RESURFACING THE SUBMERGED PAST

RESURFACING THE SUBMERGED PAST
PREHISTORIC ARCHAEOLOGY AND LANDSCAPES OF THE FLEVOLAND POLDERS, THE NETHERLANDS

J.H.M **PEETERS**, L.I. **KOOISTRA**,
D.C.M. **RAEMAEKERS**, B.I. **SMIT** & K.E **WAUGH**[†] (EDS)

© 2021 Individual authors

Authors: T. ten Anscher, J.P. Flamman, T. Hamburg, W.A.M. Hessing, L. Kooistra, L. Kubiak-Martens, J.H.M. Peeters, D.C.M. Raemaekers, B.I. Smit, K.E. Waugh† & J. Zeiler.

Editors: J.H.M Peeters, L.I. Kooistra, D.C.M. Raemaekers, B.I. Smit & K.E Waugh†

Illustrations: Cultural Heritage Agency of the Netherlands, Groningen Institute of Archaeology, Vestigia *Archeologie & Cultuurhistorie*, the authors, or otherwise noted.

English translation and final editing: K.E. Waugh† & S. McDonnell

Published by Sidestone Press, Leiden
www.sidestone.com

Lay-out & cover design: Sidestone Press

Photograph cover: Excavations at Swifterbant in 2007 by the Groningen Institute of Archaeology.

ISBN 978-94-6426-038-0 (softcover)
ISBN 978-94-6426-039-7 (hardcover)
ISBN 978-94-6426-040-3 (PDF e-book)

This research was supported by Prorail, University of Groningen, Cultural Heritage Agency of the Netherlands, Vestigia *Archeologie & Cultuurhistorie*.

Contents

List of contributors	**9**
Preface	**11**
1. Introduction of the Hanzelijn Archaeological Project	**13**
1.1 Organisation	14
1.2 The archaeological project	15
1.3 Knowledge Development Program Archaeology Hanzelijn 2012-2020	21
1.4 Some retrospective remarks	21
1.5 Acknowledgements	22
1.6 In memoriam Dr. Karen E. Waugh	22
2. The cradle of the Swifterbant culture: 50 years of archaeological investigations in the province of Flevoland	**23**
2.1 Introduction	23
2.2 History of the polders	23
2.3 Research traditions in the polder	27
2.4 Research topics and approaches	32
2.5 The positioning of the Swifterbant culture	33
2.6 Archaeology and the public	37
2.7 Conclusions	39
3. Hidden landscapes: mapping and evaluating deeply buried remains of human activity	**41**
3.1 Introduction	41
3.2 Climate, sea level rise and the structure of the subsurface	41
3.2.1 Overview of developments	41
3.2.2 Sea-level and groundwater-level rise in Flevoland	44
3.2.3 The structure of the subsurface	46
3.3 The character and quality of archaeological remains	51
3.3.1 Differences in character	52
3.3.2 Differences in preservation	54
3.4 Mapping hidden landscape units	56
3.4.1 Site versus landscape perspective	57
3.4.2 The practice of field surveys	59

3.5 The identification of sites: on statistics and indicators	61
3.5.1 The statistical uncertainty of sampling	61
3.5.2 Indicators as evidence of the presence of sites	62
3.6 Excavations: windows on the past	65
3.7 Conclusions	66

4. Exploiting a changing landscape: subsistence, habitation and skills — 67

4.1 Introduction	67
4.2 Taphonomy and analysis: the representativeness and interpretive value of find assemblages	68
4.3 Wild and domesticated mammals as sources of food	69
4.4 Fishing in a drowning landscape	73
4.5 Birds in the diet	81
4.6 Plant resources in the food economy	81
4.6.1 Wild plants	82
4.6.2 Cultivated crops	83
4.6.3 Cultivation	84
4.7 Food preparation and consumption	86
4.7.1 Animal foods	86
4.7.2 Plant food sources	91
4.7.3 Consumption	92
4.8 Resources and technology	94
4.8.1 Availability of and animal resources	94
4.8.2 Use and selection of plant resources	95
4.8.3 The utilisation of animal resources	100
4.8.4 Origin of lithic materials	103
4.8.5 Use of flint and other lithic material	106
4.8.6 Pottery production	110
4.8.7 Wood tar production?	113
4.9 Habitation patterns	115
4.9.1 Mesolithic	115
4.9.2 Swifterbant, Pre-Drouwen and Funnel Beaker occupation	119
4.9.3 Late Neolithic and Early Bronze Age	124
4.9.4 People on the move	126
4.10 Conclusions	127

5. People, ritual and meaning — 129

5.1 Introduction	129
5.2 Burial practice	129
5.2.1 Introduction	129
5.2.2 Late Mesolithic and Early Swifterbant	131
5.2.3 Classical Swifterbant and Pre-Drouwen	136
5.2.4 Late Neolithic	144
5.2.5 Conclusions	146
5.3 Other cultural practices with human bones	147
5.4 Depositions	149
5.5 Materiality	152
5.6 Conclusions	155

6. From land to water: geomorphological, hydrological and ecological developments in Flevoland from the Late Glacial to the end of the Subboreal — 157

- 6.1 Introduction — 157
- 6.2 Study of the history of the Flevoland landscape in broad outline — 159
- 6.3 Landscape dynamics in Flevoland — 164
    - 6.3.1 Late Glacial: c. 12,500 – 9800 cal. BC (Late Palaeolithic) — 164
    - 6.3.2 Preboreal: c. 9800 – 8200 cal. BC (Early and start of Middle Mesolithic) — 169
    - 6.3.3 Boreal: 8200 – 7000 cal. BC (Middle Mesolithic) — 172
    - 6.3.4 Early Atlantic : 7000 – 6000 cal. BC (Middle and early Late Mesolithic) — 175
    - 6.3.5 Middle and Late Atlantic : 6000 – 3700 cal. BC (Late Mesolithic and Early Neolithic) — 179
    - 6.3.6 Subboreal: c. 3700 – 1100 cal. BC (Middle Neolithic – Late Bronze Age) — 187
- 6.4 A new view of the landscape — 192
    - 6.4.1 Pine woodlands and heathlands in the Atlantic (7000 – 3700 cal. BC) — 192
    - 6.4.2 No salt marshes or tides in the Late Atlantic (5000 – 3700 cal. BC) — 193
    - 6.4.3 Lakes and large-scale peat accumulation in the Subboreal (circa 3700 – 1100 cal. BC) — 193
- 6.5 Three windows of observation — 194
    - 6.5.1 Zuidelijk Flevoland: Hoge Vaart-Eem microregion between 7000 and 4000 cal. BC — 194
    - 6.5.2 Oostelijk Flevoland: Swifterbant microregion between 8300 and 3700 BC — 197
    - 6.5.3 Noordoostpolder: Schokland-Urk microregion between 5000 and 1250 cal. BC — 200
- 6.6 Conclusions — 204

7. Transformations in a forager and farmer landscape: a cultural biography of prehistoric Flevoland — 205

- 7.1 Introduction — 205
- 7.2 The landscape as a source of subsistence — 206
- 7.3 Cultural structuration of the environment — 211
- 7.4 Socio-cultural relationships — 215
- 7.5 Conclusions — 219

Appendix I. Site Atlas Windows of observation: the quality, nature and context of excavated prehistoric sites in Flevoland: site atlas — 221

- I.1 Introduction — 221
- I.2 Zuidelijk Flevoland — 223
    - I.2.1 Almere – Hoge Vaart/A27 — 223
    - I.2.2 Almere – Europakwartier Site 7 — 233
    - I.2.3 Almere – Zwaanpad — 235
    - I.2.4 Zeewolde – Oz35/Oz36 — 238
- I.3 Oostelijk Flevoland — 239
    - I.3.1 Dronten N23/N307 – Site 5 — 239
    - I.3.2 Hanzelijn – Area VIII — 244
    - I.3.3 Hanzelijn – Drontermeer Tunnel (area XVI) — 246
    - I.3.4 Swifterbant Cluster — 248
- I.4 Noordoostpolder — 254
    - I.4.1 Emmeloord J97 — 254
    - I.4.2 Schokkerhaven-E170 — 257
    - I.4.3 Schokland P14 — 259
    - I.4.4 Urk – E4 — 263

Appendix II. Glossary plant species — 265

Literature — 267

List of contributors

J.H.M. Peeters
Groningen Institute of Archaeology,
University of Groningen
j.h.m.peeters@rug.nl

L.I. Kooistra
BIAX Consult
laura.i.kooistra@kpnmail.nl

D.C.M. Raemaekers
Groningen Institute of Archaeology,
University of Groningen
d.c.m.raemaekers@rug.nl

B.I. Smit
Cultural Heritage Agency of the Netherlands
b.smit@cultureelerfgoed.nl

K.E. Waugh (†)
Vestigia *Archeologie & Cultuurhistorie*

J.P. Flamman
Vestigia *Archeologie & Cultuurhistorie*
j.flamman@vestigia.nl

T. Hamburg
Archol
t.hamburg@archol.nl

W.A.M. Hessing
Vestigia *Archeologie & Cultuurhistorie*
w.hessing@vestigia.nl

L. Kubiak-Martens
BIAX Consult
kubiak@biax.nl

T. ten Anscher
RAAP Archeologisch Adviesbureau
t.ten.anscher@raap.nl

J. Zeiler
Archaeobone
abone@planet.nl

Preface

In December 2012 the Hanzelijn, a new railroad consisting of 45 kilometres of tracks and two new stations, was completed. It crosses three provinces and five municipalities and transports an average of 15,000 people every day between the cities of Lelystad and Zwolle. It reduces the time it takes to travel from Lelystad to Zwolle from 50 minutes by car to 30 minutes by train and benefits this region in more ways than one. It has provided for a better future, but what about the past?

Infrastructure, in any shape or form, has played a crucial part in the history of mankind. Knowing the routes in the surrounding landscape, which ones to use in different seasons and situations has provided humans with advantages in survival and allowed for the establishing social and economic benefits. Settlements from all periods in time and their infrastructure are therefore in a symbiotic relationship. Whether it is a worn out pathway through bushes connecting Stone Age camps, a paved road leading from Rome to Katwijk or a new railroad between Lelystad and Zwolle. Places and people need to be connected in order to communicate, survive and thrive. Routes or pathways have therefore more than once been described as the communication arteries of communities and societies both in the past and in the present.

Off course, in modern times societies focus lies towards economic growth, expanding social networks and communication. The newly build Hanzelijn fills this need by increasing mobility and accessibility between Lelystad and Zwolle and beyond. But, apart from generating this effect through use of the railroad, the construction in itself also affected the environment.

When planning the construction of railway infrastructure, the effects on the environment need to be examined and mitigated before construction can commence. Urban planning, soil contamination and unexploded ordnance, for example, are aspects that need to considered as do many other environmental issues. One aspect among these is archaeology and the careful management of potentially present archeological heritage. Archaeological remains dating to the Stone Age are present in the subsoil of Flevoland, as established, for example, by the important finds in the region of the 'Swifterbant Culture' and older (Mesolithic) finds discovered in, for example, Almere and Neolithic finds on the UNESCO World Heritage Site of Schokland.

From 2001 until 2008, a standard archaeological process was conducted prior to the construction of the Hanzelijn, attempting to find archaeological remains in the subsoil of Flevoland. This process eventually led to the discovery of, and the decision to preserve, a submerged Stone Age landscape underneath the planned railway in Flevoland. A landscape in which Mesolithic and Neolithic sites were most likely present. During these years of archaeological research, some fundamental questions were raised; for example, about the best way to locate these hidden sites, and how the investigation and preservation of sites *in situ* for this specific geological situation should commence, as well as what the effects of drainage and soil subsidence would be on archeological remains buried deeply in the soil. In 2008 it became evident that some adaptations had to be made to the initial construction design of the railway. Primary to the construction, calculations showed that the subsoil of Flevoland might not be able to carry the weight of the track and trains, with risk of subsidence and potentially even the derailing of trains. Therefore a last minute decision was made to create a solid foundation underneath the substructure of the tracks. This solution could affect archaeological remains that might be present within the preserved archaeological landscape. At this point no time was available to research the archaeological landscape beneath the planned railway. To mitigate the potential loss of important information the situation was changed into an opportunity to redirect funds and conduct research to find answers to the previous raised questions. Therefore ProRail BV, the Cultural Heritage Agency of the Netherlands and Vestigia *Archeologie & Cultuurhistorie* initiated the 'Knowledge Development Program Archaeology Hanzelijn'.

This program consisted of a wide array of archaeological research that would, for example, establish the effects of loading on the most fragile components of archaeological sites and analyze the effectiveness of different coring and remote sensing techniques to aid the discovery of these sites. All this to increase the chance of finding archaeological sites buried deep within peaty soils and to provide better means of ensuring preservation *in situ*.

The book before you is the final publication of this program. It focusses on the changes in landscape, prehistoric subsistence and socio-cultural developments during the Mesolithic and Neolithic in this wetland area. Research from sites like Almere Hoge Vaart A27, Swifterbant and Schokland-P14 is confronted with information from lesser known sites which generally have only been published in Dutch grey literature. The authors from both the University of Groningen, commercial archaeological companies and the Cultural Heritage Agency of the Netherlands have written a comprehensive overview of the way people in the past occupied and used the dynamic landscape of this region. ProRail BV, the Cultural Heritage Agency and all partners are proud of all results this program has produced.

The program's goal was to improve heritage management of the province of Flevoland and archaeological sites in similar geological conditions. Providing tools for municipalities, provinces and nations that enables them to upgrade their policies regarding archaeological heritage management.

ProRail BV and the Cultural Heritage Agency hope you will enjoy and use the results of this entire program as a connection and inspiration for the challenges of today and the safeguarding and research of our shared archaeological heritage in the future.

Jaap Balkenende
Projectmanager Hanzelijn
ProRail BV

Arjan de Zeeuw
Knowledge & Advice director
Cultural Heritage Agency of the Netherlands

Chapter 1

Introduction of the Hanzelijn Archaeological Project

K.E. Waugh (†), W.A.M. Hessing & J.P. Flamman

The *Hanzelijn* is a new, 50 km long, part of the Dutch national railway network, connecting the towns of Lelystad, Dronten, Kampen and Zwolle in the centre of The Netherlands. After actual building started in 2005, the line opened in 2012 (fig.1.1 and 1.2). Its name has been inspired by the Hanseatic League, the mercantile organisation which in Late Medieval times had been of utmost importance for the *Zuyderzee* area and member cities like Kampen and Zwolle. Design and preparation of the railway, including two new passenger stations at Dronten and Kampen-Zuid and quite extensive environmental requirements, had been started by the Dutch railway operator NS Railinfrabeheer (RIB, later becoming ProRail) at the end of the 1990's. A separate project organisation was set up at its Utrecht Head office to coordinate planning, design and contract- and environmental management. Around 2000, because of its estimated scope and complexity, it was decided that archaeology should become an integrated part of the project.[1]

Geographically, the railway crosses two distinct, and in many aspects different, landscapes. The western part is situated in the new polders of Flevoland. This former seabed of the Zuyderzee (since 1932 IJsselmeer), had been made into land -reclaimed – only in 1957. The eastern part is situated in the delta of the IJssel river, a varied landscape of sandy soils alternated by clayish deposits from the river and sea. Technically, design and construction of the track would be different in the two areas. In the east, traditional foundation and construction could be largely followed, a new bridge over the IJssel and a railway tunnel under the Drontermeer being the main challenges. In the west the subsoil was highly variable and instable (see chapter 2) over many kilometres, demanding a different substructure and embankment, also visually adapted to its flat polder surroundings.

Initial characterisation of the archaeological remains – present as well as expected – mainly followed the same distinction. In the west – 'New Land' (*Nieuwe Land*) as this part soon was called – archaeological sites either would be historic shipwrecks close to the surface, or Stone Age sites, deeply buried under marine and freshwater sediments. The east – 'Old Land' (*Oude Land*) – had seen continuous occupation from at least Neolithic times until recently. Remains from all archaeological periods could be found in, or directly under, the plough soil. Combining the technical requirements, the geological and archaeological characteristics, soon it was concluded, both areas required a different archaeological approach.

---

1 ProRail based this decision on recent experience with archaeology in two other large scale rail projects: Betuweroute (1995-2000) and the Amsterdam-Antwerp Highspeed Raillink (HSL-South 1998-2002).

Figure 1.1a-b: a) Route of Hanzelijn new railway track in the Netherlands: the town of Lelystad is situated on the western end, that of Zwolle at the eastern end and location of the railway in the Netherlands (source: a – ProRail, b – Vestigia 2008).

1.1 Organisation

In 2002 NS RailInfraBeheer contracted Vestigia *Archeologie & Cultuurhistorie* to provide the Hanzelijn Project Organisation the necessary archaeological expertise and to coordinate the archaeological process in the following years.[2] Karen Waugh (see below under 1.6) joined the project organisation and would see through the subsequent archaeological steps, almost until the very end. From the beginning close coordination and cooperation with the government, national as well as regional and local was seen as essential for RIB (ProRail). The Hanzelijn, being a national project, archaeological authority and supervision was with the National Archaeological State Service ROB (later to become the Cultural Heritage Agency of the Netherlands, RCE), but also three provinces and six were involved. Other stakeholders were two regional water authorities (*Waterschappen*), and of course a large number of individual landowners. To improve speed of action on archaeology a steering committee was set up by RIB and ROB to provide direct guidance for the archaeological project manager and speedy decision making concerning all archaeological aspects. Within the steering committee, RIB took on the role of informing other governmental organisations on issues discussed and decisions made.[3]

In the initial phase of the contract Vestigia concentrated on the planning and preparation of the contracting of the archaeological fieldwork (see below under 1.5), while ROB contributed in formulating an extensive scientific research framework, whereby the full scope of the archaeological project could be identified.[4] From the ontset the need to protect the archaeology played an important role in discussions dealing with design and construction. Archaeological sites would preferably to remain *in situ*, preserved and protected (in line with the aims of the Valletta Convention). Full excavation was seen as a last resort, if the other option was not feasible.

As little was known of the archaeology below most parts of the preferred route, the emphasis in the first years of the project was on identifying archaeologically relevant geological deposits and landforms and defining effective strategies for prospection and assessment of sites. After the so-called desktop study phase (see below), all necessary fieldwork, mainly consisting of systematic coring, trial trenching and excavation was contracted out to individual archaeological companies. Selection of these companies and their tenders was always done equally on price, as

---

2 This was the first time the archaeology of a national infrastructural project would not be managed by the government itself. The national government concluded this to be in line with its policy of privatisation and avoiding competing itself within the nascent archaeological market.

3 The Steering Committee ended its regular work in 2012, after the opening of the railway, however in a reduced setting it continued monitoring the additional archaeological programme, until 2021.

4 Peeters 2004.

Figure 1.2a-b: a) The new railway tracks in the flat landscape of *Nieuwe Land* and b) Tunnel Drontermeer (photo a-b: Henk de Jong 2010).

well as quality of proposal, and team experience. A total of 11 archaeological programs of requirements (contract specifications) and 8 evaluation and advice documents were made by Vestigia between 2002 and 2008, and 4 desktops studies and 14 fieldwork contracts were given out to seven different archaeological companies (see table 1.1).[5] At the same time some additional work was contracted out on informing the general public, although this was only given limited priority, given the uncertainty about the actual archaeological results. After the conclusion of the fieldwork phase it was decided to set up an additional synthetic research program focussing on further development of archaeological knowledge and disclosure of relevant information and best practices in the Flevoland area. This program started in 2010, the publication of this book marks its finalisation. The AKDAH program (see 1.3) was fully financed by ProRail and managed by Vestigia in close coordination with RCE.

1.2 The archaeological project

Oude Land

Within the 'Old Land' the more 'traditional' process of step-by-step assessment through desktop survey, large-scale field surveys, – mostly a combination of coring and fieldwalking -, trial trenching, evaluation and finally excavation, was followed.[6] Systematic desktop and field surveys of a 500 m wide strip over tens of kilometres in a largely unexplored landscape led to many new discoveries. Most spectacular, were the well preserved Mesolithic, Neolithic and Bronze age sites on the outcrops of the ice-pushed glacial ridge near Hattemerbroek. Due to the technical and practical challenges in the lay-out of the track, preservation and protection of newly discovered archaeological sites proved a challenge. Changing the corridor of the track or making technical adaptions to mitigate archaeological damage, although explored intensively, was not possible. However, on the

---

5 Next to these 6 additional so-called selection documents were written, prioritising discovered sites for further research or in situ conservation.

6 At a later stage a single, additional watching brief was also carried, when several extra ditches had to be dug alongside a part of the track.

Programs of requirements	2002	Peeters, J.H.M., 2002: Wetenschappelijke Uitgangspunten voor de Archeologische Monumentenzorg in het Kader van de Aanleg van de Hanzelijn (Lelystad-Zwolle), RAM 110, Amersfoort.
		Waugh, K.E. 2002: Programma van Eisen ten behoeve van het Bureauonderzoek. Vestigia rapportnummer 35, Amersfoort.
		Waugh, K.E. 2002: Programma van Eisen ten behoeve van de Bewerking van de Actuele Hoogtemeting Nederland (AHN). Vestigia rapportnummer 36, Amersfoort.
		Waugh, K.E. 2002: Programma van Eisen ten behoeve van het Geologisch Profiel. Amersfoort.
	2003	Waugh, K.E., 2003: Programma van eisen ten behoeve van Inventariserend Veldonderzoek Fase 2, Hanzelijn Oude Land. Vestigia rapport V108, Amersfoort.
	2005	Mietes E.K, K.E. Waugh, R. Schrijvers, 2005: Programma van Eisen ten behoeve van het Inventariserend Veldonderzoek Fase 2 & 3, Hanzelijn Nieuwe Land. Vestigia Rapport V181, Amersfoort.
		Waugh, K.E., W.A.M. Hessing, J.P. Flamman, E.K. Mietes en R. Schrijvers, 2005: Programma van Eisen ten behoeve van het Inventariserend Veldonderzoek Fase 3, Hanzelijn Oude Land. Vestigia rapport V204, Amersfoort.
	2006	Flamman, J.P., 2006: Programma van Eisen ten behoeve van het Definitief Archeologisch Onderzoek, Hanzelijn Oude Land. Vestigia rapport V291, Amersfoort.
	2007	Flamman, J.P., 2007: Programma van Eisen ten behoeve van de archeologische begeleiding in het project Hanzelijn. Deeltracé: Onderbouw Oude Land Vestigia rapport V418, Amersfoort.
		Quadflieg, B.I., 2007: Project Hanzelijn: Programma van Eisen ten behoeve van de uitvoering van hoog kwalitatieve boringen, Nieuwe Land. Vestigia Rapport V369 (ProRail Documentnummer BQ/HZL/20705317/20705318), Amersfoort.
		Waugh, K.E., 2007: Programma van Eisen ten behoeve van de archeologische begeleiding in het project Hanzelijn. Deeltracé: Contract Nieuwe Land. Vestigia rapport V416, Amersfoort.
		Waugh, K.E., 2007: Programma van Eisen ten behoeve van de archeologische begeleiding in het project Hanzelijn. Deeltracé: Tunnel Drontermeer. Vestigia rapport V404, Amersfoort.
	2010	Flamman, J.P., K.E. Waugh 2010: Programma Kennisontwikkeling Archeologie Hanzelijn. Programma van Eisen ten behoeve van de inventarisatie van beschikbaar bronnenmateriaal voor de Thema's 1, 2 en 3. Vestigia rapportnummer 852, Amersfoort.
Evaluation reports	2002	Vestigia *Archeologie & Cultuurhistorie*, 2002: Werkplan t.b.v. de Archeologische Begeleiding van de Definitiefase in de Hanzelijn, versie 2.0 definitief, Bunschoten.
	2003	Waugh, K.E., 2003: Werk- en Toetsingsprocedure ten behoeve van de Stuurgroep Archeologie, Project Hanzelijn. Vestigia rapport V86, Amersfoort.
		Waugh, K.E., 2003: Herziene Selectieadvies in aansluiting op het aanvullend karterend booronderzoek (Fase 1 inventariserend veldonderzoek).Vestigia rapport V107, Amersfoort
	2004	Waugh, K.E., R. Schrijvers en E. Mietes, 2004: Project Hanzelijn: Selectieadvies in aansluitring op IVO Fase 2, Oude Land. Vestigia rapport V173, Amersfoort.
	2006	Quadflieg, B.I., & Waugh, K.E., 2006: Project Hanzelijn. Selectieadvies in aansluiting op IVO Fase 2, Nieuwe Land. Vestigia Rapport V309 (ProRail Documentnummer KW/HZL/ARC/20316168/20614551), Amersfoort.
		Waugh, K.E., W.A.M. Hessing en J.P. Flamman, 2006: Project Hanzelijn: Selectieadvies in aansluiting op IVO Fase 3, Oude Land. Vestigia rapport V312, Amersfoort.
	2008	Waugh, K.E., 2008: Project Hanzelijn. Ontwerpwijziging aanleg spoorbaan tracédeel Nieuwe Land: advies over de consequenties voor archeologie. Vestigia rapport V553, Amersfoort.
	2011	Waugh, K.E., W.A.M. Hessing, M.K. Boonstra, 2011: Activiteiten met betrekking tot de afronding van Archeologie in het Project Hanzelijn. Deelproject III Afronding en Eindpresentatie Archeologie. Plan van Aanpak voor Thema 1: Publiekspresentatie, Thema 2: Evaluatie, Thema 3: Verslaglegging. Vestigia rapport V872, Amersfoort.
Archaeological reports	1999	Schute, I.A., 1999: Hanzelijn, aspectrapport archeologie; huidige situatie, autonome ontwikkeling, effectbeschrijving en effectbeoordeling. RAAP-rapport 408, Amsterdam.
	2000	NS Railinfrabeheer, 2000: Trajectnota en Milieu-effectrapport Hanzelijn, Utrecht.
	2002	Gouw, M.J.P., W.A.M. Hessing, E.K. Mietes en K.E. Waugh, P. van der Kroft, 2002: Archeologische begeleiding van de Definitieve fase van het project Hanzelijn, Rapportage SAI-2 Fase, deel 1: Archeologisch en Geologisch Bureau onderzoek. Vestigia rapportnummer 45, Amersfoort.
	2003	Leijnse, K, 2003: Hanzelijn tracédeel Oude Land. Een inventariserend archeologisch onderzoek: aanvullingen op IVO Fase 1. RAAP-rapport 953, Amsterdam.
		Müller, A., 2003: Hanzelijn tracédeel Oude Land. Een inventariserend archeologisch onderzoek. RAAP-rapport 869, Amsterdam.
		Müller, A. & K. Leijnse, 2003: Hanzelijn tracédeel Nieuwe Land. Een inventariserend archeologisch onderzoek. RAAP-rapport 932, Amsterdam.
		Vos, P.C., 2003: Geologisch profiel Hanzelijntracé. Geologisch onderzoek ten behoeve van de archeologische bureaustudie Nieuwe Land (Flevoland) en Oude Land (Kamperveen). TNO-rapport NITG 03-006-B, Utrecht.
	2004	Leijnse, K, 2004: Hanzelijn tracédeel Nieuwe Land. Een inventariserend archeologisch onderzoek: aanvullingen op IVO Fase 1. RAAP-rapport 1003, Amsterdam.
		Roller, G.J. de, 2004: Hanzelijn tracédeel Oude Land, inventariserend archeologisch onderzoek Fase 2. ARC-publicaties 105, Groningen.
		Vos, P.C., 2004: Detaillering geologisch profiel en aanvullende rapportage Hanzelijntracé Nieuwe Land, ProRail, Utrecht.
	2006	Colard, G., en M.J. den Uil, 2006: Hanzelijn: Archeologie Hattem – Onderzoek naar de optredende spanningen, dwarskrachten en vervormingen met EEM (Plaxis). Memo Holland Railconsult GMV-GC-060007272.
		Hamburg, T., en S. Knippenberg, 2006: Steentijd op het Spoor. Proefsleuven op drie locaties binnen het tracé van de Hanzelijn 'Oude Land', Archol rapport 54, Leiden.
		Leijnse, K, 2006: Hanzelijn tracédeel Nieuwe Land. Archeologisch vooronderzoek: een inventariserend veldonderzoek – IVO fase 1 (afronding) & IVO fase 2. RAAP-rapport 1305, Amsterdam.
		Tol, A.J., 2006a: Hanzelijn, tracédeel Oude Land; een inventariserend veldonderzoek: afronding IVO Fase 1. RAAP-rapport 1303, Amsterdam.
		Tol, A.J., 2006b: Hanzelijn, tracédeel Oude Land; een inventariserend veldonderzoek: IVO Fase 3 deelgebied A. RAAP-rapport 1304, Amsterdam.
	2008	Corver, B.A. 2008: Project Hanzelijn: 'Het Oenen' – De Slaper nr. 2 te Kamperveen, gemeente Kampen. Een Archeologische Begeleiding. ADC Rapport 1102, Amersfoort.

	2009	Prangsma, N. M. & D.A. Gerrets, 2009: Hanzelijn Tunnel Drontermeer: verbinding tussen Oude en Nieuwe Land. Een Archeologische Begeleiding bij de Sallanddijk en een compenserend archeologisch onderzoek in gebied XVI. ADC Rapport 1601, Amersfoort.
	2010	Hamburg, T., 2010: Archeologische Begeleiding Hanzelijn – Oude Land. Archol Rapport 131, Leiden.
		Osinga, M. & J.J. Hekman, 2010: Archeologisch onderzoek Hanzelijn deelgebieden XIV en XV, archeologische begeleiding. Grontmij Archeologische Rapporten 740, Assen.
	2011	Kerkhoven, A., 2011: Compacte analyse verzamelde gegevens onderdeel Archeologie. Interne memo ihkv Programma Kennisontwikkeling Archeologie Hanzelijn (PAKH), 21 oktober 2011.
		Kooistra, L.I. & H. van Haaster, 2011: Inventarisatie van bio-archeologisch onderzoek in het kader van het Programma Kennisontwikkeling Archeologie Hanzelijn (PKAH). BIAXaal 554, Zaandam.
		Lohof, E., T. Hamburg & J. Flamman (red.) 2011, Steentijd opgespoord. Archeologisch onderzoek in het tracé van de Hanzelijn – Oude Land, Archol rapport 138/ ADC-rapport 2576, Leiden/ Amersfoort.
		Vorenhout, M., 2011a: Evaluatie invoer database PKAH. Korte notitie MVH Consult n.a.v. literatuur database bestaande onderzoeken binnen in situ behoud. Interne memo ihkv Programma Kennisontwikkeling Archeologie Hanzelijn (PAKH), 15 december 2011.
	2014	Hamburg, T., A. Tol, J. de Moor & Y. Lammers-Keijsers, 2014: Afgedekt verleden. Opsporing, waardering en selectie van prehistorische archeologische vindplaatsen in Flevoland. Programma Kennisontwikkeling Archeologie Hanzelijn (Thema 1B), Leiden/Amersfoort (Archol-rapport 244/Earth Integrated Archaeology rapporten 49).
		Muller, A., H. Meerten, R.B.J. Brinkgreve, & D.J.M. Ngan-Tillard, 2014. Flevoland kennisontwikkeling programma archeologie Hanzelijn: Mogelijkheden tot in-situ conservering van begraven archeologische landschappen, Deelonderzoek 2, De invloed van tijdelijke en permanente afdekkingen of ophoging op maaiveld op de conservering van archeologische vindplaatsen in de ondergrond, Amersfoort.
		Vissers, M.J., S. van Asselen & J.J. Hekman, 2014: Programma Kennisontwikkeling Archeologie Hanzelijn, Thema 2A: veranderingen in de waterhuishouding gerelateerd aan bodemeigenschappen en de gevolgen daarvan voor de conservering van afgedekte archeologische vindplaatsen in Flevoland, De Bilt (Grontmij Archeologische Rapport 1314).
	2018	Heeringen, R.M. van, R. Schrijvers, K.E. Waugh, 2018: Handreiking prospectief onderzoek in Flevoland voor het opsporen en waarderen van vindplaatsen uit de vroege prehistorie. Vestigia rapport V1372, Amersfoort.
Syntheses	2021	Peeters, J.H.M, L.I. Kooistra, D.C.M. Raemaekers, B.I. Smit & K.E. Waugh 2021: Resurfacing the submerged past. Prehistoric archaeology and landscapes of the Flevoland Polders, the Netherlands, Leiden.

Table 1.1 Products of the Hanzelijn Archaeological Project and Knowledge Development Program Archaeology Hanzelijn.

Oude Land protection of 70,000 m² of archaeological sites could be realized, mainly by shifting a number of planned interventions alongside the track to other areas, and using superfluous ground to cover threatened sites directly beneath the plough soil.

In 2006 and 2007, circa 35,000 m² were excavated in the Hattemerbroek area and three smaller locations, De Slaper, Oenen and De Enk-Zuid. All these sites produced a vast amount of new evidence on the Early-Prehistoric (8800 – 5000 BC) and Late-Prehistoric (3200 – 400 BC) occupation of the Oude Land area (fig. 1.3). The final publication of all results appeared in 2011.[7] In addition, as a result of the extensive field surveys of the Hanzelijn campaign, several other archaeological sites were discovered outside the railway track itself. One of these turned out to be situated in the middle of a large industrial estate at a short distance of Hattemerbroek. This early prehistoric site was excavated by the same team in 2007, using the specifications and experience gained in 2006-2007 and fully published in 2011, making comparative studies possible.[8]

Nieuwe Land

As information on the geology and archaeology in the 'New Land' in 2002 was limited, archaeological assessment of the proposed corridor of the Railway had to start with cataloguing relevant information from a wide array of sources. All available geological data collected by the various research institutes and government services during the reclamation, cultivation and parcelling of the new polder in 1960s and 1970s was collected. TNO/NITG (now Deltares and TNO), the former Dutch research institute for geosciences, used this information to create a new model of the geological and soil substrata. This model was subsequently used by Vestigia to make detailed characterization maps of the proposed rail track and its surroundings and predict the most promising areas and geological deposits containing archaeological sites. In Flevoland, the focus during the prospection phase was on identifying various levels of potential prehistoric occupation and the effects of the rise in sea levels during the prehistoric period, causing a gradually drowning landscape. Because of the depths of the relevant deposits – up to 6 metres below surface – trial trenching was not possible, but corings showed the prehistoric landscape well-preserved under the many later layers of maritime and freshwater clay and peat (fig. 1.4).

In the New Land the depths of the archaeological layers led to the assumption that preservation in situ would be technically feasible in most areas. Before construction could begin, a detailed geotechnical analysis of the deformation and stability of the underlying soil structure was carried out in order to calculate the probability of compaction, displacement and deformation of the archaeological layers under the proposed infrastructure. In the meantime prospection of prehistoric sites in the deeply buried landscape by coring continued, identifying several promising locations. However when the geotechnical analysis and additional design adaptations were positive, it

---

7   Lohof, Hamburg & Flamman (eds) 2011.
8   Hamburg, T., E. Lohof & B. Quadflieg (eds) 2011.

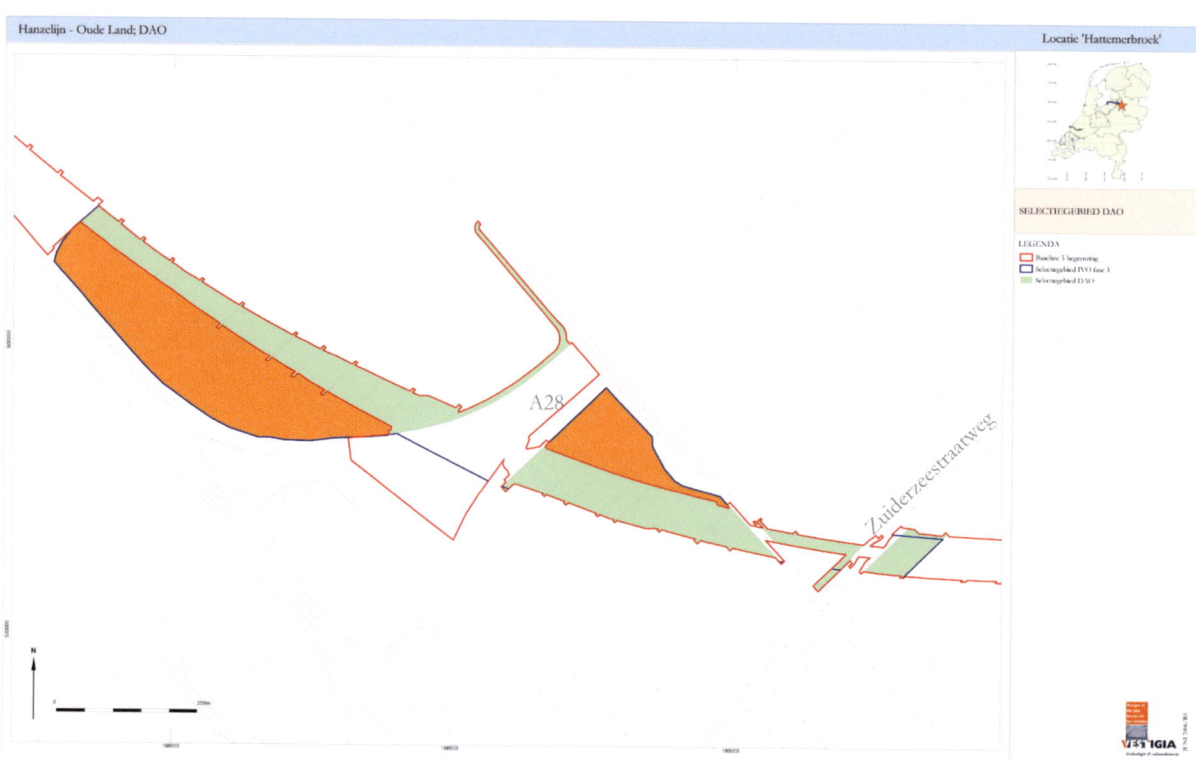

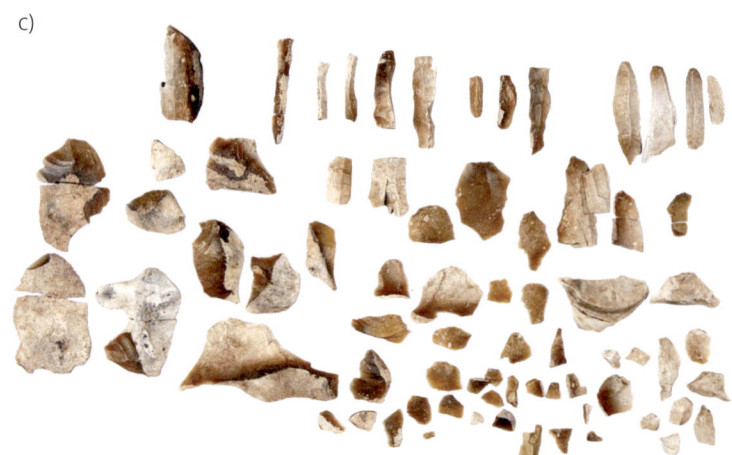

Figure 1.3a-c: Lay-out of the new railway near the motorway junction Hattemerbroek (A28/A50), showing a) the excavated areas in green and *in situ* preservation in orange; b) shows a Late Neolithic Bell beaker from one of the burials on this location; c) shows part of a Late Paleolithic flint assemblage found here as well (map: a Vestigia 2010, photo: b-c Archol 2011).

Figure 1.4a-b: a) Mechanic coring using aqualock technique at *Nieuwe Land*; b) coring grid and results showing relief of buried Pleistocene surface and possible archaeological site (red) just outside the actual track (photo a-b: Vestigia 2012).

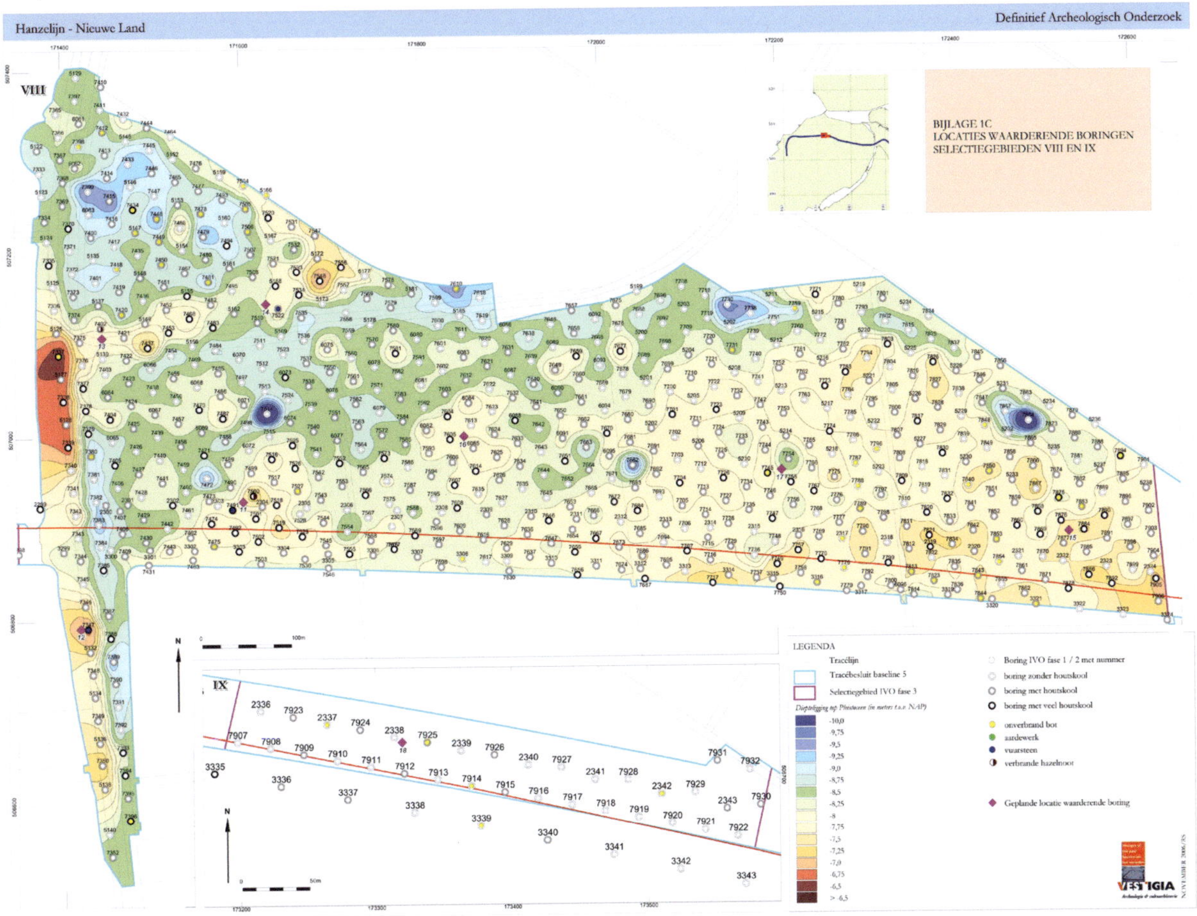

Figure 1.4 c) Sequence of aqualock cores showing continuous sedimentation since last Ice Age below *Nieuwe Land* (photo: Vestigia 2012).

was decided to refrain from further archaeological evaluation in those areas, where in situ preservation could be achieved. The necessary archaeological requirements were incorporated in the final design and building contract for a nearly 10 kilometers long stretch of the railway, west of Dronten.

Just before construction work in the New land area in 2010 was supposed to start, unexpected new ground stability problems became apparent. As a consequence, before construction could begin, a large part of the underground corridor west of Dronten, would have to be excavated down to Pleistocene sand level (6-8 metres below surface) anyway. Adding to the setback, planning and contractual agreements at this stage allowed time for neither a restart of the archaeological process, nor large-scale excavations. These last minute alterations meant that several significant areas previously selected for *in situ* preservation along the corridor, could now not be further explored nor protected. The project, where the focus of the developer and the commissioning authority had been from the beginning on safeguarding the potential archaeological resource, was faced with an insoluble dilemma.

Extensive consultation, deliberation and negotiation on the highest levels between the national Department of Transport, Public Works and Water Management and the Department of Education, Culture and Science resulted in a compromise to turn the negative into something positive. ProRail in collaboration with RCE were asked to set up an additional archaeological research program to improve and facilitate heritage management and future archaeological research in the Flevoland area. The *Knowledge Development Program Archaeology Hanzelijn* started in 2012 and was fully funded by ProRail and will be further explained in the next section. In the meantime archaeological work along other parts of the corridor continued and resulted in several small scale excavations and watching briefs and the in situ preservation of a historical ship wreck located next to the new Drontermeer tunnel. Also in New Land the Hanzelijn surveys directly lead into other archaeological interventions. A newly discovered early prehistoric site, happened to be situated at a newly planned intersection of the N23 / N307 road west of Dronten. Large-scale excavations of what proved to be an extensive and well-preserved Mesolithic site (8300-5000 BC) followed in 2013, underlining the archaeological potential of the area.[9]

1.3 Knowledge Development Program Archaeology Hanzelijn 2012-2020

In addition to the agreed developer-bound archaeological mitigation and research investments, in 2012 a research program was set up aimed at the improvement of archaeological survey and research within Flevoland and to a better understanding of environmental aspects influencing in situ conservation. An additional budget of 2 million euro was allocated to the program. Research design and project management were done by Vestigia. Selected parties were asked to hand in research proposals and were subsequently contracted.[10] The project focussed on four specific subjects:

1. Improvement of prospection, evaluation and selection procedures for (deeply) buried prehistoric landscapes and archaeological sites;
2. A better understanding of soil and hydraulic conditions in Flevoland in relation to the possibilities and limitations of executing in situ conservation, and the monitoring of the long term effects;
3. A better understanding of the effects of covering archaeological sites, either temporarily or permanently in order to successfully achieve in situ conservation;
4. A new analysis of the evidence for the earliest occupation of Flevoland.

Part 1 of the program was carried out in three phases: Under 1A a catalogue of all relevant primary sources on paleo-ecological, geological and archaeological data in Flevoland was composed in 2011.[11] Under 1B a detailed an critical evaluation of current prospection techniques used in Flevoland was published in 2014.[12] Under 1C a guideline (best practises) for the prospection and evaluation of early prehistoric sites in Flevoland was published in 2018.[13] Part 2 resulted in a report on the effects of groundwater fluctuations on the preservation of buried archaeological sites in Flevoland in 2014.[14] Part 3 resulted in a technical study on the possibilities of in-situ conservation through ground covering of archaeological sites, also published in 2014.[15] Under Part 4 the publication of Ten Anschers PhD thesis on Early Neolithic occupation of the northern part of Flevoland, was seen as an essential missing element, before further analyses could be made. Therefore the program financially contributed towards the final stages of this substantial work.[16] The current publication before you forms the apotheosis of Part 4, and in a way also of the Knowledge Development Program Archaeology Hanzelijn, and the Hanzelijn project in its entirety.

1.4 Some retrospective remarks

As an infrastructural building project the Hanzelijn was a success. Commissioning, construction and opening of the line was realised in a relative short time span. The question of the success archaeologically is more difficult to answer. From the start in situ preservation of archaeological sites had been emphasized. After initial successes, in the end a large concession on this had to be made. In contrast, looking especially at the level of archaeological knowledge and access to relevant information at the start of the project, substantial progress has been made, especially concerning the practical challenges of in-situ conservation, and most importantly the Early Prehistoric occupation of this part of the Netherlands. Hopefully this book will underline the importance of the Flevoland area as a unique archaeological landscape.

The Knowledge Development Program Archaeology Hanzelijn started in 2012, now in 2021 it is brought to closure marked by this syntheses. The work on this synthesis started around 2014 with the main body of work done between 2016 and 2017. Finalising the book and get everything ready for publication took more time than expected, however. Ongoing research elsewhere, both in and outside Flevoland, has returned exciting results,

---

9 Hamburg, Müller & Quadflieg 2013.
10 It was seen as preferable that academic institutions as well as archaeological companies would form strategic alliances to tender these contracts. This procedure has been largely successful.

11 Kerkhoven 2012; Kooistra & van Haaster 2011; Vorenhout 2011.
12 Hamburg, Tol, de Moor & Lammers-Keijsers 2014 (eds).
13 Van Heeringen, Schrijvers & Waugh 2018.
14 Vissers, Van Asselen & Hekman 2014.
15 Muller, van Meerten, Brinkgreve & Ngan-Tillard 2014.
16 Ten Anscher 2012.

Figure 1.5: Dr. Karen E. Waugh (8th of March 1963 – 16th of July 2019) (photo: N. Liskaljet 2018).

which the authors would liked to have integrated in the various chapters. However, as this would have required substantial reworking of the text, at least of some parts, and more time taken, it was decided to restrict this to the inclusion of footnotes where appropriate. Despite this situation, the authors hope that this book provides a valuable source for the international professional audience interested in the Early Prehistory of NW Europe, as it brings together a rich body of data and information from published work and 'grey literature' predominantly written in Dutch.

1.5 Acknowledgements

During the last two decades many organisations and persons have contributed to the success of the Hanzelijn Archaeological Project. Our gratitude goes out to all authors, researchers, colleagues and others involved in this program, its results and outreach. Contributions of a number of them cannot remain unmentioned: from (RIB) ProRail J.H.P. van den Berge, H. van Helvoort, K. Renzen, H. Hartgers and J. Huisman were active members of the Steering Committee. From (ROB) RCE these were subsequently H. de Haan, P. Schaap, M. Stafleu, H. Peeters, B.I. Smit. H. Peeters and B.I. Smit also added their extensive archaeological expertise to many phases of the project. The Vestigia project team over the years consisted mainly of K.E. Waugh, J.P. Flamman, B. Quadflieg, R. Schrijvers, E. Mietes, R. van Heeringen and M. Gouw. Archaeological companies involved in surveys, fieldwork and synthetic research were ADC, ARC, Archol, Grontmij and RAAP. Specialist additional services and advise were provided by NITG-TNO, Deltares, TU Delft, Grontmij/Sweco. Companies and organisations who were engaged in the Knowledge Development Program Archaeology Hanzelijn were ADC, Archeobone, Archol, BIAX Consult, Earth integrated archaeology, Groningen Archaeological Institute of the University Groningen , Grontmij/Sweco, RAAP, Transect, Technical University Delft, the Cultural Heritage Agency of the Netherlands and Vestigia *Archeologie & Cultuurhistorie*.

1.6 In memoriam Dr. Karen E. Waugh

Karen Waugh unexpectedly passed away in July 2019, following a short illness. From the very beginning in 2002, she was the central figure in the project, providing endurance, stability and continuity to see good archaeological values realized for a fair prize. Running the project, she was analytical, business minded and always looking for answers. Coming from abroad, she became genuinely interested in the landscape and the archaeology, especially of the new polders. She will be foremost remembered for her warm personality and empathy when obstacles were encountered, and conflicts needed to be resolved. It is utterly sad she has not been able to finish the publication, whereat she had been working almost to the end. For this reason, ProRail, the Cultural Heritage Agency of the Netherlands, University of Groningen, Vestigia and the authors have dedicated this book to her.

Chapter 2

The cradle of the Swifterbant culture

50 years of archaeological investigations in the province of Flevoland

D.C.M. Raemaekers & J.H.M. Peeters

2.1 Introduction
This chapter considers the creation of the polders and the advances in thinking about Flevoland as a former seabed. It looks at the history of archaeological research into prehistoric Flevoland since the creation of the polders, the discovery of the 'Swifterbant culture' and the significance of the research carried out for the development of the discipline and for the image of the region's prehistoric past, both in the Netherlands and internationally.

2.2 History of the polders
The history of prehistoric archaeology in Flevoland begins in the 19th century, not because of a sudden enthusiasm for archaeological research, but because the dire environmental conditions in the region called for drastic measures. At the time, the territory that is now the province of Flevoland was part of the seabed of the "southern sea" or Zuiderzee (Zuyderzee), a shallow inlet or bay of the North Sea which connected Amsterdam and a number of market and fishing ports around its coast (including Enkhuizen, Hoorn, Urk and Marken) with the Waddenzee (the inland sea between the Wadden islands and the North Sea on the one hand and, on the other the mainland of The Netherlands, Germany and Denmark) and from there with the international maritime world. After repeated and devastating flooding of the coastal towns and villages, national politicians decided that a permanent solution needed to be found for the problem. The first decision was to evacuate Schokland in the Zuiderzee, because repeated storm surges had eroded away large sections of the long, narrow island. A Royal Decree of 1859 ordered its population to be relocated to a number of towns and villages along the Zuiderzee coast and all buildings on the island to be demolished to discourage them from returning.[17] In the period following this, various plans were put forward to address the problem of flooding. In the end, it was Dr. Cornelis Lely who came up with a plan that passed as an Act in Parliament in 1918 (fig. 2.1). This plan would provide the blueprint for later civil engineering projects.[18] There were two parts to his plan. First, a dyke (Afsluitdijk) was to be constructed between the provinces of Noord-Holland and Friesland to close off the Zuiderzee and to prevent any further impact from the sea on the region. With the completion of this barrier in 1932, the Zuiderzee became the IJsselmeer, a non-tidal, shallow, inland freshwater lake. The

---

17 E.g. Geurts 1991.
18 E.g. Geurts 1991.

Figure 2.1: Original sketch for the planning of the IJsselmeer polders by Ing. Cornelis Lely. In red, the outline of the new polders, with the phasing schematically indicated at the bottom of the map (source: PDB Flevoland/Erfgoedpark Batavialand).

Figure 2.2: Pioneers in the polder: geological survey (source: PDB Flevoland/ Erfgoedpark Batavialand).

second phase was the creation of several large polders to increase the amount of agricultural farmland available to feed the rapidly growing Dutch population.[19] A large dam was built around the area of the intended polder, and the water within it was pumped away. The former seabed of the IJsselmeer thus became the surface of the new polders, laying several metres below sea level.

The polders were created one by one, each new reclamation project drawing on expertise gained in the previous polder. A special government civil engineering agency – the *Rijksdienst voor de IJsselmeerpolders* (RIJP) – was established for the purpose. The methods used for land reclamation were first tested on the 'pilot polder' at Andijk (Noord-Holland, 1927), after which the first large-scale polder, the Wieringermeerpolder (1930), was created in the far north of the province of Noord-Holland. Subsequently the first polder, the Noordoostpolder (1942), was created in what would later be the territory of the new province of Flevoland. It became clear that the creation of the new polders was having a major negative impact on groundwater drainage in the adjacent areas of old land. In order to curb this effect, an elongated, narrow, border lake, known as the Randmeer, was created between the old and new lands during the reclamation of Oostelijk Flevoland, the second Flevopolder (1957). The purpose of this border lake was to isolate the management of water on the new polder lands from that of the neighbouring old lands.[20] Zuidelijk Flevoland (1968) was the final polder to be created. By this time, farming had become a less significant factor in the Dutch economy, but society was increasingly becoming aware of the important nature conservation potential of the remaining part of the IJsselmeer. The new polder of Zuidelijk Flevoland also provided space for the development of the new town of Almere, built to accommodate the population overspill from Amsterdam.

Given the originally intended use of the polders as farmland, investigations were essential to establish the suitability of the soil structure of the land to be reclaimed (fig. 2.2). These investigations didn't wait, however, until after the polders were created. Even before reclamation work began, borehole samples were being collected from the seabed using boats. Analysis of the soil continued on a large scale after the polders were complete. This was carried out by recording the stratigraphic soil horizons exposed in the profiles of recently dug ditches (fig. 2.3). This work produced unique maps illustrating a 'profile code chart' for each ditch (fig. 2.4). This information was then made available to the pioneer farmers who were starting out on the new land, so that they could decide which crops could best be grown where.

The information was also used for scientific publications in which the development of the Holocene

---

19 The First World War – in which the Netherlands was neutral – was an important factor in the plans. Rather than annexing land by force to feed its population, the Netherlands decided to reclaim land from the sea.

20 The creation of the polders had a huge impact on hydrological conditions in the region, and therefore also on the preservation conditions for organic archaeological remains (see Appendix).

Figure 2.3: Pioneers in the polder: geological observations of sections of newly dug ditches (source: PDB Flevoland/Erfgoedpark Batavialand).

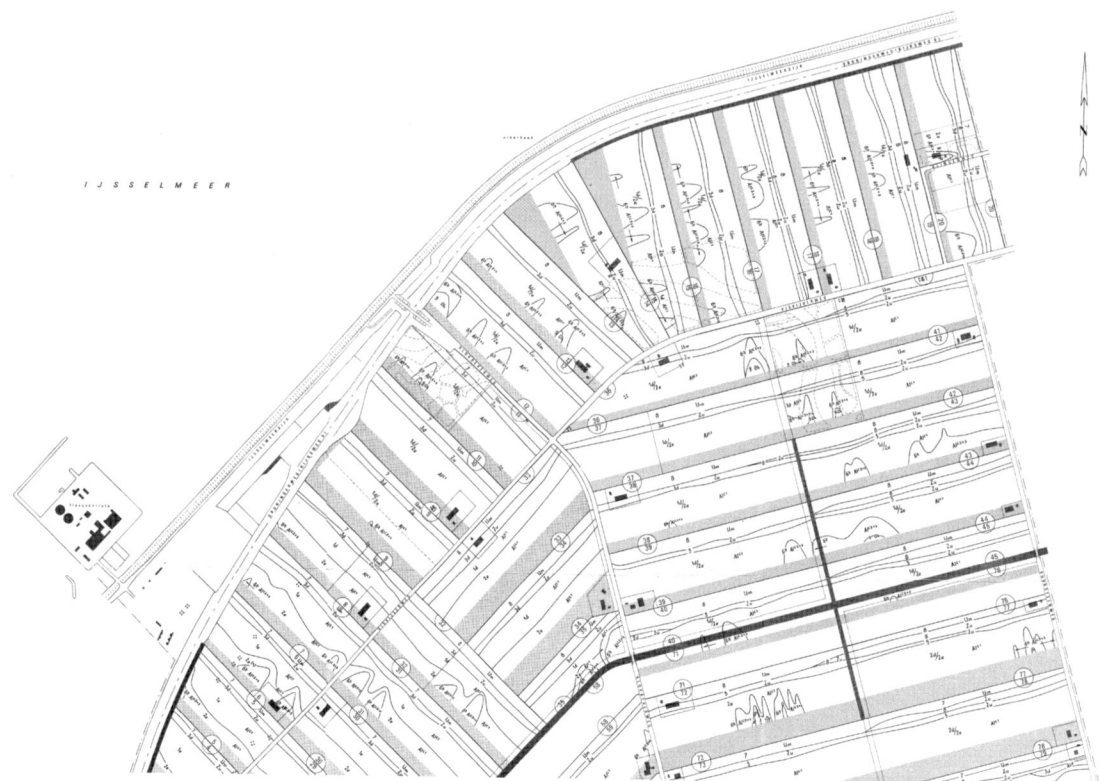

Figure 2.4: Examples of the codes used to classify the geological units discerned during the geological surveys. The section provide a general picture of the geological stratigraphy, including gully incisions (source: Bodemkundige Code- en Profielenkaart van Oostelijk Flevoland, Sectie G). The section in this figure corresponds with the area shown in figure 2.5.

landscape could be described in considerable detail thanks to the analysis of the core samples from the many boreholes and the recorded profile code charts. The physical-geographical evidence for land formation processes in the Noordoostpolder was the basis for Wiggers' doctoral thesis (1955). This became an important source of information for archaeological research in this part of Flevoland. The subsurface of Oostelijk Flevoland was described in detail by Ente *et al.* (1986), while Menke *et al.* (1998) focused on Zuidelijk Flevoland.

The RIJP also regarded archaeology as important. An archaeological survey of the Wieringermeerpolder provided the basis for Braat's doctoral thesis (1932). Braat worked at the National Museum of Antiquities in Leiden (RMO: *Rijksmuseum voor Oudheden*), one of the two research institutes for archaeology in the Netherlands at the time. The other, the Biological-Archaeological Institute (BAI) at Groningen University (BAI: *Biologisch-Archaeologisch Instituut*)[21], was involved in an archaeological survey and some small-scale excavations in the Noordoostpolder. During the Second World War, P.J.R. Modderman worked on the three topics that would come to characterise the archaeology of Flevoland: prehistoric remains from the period before the region became submerged under water, shipwrecks from the time of the Zuiderzee and the Medieval occupation of Schokland. Modderman would be the first scholar to be awarded a doctoral degree in the archaeology of Flevoland.[22] G.D. van der Heide, who worked for the RIJP from 1949 to 1974, made an important contribution to research on the prehistory of Flevoland with his publications on the Noordoostpolder[23] and Oostelijk Flevoland. It was van der Heide who began researching the sites at Swifterbant (from 1962).[24]

2.3 Research traditions in the polder

A number of bodies have contributed to the development of prehistoric archaeology in Flevoland. From a historical perspective, it is important to distinguish between three phases, each of which corresponded to a phase in the social development of the Netherlands. The first phase was dominated by a central government that was keen to exercise control over developments in Flevoland. The political idealism of a 'socially-engineered' society was a dominant principle during this period. The second phase commenced in 1972, when the BAI in Groningen took over the archaeological investigations in Swifterbant from the RIJP. Thirdly, the implementation of the Valletta Convention (1992) introduced market forces into archaeological practice in the Netherlands. As a result, the various public authorities stepped back from providing archaeological fieldwork services and left this task to the growing number of commercial companies. In Flevoland, this third phase can be seen to start with the A27-Hoge Vaart excavation, with a large budget supplied by the developer (the Ministry of Transport, Public Works and Water), although the work was still carried out by the *Rijksdienst voor Oudheidkundig Bodemonderszoek* (ROB, the State Service for Archaeological Investigations)[25]. The third phase culminated with the appointment of a municipal archaeologist in Almere in 2000.

The RIJP and academic curiosity

The first organisation to carry out archaeological investigations in Flevoland has already been mentioned: the *Rijksdienst voor de IJsselmeerpolders*, or RIJP. Initial investigations concentrated on the archaeology of the Wieringermeerpolder. After the creation of the Noordoostpolder, the RIJP relocated there and the archaeological finds from the Wieringermeerpolder investigations were deposited in the provincial depot for archaeological finds from Flevoland. G.D. van der Heide continued to work for the RIJP for many years, playing an important role in Flevoland archaeology. It soon became clear that the investigations at Swifterbant had uncovered important new finds from the natural levees and adjacent stream channels: a cultural group for which no evidence had previously been recorded. Van der Heide revealed the finds in several publications.[26] It is important to note that he did not target his findings exclusively at academics; he also produced several publications for the broader public. In this sense, he contributed to another key feature of Flevoland archaeology: public awareness (see below). Under the direction of the RIJP, several of the sites at Swifterbant were set aside from agricultural activities, and were instead protected *in situ* within a reconstructed natural environment: the levees identified under the surface were grassed over and the course of stream channels and the marshlands behind the levees were planted with trees (fig. 2.5). Visitors to the area today are thus able to experience to some extent the scale of the prehistoric landscape. Since 2017 several of these sites have been listed as archaeological sites of national importance (national archaeological monuments).

The BAI, in the person of P.J.R. Modderman, was involved in the archaeology of Flevoland at an early stage, although after Modderman completed his doctorate in 1945, the institute would not play a role in Flevoland archaeology again until 1972. In the summers between

---

21 Predecessor of the Groningen Institute for Archaeology of the University of Groningen.
22 Modderman 1945.
23 Van der Heide 1950, 1951, 1955; Van der Heide & Wiggers 1954.
24 Van der Heide 1965a, 1965b.
25 Predecessor of the Cultural Heritage Agency of the Netherlands.
26 Van der Heide 1965a, 1965b.

Figure 2.5: Aerial view of the area west of Swifterbant. The rectangular parcels are crosscut by sinuous patterns indicative of prehistoric rivergullies and banks. Modern land use has been adapted to the preserved palaeolandscape and archaeological heritage as demonstrated by several excavations in the area (source: Google Earth).

Figure 2.6: Pioneers in the polder: excavation near Swifterbant by the Biologisch-Archaeologisch Instituut of the University of Groningen in 1975 (source: PDB Flevoland/Erfgoedpark Batavialand).

1972 and 1979 the BAI carried out seasons of excavations at various locations in the Swifterbant region under the direction of J.D. van der Waals (fig. 2.6). Along with fellow Groningen archaeologist H.T. Waterbolk, Van der Waals set out a specific research programme for Swifterbant, an important innovation compared to the RIJP investigations, and a first for Dutch archaeology.[27] The programme focused on four research themes. The first of these was to position the occupation remains at Swifterbant in a chronological sense, between the Mesolithic and the Funnelbeaker culture (*Trechterbeker cultuur*,TRB), and in a geographical sense, as the link between the Mediterranean region and southern Scandinavia. The second focus aimed at reconstructing the landscape of the region and providing a model for how it had been used. The third focus built on this: research would be carried out to ascertain the times of year when the inhabitants of Swifterbant exploited the micro-region, and the seasons when other parts of the wetlands in the western Netherlands or, alternatively, the dry Pleistocene areas were exploited. The fourth research focus concerned the transition from forager or hunter-gatherer to food-producing subsistence.

Three American and Canadian archaeologists were involved in the BAI research: T.D. Price, R. Whallon and C. Meiklejohn. Price and Whallon each carried out separate fieldwork on a river dune, where they expected to find evidence for the transition from a hunter-gatherer to farming subsistence. Though Price's study was published,[28] it did not produce the desired results. This was due to the absence of a clear stratigraphy and the poor preservation conditions which meant that only flint, lithics and pottery were found. Whallon's study was only published in the form of a preliminary report.[29] Meiklejohn and fellow Utrecht anthropobiologist T.S. Constandse-Westermann worked on the human skeletal material from Swifterbant.[30] The contributions made by the American archaeologists extended further than their excavations and the subsequent analysis, however. In the American academic tradition, prehistoric archaeology and cultural anthropology were much more closely linked than in the Dutch tradition, which had strong roots in biology and earth sciences. Though this did not necessarily give rise to new research methods, it did raise new research questions. The Americans introduced the idea that fieldwork could be used to test hypotheses concerning human behaviour, in line with the theoretical developments in processual archaeology (see also 2.4).

The vast scale of the BAI's Swifterbant project eventually meant that it was not published in its entirety. After a series of 14 'Swifterbant contributions' to the journal *Helinium* (1976-1985) and four 'Final Reports on Swifterbant' in the journal *Palaeohistoria* (1978-1981), several material assemblages remained unpublished.[31]

After the BAI's final season at Swifterbant in 1979, the University of Amsterdam's Institute for Prehistory and Protohistory (A.E. van Giffen *Instituut voor Prae- en Protohistorie*, IPP) took up the baton. Between 1984 and 1990, research concentrated on the Noordoostpolder, mainly on the site of Schokland-P14 (fig. 2.7). The 'Wet Heart Project', as it was known, was led by J.A. Bakker and, like the Swifterbant project, had a research programme with a clearly processual approach.[32] Ten Anscher distinguished nine objectives within the programme that can be grouped into three topics:[33] The first was to build a chronology of occupation at Schokland-P14 based on the typology of the pottery; the second was to place the pottery assemblages in a broader geographical framework; the third was to consider the occupation history of the site in relation to landscape developments. Three PhD students would carry out research for their doctoral theses, focusing respectively on the landscape, the ecological archaeology and the many cultural remains at Schokland-P14.[34] This last study, in particular, turned out to be much more complex than originally envisaged. For example, during the excavation it seemed that the excavated fill from a gully contained finds dating to a late Swifterbant phase,[35] but when a series of ^{14}C dates became available, it became clear that the evidence for the chronological range represented in the gully fill was much greater. As a result, the relationship between Swifterbant and the Funnelbeaker culture (TRB) had to be reconsidered (see also 2.4).

Eventually, after a break of exactly 25 years, the Groningen Institute of Archaeology (GIA; formerly BAI) picked up the thread again with a small-scale excavation at Swifterbant led by D.C.M. Raemaekers. In the context of this 'New Swifterbant Project', annual fieldwork campaigns took place between 2004 and 2010, involving students and amateurs from the National Amateur Archaeology Association (AWN: *Archeologische Werkgemeenschap Nederland*) (fig. 2.8). This project was also based on a processual research programme concentrating on three

---

27 Van der Waals & Waterbolk 1976.
28 Price 1981.
29 Whallon & Price 1976.
30 Constandse-Westermann & Meiklejohn 1979; Meiklejhon & Constandse-Westermann 1978.

---

31 Studies of material from Swifterbant have fortunately been published in later periods Zeiler 1997 (animal bone); De Roever 2004 (ceramics); Devriendt 2013 (lithic and stone material).
32 Ten Anscher & Gehasse 1993; Ten Anscher *et al.* 1993; Ten Anscher 2012, 31-40.
33 Ten Anscher 2012: 32.
34 Gotjé 1993 (landscape); Gehasse 1995 (ecological archaeology); Ten Anscher 2012 (cultural remains).
35 Hogestijn 1990; Ten Anscher *et al.* 1993.

Figure 2.7: Work in progress, archaeological research at Schokland-P14 (1987) by Dr Jan-Albert Bakker and (then student) Theo ten Ascher, A.E. of the Giffen Instituut voor Prae- en Protohistorie, University of Amsterdam (source: Ten Ascher 2012).

Figure 2.8: Work in progress, archaeological research at Swifterbant-S25 in 2009 by the Groningen Institute of Archaeology (© University of Groningen, Groningen Institute of Archaeology).

Figure 2.9: The state secretary of Education, Culture and Science Aad Nuis (second on the right) is guided by Willem-Jan Hogestijn at the excavation of Hoge Vaart A27. This excavation played an important role in the implementation of the Valletta treaty and consecutive development of archaeological order in the Netherlands (photo: Cultural Heritage Agency of the Netherlands).

related themes and objectives:[36] firstly, to generate a broader insight into landscape use outside the settlements; secondly, to obtain a clearer picture of the functional differences between the various sites in order to gain a better understanding of the relationship between them; thirdly, research focused on the diachronous developments: from Late Mesolithic hunter-gatherers, by way of semi-agricultural Swifterbant communities to fully-fledged farming communities. The fieldwork was published in site reports,[37] while studies derived from the results were published in the form of academic articles.[38] The research also led to three PhD theses on the lithic and flint assemblages, the landscape and exploitation, and on aspects related to archaeological heritage management.[39]

New statutory frameworks

Around the turn of the millennium the context in which archaeological research was to be carried out fundamentally changed. This change included steps towards the decentralisation of curatorial responsibilities from national to regional and local authority level. When Flevoland became the twelfth province of the Netherlands in 1986, a provincial archaeologist was appointed to meet the provincial authority's responsibilities in this regard. The first provincial archaeologist, J.W. Hogestijn, was appointed in 1989. He had previously been involved in the IPP's 'Wet Heart Project'. Hogestijn, working under the auspices of the ROB, carried out several small-scale excavations on Mesolithic and Neolithic sites in the Noordoostpolder. In 1994 he initiated the large-scale excavation at Hoge Vaart-A27 on behalf of the ROB (fig. 2.9). The excavation was carried out as a consequence of construction work for the final section of the A27 motorway across the polder. It was the first Flevoland excavation with a large budget, provided by the Ministry of Transport, Public Works and Water Management. In contrast to the small-scale university excavations, an extensive area of 8600 m^2 containing evidence of occupation dating to the Mesolithic and the Swifterbant culture was excavated by paid staff – some of them on a job creation scheme – rather than by students and volunteers. The project was long-running, with fieldwork being carried out continuously from mid-1994 to early 1997. The final publication appeared in 2001, presenting a complete overview of all the results.[40]

The excavations on the Hoge Vaart-A27 site took place on the eve of the implementation of the Valletta Convention in Dutch legislation, and was literally the backdrop to the political debate as to what form this implementation should take. The context in which the excavation took place was completely new for The Netherlands: paid workers employed on fixed-term contracts, a project-based management approach and the use of an automated finds registration and processing system. This laid the foundations for procedures and methodologies for the study of prehistoric sites that would be adopted and further developed in subsequent large-scale commercial projects, such as the excavation of the Mesolithic and Early

---

36 Raemakers *et al.* 2005.
37 Prummel *et al.* 2009; Raemaekers *et al.* 2014.
38 Huisman *et al.* 2008; Schepers *et al.* 2013; Raemaekers *et al.* 2013; Huisman & Raemaekers 2014; Raemaekers 2015.
39 Devriendt 2013; Schepers 2014; Woltinge 2020.

40 Hogestijn & Peeters (eds) 2001.

Neolithic sites at Hardinxveld-Giessendam, carried out in connection with the construction of the Betuwe freight rail link from Rotterdam to the German border.[41]

The Hoge Vaart-A27 excavation also raised awareness of the fact that important remains of prehistoric occupation were preserved in Zuidelijk Flevoland and that archaeology should therefore be recognised as a material consideration in major spatial developments, particularly in the municipality of Almere. From the very beginning, the local authority in Almere had been instructed to focus on accommodating a rapidly growing population. With that in mind, it is quite remarkable that the local authority decided that archaeological heritage management should be a permanent consideration in the planning procedures as it set about building the approximately 3000 hectares of new developments. In 2000 a municipal archaeologist, J.W. Hogestijn (the former provincial archaeologist for Flevoland), was appointed in Almere to shape the new policy. Almere is currently the only municipality in Flevoland with its own archaeological department. The five other municipalities rely on the expertise of the provincial authority's archaeology and heritage management support service (*Steunpunt Archeologie en Monumenten*).

Currently, most archaeological research in Flevoland is carried out under the Heritage Act (2016). A considerable proportion of the fieldwork comprises geoarchaeological borehole evaluation surveys. Only a small proportion of the sites discovered have subsequently been excavated, for example along the route of the Hanzelijn rail link and the N23 trunk road (see Appendix for an archaeological description of these sites). Small-scale investigations are also carried out, for instance by the Cultural Heritage Agency of the Netherlands (RCE:*Rijksdienst voor het Cultureel Erfgoed*, the former ROB),for the purposes of archaeological heritage management, to establish and monitor the preservation conditions at sites with scheduled protected monument status (or protection pending). Fieldwork performed in a purely academic context is limited and is always small scale, as in the case of the 'New Swifterbant Project' mentioned above.

2.4 Research topics and approaches

The narratives presented in this account of the history of archaeological research in Flevoland are all very similar. This history cannot readily be divided into paradigmatic research traditions. In the Flevoland context, it is better to regard these traditions as more or less parallel research perspectives. The culture-historical approach to Flevoland archaeology began with Modderman's 1945 survey, and continued right through to Ten Anscher's thesis in 2012. This tradition, with its emphasis on identifying units in time and space (archaeological cultures), provides the framework for other perspectives, but the short history of research in Flevoland means that this framework is not yet fully defined (see 2.5). Alongside the culture-historical approach, a processual tradition developed out of the BAI research from the 1970s onwards. This focused particularly on development processes (Neolithisation) and the relationship between landscape and occupation. This processual tradition also persists to this day, as evidenced by the research questions used in large-scale developer-funded investigations. Post-processual approaches are rare. This is undoubtedly connected with the fact that the buried prehistoric landscapes of Flevoland are not easily 'experienced'. Flevoland archaeology does however provide scope for a post-processual approach: studies that explore the conceptual relationships between different forms of material culture. Examples include the study of the use of exotic raw materials (Chapter 4), deposition patterns and deliberate fragmentation (Chapter 5).

The different research traditions share a number of research themes, including those of Neolithisation and adaptation, iconic themes in the study of the archaeology in Flevoland. The Flevoland wetlands also play an important role in the debate on the interpretation of wetland habitation. Did these communities have a wetland 'world view',[42] or should the occupation remains simply be interpreted as the particularly well-preserved remains of communities that engaged in the same spectrum of activities in other parts of the Netherlands, but for which there is less well-preserved archaeological evidence? This debate focuses on the potential (or lack of it) for cultivation and arable farming in the marshlands. The excavation of a cultivated field at Swifterbant-S4 has not resolved the debate, but has changed its character. Now that evidence for local arable farming has been established, the debate has turned to the nutritional importance of this activity. Were the occupants of Swifterbant-S4 hunting farmers or hunter-gatherers who occasionally grew crops on a small scale? (See also Chapter 4).

As remarked on above, the discoveries at Swifterbant are notable for the involvement of American and Canadian archaeologists in the research. Their input provided a catalyst for a more anthropological approach to the investigation of the Mesolithic and, more especially, the Late Palaeolithic. Another contribution to this development came from the excavations of the Late Mesolithic site on the shore of the Bergumermeer (province of Friesland, 1971-1974) led by American R.R. Newell who also worked for the BAI. Newell never actually excavated in Flevoland, but was involved in research on contemporary sites in the

---

41 Louwe Kooijmans 2001a, 2001b, 2003.

42 See Amkreutz 2013, 435.

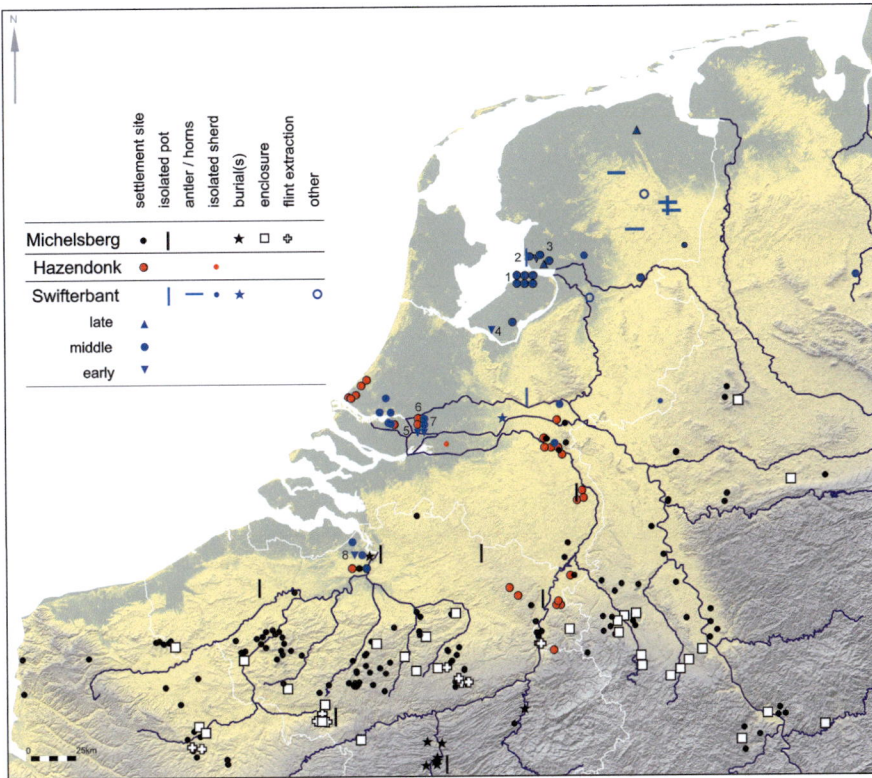

Figure 2.10: Distribution of some key sites of the Swifterbant culture, and contemporaneous traditions (1: Swifterbant, 2: Urk, 3: Emmeloord, 4: Almere Hoge Vaart/A27, 5: De Bruin, 6: Brandwijk; 7: Polderweg, 8: Doel (source: Devriendt 2014 figure 2.1).

southern Netherlands.[43] It may be that the introduction of an anthropological approach to the study of the Stone Age in the Netherlands had more to do with the particular constellation of collaboration that existed between the researchers themselves: Newell (an archaeologist) worked with Constandse-Westermann (an anthropobiologist), who in turn worked with Meiklejohn (an anthropobiologist and anthropologist), while Price and Whallon maintained contacts with the BAI during their excavations at Swifterbant and Havelte (in Drenthe).[44] This contact and collaboration meant that students also became involved in research with an anthropological or ethno-archaeological perspective. Interest in this approach continued to grow in the 1980s, as reflected, for example, in the creation of a national organisation for ethno-archaeological research (*Stichting voor Ethno-Archeologisch Onderzoek Nederland*).

Anthropological models are not generally explicitly reflected in the research designs applied to Stone Age research in the Netherlands. However, research carried out at the Ahrensburg site of Vessem by Arts and Deeben (who had close links to Newell), is an exception – the influence of anthropological concepts and explanatory frameworks inspired by ethnography is clear.[45] This is particularly evident in the interpretation of find concentrations in terms of activity areas, the reconstruction of mobility patterns and settlement systems, and the functional interpretation of lithic tools on the basis of use-wear analysis. When it comes to this last field of research, there is yet another American connection. P.A. Bienenfeld carried out the first use-wear analysis on artefacts from Swifterbant as part of her doctoral studies at the State University of New York. In addition, A.L. van Gijn, who had a Bachelor's degree from Washington State University, went on to study anthropology at the University of Groningen followed by a doctoral degree at the University of Leiden which was also based on a use-wear analysis of artefacts. Several international lines do seem to come together at the BAI and in Flevoland, albeit entirely unintentionally.

2.5 The positioning of the Swifterbant culture

One important international contribution that was to emerge from Flevoland archaeology was the Swifterbant culture. Until the discoveries made by the RIJP at Swifterbant in the early 1960s, there had been a large gap in the chronological and spatial model for the Neolithic in the Low Countries. This gap lay roughly between the Linear Bandkeramik (or LBK) of southern Limburg and Belgium (5300-5000 cal. BC), the flint mines at Rijckholt and Spiennes (from 4600 cal. BC) and the Vlaardingen-Stein-Funnelbeaker horizon (from 3400 cal. BC). This gap

---

43 Newell & Vroomans 1972; Newell 1980; Niekus 2014: Niekus, Jelsma & Luinge 2018.
44 Price 1980; Price, Whallon & Chappel 1974.
45 Arts & Deeben 1981.

has now been largely filled by the Swifterbant culture (fig. 2.10). The sites at Swifterbant not only became the type sites for the naming of the culture, but also provided evidence for a new pottery group that emerged around 4300-4000 cal. BC, referred to here as Classical Swifterbant.

The discovery of the sites at Swifterbant prompted a debate about the relationship between the pottery found there and the Ertebølle Culture of southern Scandinavia. The Ertebølle Culture has a much longer research history, starting in the 19th century. Research has established that this was a Late Mesolithic tradition with an aceramic phase (5400-4800 cal. BC) and a ceramic phase (4800-4000 cal. BC).[46] The *køkkenmøddinger* (shell middens), comprising heaps of shell waste and found at sites near the coast, are typical of this culture.[47] Thanks to its long research history, the Ertebølle culture enjoys broad international recognition, and most overviews of European prehistory contain a section on it.[48] Knowledge of the Ertebølle culture within the archaeological profession has not only relied on the dissemination of literature; personal experience has also been a factor. H.T. Waterbolk – who would later become Professor of Archaeology in Groningen – studied for a time in Denmark, and had a detailed knowledge of Danish prehistory. When potsherds from an unknown culture were found in association with a heap of shell waste at Haamstede (Zeeland) in 1957, Waterbolk published an article entitled 'Ertebølle culture in the Netherlands?'.[49] Several years later, however, the finds were attributed to the Vlaardingen group.[50]

Research around Swifterbant revealed for the second time an unknown cultural group in a wetland context. Again, parallels were drawn with the Ertebølle culture.[51] In 1979, J.P. de Roever published her article 'Swifterbant – Dutch Ertebølle?' arguing for a link with the Ertebølle culture – a source that is often cited in the international literature. As a result, the Swifterbant culture was often presented as a 'side-show' to the Ertebølle, which meant that its own identity and its role in the development of the Funnelbeaker culture remained somewhat overlooked. In the 1990s a countermovement emerged, emphasising the differences between Swifterbant and Ertebølle.[52] Kampffmeyer lamented that: 'Because of their point-based pottery, the sites from the Dutch Early Neolithic in the Rhine-Meuse Delta and the IJsselmeerpolders near Swifterbant share with Hüde I the fate of frequently being interpreted in relation to the Danish Ertebølle culture.'[53] In emphasising the differences, researchers did not simply focus on a comparison of pottery typology. It became clear, for example, that there were also no similarities to be found when comparing the flint assemblages from the two cultures.[54] More importantly, while the Ertebølle culture steadfastly maintained a Mesolithic mode of subsistence, the people of Swifterbant gradually mastered livestock and arable farming, which points to a fundamentally different world view.[55]

Interestingly, there is wide variation in Swifterbant pottery from one site to another. This has led in the past to a heated debate on the cultural affinity of, on the one hand, the archaeological remains from Swifterbant in the Flevopolders, and on the other, the remains from the Hazendonk site in the river area of the central Netherlands.[56] Louwe Kooijmans,[57] who excavated Hazendonk (1967-1976), compared the pottery from the two sites and concluded that '...there remains the distinct difference in pottery style [between Swifterbant and Hazendonk 1], a greater difference than that between Swifterbant and the considerably more remote Ertebølle culture s.s. The decorated pottery of Hazendonk 1 has its best counterparts at Hamburg-Boberg site 15'.[58] De Roever did not agree with these conclusions, and included the Hazendonk pottery within the range of variation in the Swifterbant pottery: 'In my opinion this [Hazendonk] pottery also falls within the range of variety of the Swifterbant ware'.[59] The debate was resolved by describing all the pottery in the two regions within a single system.[60] This revealed the major morphological and technological similarities, and highlighted the fact that the differences lay mainly in the frequency with which certain specific decorative motifs were used.

The finds at Swifterbant also prompted a debate, which continues to this day, about the cultural relationship between the Swifterbant culture and the Funnelbeaker culture. In the 1970s and 1980s this debate was dominated by the idea that the Funnelbeaker culture was a new cultural phenomenon in The Netherlands, whose emergence was not related to the Swifterbant culture. Arguments in support of this view, which has been dubbed the 'discontinuity model',[61] were based on the absence of Swifterbant sites in the area of the younger Funnelbeaker culture (wetlands versus dry areas), the large cultural

---

46 Andersen 2010.
47 Andersen 2000.
48 E.g. Champion *et al.* 1984, 101; Cunliffe 2008, 126-127; Whittle 1996, 154-155, 192.
49 Waterbolk 1957.
50 Van Regteren Altena *et al.* 1962.
51 E.g. Van der Waals 1972; Louwe Kooijmans 1976.
52 E.g. Ten Anscher 2012; Raemaekers 1999.

53 Kampffmeyer 1991: 159; translated by first author.
54 Deckers 1982: 38. Stapel 1991.
55 Raemaekers 1997.
56 Louwe Kooijmans 1976; De Roever 1979.
57 Curator at the National Museum of Antiquities at the time.
58 Louwe Kooijmans 1976, 259.
59 De Roever 1979, 25. See also De Roever 2004, 145-147.
60 Raemaekers 1999.
61 Raemaekers 2015.

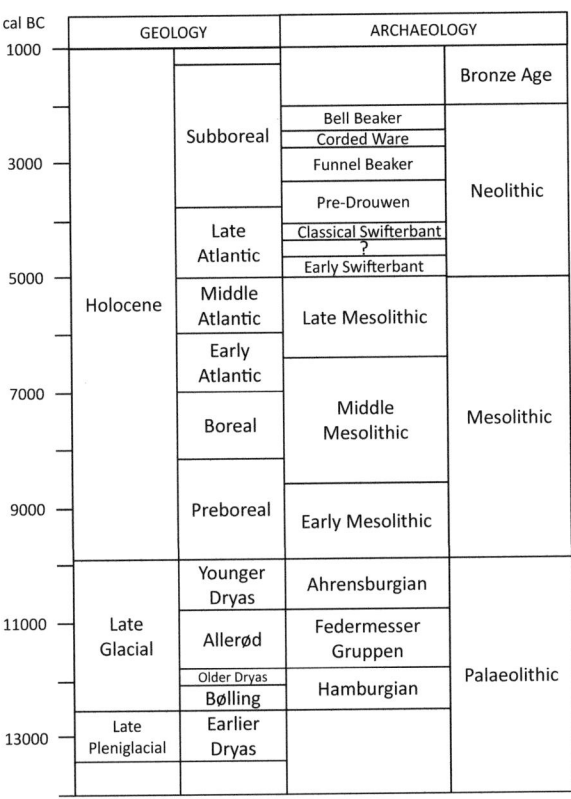

Figure 2.11 Chronology and phasing of geological and archaeological periods as used in this book.

differences in material culture and funerary practices, and the major difference in age: there appeared to be a hiatus in occupation from 4000 to 3400 cal. BC.[62]

A second model, the 'continuity model', emerged thanks to the finds made in the 'Wet Heart Project' from 1986 onwards. In this project, find assemblages excavated at Schokkerhaven-E170 and Schokland-P14, both in the Noordoostpolder, were found to occupy an intermediate place in terms of both time and characteristics between the levee sites at Swifterbant and the Funnelbeaker culture. This model assumed a continuous cultural development from the Swifterbant culture through to the Funnelbeaker culture. The intervening phase has long been known as Late Swifterbant. This perpetuated the idea both in The Netherlands and internationally that the western group of the Funnelbeaker culture was actually a late development of the culture. More recently, the term 'Pre-Drouwen' has been introduced to replace 'Late Swifterbant', in order to emphasise the continuity with the later Drouwen Funnelbeaker culture and to make it clear that developments towards the Funnelbeaker culture also began much earlier than 3400 cal. BC in The Netherlands.[63] The term 'Pre-Drouwen 'is also used in this book (fig. 2.11).

The debate about the distinctiveness of the Swifterbant culture was conducted – and indeed is still being conducted – not only in reference to the Ertebølle and Funnelbeaker cultures, but also to the Linear Bandkeramik. With the levee settlements at Swifterbant dating to between 4300 and 4000 cal. BC, there was of course a hiatus of 700 years between the end of the Linear Bandkeramik and Swifterbant. In view of the reconstructed date for the inundation of the Hoge Vaart-A27 site, there was a secret hope that activity on the site would fall in the hiatus. This did indeed turn out to be the case, and furthermore, pottery was found that could be attributed to the Swifterbant culture. Shortly afterwards the sites at Hardinxveld-Giessendam in the river area of the central Netherlands were investigated and two new data points could be added to the hiatus. This resulted in the definition of an Early Swifterbant phase that could be placed between 5000 and 4650 cal. BC, thus predating the levee settlements at Swifterbant itself. This Early Swifterbant phase, which could be interpreted as ceramic Late Mesolithic, does not align seamlessly with the Classical Swifterbant period on the basis of current data, and it is still difficult to place it in the greater geographical picture of cultural variation (fig. 2.12).

On the basis of current knowledge, we can interpret the Swifterbant culture as being representative of an independent tradition with its roots in a Mesolithic hunter-gatherer subsistence lifestyle (early phase), to which new elements were added (classical phase) and after which it evolved towards the Funnelbeaker culture (Pre-Drouwen). The geographical extent of the Swifterbant culture is not entirely clear at this stage. The proposed north-eastern boundary in Lower Saxony in Germany is difficult to define because of the almost complete lack of find assemblages. The only finds discovered are generally non-diagnostic pots that could be interpreted as Swifterbant, but could just as easily be given another cultural interpretation.[64] The most important site in Lower Saxony is Hüde I, on the Dümmermeer lake. Some of the pottery from this site clearly belongs to the Swifterbant culture, but other cultural groups are also represented. The south-western boundary has been provisionally defined by a number of sites near Antwerp in Belgium.[65] The area between the wetlands and the Central European loess region has produced remarkably few sites that can be placed with

---

62 E.g. Waterbolk 1985; Fokkens 1998, 96-100.

63 Ten Anscher 2012. The term was proposed by Bakker (1979: 115) in a description of several pottery finds that possibly predate the construction of the megalithic tombs. The author then concludes that these finds are not older than the Drouwen TRB.

64 Raemaekers 1999; Ten Anscher 2012.

65 For example Crombé et al. 2015.

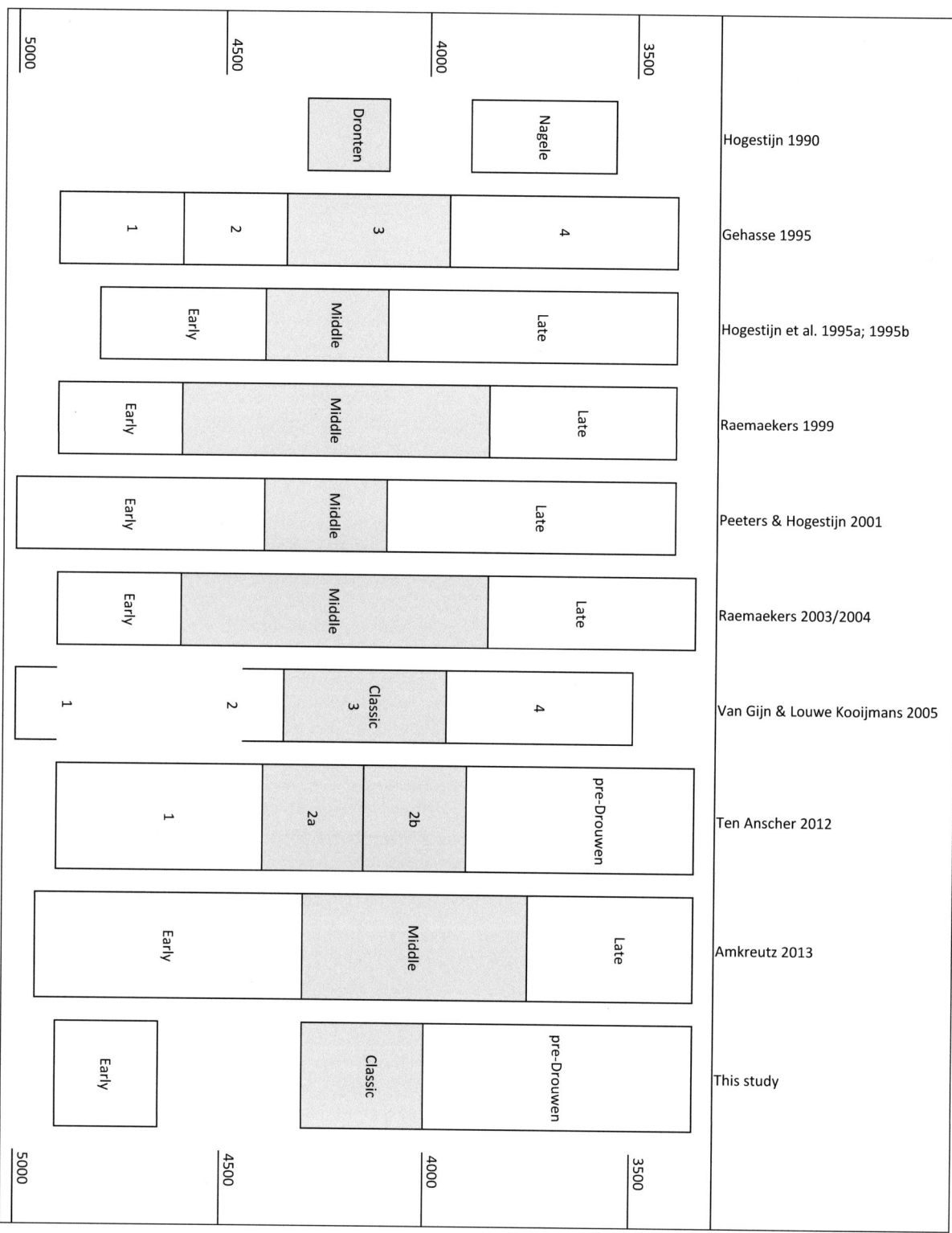

Figure 2.12: The chronological phasing of the Swifterbant culture has been changed several times during the research history, based on new finds, changing ideas about pottery characteristics and, to a lesser extent, other find categories.

Figure 2.13: Impression of the former exhibition OER! in Museum Nieuwland (Lelystad) in 2016. This exhibition highlighted the Swifterbantculture as icon of the prehistory of the province of Flevoland (source: PDB Flevoland/ Erfgoedpark Batavialand).

certainty in the time slot attributed to the Swifterbant Culture. The poor preservation conditions mean that even pottery has not generally survived, if indeed it was ever present. This makes it difficult, if not impossible, to define a regional boundary in this area, even a sketchy one.

2.6 Archaeology and the public

In academic terms, the prehistoric archaeology of Flevoland – with the Swifterbant culture at its core – has played a key role in defining the image of The Netherlands' prehistoric past. Equally important, however, is the fact that the country's youngest province has always focused on communicating with the public. This remains crucial, as few people realise that this tract of land, so recently reclaimed from the sea, has a long history of occupation before it transformed into a maritime landscape characterised by seafaring and fishing. People still react with surprise when they hear accounts of prehistoric hunters and early farmers who lived in a landscape whose remains now lie deeply buried beneath several metres of clay and peat, under the shipwrecks that are now visible, or have been made visible on the surface of the land that used to be the former sea bed.

As mentioned in section 2.3.1, Van der Heide had already begun publishing articles intended for the general public. In addition, accessible articles about new finds and research were published every year for interested amateurs in cultural heritage yearbooks and, from 1990, in the series *Cultuur Historisch Jaarboek voor Flevoland*. The Swifterbant culture was the main subject of a book intended for the public, paid for by a subsidy from the provincial authorities of Flevoland and published in 2004.[66] Peeters' doctoral thesis (2007) on the site of Hoge Vaart-A27 in Almere was also presented in the form of an accessible printed lecture.[67]

Exhibitions were staged at various locations, firstly at Museum Schokland and then at the Nieuwland Poldermuseum, where the prehistoric occupation history of Flevoland became a prominent feature in the permanent display. The first section of the permanent display that visitors encounter on entering the museum is devoted to the Swifterbant Culture. In an exhibit entitled *OER!*,[68] archaeological finds are displayed alongside monitors showing 'ethnographic images' of a living experiment conducted in the 1970s (see below) thereby bringing the objects to life (fig. 2.13).

In the context of the Hoge Vaart-A27 excavation and also during the more recent excavation at Dronten-N23, a great deal of attention was paid to engaging with the public. During construction work on the motorway, an exhibition at the visitor centre showed the progress of the archaeological excavation. A specially appointed communications officer gave weekly tours for members of the public, stakeholders and schools. School children could be given lessons on the province's prehistoric past using a specially designed education pack. A 20-minute documentary entitled 'Tarzan's Black Box' was broadcast on national television.

Particularly interesting is the role of experimental archaeology in Flevoland. In the 1970s Roelof Horreüs de Haas set up a living experiment with his family and friends

---

66   Peeters, Hogestijn & Holleman *et al.* 2004.
67   Peeters 2008.
68   PRIMAL!

Figure 2.14: Impression of the 'living experiment' of Roelof Horreüs de Haas (source: Horreüs de Haas & Horreüs de Haas 1984).

in a fairly wild and overgrown area of Flevoland (fig. 2.14). It was inspired largely by the archaeological fieldwork at Swifterbant. Although the experiment was not set up with any scientific goals in mind, the participants tried to 'survive' without any modern aids. The experiment went well until one of the children fell ill and the experiment had to be abandoned.[69] In 1987 an organisation specially established for the purpose (*Stichting Prehistorische Nederzetting Flevoland*) set up a replica prehistoric settlement in Natuurpark Lelystad. Schools and other groups still go there to experience how prehistoric people lived for a day, or longer.[70] First-year archaeology students from Groningen University also spend a weekend there before attending their first lectures. In 2005 a group of people interested in learning prehistoric skills spent four weeks at a Mesolithic camp they had set up at Oosterwold near Zeewolde.[71] Based on these experiences, Leiden University began in 2012 on a scientifically-accurate reconstruction of a Neolithic farm. This project focused

---

69   Horreüs de Haas & Horreüs de Haas 1984.

70   Van Betuw-Demon 1997; http://www.spnf.net/index.ph The location is now known as Swifterkamp.

71   Pomstra & Olthof 2006.

mainly on understanding how the Neolithic tools would have been used and on the decay process of the structure.[72]

Finally, it is important to note that the provincial depot for the archaeological archive (*Provinciaal Depot voor Bodemvondsten*), originally part of the *Nieuwland Erfgoed Centrum*, and since 2017 part of *Stichting Erfgoedpark Batavialand* plays a key role in involving amateur archaeologists (from the AWN) in fieldwork activities, including an annual field school, fieldwalking surveys and finds documentation. The manager of the depot also runs a weekly 'surgery' where members of the public can bring objects for identification or simply ask questions. These surgeries regularly unearth new, sometimes surprising, discoveries, all of which ultimately contribute towards our efforts to create a picture of Flevoland's past.

2.7 Conclusions

Research into the prehistoric archaeology of Flevoland may have only a short history, but this history clearly has its own character. It is important to note that it is actually the history of the reclamation of the Flevopolders that made the archaeological research possible. The good state of preservation of archaeological remains that over time had been buried under deep layers of deposited sediments, provided a good basis for the processual research tradition that gained influence in The Netherlands from the 1970s onwards. The archaeological remains provided important material evidence for research into the main processual themes: the relationship between humans and the landscape and the development of occupation, with a particular focus on the transition from a hunter-gatherer subsistence to a farming community.

Research into the Swifterbant culture has defined the image of the prehistoric archaeology of Flevoland in several ways. Firstly, the research connected with a global research theme: the emergence of farming communities. This almost automatically ensured that the results would reach an international audience. Secondly, the research attracted a number of academics from the United States and Canada. Their academic background, in which archaeology and ethnography were more closely connected than in the European academic tradition, left a clear mark on Dutch Archaeology. Furthermore, these researchers moved on after a while, taking their experiences in The Netherlands with them, and becoming important advocates for the international importance of the Swifterbant culture. Thirdly, research on the Swifterbant culture made an important contribution to the phasing of the Neolithic in the Low Countries. Prior to the excavations at Swifterbant, archaeological cultures like the Funnelbeaker and the Vlaardingen/Stein group did not appear to have any roots in our region.

Another vital aspect of Flevoland archaeology is that from the very beginning it was not merely about scientific research, but about communicating new insights and understanding to the broader public. Hopefully, this publication will also contribute to those efforts.

---

72 Van Diepen 2013; Pomstra & Van Gijn 2013. Zie ook http://experiment-horsterwold.nl/.

Chapter 3

Hidden landscapes
Mapping and evaluating deeply buried remains of human activity

J.H.M Peeters & B.I. Smit

3.1 Introduction

The previous chapter has explained how research into the remains of prehistoric occupation in Flevoland took shape over the years. The chance discovery of the first artefacts was linked to an extensive soil mapping survey in the 'New Land' to assess the suitability and potential of the soil for development as agricultural land. The archaeological investigations carried out since have always been closely linked to soil and geological mapping and have even included these as part of fieldwork programmes. Research into the soil and lithostratigraphic characteristics of the region remains crucial when trying to understand the archaeology of the area and interpret the remains of prehistoric occupation.

The patterns found in the archaeological data cannot be seen in isolation from developments in the palaeolandscape and the lithostratigraphy of the region. Climate change and associated sea-level rise since the end of the last ice age, the Weichselian glacial, were the main drivers of these developments. The processes linked to these developments were caused by a complex sequence of sedimentation and erosion which resulted in the formation of the subsurface as we now know it. These landscape formation processes not only had an impact on the geographical location of sites, but also on the preservation of remains from the prehistoric past. After the initial formation of the record by human activity – depositional processes, in other words – post-depositional processes determine what the archaeological landscape ultimately contains. These characteristics in turn have a great impact on the potential for detecting evidence for prehistoric activity.

This chapter explains factors that are important for an understanding of how the archaeological record in Flevoland as we now know it was formed and how the characteristics of buried landscapes (see Chapter 6), and the remains of human activity found in them, impact on the results of archaeological prospection in the region.

3.2 Climate, sea level rise and the structure of the subsurface

3.2.1 Overview of developments

Since the end of the Last Glacial Maximum (LGM) some 20,000 years ago, the temperature has risen on a global scale. This was, however, no rectilinear event: temperature fluctuations during the Late Glacial, in the period immediately preceding the Holocene, produced rapid and relatively abrupt shifts between warm (interstadial) and cold (stadial) phases. The sea level was tens of metres lower than it is today – during the LGM some 120 m

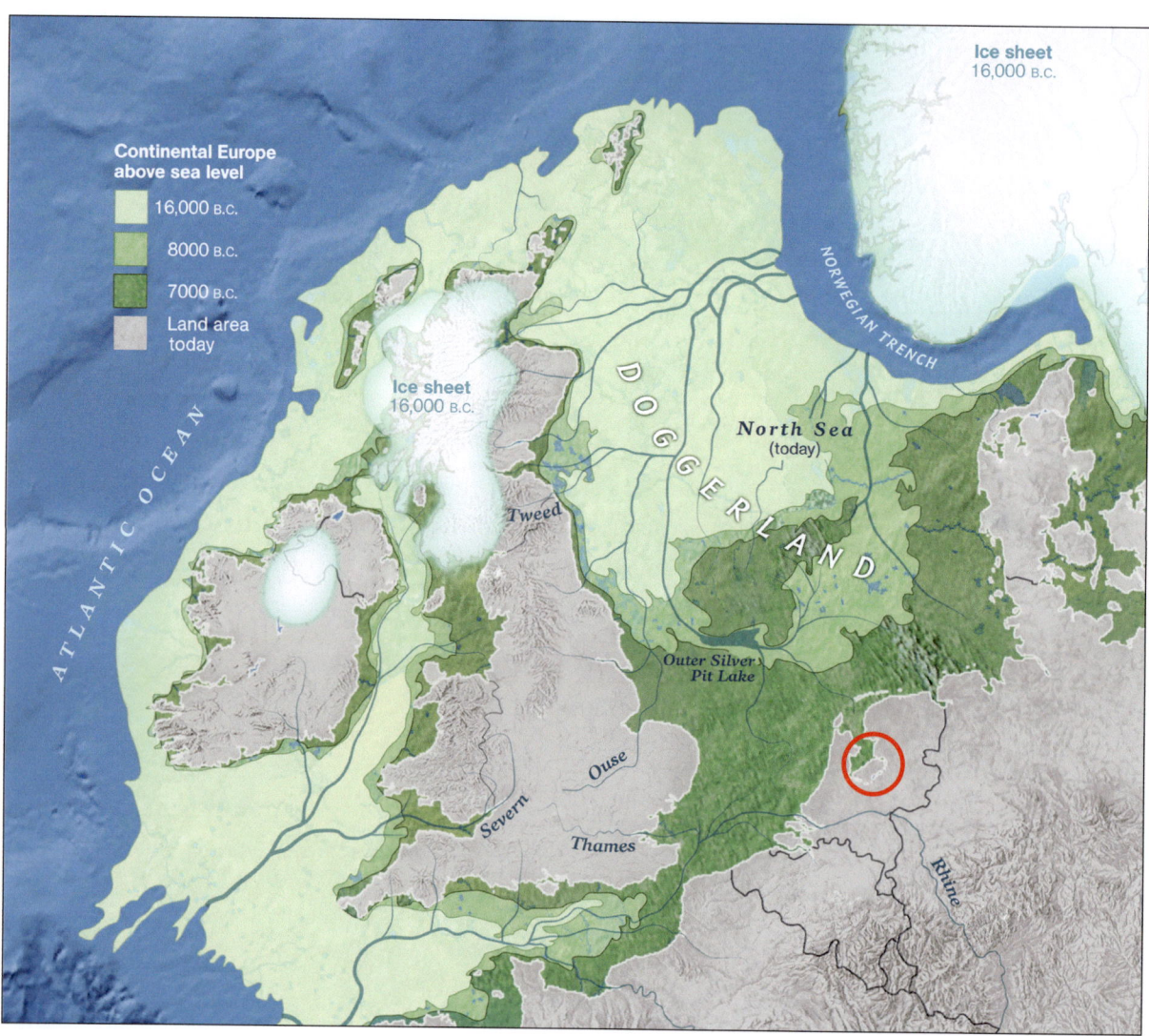

Figure 3.1: Flevoland positioned in the north-western European landscape at the end of the Late Glacial Maximum. The landscape in the present North Sea Basin would gradually drown, due to sea-level rise, leading the region of present-day Flevoland to progressively be closer to the coastline (reproduced with permission of National Geographic).

lower – and the North Sea did not exist. At the end of the Pleistocene the relative sea level was still some 70 m below where it is today and the southern North Sea was still dry land. The region that became Flevoland thus lay inland, at a great distance from the former coastline, which ran from southern Scotland, along Dogger Bank to northern Denmark (fig. 3.1). However, under the influence of a structural rise in temperature, the ice caps rapidly melted and the sea level rose, causing the coastline to shift rapidly.[74] This rapid landward retreat, or transgression, of the coastline occurred particularly towards the Netherlands, where there was only a slight incline in the land surface.

During the Early Holocene, between c. 11,700 and 6800 cal. BC, rising sea levels meant that Britain was cut off from the northwest European continent when a connection formed between the northern sea and the sea arm that flowed in a northerly direction via the English Channel and the Straits of Dover. Importantly, this connection allowed a marine environment with a steadily stronger tidal regime to develop. As the North Sea expanded, Dogger Bank would have remained visible as an island between the Netherlands and England for some time. Between 6500 and 6300 cal. BC the sea level rose 4 m, from approx. 19.5 to 15.5 m -NAP, due to a combination of factors: a structural sea level rise of 2 m, and an extra rise of 2 m caused by the rapid drainage of Lake Agassiz in

---

74 Cohen *et al.* 2017.

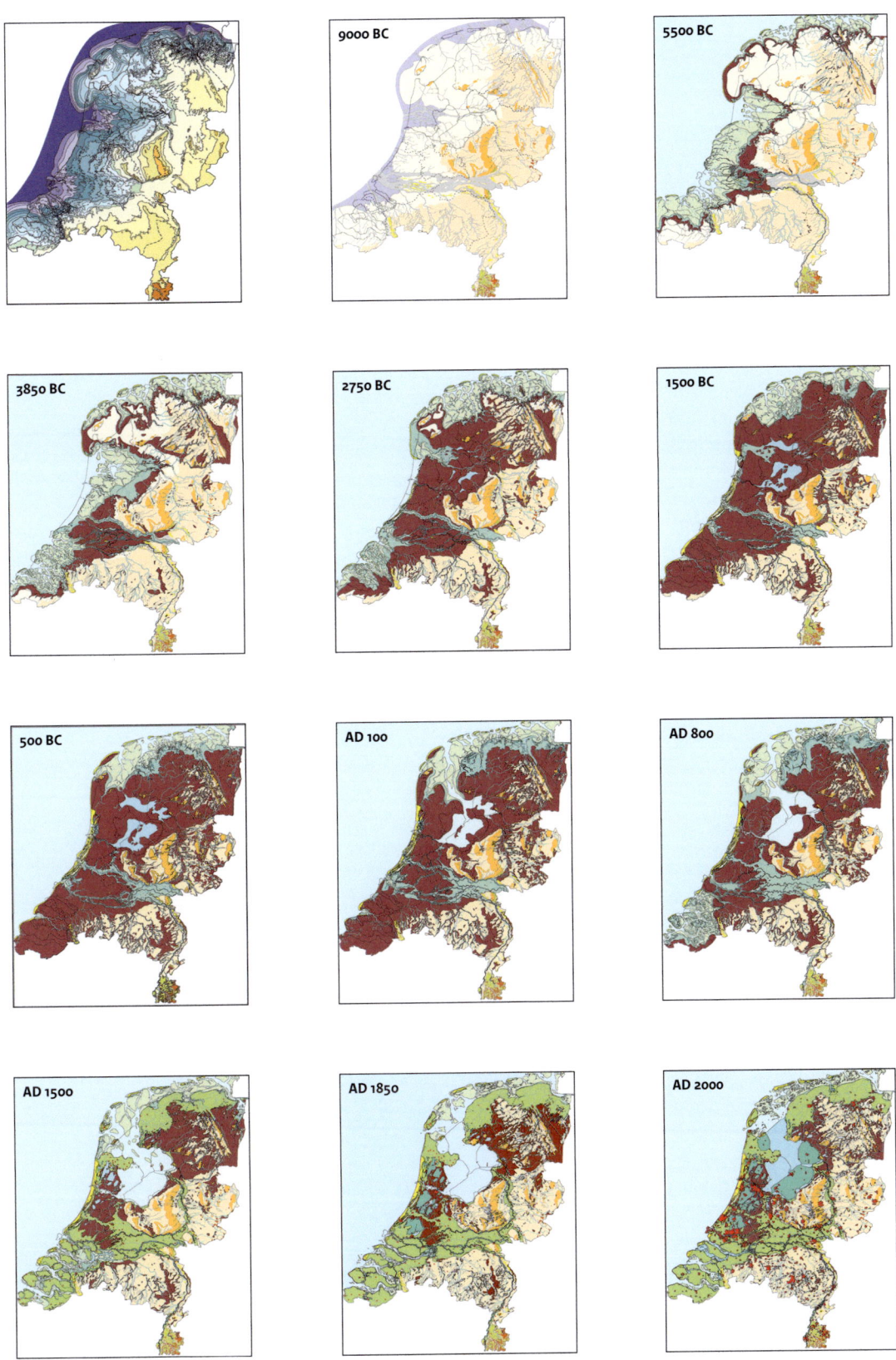

Figure 3.2: The palaeogeographical development of the Netherlands demonstrates how the Flevoland area was positioned in large-scale landscape change during the Holocene (after Vos & De Vries 2018).

Canada.[75] Dogger Bank was submerged shortly afterwards, by which time the eastern North Sea coast was roughly in the position of the current Dutch coastline.

Along the coast of the western Netherlands the inundation process and coastal development were strongly influenced by two important river systems. Between Rotterdam and Leiden, the Rhine-Meuse system formed an estuary in which, during the Atlantic, sedimentary regimes more or less kept pace with the decrease in sea level rise.[76] As a result, the coastline stabilised and, during the Subboreal, the Rhine was even able to develop a small delta. A second estuary was located slightly further to the north, between Haarlem and Alkmaar. A number of smaller rivers – the Overijsselse Vecht, the Hunnepe and the Eem, the three main rivers in prehistoric Flevoland – flowed through this estuary into the sea (fig. 3.2). There was limited sediment influx here during the Atlantic, which meant that the wide estuary remained intact and open for longer. Despite the further rise in sea level, Flevoland encountered virtually no direct effects from the sea via this estuary. The expansion of the peat and the marine influx of sand along the coast caused the system to become blocked in the Subboreal, which meant that the coastline remained a good distance from Flevoland. An important opening in the coastline at Bergen silted up around 500 cal. BC. This meant that the Overijsselse Vecht and Hunnepe were no longer able to drain into the sea at this location. A gap in the coastline at Castricum also silted up between 200 and 100 cal. BC.[77] As a consequence of the poorer drainage conditions, a large complex of lakes known collectively as Lake Flevo, was formed. At a later period, in the Subatlantic, these lakes developed into Lake Almere, which remained connected with the sea to the north. The enlargement of this sea connection as a consequence of a series of floods, led to the formation of the Zuiderzee (c. AD 1250) (fig. 3.2).

3.2.2 Sea-level and groundwater-level rise in Flevoland

To gain a better understanding of the properties of the subsurface in Flevoland and the archaeological remains within it, we need to take a closer look at the subsequent rise of the groundwater level in the light of sea-level rise and the general palaeogeographical developments described above. Firstly, it is important to note that relative sea-level rise along the Dutch coast varied – and indeed still does – due to differences in land subsidence resulting from tectonic and glacio-isostatic factors.[78] Differences in soil compaction and sediment load also played an important role. Land subsidence gradually increased from the south (Flanders, Zeeland) towards the north (western and northern Netherlands), which meant that the groundwater-level curve in the north was steeper than that in the south (fig. 3.3).[79] This was particularly the case before 7000-6500 cal. BC as a large expanse of land ice was still melting and the amount of ocean water was increasing. Thereafter the curves flattened and regional differences were caused mainly by variations in isostacy, subsoil compaction and sediment load.

Reconstructions of relative sea level rise have always played an important role in archaeological research in Flevoland, providing insight into what parts of the province were habitable in which periods, for example. Until the 1980s, average sea-level curves were based on time-depth data from the river area of the central Netherlands and the North Sea basin.[80] New dates from basal peat samples from the Noordoostpolder have, however, produced a groundwater-level curve that deviates from the sea-level curve generally used in the past. The new curve suggests that the groundwater level in the Noordoostpolder lay below the average sea level from c. 4800 cal. BC onwards.[81] Further research has shown that the data for the Noordoostpolder is reliable and that the deviation from the average sea-level curve is probably the result of contaminated peat samples.[82]

The curve for the Noordoostpolder and a curve derived for Zuidelijk Flevoland both provide a fairly good picture of structural groundwater-level rise and the expansion of gradually waterlogged conditions across the landscape which were related to relative sea-level rise (fig. 3.4).[83] It is clear that this process began fairly late. While the sea level was at approx. 15.5 m -NAP around 6300 cal. BC, it would take another few hundred years before the lowest-lying parts of Flevoland, which were approx. 13 m -NAP, really did become waterlogged. In the preceding period any groundwater-level rise was due to local circumstances and variables, *i.e.* the potential for the water to drain away via the soil or via streams and rivers. Poorly-drained depressions and relict river and stream meanders (cut offs) filled with stagnant water. The degree to which this

---

75 Hijma & Cohen 2010.
76 The most important factor in the relative rise of the sea level along the Dutch coast was now land subsidence, rather than the melting of the ice caps.
77 Vos 2015.
78 In the Weichselian, The Netherlands was the 'forebulge' of the terrestrial ice cap that lay over southern Scandinavia and northern Germany. As the ice cap melted the forebulge slowly subsided as the areas previously covered by land ice 'sprang back'. This phenomenon, known as post-glacial rebound, is part of the process known as glacio-isostasy.
79 Kiden, Denys & Johnston 2002.
80 Jelgersma 1979; Van de Plassche 1982.
81 Gotjé 1993.
82 Van de Plassche *et al.* 2005.
83 Makaske *et al.* 2002, 2003; Van de Plassche *et al.* 2005.

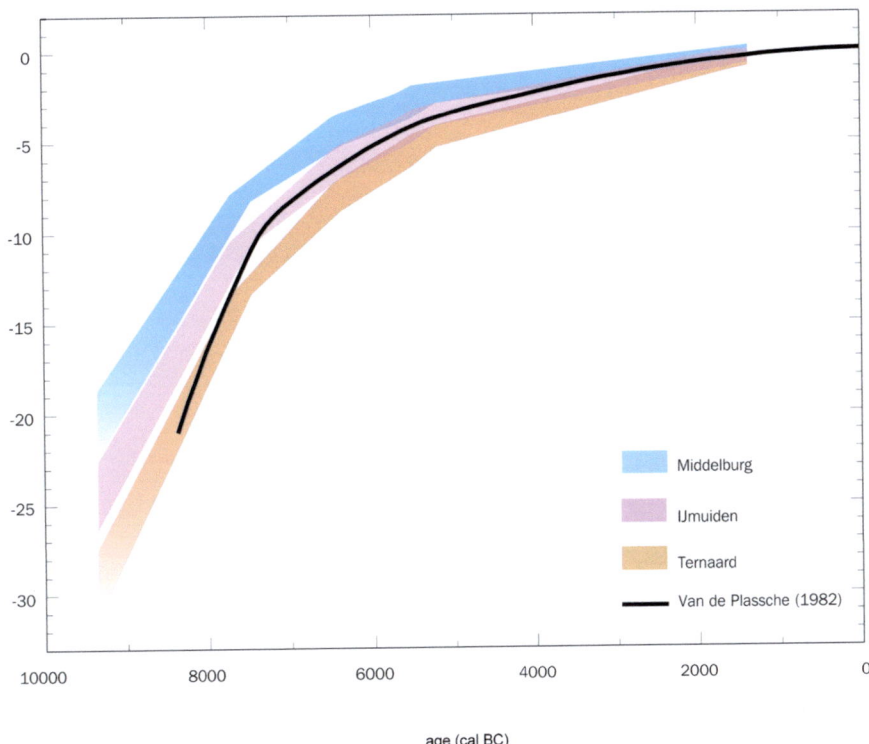

Figure 3.3: Differences in relative sea-level rise along the Dutch coast during the Holocene. The time-depth curves for Middelburg (South-West), IJmuiden (North-West) and Ternaard (North) are variably positioned relative to the average sea-level curve of Van der Plassche. The differences in the earlier half of the Holocene are mainly due to changing forebulge effects, whereas the convergence in the later half corresponds to decreasing forebulge effects and lower absolute sea-level rise (source: Vos & Kiden 2005).

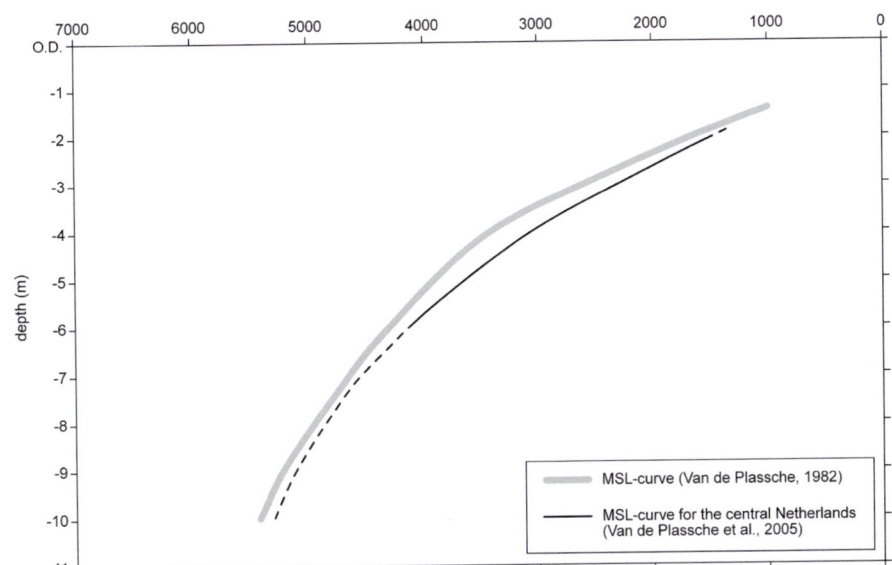

Figure 3.4: Relative sea-level curve for Flevoland, which is slightly lower than the a relative sea-level curve for the Netherlands (adapted from van de Plassche et al. 2005).

occurred also depended on the specific climate conditions and the capacity of surrounding vegetation to buffer water.

From ca. 6000 cal. BC, when the entire region became structurally waterlogged, the average groundwater level rose gradually, starting in the west and progressing towards the east. The process began in the river valleys of the Eem in Zuidelijk Flevoland, the Hunnepe in Oostelijk Flevoland and the Overijsselse Vecht in the Noordoostpolder. Within the territory of present-day Flevoland these river valleys were separated by extensive areas of undulating coversand (fig. 3.5). Again, localised higher groundwater levels occurred as a result of localised conditions. This could easily be confirmed in Zuidelijk Flevoland, where local underwater deposits turned out to lie 1.5-2 m above the regional groundwater-level curve – between the dates 5400 and 5000 cal. BC the local groundwater level probably lay between 2 -3 m above the average groundwater level.[84] After that date, the local and the average groundwater levels slowly began to converge; the local regime began to

---

84 Peeters 2007, 54-55.

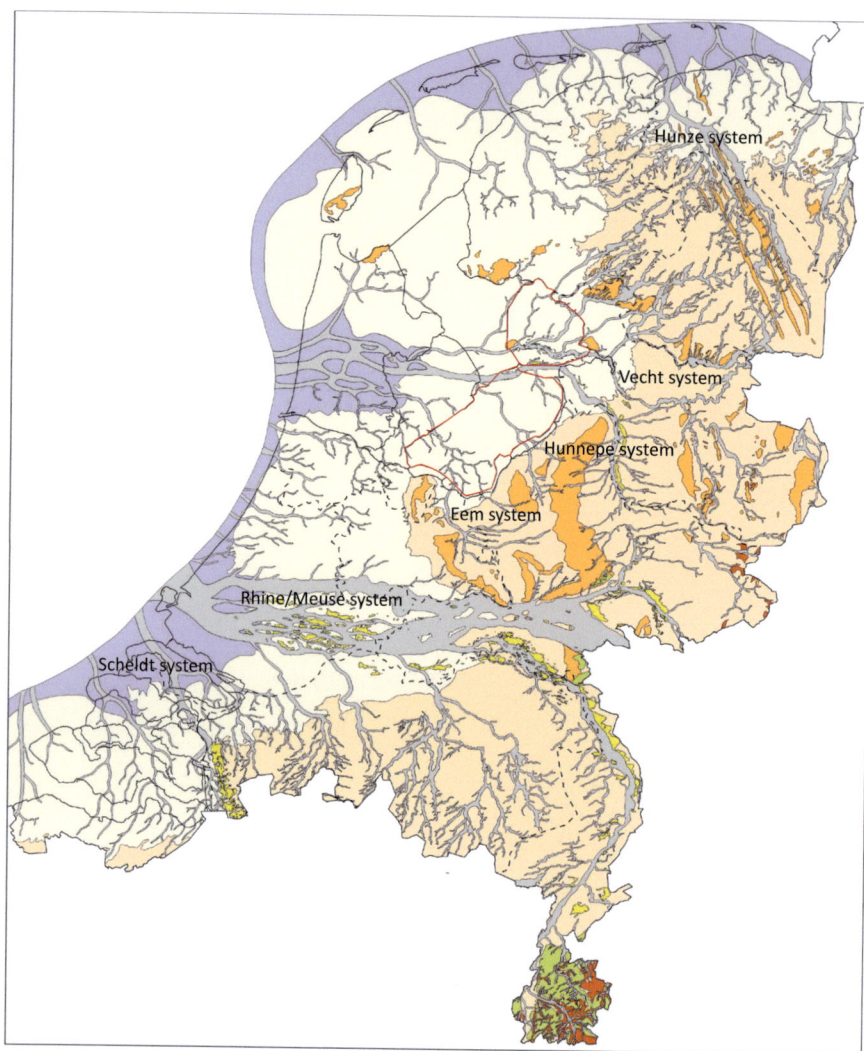

Figure 3.5: Elevation model of the Pleistocene surface and drainage systems of the Netherlands. The drainage of the Flevoland area (central in the figure) was determined by the rivers Eem, Hunnepe, and Vecht (after Vos 2015).

make way for the regional regime, which was controlled primarily by sea-level rise. Such a process eventually took place over the whole of Flevoland, with localised hydrological conditions ultimately being absorbed into the regional regime.

A simple computer model shows that, around 4000 cal. BC, the majority of the region might have been dominated by marsh vegetation and open water.[85] We must not, however, forget that the landscape grew more waterlogged as a result of constant interaction between local and regional conditions and processes. The different characteristics of the river systems (Overijsselse Vecht, Hunnepe, Eem) that drained the region may have played a key role in this. It is, for example, striking that there is little evidence of any structural influence from the sea, despite the fact that the region lay on the edge of the coastal zone and was connected to the sea. The same computer model systematically shows large areas of open water from 4300 cal. BC onwards, but it is likely that this was not actually the case, at least not on such a large scale.[86] A considerable proportion of this modelled open water was probably peat bog, which may have helped buffer any marine influence, certainly as the coastline gradually closed in the Subboreal. The deteriorating drainage ability of the rivers would, however, have caused large, deep lakes to form, meaning there would have been more and more open water during the Subboreal and Subatlantic (Lake Flevo complex followed by Lake Almere).

3.2.3 The structure of the subsurface

It will be clear from the above that the changing palaeogeographical situation and hydrological conditions since the beginning of the Holocene influenced the

---

85 Peeters 2007, 56-74.

86 The relative proportion of open water decreases as more sedimentation of clay is introduced into the model (Peeters & Romeijn 2016).

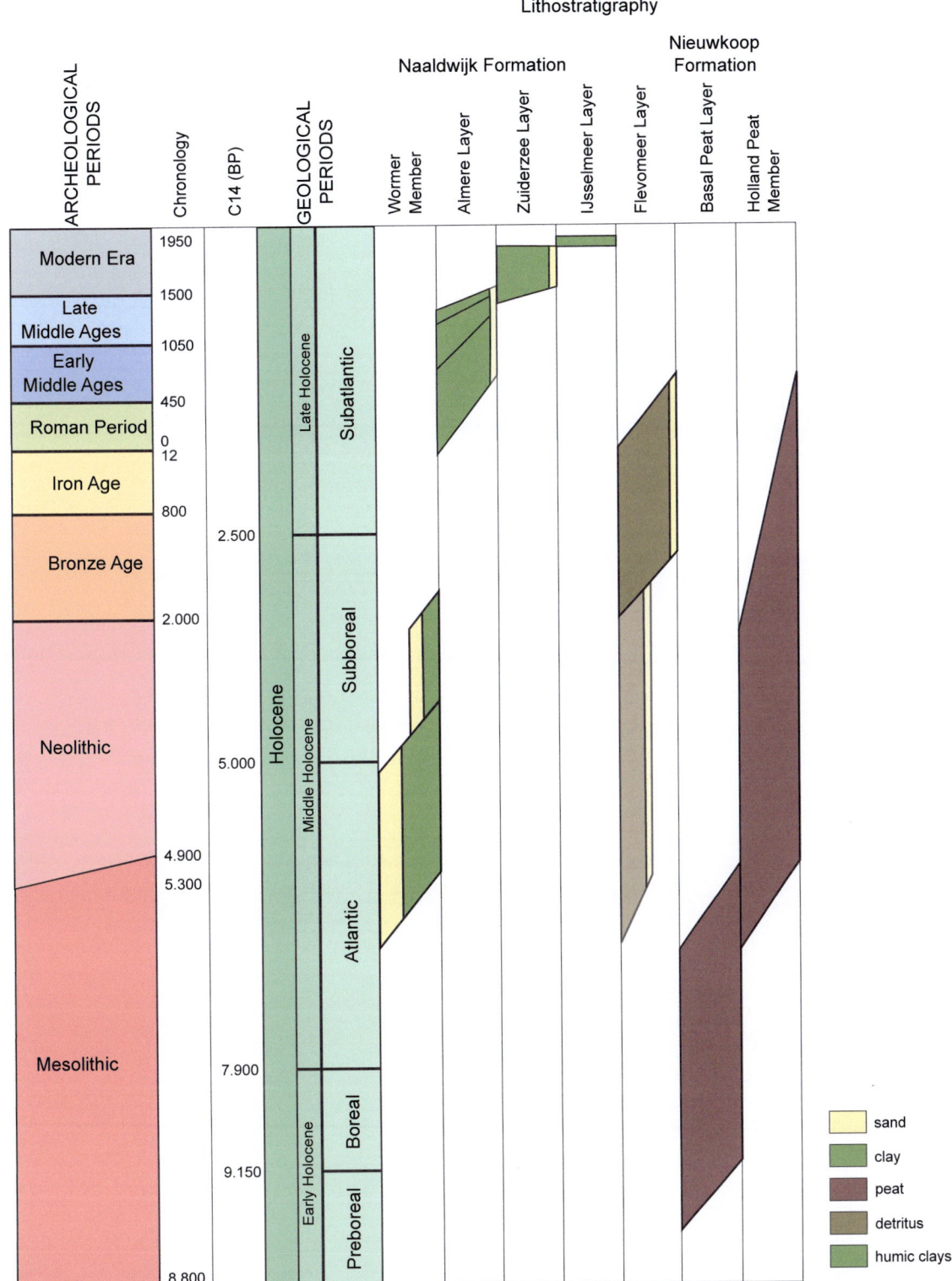

Figure 3.6: Lithostratigraphical table with lithothstratigraphical units relevant to Flevoland.

HIDDEN LANDSCAPES 47

formation of prehistoric landscapes in Flevoland to varying degrees. Between 11,700 and 6300 cal. BC, climatological and local hydrological processes were responsible for developments in the landscape. After this period, regional processes driven by sea-level rise and coastal development took over. These contrasting processes are clearly recognisable in the structure of the subsurface in Flevoland (fig. 3.6).

General structure

To properly understand the processes connected with lithostratigraphic formation in Flevoland, we need to go further back in time to the Saalian glaciation. During this ice age, when the land ice extended halfway across The Netherlands, geomorphological structures formed, such as ice-pushed ridges, glacial basins and moraine plateaus. These would have a major impact on successive landscape development. Ice-pushed ridges formed mainly to the south and east of Flevoland (Utrechtse Heuvelrug ridge, Veluwe), whereby thick blocks of Middle Pleistocene sediment (mainly river deposits) on the edge of glacier termini were tilted by the shearing action and weight of the advancing glacier. The asymmetrical escarpments that formed in this way reached heights of approx.120 m above the current sea level. Beneath the glacier termini, ice-scoured basins formed under the pressure of the ice cap and the mechanical shifting of sediment by the moving ice. Beneath the southern half of Flevoland, a glacial basin formed in this way reached a depth of approx.120 m below the current sea level. The basin grew progressively shallower towards the north, eventually transitioning to a plain consisting of ground moraine (glacial till). In the Noordoostpolder the glacial till was slightly pushed up during a static period in the retreat of the land ice. From the northern and north-eastern boundary of Flevoland the ground moraine extended into the Drenthe-Friesland till plateau.

The large glacial basin beneath Flevoland filled up with thick layers of sediment from the end of the Saalian, continuing into the Weichselian. As the ice cap melted at the end of the Saalian, large quantities of sand and gravel were washed away with the meltwater. During the Eemian, when temperatures rose sharply, part of Flevoland was flooded by the sea which deposited marine sands and clay sediments, whilst in the lower-lying areas, peat began to develop. In the Early Weichselian (Early Glacial), when the sea level began to fall again, alluvial deposits of river sand and gravel were deposited by the Rhine. During the Middle Weichselian (Pleniglacial) the Rhine changed course and only smaller river systems (the Overijsselse Vecht, Hunnepe, Eem) remained active. These rivers transported significantly less sediment. Large quantities of eolian coversand were deposited outside the river floodplains.

The Pleistocene deposits that lay on the surface at the end of the Weichselian provided the foundation for developments in the Holocene. In higher parts of the landscape these Pleistocene deposits mainly consisted of Late Glacial coversand and glacial till from the Saalian. In the river valleys these comprised river sands and clay on which wind-blown sandy river dunes formed in the Younger Dryas. The formation of river dunes may have continued into the Early Holocene.[87] This development, however, was constrained by an increasingly dense vegetation cover, combined with a rise in the annual average temperature. In the Preboreal, marshes in which peat developed evolved in the lower parts of the landscape. The formation of this peat was not connected with the rise in sea level, but with high local groundwater levels. Preboreal peat layers are therefore not found over large contiguous areas. In the Boreal, too, there appears to have been limited peat formation or sedimentation by other water-related deposits. This might be connected with the decreasing groundwater levels in the region. There is evidence of peat formation only in the Noordoostpolder, but this was probably caused by the layers of glacial till in the subsurface, which hampered drainage.

It was not until the Atlantic that sedimentation and erosion processes were directly affected by structural sea-level rise. In the Early Atlantic the increasingly waterlogged conditions in the region led to localised higher groundwater levels, prompting the deposition of humus-rich material (detritus) under water. Around the transition from the Early to the Middle Atlantic the gradual expansion of the wetland environment led to peat formation (Basal Peat Bed) over extensive contiguous areas. Starting in the west, the basal peat (sedge and woodland peat) gradually crept eastward towards higher parts of the coversand landscape.

We generally see peat mires spreading to the east, with little erosion at first. Only in the river valleys, where the transgressive sea periodically exerted its influence, is there evidence for the clearing out of older sediments. Clay was deposited in new and 'cleared' channels. The limited influence of the sea meant that the dynamics were fairly restricted, and stagnation probably set in quite quickly, as a result of which detritus was once more deposited, followed by peat growth and land accretion in channels and meander cut offs. This certainly occurred in Zuidelijk Flevoland, where the Eem, a river with a low dynamic, was the most important drainage system. The wider river valley of the Hunnepe and Overijsselse Vecht, between Oostelijk Flevoland and the Noordoostpolder, provided

---

87 Evidence from the river area of the central Netherlands suggests that river dunes also formed in the Preboreal (and possibly the Boreal). We do not yet have reliable dates for the formation of river dunes in Flevoland.

the space for a higher flow dynamic. Here too, however, erosion was limited to the river valleys and particularly the stream channels.

The limited influence of the sea in Flevoland decreased even further with the closing up of the coastline of the western Netherlands in the Subboreal. Large scale peat formation did occur (Holland Peat Member), but the deteriorating drainage conditions led to the simultaneous formation of large, shallow lakes. It was the presence of these large lakes that led to the erosion of the peat landscape. The seasonal increase in storms led to the erosion of the eastern lake banks and the gradual merging of the lakes, a process that led to the emergence of Lake Flevo. The organic material from the eroded peat was deposited on the lake bed as detritus gyttja. Eventually, not even the highest parts of the Pleistocene landscape (mainly in the eastern half of Oostelijk and Zuidelijk Flevoland) were left unaffected by large-scale erosion of the peat landscape. As a new connection with the sea formed (this time from the north) and Lake Almere and, subsequently, the Zuiderzee were created, the dynamics of the area increased and eventually the whole of Flevoland was submerged. In this phase, almost the entire Subboreal and Subatlantic landscape eroded away, and the remaining older landscape features were covered with layers of detritus gyttja, locally-deposited Pleistocene sand and marine clay.

Sub-regional similarities and differences

Based on the previous section, we must conclude that the lithostratigraphical structure of the subsurface in Flevoland no longer features a 'complete' sequence of the lithological layers originally deposited. Certainly the Subboreal and Subatlantic peat layers have almost completely eroded away. On the other hand, a substantial part of the Atlantic sequence has remained preserved, particularly in the western part of Flevoland and in river valleys. The Early Holocene (Preboreal, Boreal) sequence has also remained fairly well preserved, although this phase is characterised by relatively little sedimentation. Differences in the formation of the landscape mean, however, that the structure of the subsurface in the three regions (Zuidelijk Flevoland, Oosterlijk Flevoland, Noordoostpolder) is not identical. A more detailed description of the situation in each region is therefore given below, on the basis of representative profiles.

Zuidelijk Flevoland

Zuidelijk Flevoland lies in the deepest part of the glacial basin that continues southward into the Gelderse Vallei, which itself is bordered by ice-pushed ridges (Utrechtse Heuvelrug, southern Veluwe) (fig. 3.7). The Pleistocene surface in Zuidelijk Flevoland consists mainly of aeolian coversands (Boxtel Formation; Wierden Member, with local river sands (Kreftenheye Formation). Late Glacial soils (Bølling, Allerød) and thin layers of peat (Allerød) occur in the coversands. The undulating coversand landscape is cross-cut by the river Eem, which was the most important drainage system for the Gelderse Vallei, and by a number of streams. River dunes have not been identified with any certainty in this area. Localised peat developed in poorly drained, low-lying depressions during the Preboreal. There is no evidence for peat formation in the Boreal. The progressive formation of basal peat (Nieuwkoop Formation; Basal Peat Layer) in the Atlantic was preceded by underwater deposition (detritus). During the Middle and Late Atlantic the peat (Nieuwkoop Formation; Holland Peat Member gradually advanced to the higher parts of the landscape. Under the influence of the sea (which was still some distance away, however), temporarily increased dynamics in the Eem fluvial system caused the clearing out of some old meanders. There was also incidental disturbance of some peat layers on higher ground. Clay (Naaldwijk Formation; Wormer Member) was rapidly deposited in the main channel of the Eem and in its tributaries (possibly over one or two centuries). This process culminated in stagnation and detritus deposition, followed by peat formation (Nieuwkoop Formation; Holland Peat Member. In most parts of the area, the sequence was cut off by the Flevomeer Layer (Nieuwkoop Formation), which consists of lacustrine detritus/gyttja. On top of this we find deposits from the Almere Layer and Zuiderzee Layer (Naaldwijk Formation), with intercalated horizons of mineral and detritus deposits. The Zuiderzee Layer can lie directly on the Pleistocene sand in the southeastern part of Zuidelijk Flevoland. The distinct boundaries between the layers suggests that occasional erosion did occur.

Oostelijk Flevoland

Under Oostelijk Flevoland, the glacial basin rapidly ascends towards the north (fig. 3.8). The old Pleistocene surface in this region consists overwhelmingly of coversand (Boxtel Formation; Wierden Member) and river sands in the Hunnepe valley (Kreftenheye Formation). Allerød soils and thin horizons of peat occur through the coversands. In contrast to Zuidelijk Flevoland, river dunes are found here. The area was transected from east to west by the Hunnepe, a fairly small river system that drained water from the coversand landscape in the eastern Netherlands and bordering parts of Germany. It is not clear whether the Hunnepe was actually connected to the Overijsselse Vecht, which flowed slightly further to the north. Though we have no direct evidence for peat formation in the Preboreal and Boreal (Nieuwkoop Formation; Basal Peat Layer), it is likely that there was localised peat growth. Structural expansion of the peat beds occurred during the Atlantic (Nieuwkoop Formation; Holland Peat Member). An anastomosing

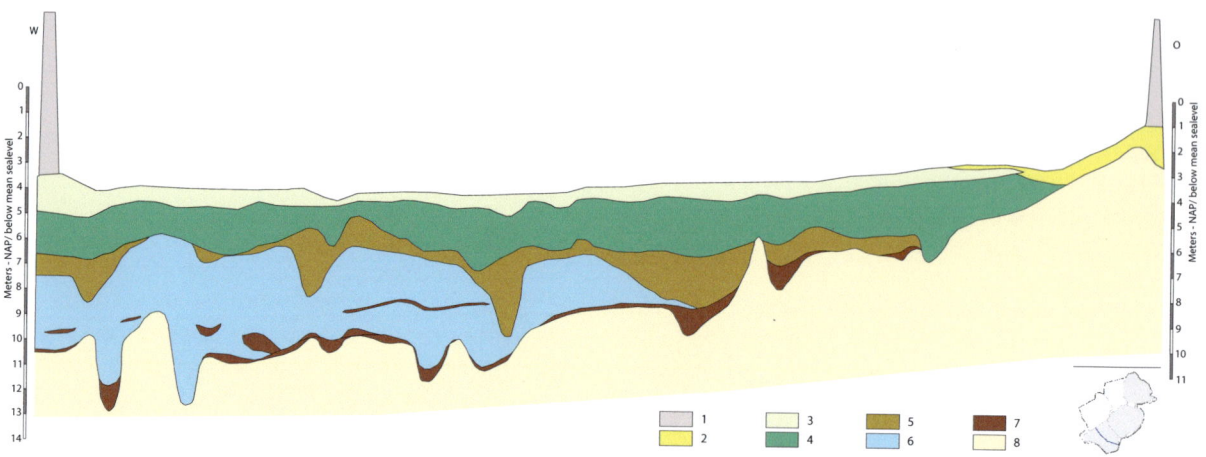

Figure 3.7: Typical longitudinal section through Zuidelijk Flevoland. Legend 1 Dike; 2 redeposited Pleistocene sand; 3 IJsselmeer Layer / Zuiderzee Layer (Naaldwijk Formation-Walcheren Member); 4 Almere Layer (Naaldwijk Formation-Walcheren Member); 5 Flevomeer Layer (Naaldwijk Formation-Walcheren Member); 6 Wormer Member (Naaldwijk Formation); 7 Holland Peat Member/Basal Peat Layer (Nieuwkoop Formation); 8 Pleistocene sand (Boxtel Formation/Kreftenheye Formation) (adapted from DINOloket).

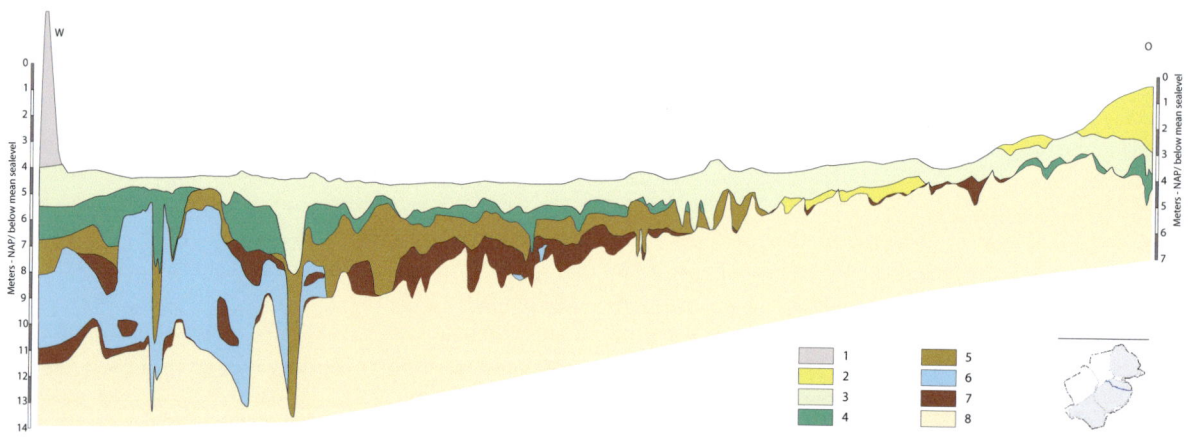

Figure 3.8: Typical longitudinal section through Oostelijk Flevoland. Legend 1 Dike; 2 redeposited Pleistocene sand; 3 IJsselmeer Layer/Zuiderzee Layer (Naaldwijk Formation-Walcheren Member); 4 Almere Layer (Naaldwijk Formation-Walcheren Member); 5 Flevomeer Layer (Naaldwijk Formation-Walcheren Member); 6 Wormer Member (Naaldwijk Formation); 7 Holland Peat Member/Basal Peat Layer (Nieuwkoop Formation); 8 Pleistocene sand (Boxtel Formation/Kreftenheye Formation) (adapted from DINOloket, Vos & Van Gessel 2004, Ente 1976).

system of fairly small river channels developed in the Hunnepe river basin during the Late Atlantic, depositing clay on the floodplain (Naaldwijk Formation; Wormer Member). The clay deposits lay intercalated through the Holland Peat Member. The Atlantic deposits are delineated by further deposits of the Flevomeer or Almere Layer, above which are Zuiderzee Layer deposits.

Noordoostpolder

In the southern part of the Noordoostpolder glacial till ridges occur at Urk, Schokland and De Voorst (fig. 3.9). Another ridge top at Tollebeek would have been present in prehistory but has been removed in recent times. The rest of the old Pleistocene surface, which increases in elevation in a north to north-easterly direction, consists of coversands intercalated with Late Glacial peat horizons (Allerød). River sands and clay (Kreftenheye Formation; Wijchen Member) have been found in the stream valley of the Overijsselse Vecht, in the southern Noordoostpolder. River dunes occur frequently in and alongside the floodplain, some forming large complexes, such as the one to the south of the former island of Schokland. The

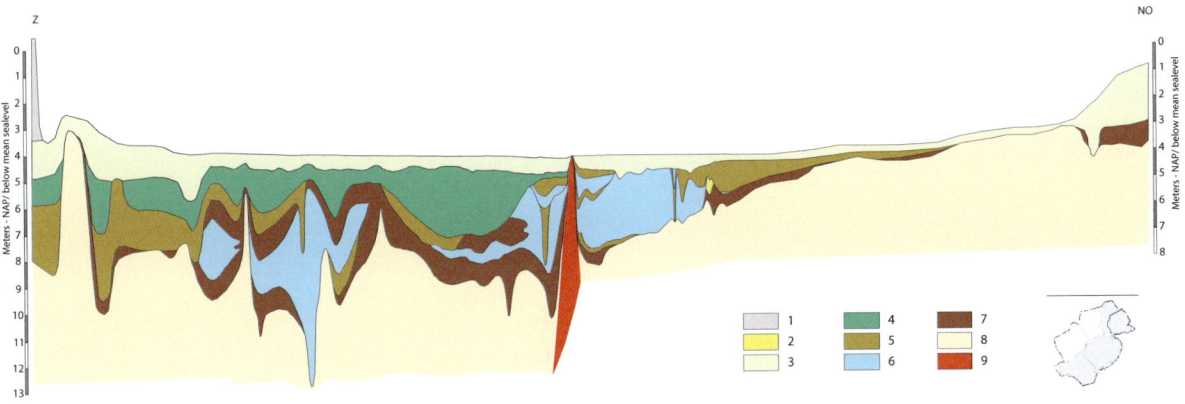

Figure 3.9: Typical longitudinal section through the Noordoostpolder. Legend 1 Dike; 2 redeposited till; 3 IJsselmeer Layer/Zuiderzee Layer (Naaldwijk Formation-Walcheren Member); 4 Almere Layer (Naaldwijk Formation-Walcheren Member); 5 Flevomeer Layer (Naaldwijk Formation-Walcheren Member); 6 Wormer Member (Naaldwijk Formation); 7 Holland Peat Member/Basal Peat Layer (Nieuwkoop Formation); 8 Pleistocene sand (Boxtel Formation/Kreftenheye Formation); 9 Pleistocene till (Drenthe Formation) (adapted from DINOloket, Wiggers 1955).

formation of basal peat in the Holocene began in lower-lying parts of the river valley. During the Atlantic, peat began to form as the marshes expanded (Nieuwkoop Formation; Holland Peat Member). Clay sediments were deposited in the stream channels intercalated between the peat horizons. These clay layers, fan out against the flanks of river dunes. Around 3700 cal. BC, at the start of the Subboreal, part of the Overijsselse Vecht valley to the south of Urk became blocked by peat growth as a consequence of the closing off of the sea inlet at Castricum. The Overijsselse Vecht changed course to the north of Urk, which led to the erosion of previously-developed peat and deposited detritus. This activity also created a large lake whose size and position changed in the course of the Subboreal. More and more raised bogs developed in this poorly drained area. Towards the end of the Subboreal and in the Subatlantic, however, lots of open water was again present in the landscape, incurring large-scale erosion of the older peat deposits. Eventually the Noordoostpolder, apart from the highest points (Urk and Schokland),would become submerged and as a result covered with a layer of marine Zuiderzee deposits (Naaldwijk Formation).

Diversity in peat

As Flevoland became increasingly saturated, extensive peat bogs or mires formed, as described in the previous section. There were of course different environments, depending on the specific localised circumstances.[88] The earliest peat, which developed on a more localised scale, comprised sedge, birch and willow.[89] During the course of the Early and Middle Holocene larger peat environments developed, with more diverse characteristics. The widespread saturated conditions caused by structural water level rise resulted in the expansion of eutrophic marsh woodland (carrs, riparian forest) where willow and alder grew, as well as oak. Mesotrophic to eutrophic sedge and reed peat also occurred on a large scale. Oligotrophic peat environments, particularly raised bogs, also developed under specific hydrological conditions and in places with a particular soil chemistry. This initially occurred on a small scale, but as a result of increasing acidification caused by poor drainage, such environments gained more ground. The peat beds found in Flevoland therefore generally contain several distinct types of peat: sedge peat, reed peat, woodland peat (willow carr, alder carr, birch carr) and sphagnum peat (raised bogs). These types of peat not only have a bearing on the reconstruction of the palaeolandscape, but also on the preservation conditions for various categories of archaeological material.

3.3 The character and quality of archaeological remains

Prehistoric archaeological remains in Flevoland are generally '*low resolution phenomena*', mostly buried under sediment, which hampers an easy detection of these remains.[90] The Neolithic sites at Swifterbant discovered when ditches were dug to drain the new polders (see Chapter 2) are mostly situated fairly near the surface. These sites, some of which were found in clayey levee deposits, are characterised by the presence of archaeological material such as pottery sherds and a dark-

---

88   Gotjé 1993.
89   Gotjé 2001; Spek, Bisdom & Van Smeerdijk 2001b.

90   Peeters *et al.* 2002, 83.

coloured 'cultural layer' that is easily distinguished in borehole core samples. The delimitation of this layer was generally assumed to coincide with that of the settlement. Recent research has, however, established that this layer should in fact be seen as a waste layer (midden) that may represent only a portion of the original occupied area.[91] Over the years, archaeological investigations in Flevoland have revealed a large variety of prehistoric archaeological phenomena in the buried landscape. Many Mesolithic sites, for example, consist of scatters of flint of variable density, while some locations are almost devoid of flint artefacts, but have many pit hearths. Such archaeological remains are less easy to identify in a core sample. In other situations, there may be highly localised phenomena that are impossible to detect in borehole surveys, such as fish traps in clayey channel fill. Furthermore, most sites are in the upper part of the Pleistocene sand and were not covered with sediment until some time after habitation, so very few, if any, organic remains have been preserved. To better understand the current picture of prehistoric Flevoland outlined in this publication, the section below shall therefore take a closer look at these aspects.

3.3.1 Differences in character

The previous sections roughly describe how the landscape in the area that we now refer to as Flevoland has changed since the last part of the Pleistocene. A broad range of landscape-forming processes not only affected the degree to which evidence of these old landscapes has been preserved, but above all affected the possibilities this changing landscape offered its human inhabitants. The activities of prehistoric humans in Flevoland caused the initial formation of the buried archaeological record. Attempts to understand the relationship between humans and the landscape – or rather the place of humans *in* the landscape – is perhaps one of the most important focuses of the archaeological research that has taken place in this region so far. However, we still do not know very much about what humans actually did. Nevertheless, current data does allow us to indicate a *minimum* variation in human activities, and to define the character of the associated archaeological phenomena.

To begin with, we must acknowledge that not all prehistoric human activity in this area was related to *wetlands*. The general description of landscape developments given above suggests, certainly for the Early Holocene, that there was no structural wetland expansion. Although wetland areas did of course exist, not all activities took place in such a context. This also means that many sites that 'drowned' in the course of the Holocene must primarily be placed in a 'dry context'. The wet conditions arose later in those locations. However, there were certainly also *wetland sites* in this period, locations where activities were directly related to, or took place in, wet environments. As indicated in the previous section, structural wetland expansion occurred during the Atlantic, so Late Mesolithic and more recent sites must be understood largely in a wet context. We should also note that, in the event of a long-term history of occupation, a dry context for activities might well have evolved to a wet context.

The extent to which the environmental context evolved had a significant impact on the nature or character of the archaeological phenomena that we find in Flevoland. Locations where activities left behind some – mainly non-perishable – remains tend to be characterised by a scattering of artefacts, in many cases of worked flint. Depending on the nature of the activities and the duration of site use, there may be a thin spread of material (a low-density distribution) over a relatively small area, or an extensive, thicker spread (high-density distribution) of material. The nature of the activities themselves primarily determines what materials were left behind: the production of a few flint blades leaves behind different evidence than the slaughter of a shot red deer, or a catch of fish with the aid of a fish trap. At the level of a single 'event', the archaeological record will generally be fairly limited, but as the number of events within a certain area increases, accumulation occurs and a palimpsest develops in the archaeological record.[92] Also in the case of a complex of different but simultaneously carried out activities, one would expect a larger (more varied) archaeological record.

With this in mind, the archaeological data that we currently have from Flevoland displays a large degree of variation. Small concentrations of artefacts that are probably related to 'brief' moments of activity have been found at various sites, such as Almere-Zwaanpad and Hoge Vaart-A27 (northern concentration). Sizable palimpsests have been found at locations used over a long period of time, such as Schokland-P14, Urk-E4, Dronten-N23 and Hoge Vaart-A27. These sites do not always feature very large quantities of material remains. Sometimes there are clusters of features such as pit hearths, dozens or even hundreds of which are sometimes found on one site, as at Hanzelijn-Drontermeer and Dronten-N23. There are also function-specific phenomena, such as fish weirs and fish traps found in channels filled with clay, as at Emmeloord-J97 and Hoge Vaart-A27 (phase 4). In many cases, the nature of the evidence of activity also varies widely, perhaps because the nature of activities at a specific location changed over time. In addition, the nature of the archaeological record changes from one site to another as

---

91 Huisman & Raemaekers 2014.

92 Bailey 2007.

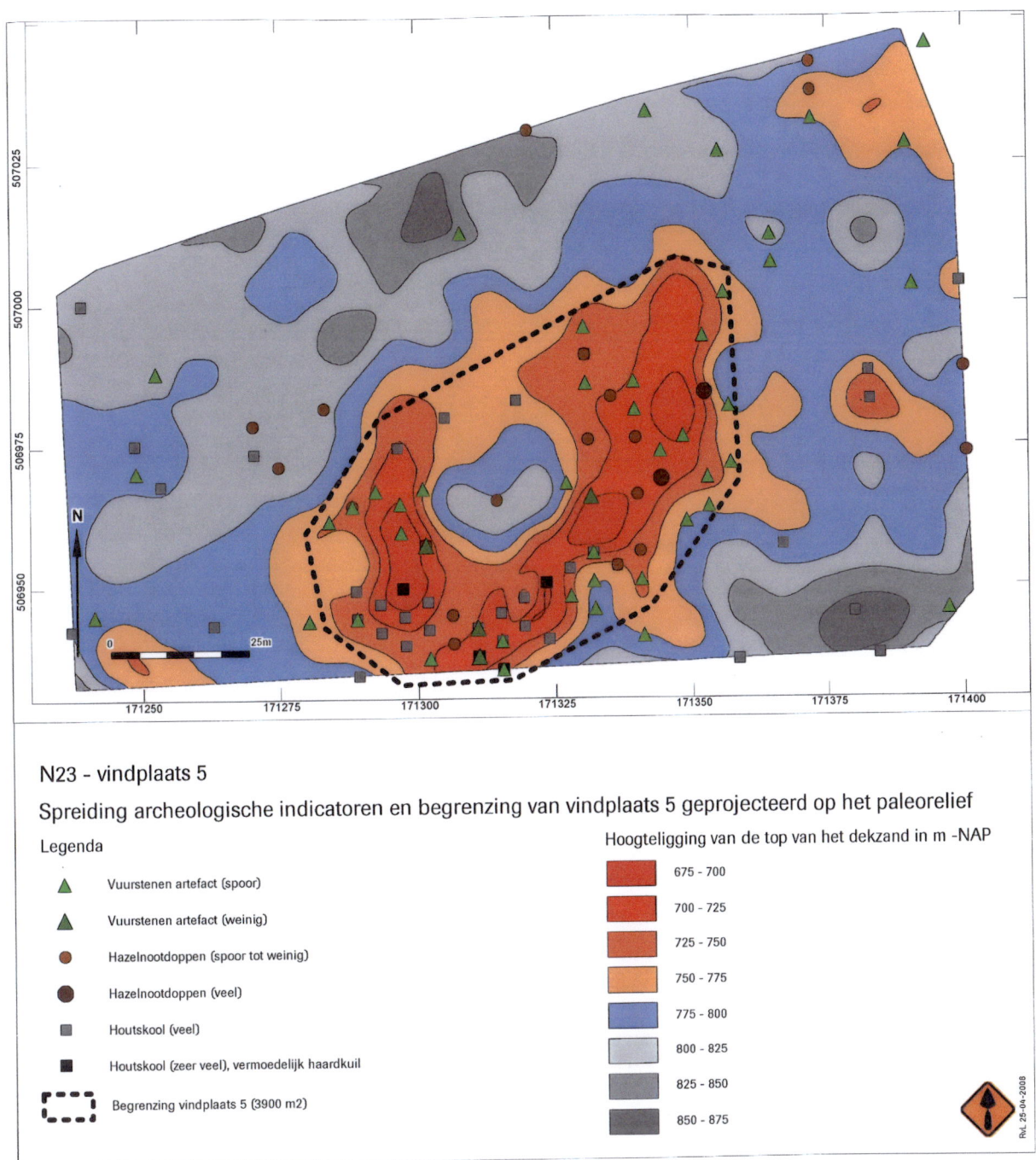

Figure 3.10: Pleistocene surface model of and delimitation (dashed line) of Dronten-N23, showing the distribution of anthropogenic indicators discovered during borehole surveys. Legend triangles: lithic artefacts; circles: charred hazelnut shell; rectangles: charcoal (source: Van Lil 2008). Elevation intervals in centimeters below Dutch O.D.

a result of differences in the contexts of usage, as at the sites near Swifterbant.

Although archaeology is perhaps rather fixated on the material remains of prehistoric activity, attention doesn't solely concentrate on this aspect. Indirect evidence of human activity can, for example, be found in pollen diagrams, which can show signs of human influence in the vegetation, sometimes in combination with the presence of charcoal particles, parasites and dung mould. One example is Hanzelijn Area VIII, near the large Mesolithic site of Dronten-N23 (fig. 3.10).[93] The presence of burnt

---

93 De Moor *et al.* 2009.

reeds in clay layers can also indicate human activity, as found in borehole core samples near Almere, where burnt plant remains were found at several levels in channel fill.[94] The localised presence of colluvial deposits of displaced sand that has slid down slopes – the site of Hoge Vaart-A27 is an example – could be connected with human (or animal) activity, such as treading. Sometimes footprints and hoofprints are even found, as at Schokland-P14. These are also archaeologically relevant phenomena that can provide information about specific forms of landscape use. Such phenomena are however spatially diffuse, which makes it difficult to refer to them as 'sites', and their research requires a different approach.

3.3.2 Differences in preservation

The prehistoric occupation history of Flevoland covers a very long timespan. Although the focus of this publication is the Holocene, there is certainly also evidence for human activity in the Late Palaeolithic and even the Middle Palaeolithic.[95] Despite the fact that these occupation remains have been covered by younger sediment, the preservation conditions are highly variable. Sediment cover is no guarantee that archaeological remains will be well preserved. As already stated, favourable preservation conditions depend on the geogenetic properties of the area, as well as on processes associated with the reclamation of the polders.

Many known sites in Flevoland are situated on the stable surfaces in the former Pleistocene and Early Holocene sandy landscape, particularly on the relatively speaking higher elevations (coversand ridges and hillocks and on river dunes), when compared to the surrounding area. As we have seen, it took some time, not until the Late Mesolithic, before the large-scale expansion of wetter conditions became apparent in the area. In preceding periods, waterlogged conditions which were conducive to the preservation of unburnt organic remains would only have occurred on a more localised scale, probably only in river and stream valleys, and in depressions with a poorly permeable subsurface. This means that any unburnt organic remains on the higher, dry landscape elements would have disappeared before those parts of the landscape also became saturated. This process is of course not only connected with the groundwater level and sedimentation, but also with soil formation, during which leaching would have led to decalcification. Biological decomposition of unburnt organic remains would also have played an important role. As a result, only burnt remains – generally of vegetation – would have had a reasonable chance of surviving as the landscape became progressively wetter. In this sense Flevoland is no different from other parts of

Figure: 3.11 Poorly preserved skeletal remains in inhumation graves at Urk-E4. Although the grave on the right contained somewhat better preserved remains during excavation, these could not be salvaged due to the poor preservation (photo: Cultural Heritage Agency of the Netherlands/Ton Penders).

the Netherlands, where remains of prehistoric occupation are also found on or near the current surface.

Conditions for the preservation of unburnt organic remains began to improve as the area became wetter in the Late Mesolithic (section 3.2.2), causing peat to form on a large scale, and organogenic sediment (detritus) to be deposited under water. As a consequence, there is a greater chance of finding well-preserved remains not only at or near locations that were inhabited for longer periods (such as on the levees), but also in parts of the landscape where people were present for shorter periods, or where specific activities took place (along water channels, for example). As wetland-related sedimentation increased, the survival chances of other, not directly archaeological, indicators of human activity increased. These include indications in pollen diagrams for the burning of vegetation. Generally speaking, remains that immediately ended up in peat or clay deposits (Nieuwkoop Formation and Naaldwijk Formation) are better preserved than remains left behind on the surface of the sand. The underrepresentation of unburnt bone remains on 'drowned' sandy ridges where

---

94 Woltinge 2009.
95 Hogestijn, 1986; Johansen, Niekus & Stapert 2008; Ten Anscher 2012.

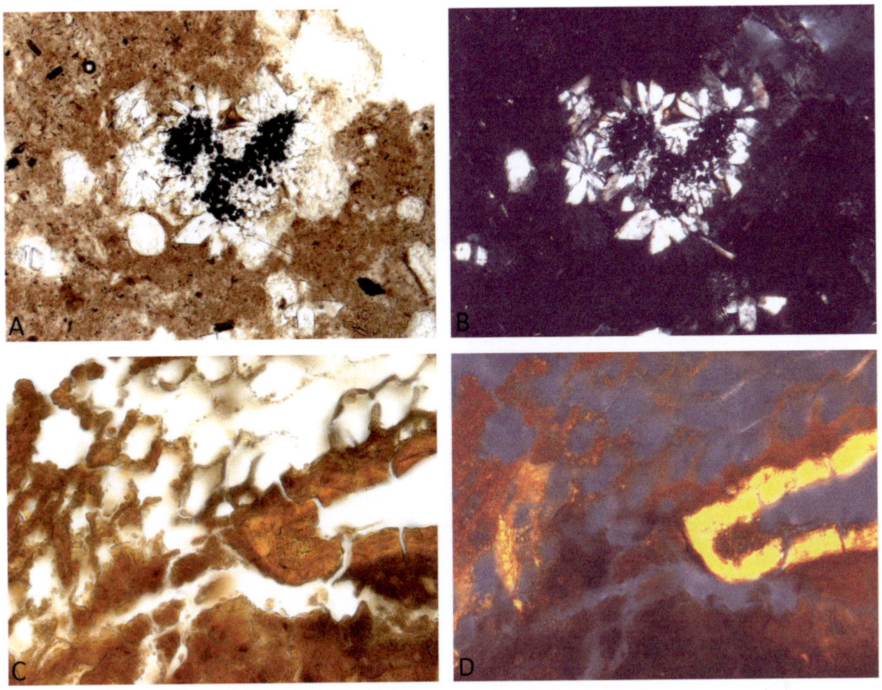

Figure 3.12: Degradation features observed in micromorphological thin sections from the Schokland-J112 site. A-B: Framboidal pyrite surrounded by crown-shaped gypsum crystals. C-D: Wood-cell walls impregnated or replaced by iron oxides. A and C were made using plane polarized light (PPL); B and D using crossed polarizers (XPL) (photo: Cultural Heritage Agency of the Netherlands).

Figure 3.13: Preservation of bone material discovered in Flevoland differs to a great degree. Top: a well preserved bone chisel; below: a poorly preserved antler (*Cervus elaphus*) mattock. Both discovered at Emmeloord-J97 (photo: Dick Velthuizen).

skeletal remains have nevertheless been found in graves – as, for example, at Swifterbant S21-23 and Dronten-N23 – could therefore be explained by the dating of the burials, which appear to be relatively young and come from a phase when the sandy ridges were already fairly wet.

Although the development and expansion of full-blown wetland conditions may have been favourable for the preservation of occupation remains – not only organic remains, but also for example erosion-sensitive pottery – we must also consider the other side of the argument. The progressively wetter conditions and the periodic increase in the dynamism of stream channels would have resulted at the same time in erosion on various spatial scales. Previously deposited layers of sediment in river and stream valleys were washed away in places, and major erosion could occur on lake beds and around shores. As a result, the majority of the late prehistoric – and in Flevoland also historic – landscape seems to have disappeared. Evidence of Bell Beaker culture occupation, for example, has only been preserved on a local scale – the Schokland-P14 site, with features from this phase of occupation, is exceptional. Various pieces of evidence, such as stray finds of Bell Beaker pottery sherds at sites in the Noordoostpolder and Oostelijk Flevoland suggest, however, that occupation in this period may have been more structural and widespread and what has been found by archaeologists is merely the residue of an eroded landscape.[96]

The 'preservation potential' of the subsurface in Flevoland doesn't only depend on the specific circumstances in which sedimentation and erosion occurred in relation to habitation phases. The process of wetland formation under the influence of sea-level rise and deteriorating drainage conditions in the hinterland was accompanied by strong groundwater fluctuations which accelerated the degradation of soils and caused shifts in the reduction-oxidation boundary. It is entirely possible that vulnerable categories of material degraded rapidly in such conditions; the poor conservation of skeletal remains in Neolithic inhumation graves could, for example, have been as a result of these conditions (fig. 3.11).

The processes set in motion by the creation of polders in the region also affected the preservation potential of the subsurface. Such a process introduces radical changes in the hydrological conditions. New groundwater flows (seepage), for example, cause geochemical changes, which in turn cause the polders to become brackish. Lowering the water level to enable cultivation also causes oxidation and, for instance, the formation of pyrite crystals in sulphur-rich soil horizons (fig. 3.12). Research has shown that lowering the water level has a very negative impact on the physical quality of wood remains, for example. The condition of skeletal remains in graves, as mentioned above, may also be connected with the polder reclamation processes. At the same time, there are major differences. For example, bone remains from site A – *e.g.* Emmeloord-J97 (see Appendix) – are of excellent average quality, while those from other sites – *e.g.* Hoge Vaart-A27 or Dronten-N23 (see chapter 3) – are poorly preserved (fig. 3.13). In these cases, differences in preservation have been determined by cultural deposition processes in the past, combined with local soil properties and developments in the soil over time.

Another factor connected with the creation of the polders that has a bearing on the preservation of archaeological sites is differential settling resulting from the compaction of clastic layers and peat. This process eventually brings archaeological remains within the reach of the plough, or causes cultural layers to end up in an oxidizing environment as the water level falls.[97] This problem clearly exists in the Noordoostpolder and the eastern edge of Oostelijk and Zuidelijk Flevoland, where the Pleistocene surface lies at or near the current surface. If no further measures are taken with respect to the decreasing water level, differential settling can have major consequences for many archaeological sites. A groundwater buffer has been created around the Schokland-P14 site, which is part of the UNESCO World Heritage Site of Schokland. This keeps the groundwater level relatively high on a local scale, thus preserving the archaeological remains on the site. Recently the Dutch government has decided that the area providing the groundwater buffer will be enlarged to protect a larger area with archaeological remains.

3.4 Mapping hidden landscape units

It will be clear from the previous sections that the properties of the buried archaeological record in Flevoland depend heavily on the geological developments and processes associated with the creation of the polders in the 20th century. While the discovery of sites close to the surface during soil mapping surveys was a matter of coincidence in the initial period after land reclamation was complete, it is clear that a more systematic mapping of prehistoric habitation remains needs to rely on more than the inspection of the sides of ditches. Geological developments in the area have also covered many archaeological remains with layers of sediment that can be as much as several metres thick, which means such sites cannot be identified on the present surface level. There are only a few places where sites are discovered at surface level. This is, for example, only the case for archaeological remains situated on the higher sand ridges or on river dunes and which are now (partly through settling) being destroyed by ploughing. Most archaeological remains are located more than 2 m under the present surface. Their density is generally low and their spatial distribution variable. Furthermore, it is not always possible to categorically identify remains as archaeological artefacts.[98] As a result, specific strategies have been developed for the mapping and evaluation of occupation remains. These strategies have a strong bias towards borehole surveys with a focus on the recording of the palaeolandscape characteristics of buried landscape units, and the extrapolation of observations, whether statistically substantiated or not. Models based on the use of the landscape by hunter-gatherers and early farmers are also used. The depth of the archaeological remains and the resulting high costs of excavation, coupled with efforts to preserve remains *in situ*, means that excavation is avoided where possible. As a result, our picture of prehistoric Flevoland is defined by a limited number of 'windows' within which detailed information has been gathered, as well as a large number of observations of more limited resolution.

---

96 Raemaekers & Hogestijn 2008.

97 On the other hand, differential settling does make buried channels and levees visible at ground surface, allowing them to be mapped more easily (Dresscher & Raemaekers 2010; Van Heeringen *et al.* 2014).

98 Peeters *et al.* 2002, 83; Wansleeben & Laan 2012a/b; Van de Geer 2014.

3.4.1 Site versus landscape perspective

A greater insight into the nature and quality of archaeological remains of prehistoric habitation in the region means that the general approach to research has shifted from a focus on sites to a landscape perspective. This shift might not stand in isolation, as it seems to fit in a national trend in Dutch archaeological heritage management (see Chapter 1). This more landscape-oriented approach focuses on identifying and explaining human activities in relation to the dynamics of the landscape, which means that activities that took place 'off-site' are also the subject of study. However, the choice of a site- or landscape-oriented perspective has implications for the approach adopted in research in terms of the potential for detecting and interpreting traces of human activity.

If the focus is on the 'site' as a spatially defined assemblage of habitation remains, strategies are needed that primarily ensure the recovery of material remains (artefacts, food remains). One clear example of this is the research carried out in the 1970s on Mesolithic habitation remains on the river dunes near Swifterbant, which had as its main objective: 'the location and recovery of an intact Mesolithic occupation surface'.[99] The discovery of Neolithic levee settlements at Swifterbant on the basis of the identification of dark 'cultural layers' or middens in borehole core samples and ditch profiles also fits with this perspective. The result of this approach is that the majority of sites discovered (and investigated) are those where the density of archaeological remains is relatively high and site classifications emerge based on quantitative (numbers, densities) and qualitative (artefact type) differences in find assemblages.[100] This generally results in a distinction between, for example, 'domestic sites' and 'industrial sites', or habitation sites (settlements), base camps and specialised or special-purpose camps.

The landscape approach encompasses *all* recognisable evidence of human behaviour in different parts of the landscape. This includes 'traditional' sites, but also isolated archaeological remains and more diffuse phenomena that can be linked to human activity.[101] In this perspective, it has been proposed that the term 'site' be used to denote any location where evidence of human activity can be found, irrespective of any quantitative factors.[102] The discovery of a felled tree, or evidence in a pollen diagram of the deliberate burning of vegetation also constitute archaeological information that requires investigation. Since the archaeological remains in Flevoland can be found in a context rich with palaeoecological information, as a result of the inundation of the landscape, it is particularly important to consider these aspects in any research.[103] Section 3.5.2, however, shows that this broader definition of what a site 'is' does not enjoy universal support.

The specific situation in Flevoland means that, in all cases, the amount of information available is limited. Location choice models used to detect occupation remains are therefore based largely on information from other parts of the Netherlands.[104] The morphology of the Pleistocene surface plays a key role in these models, and therefore also in Flevoland. Given the frequent occurrence of sites on higher ridges in the landscape and in gradient zones, the significance of such zones are greatly emphasised in the models.[105] Partly as a result of the difficulty involved in discovering buried archaeological remains, other parts of the former palaeolandscape have never, or hardly ever been investigated. Apart from the problem of the differential preservation of archaeological remains, it is, therefore, also by no means certain that the currently known sites portray a balanced picture of Flevoland's prehistoric past.

The observation above is important if we want to understand human behaviour in the past. At the same time, we must realise that we will never have a 'neutral' or fully 'representative' sample. Our frames of reference may be extremely narrow, which not only makes it difficult to get to grips with the variation hidden in the archaeological data itself, but also makes it difficult to develop an understanding of the potential variation from a more model-based perspective – whether this is of an analogue or computational character.[106] There are many well-researched sites and other records (*e.g.* palaeoecological) in Flevoland (see Appendix I) that allow us to sketch a coherent, albeit fragmented picture of the province's prehistoric occupation history. But the composition of the sample is heavily influenced by factors – post-depositional processes, research interests, generalised representations – that have nothing to do with human behaviour in prehistory.

Consequently, the archaeological data cannot be used as reliable input for predictive models, or as a dataset for a neutral assessment of predictive models as part of field surveys and assessments.[107] A statistical approach is preferable in this respect, although this too is subject to certain assumptions and depends on the objective. In the municipality of Almere, for example, a system is used

---

99 Price 1981, 102.
100 Newell 1973; Price 1978; 1980; Stapert 1985; Arts 1988; Wansleeben & Verhart 1990
101 Zvelebil *et al.* 1992.
102 Peeters 2007, 27.

103 Peeters 2007, 28.
104 Models by Arts (1985) and Groenendijk (1997), among others; these are heavily based on the ideas of Newell (1973) and Price (1978).
105 For gradient zones, see the discussion in Smit 2010.
106 Peeters 2007, 2010.
107 Brouwer Burg, Peeters & Lovis 2016.

Figure 3.14: Impression of techniques used during borehole surveys in Flevoland. Top left: manual auger; top right: mechanic sonic aqualock drill; bottom left: mechanic Avegaar auger; bottom right: detail of an Avegaar sample (source: De Boer & Lesparre-De Waal 2012; Hamburg et al. 2014).

whereby 'the landscape' – in actual fact the relief of the Pleistocene surface – plays an important role, and whereby zones differentiated by borehole surveys (high-lying, low-lying, slopes) are further explored for archaeological indicators on the basis of random sampling. No predictions are made beforehand as to what archaeological remains are likely to be found.

This does not, however, mean that it is not useful to develop and use models. They are useful as an heuristic instrument in a more explorative context in order to develop an understanding of (causal) relationships between the initial formation of the archaeological record by human behaviour and the factors that we as researchers *think* have influenced that behaviour.[108] Models can thus be regarded as 'thought experiments' that on the one hand give an indication of the behavioural variation created by adjustments to variables and parameters, and on the other hand that produce hypotheses that can, to some extent, be tested on the basis of the archaeology.[109]

---

108 There is essentially no difference in the structure of 'computational' and 'analogue' models, albeit that the former requires the formalisation of 'model behaviour'.
109 See Brouwer Burg, Peeters & Lovis 2016.

Figure 3.15: Archaeological remains and anthropogenic indicators discovered in borehole samples are usually very small. However the borehole samples on Almere Hoge Vaart-A27 were extremely rich. a) fragments of burnt bone (largest fragments are about 6 mm); b) lithic microdebitage (largest fragment is about 10 mm).

3.4.2 The practice of field surveys

After the 'academic' period in the 1970s and 80s, archaeological research in the province of Flevoland became mainly driven by more methodically-oriented questions and excavations – by commercial companies – as part of the implementation of the new heritage legislation (Valletta Convention). These days, some 95% of all archaeological research is performed in the context of archaeological heritage management in the Netherlands, in which the primary aim is preservation *in situ*. Most research therefore focuses on gathering information on the physical and intrinsic quality of a 'site', which still clearly speaks of a site-oriented approach. Explaining human behaviour in the past is in fact a secondary consideration. Nevertheless, determining the 'intrinsic quality' of the phenomena encountered does refer to human behaviour, as it concerns the question of what has been found and what these remains can be associated with.

These questions are difficult to answer in the context of Flevoland where it is usually only possible to carry out borehole surveys. Remains tend to be located at great depth, and below groundwater level, which makes excavation extremely costly. This means that, on the basis of a very limited spatial sample, conclusions have to be drawn on aspects that are difficult to determine, such as the size of a site, the degree of preservation of materials actually or potentially present and of the site as a whole, as well as the character and age of the site. Geophysical prospection methods are rarely used to map the prehistoric landscape and identify occupation remains.[110] Research has shown that there are too many variables – *e.g.* brackish-saline groundwater and depth – that affect magnetic or resistivity measurements, making results unreliable. For the time being, it seems that electromagnetic surveys are only suitable for mapping the relief of the Pleistocene sandy surface, though verification using boreholes and core penetration tests remains necessary.[111]

A lot of research is thus necessarily restricted to taking soil samples from borehole cores. Boreholes were made manually until the beginning of this century (fig. 3.14).

---

110 Visser, Gaffney & Hessing 2011.
111 Hamburg *et al.* 2014.

This meant that the intensity of a borehole survey also remained limited. As a result, the chances of even finding archaeological remains were extremely small. Experiments involving the development and application of various mechanical coring techniques were therefore carried out in Flevoland, particularly in Almere, where extensive urban expansion was taking place. In recent years, various techniques have been used to map the archaeological landscape at various stages of investigations. These techniques focus not only on archaeological remains, but also on palaeolandscape elements in particular that are implicitly or explicitly identified as a factor when research strategies are devised.

Archaeological field surveys in Flevoland, just as elsewhere in the Netherlands, are carried out in phases. Prior to the actual fieldwork, a desk-based assessment is carried out in order to determine as far as possible from existing records, the nature extent and significance of the lithostratigraohical and cultural-historic environment within an area specified for development. For the past few years, precise elevation measurements of the current land surface have played an important role.[112] Differential settling of deposits produces a pattern of relief in the current surface – albeit limited in absolute terms – that enables the identification of all kinds of palaeolandscape structures that are not buried too deeply under the surface. Geostatistical analysis of elevation data is particularly effective for revealing structures that would otherwise barely be visible.[113] Such information is ideal input for a survey strategy.

The first phase of fieldwork, might be regarded as more an exploratory geoarchaeological survey because it generally focuses on the characterisation of the relief of the Pleistocene surface and the stratigraphy using sample boreholes. A mechanical sonic core drill equipped with an aqualock ('Sonic aqualock') (fig. 3.14) is often used for large-scale surveys. This mechanical device, developed specially for archaeological research, uses high-frequency vibrations to drill a core sampler into the soil and extract a two-metre core. The advantage of this is that relatively intact, continuous cores can be obtained quickly, and then described and sampled for various purposes.[114] The evidence for lithostratigraphy visible in the core can be used as a basis for maps indicating the landforms present in the subsurface, as well as zones where erosion has occurred. Pedological characteristics are studied in order to verify the latter. Soil properties such as the intactness of podzol profiles in Pleistocene coversand, or the initial soil formation in Holocene levees along river channels can be used to determine erosion. The data provides a basis for a geoarchaeological model that is used to target the search for archaeological indicators.

The second phase of fieldwork – archaeological mapping and assessment of the potential of a location – is primarily focused on detecting archaeological indicators such as flint artefacts, pottery, charcoal and charred fragments of hazelnut shells (fig. 3.15). Evidence may already have been found in the first fieldwork phase if samples from cores have been sieved. In this second phase, the sonic AquaLock and screw augers (*Avegaar*) with a larger sample volume are used to take core samples. In most cases, these samples are taken from the top of the Pleistocene sand, where a significant proportion of archaeological indicators are located. The sand is sieved through a mesh ranging from 4 to 1 mm. Depending on the results, further investigation may take place on the basis of a more closely-spaced borehole sampling grid, or alternatively, larger sample volumes taken.

If a site is found – *i.e.* if several adjacent boreholes yield archaeological indicators – a third phase of fieldwork will generally be initiated to gather more information about the nature of the site for the purposes of an archaeological assessment or validation of the remains. This might entail collecting more artefacts for the purposes of chronological or cultural interpretation, for example. But it might also involve further 'contextualisation', focusing on the relationship between the palaeolandscape and the remains of human activity that have been found. Wherever possible, this research is carried out by means of mechanical coring in order to produce high-quality, fairly undisturbed core samples, known as *Begemann* samples (fig. 3.14). The cores are then sampled for use in further specialist analyses, such as dating, vegetation analysis and soil micromorphological analysis.[115]

We must bear in mind that although borehole surveys allow us to localise places where archaeological remains are present in the subsurface, it is not possible to draw far-reaching conclusions about the nature and size of clusters of archaeological remains.[116] The age of occupation remains can of course be at least partly determined by ^{14}C dating of carbonised organic remains, such as hazelnut shells. Depth in relation to groundwater-level curves (see section 3.2.2) can be used to approximate age relative to the 'inundation' of a location, which gives a minimum age for the occupation remains. However, some form of excavation will always be necessary in order to carry out a final assessment of the remains found (settlement, extraction camp etc.). Given the depth of many sites, this

---

112 The Elevation Map of the Netherlands (AHN). Measurements are obtained by laser altimetry with a grid/raster of 0.5 m² (horizontally) and and a 5 cm margin of error for the elevation.
113 Hamburg *et al.* 2014; Van Heeringen *et al.* 2014.
114 It has been found that elevation differences caused by compaction of clastic sediment must be taken into account while boring.
115 Hamburg *et al.* 2014.
116 Hamburg *et al.* 2014; Smith & Hogestijn 2013.

rarely happens. Excavation is often hampered by technical and budgetary constraints. This is especially the case if sheet piling has to be installed, deep layers of topsoil have to be mechanically removed, and groundwater has to be pumped away. Only if archaeological remains are closer to the surface, as at Swifterbant and locations in the Noordoostpolder,[117] is it possible to dig trial trenches that can be kept dry using a simple pump. The assessment of deeper buried sites must therefore rely on knowledge from excavations at other locations or *expert judgment*.[118]

3.5 The identification of sites: on statistics and indicators

As indicated above, archaeological mapping surveys focus on locating sites. The majority of investigations carried out under current heritage legislation involve little more than estimating the 'risk' that archaeological remains are present and might potentially be disturbed by planned interventions. In the Netherlands, mainly as a consequence of fieldwork carried out in Flevoland, a great deal of attention has been given to ensuring such estimations are as accurate as possible whilst establishing the most suitable detection methods and strategies at the lowest cost,. As part of this process, it has always been assumed that a site is a unit that can be spatially and temporally defined. This principle, although a matter of definition is, however, debatable.[119]

It goes without saying that borehole surveys only recover small samples of lithological units in the subsurface. Since lithological units can be linked to geogenetic processes that are associated with the dynamics of the landscape, a lot of attention has understandably been given to the identification of palaeolandscape zones in which the presence of archaeological remains is most likely. But the question remains, of course, how we can precisely determine whether and where a site is present in that buried landscape? In other words: how can we tell from small samples (*i.e.* core samples) whether or not there is evidence for a site? In some respects, this problem is related to statistical factors – how great is the chance that a borehole will be made in a site (intersection probability) and that there will be an archaeological indicator in that core sample (detection probability)? The problem is, however, also a qualitative one. On what basis do we determine whether we have an archaeological site or not- assuming we have indicators of human activity – and how do we determine that the site is worth preserving *in situ* or excavating?

3.5.1 The statistical uncertainty of sampling

The debate about 'optimum' methods and strategies of detection is primarily a quantitative matter, targeted on sampling an invisible and unknown 'archaeological landscape' in which potential traces of human activity – archaeological indicators – are present. As we saw in section 3.3.1, the nature of the archaeologically detectable deposits left by human activity in prehistoric Flevoland is highly diverse, but in this context it is the variation in the spatial distribution and density of archaeological remains that is of particular importance. The quantitative approach to the issue has been inspired above all by similar research into the effectiveness of shovel test pits to detect sites using trial trenches.[120]

Discovery probability, intersection probability and detection probability play a key role in the debate. The likelihood that a site will be discovered depends on the distance between boreholes – the borehole grid – relative to the size of the site. The chance that a borehole will be made within the site depends above all on the configuration of the boring grid – the positioning of boreholes relative to each other – in relation to the overall shape of the site. The probability that an archaeological indicator will actually be found in the core (detection probability) depends on the distribution and density of remains within the site, the size of the remains and the diameter of the borehole.[121]

This theoretical model makes several, not unproblematic, assumptions. Where there is a continuous layer, such as an anthropogenic layer recognisable over a certain area, then the model theoretically works perfectly, but only so long as the layer is recognisable as such in a core sample. This is the case with regard to the levee sites (midden deposits) at Swifterbant. In the absence of such a recognisable anthropogenic layer, however, major problems can arise. In most cases, what we have is a scattering of occupation remains that are distributed neither evenly or randomly over a surface; there is no neat Poisson normal distribution.[122] There is always a continuum in which variable densities or clusters of material occur. Empty zones, or zones with very low densities transition into zones with higher densities; boundaries cannot in fact be drawn with any accuracy and are necessarily arbitrary.[123] This implies that the archaeological reality does not in fact consist of average site characteristics, but a continuum that is differentiated to a greater or lesser extent.[124] The statistical approach that has come to be

---

117 Ten Anscher 2012; De Roever 2004; Raemaekers 1999.
118 Hamburg, Müller & Quadflieg 2012, 21-24; Hamburg *et al.* 2001, 7.
119 See for example Foley 1981; Binford 1992; Dunnell 1992.

120 Krakker *et al.* 1983; Kintingh 1988.
121 Tol *et al.* 2004; Verhagen & Tol 2004; Verhagen *et al.* 2011, 2013.
122 Peeters 2007, 266.
123 The variation in the distribution of artefacts can perhaps best be characterised as 'scatters and patches' (Isaac 1981), or as a 'lithic landscape' (Zvelebil, Green & Macklin 1992). These views are close to the 'off-site' concept proposed by Foley (1981).
124 Peeters 2007, 266.

applied over the years in the practice of archaeological survey is however based on average characteristics.[125] It has therefore been shown in practice that such a model does not always work, as on closer inspection many sites turn out not to have been recognised.

Recent statistical studies have acknowledged the importance of inherent variability of distribution.[126] The probability that a site will be effectively 'spotted' can apparently be best improved by taking larger volumes of core samples, which is perhaps less feasible in considerable tracts of Flevoland in view of the great depth at which cores have to be taken. Another factor that has a positive impact on the detection probability is the use of a fine mesh (1 mm) for sieving soil samples out of the cores. This makes it clear that a largely theoretical, statistical model is of limited value in defining prospective studies, but that statistically useful features – such as fractal distributions[127] – can be identified by analysis of 'actual' distribution patterns.

Essentially, therefore, discovering prehistoric sites with borehole surveys, particularly in a region like Flevoland, is extremely problematic. The best chances of discovery apply to sites characterised by an anthropogenic layer, such as the Neolithic levee settlements at Swifterbant. Another specific factor applying to the Swifterbant levee sites lies in the fact that the levees are relatively limited in size. A levee first actually needs to be discovered before a follow-up study, the search for archaeological indicators, can be restricted to the known area of the levee. The disadvantage of this approach is of course that this gives us little insight into Swifterbant activity that occurred in other parts of the landscape. In the statistical perspective outlined above, large find assemblages, in which the density of occupation remains is so great that there is almost a continuous layer, do have a reasonable chance of being discovered. The Hoge Vaart-A27 and Dronten-N23 sites are such examples. However, in many cases material is scattered more thinly and from a statistical perspective the probability of discovery relies mainly on chance. In this respect, the result of intensive borehole surveys around Almere may be the writing on the wall: if the discovery of sites relies largely on chance, but archaeological remains are nevertheless regularly discovered in borehole cores, then we have a picture of only a fraction of the archaeological (lithic) landscape.[128]

3.5.2 Indicators as evidence of the presence of sites

The potential for discovering archaeologically important locations depends not only on the quantitative problem discussed above. Recognisability is also a crucial factor. How can we be sure we are dealing with an archaeological site? We need to find phenomena associated with human activity (see section 3.3) which can be recognised in a borehole core: archaeological indicators, in other words.

It is of course important to determine which indicators can be regarded as anthropogenic and which cannot. For the prospection of buried sites in Flevoland, the primary focus is on the presence of worked stone and flint, pottery, burnt botanical remains, burnt bone and charcoal in core samples.[129] Although worked flint is not regarded as especially problematic, a correct identification is not always obvious. In most cases we find 'microdebitage': splinters that are the residue of flint flaking. Such small material can, however, easily be transported by water and wind. Furthermore, splinters can also form naturally, which cannot always be readily distinguished from anthropogenic material. Such problems are particularly prevalent in areas where glacial till occurs at or near the surface.

Charcoal is by definition regarded as a secondary indicator because it can also have natural origins; though when found in the proximity of other indicators that can be linked with more certainty to human activity, charcoal can be regarded as an archaeological indicator. The presence of unburnt fish remains in core samples – mainly perch scales and pike teeth – has prompted a great deal of discussion. Can they be regarded as archaeological indicators or not? It is currently assumed that these are remains deposited in natural circumstances. Other possible evidence of human activity, such as indications of human intervention in the vegetation, rarely play a role in the discovery of sites. Moreover, palaeoecological analysis is generally only used during excavations of identified sites, rather than at the survey stage.

At the same time, it is important to consider the fact that the research perspective is of at least equal importance. Despite the adoption of a certain 'landscape perspective' (see section 3.4.1), we are forced to conclude that archaeological survey in Flevoland is principally targeted at identifying sites, *i.e.* locations with clear archaeological indicators. As a consequence of research tradition and approaches to heritage management based largely on the site perspective, a broader approach based on the relationship between human behaviour and the dynamics of landscapes in prehistory[130] has made little

---

125 Verhagen & Tol 2004.
126 Verhagen *et al.* 2011, 2013; Smith & Hogestijn 2013; Smith 2013; Wansleeben & Laan 2012b.
127 See Peeters 2007, 167; Verhagen *et al.* 2013.
128 For example research like Nales (2010) and Warning, Smit & Visscher (2009) amongst many other archaeological heritage management reports on the archaeology of Almere, mainly in Dutch.

129 Hamburg *et al.* 2014.
130 A proposal for a research strategy geared to the integrated assessment of archaeological and palaeolandscape data was presented by Peeters (2007, 270-277).

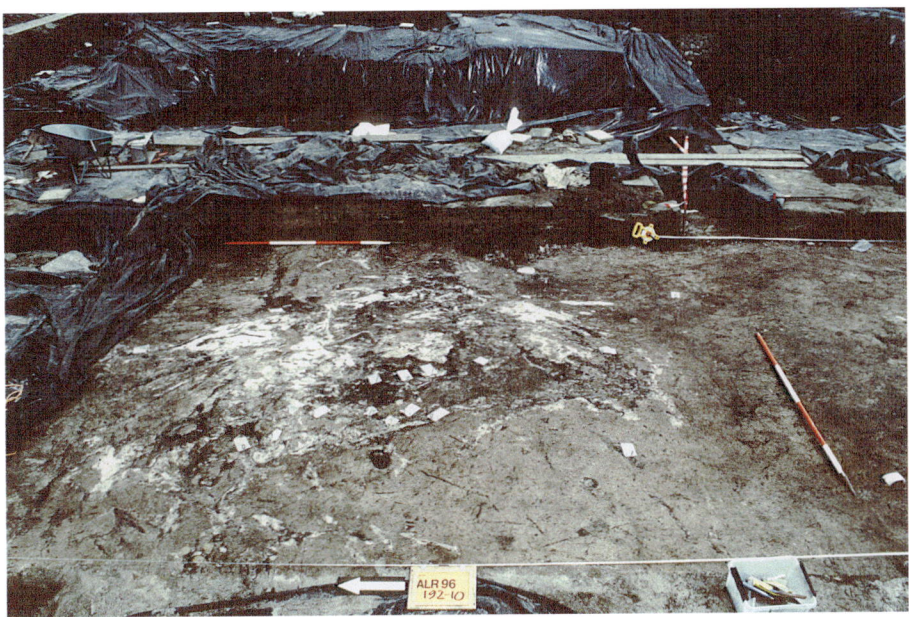

Figure 3.16: Washed sand deposited around prehistoric tree roots at Almere Hoge Vaart-A27. Although a common feature in prehistoric wetland environments, these phenomena have not resulted in large scale disturbance of the archaeological levels. These localized phenomena are characteristic in dynamic environments where peats are formed (photo: Cultural Heritage Agency of the Netherlands).

Figure 3.17: During the construction of the Hanzelijn railway near the Drontermeer observations permitted to gain insight into the recent formation of soil horizons due to subsurface transportation of humic elements. The picture shows the relationship between the depth of a recent drainage pipe and a humus-rich horizon (photo: J.H.M. Peeters).

headway.[131] Picturing 'the landscape' in the evaluation phase of research rarely constitutes anything more than mapping the Pleistocene surface and distinguishing zones on the basis of relief. The focus remains on identifying the presence or absence of archaeological indicators in the area of interest. Further investigations are generally carried out only at locations where the density of indicators is greatest, with the greatest weight being attributed to worked flint, pottery and burnt hazelnut shells. Clearly, however, this overlooks all kinds of anthropogenic phenomena. Not everyone recognises the potential importance of layers containing burnt reed remains in the clay layer in a channel. Wooden fish weirs leave no trace in core samples unless, with some luck, the wickerwork or a fish trap also happens to be intersected.

Another qualitative aspect that is generally considered in evaluation research is information on the preservation of the landscape in which archaeological remains are found. This is not in fact a matter of the intactness of 'the landscape' or former land forms, which is after all a set of temporally and spatially defined contextual factors, but of the intactness of the assumed prehistoric surface. This is generally determined on the basis of macroscopic soil phenomena visible in cores, although micromorphological research has also recently been used to establish the intactness of a prehistoric surface.[132] The dilemma in studies of this kind is the extent to which local observation can be extrapolated to larger areas. There is after all a great deal of variation in terms of the spatial scale of erosion processes. Some large-scale processes, as described in section 3.2, have had a major impact on the intactness of old prehistoric ground surfaces (see also section 3.3.2). Other erosion processes are more localised and are generally associated with disturbance to vegetation cover, or the dynamics of flowing water. Archaeological investigations in Flevoland have made it clear that the 'intactness' of soils can be spatially variable and temporally differentiated, even at the level of the individual site.[133] The presence of a disturbed or only partially intact prehistoric surface does not therefore automatically mean that no important archaeological remains are present.

It is therefore important to understand the spatial and temporal scale on which erosion (and sedimentation) occurred in relation to the human activity that took place. At Hoge Vaart-A27, for example, it proved possible to conclude that some of the prehistoric activity occurred in a landscape context where there was continuous localised, small-scale erosion – that is to say, the washing away of earth around tree roots (fig. 3.16) – in combination with the deposition of peat.[134] Yet despite this dynamic context, all kinds of archaeological phenomena remained intact. At Dronten-N23 and Hanzelijn Area VIII, coversand was found to have washed away in a manner similar to that in Hoge Vaart-A27, but when these locations were investigated it was not possible to say whether this occurred before or after occupation.[135] From an explicitly geological perspective, identifying these signs of erosion in cores could have led to the conclusion that little of the surface was intact, and that there was therefore little hope of uncovering a well-preserved site. From an archaeological perspective, however, things

---

131 Hamburg *et al.* 2014; however see also Van Heeringen, Schrijvers & Waugh 2018.

132 Hamburg *et al.* 2012; Hamburg *et al.* 2014. Micromorphology has in fact been used previously in the context of an excavation to ascertain how soil profiles were formed (Spek, Bisdom & Van Smeerdijk 2001a; Exaltus 2001; Peeters & Hogestijn 2001).

133 Hamburg, Müller & Quadflieg 2012; Hogestijn & Peeters 2001.

134 Spek, Bisdom & Van Smeerdijk 2001b.

135 De Moor *et al.* 2009; De Moor 2012.

can look very different. Many undisturbed phenomena may still be present, so that a site might seem relatively well-preserved. This kind of interpretive dilemma, the essence of which lies in the problem of linking different scales, is a common occurrence in Flevoland.[136] The interpretation of geological phenomena in relation to archaeological indicators can have a major bearing on the identification of archaeological sites. From an archaeological perspective, it is easy to overestimate the impact of erosion on any archaeological sites that may potentially be present. Perfectly preserved sites are more the exception than the rule. Furthermore, areas that have been subjected to erosion might still contain archaeological phenomena that can help us understand how the landscape was used in prehistory.

On the other hand, problems can occur when it comes to defining a chronological framework for pedogenic phenomena. The reclamation of the polders in Flevoland and the intensive drainage of the region caused humus particles to be transported both horizontally and vertically on a large scale, creating podzol-like profiles. In cores, such profiles can easily be interpreted as intact soils formed in prehistory. Along the Hanzelijn route to the east of Dronten, for example, intact podzol soils in Pleistocene coversand were documented in a prospective borehole survey in a zone where many core samples also contained charcoal.[137] However, during the construction works for the railway track it was found that a humus- and iron-rich horizon in the coversand was related to the lateral flow of groundwater flow towards modern drainage pipes (fig. 3.17). The humus and iron particles originated in the sediment layer above. This was largely made up of disintegrated peat (coarse detritus). Micromorphological analysis also made it clear that the coversand had been levelled off, which could barely be seen in macroscopic analysis because of the more recent soil formation.[138] In fact, several dozen Mesolithic pit hearths were found here, so despite the erosion of the prehistoric surface, some deeper soil features were still present.

3.6 Excavations: windows on the past

The landscape development described in the previous sections, the way humans exploited and influenced their environment in the past, and the procedures used in archaeological research have given rise to a unique dataset of archaeological information. The character of this dataset varies, however, ranging from isolated surface finds or individual borehole core samples containing an archaeological indicator, to fully-scale, complex excavations. The discussion in the following chapters is based on a selection of more closely investigated locations. The 12 sites in question are presented in Appendix I. Research at these locations has been carried out over the past 30-40 years. These sites are described as 'windows' on the post-glacial occupation history of Flevoland from the Mesolithic to the Middle Bronze Age. Research at these locations is characterised both by systematic research into the archaeological remains and extensive multidisciplinary analysis of the stratigraphical and landscape context in which these finds were made.

Four sites have been selected from each polder. These are related to human activity in parts of the landscape that had their own specific characteristics in prehistory. Furthermore, the sites represent activities that took place on the Pleistocene subsurface and on landforms that developed during the Holocene. The landscape in Zuidelijk Flevoland was dominated by the Eem river system, in Oostelijk Flevoland by the Hunnepe river system, and in the Noordoostpolder by the Overijsselse Vecht system. Since these systems developed differently and had different dynamics, the archaeological record in these areas is varied. When this variation is considered as a whole, however, along with detailed information from other sites, it is possible to outline a coherent picture of the prehistoric occupation of Flevoland.

The windows selected for the Noordoostpolder are: Emmeloord-J97, Schokkerhaven-E170, Schokland-P14, Urk-E4. In Oostelijk Flevoland these are: Dronten-N23/N307. Hanzelijn Area VIII, Hanzelijn Tunnel Drontermeer and the well-known Swifterbant cluster. From Zuidelijk Flevoland: Almere Hoge Vaart-A27, Almere Europakwartier site 7, Almere Zwaanpad and Zeewolde-Oz35/Oz36. An excavation site is included for each polder where research campaigns were conducted over several years and provided the basis for one or more doctoral theses: Schokland-P14, the Swifterbant cluster and Almere Hoge Vaart-A27.[139]

In most cases several successive research campaigns were carried out at the sites, each phase selecting particular parts of the site. After a phase involving several borehole surveys, each site was excavated, with the exception of Hanzelijn Area VIII. The excavations were carried out in pits and trenches. In most cases archaeological finds were excavated in a grid system of units (1x1 m or 50x50 cm), and the find level was sieved through 2-4 mm meshes. Larger artefacts were sometimes collected by hand. Features were excavated as separate entities and generally sampled for specialist research (including dating, botanical, micromorphological, zoological, geochemical research).

---

136 Leijnse 2006.
137 See for example Makaske *et al.* 2002.
138 Van Zijverden 2009.

139 Schokland P14: Ten Anscher 2012; Gehasse 1995; Swifterbantcluster: Raemaekers & de Roever 2020; De Roever 2004; Raemaekers 1999; Devriendt 2013; Almere Hoge Vaart-A27: Peeters 2007.

The archaeological remains found were also subject to specialist analysis (including residue analysis, use-wear analysis etc), as well as research into their landscape context (including pedological, physical-geographical analysis). The Appendix of Sites describes the sites in terms of their geographical location, the research methods used, the geological and pedological context, dates, find material, features, cultural context, palaeoecological and geographical context and taphonomic factors.

In this book we also make reference to sites that are not listed in the Appendix. These were all investigated on a smaller scale, and provide details that contribute to the overall picture. Locations where ecological research has been carried out have also been referred to where relevant, as a source for detailed descriptions of the landscape.[140] As a result, the synthesised information presented in the following chapters is logically based on a heterogeneous set of data.

3.7 Conclusions

The above discussion of the geological development of Flevoland, the properties of remains of prehistoric occupation and the potential for discovering and identifying those traces has shown that archaeologists encounter many practical problems. The complex dynamics of the palaeolandscape resulted in a differentiated subsurface structure and development of geological phenomena that hamper the interpretation of evidence of prehistoric activity. The variable depths at which remains of human activity are found means that it is not possible to use the same methods everywhere to detect and assess sites.[141] Furthermore, the effectiveness of most detection methods is limited because of the specific properties of the archaeological remains. As a result of the perfectly understandable choice of borehole surveys as a research tool, bearing in mind the limitations of this method, archaeological remains that are potentially present are difficult to discover and assess on the basis of their intrinsic properties. Our present picture of the archaeological landscape probably does not do justice to the wide variety of ways in which landscapes were used in prehistory. It is inevitably the larger sites that have been discovered and investigated, that form the most important basis for our image of prehistoric Flevoland.

It is also clear that researchers are ambivalent when it comes to adopting an overtly landscape archaeology perspective. If we look at the focus of most survey research, we find a great emphasis on discovering sites in the traditional archaeological sense. However, human activity can take forms that are not so easy to capture using the site concept, which is still the starting point for the majority of investigations.[142] There are certainly major methodological issues that must be addressed: the borehole surveys that are still the dominant method for prospecting hidden landscapes have their limitations. But at least as important is the fact that a change of attitude is required in order to broaden the focus of research. A truly landscape-oriented approach that aims to enhance our understanding of human activity in a dynamic prehistoric landscape that has changed dramatically over time will require us investigate other locations than only those where a large amount of occupation remains have been left behind. New and different prospection techniques will also have to be used and developed if we are eventually to ascertain the true variation of Flevoland's buried archaeological resource.

---

140 See chapter 6.
141 See also Van Heeringen, Schrijver & Waugh 2018; Hamburg et al. 2014.
142 Examples of investigations where aspects such as this have played a role are those associated with the Hanzelijn rail link and Lelystad-Kotterbos (Gerrets, Opbroek & Williams 2012; Lohof, Hamburg & Flamman 2011; De Moor et al. 2009; Van Heeringen et al. 2014).

Chapter 4

Exploiting a changing landscape: subsistence, habitation and skills

J.H.M. Peeters, T. ten Anscher, L.I. Kooistra,
L. Kubiak-Martens & J. Zeiler

4.1 Introduction

The sites investigated in Flevoland have yielded large quantities of data that provide an insight into how its prehistoric inhabitants exploited the landscape. As explained in Chapter 3, and explored in more detail in Chapter 6, dramatic changes occurred in the landscape during the Holocene, mainly as a result of progressively wetter conditions. These changes had a direct impact on people's ability to obtain basic necessities and on the habitability of the region.

One vital prerequisite for life was, and is, of course the availability of food. However, the availability of food sources in the region varied on different spatial and temporal scales. In the long term (centuries, millennia) there was gradual, large-scale change in the landscape, under the influence of climatological, hydrological and pedological conditions. In the shorter term (months, human lifetimes) changes were linked above all to the seasons and to short-lived events in the landscape. Given the large-scale transformation of the Flevoland landscape over the long term, whereby in the course of the Holocene the region changed from a dry, forested hinterland into an ecologically differentiated zone dominated by water behind the coastal beach barrier (see Chapters 3 and 6),[143] we might expect the potential food supply to become more differentiated over the course of prehistory. There may have been a shift in the relative availability of food sources. Cultural developments would also have played a role in the food supply, on the one hand as a consequence of cultural choices determining the exploitation of certain food sources, and on the other due to the introduction of new food sources.

Besides food, the landscape also provided other resources that were important for human life. Wood and other plant resources were needed for building shelters and other structures and devices, such as fish traps. Various raw materials, including wood, reeds, flint, bone, antler, hides and amber, were used in the manufacture of a broad range of tools, utensils, clothing and jewellery. A considerable proportion of these raw materials were available and could be exploited in the region, a small proportion, however, came from regions outside Flevoland. The choice of raw material was not only limited to those that could be used directly. To produce pottery, for example, various primary raw materials – clay and some form of tempering – first had to be combined to produce a workable mass. Adhesives, such as wood tar used to haft flint tips to arrow shafts, had to be distilled from raw materials such as wood or bark. Such processes concern aspects of

---

143   Gotjé 1993; Peeters 2007; Ten Anscher 2012, 507-536.

transformation technologies, which involve high levels of knowledge and skill.

The development of the palaeolandscape in the region also gives us an insight into the relationship between landscape characteristics and habitation. During the course of the Mesolithic and Neolithic the landscape not only changed radically in terms of the availability of all sorts of raw materials, but it also changed in a spatial sense (see also Chapter 6). Suitable locations for temporary encampments, or for more or less permanently occupied settlements, were not always available: some locations disappeared, while other new ones were created. River channels and streams changed course, lakes and pools expanded or disappeared. The dynamics of the landscape had an impact on the choice of location and variability of settlements and other sites in the landscape.

This chapter considers what the archaeological data tells us about the exploitation of food and non-food resources in a diachronous perspective. Links will be made to a certain extent with the dynamics of the landscape, which are examined in detail in Chapter 6. The diachronous relationships between the economic exploitation of resources and developments in the landscape will be discussed in more depth in Chapters 6 and 7. The quantities of animal and plant remains collected from excavations in Flevoland provide a relatively large amount of information about the importance of plants and animals as food sources, the choices that the prehistoric inhabitants of Flevoland made as regards their exploitation, and the ecological zones that were important for the food supply. We also have a relatively large amount of information on the availability and use of non-food resources. Before we examine this, however, we shall first briefly consider the extent to which these are representative.

4.2 Taphonomy and analysis: the representativeness and interpretive value of find assemblages

The sections below show that, on the whole, the set of data from Flevoland suggests that a broad range of plant and animal resources were exploited. We must, however, ensure that we take into account the possibility that this picture is distorted. As discussed in Chapter 2, geological, hydrological, biological and anthropogenic processes in the region impacted in various ways on the structure of the substrate and the archaeological remains present in different lithological units. The buried archaeological record as we know it is therefore the result of a complex series of depositional and post-depositional processes.

Remains from various periods of prehistoric activity have not always been exposed to the same taphonomic processes. The preservation of organic remains from different periods is, for instance, particularly variable. Waterlogged plant and animal remains from before the Mesolithic tend, in general, to be underrepresented at the sites investigated. The Early Neolithic, on the other hand, is well represented, while later periods of prehistory are again underrepresented. As a result, any diachronous analysis of the exploitation of the dynamic landscape cannot be based on data that are comparable in qualitative and quantitative terms. Furthermore, the sites described in the site atlas (Appendix I) have not all been investigated in the same way or with the same intensity, which also leads to quantitative and qualitative distortions when comparing data.

One problem concerns the different way in which zoological and botanical remains have been collected. In earlier excavations, in particular, this was done mainly by hand. As a result, only large fragments of skeletons were recovered and smaller remains – not only those of birds and fish, but also of small mammals and reptiles – would usually have been missed. Botanical remains were rarely collected systematically over a contiguous area, which means that data generally refers only to specific sampling locations. As a result, the picture of local flora and fauna might not only be incomplete, but could result in a distorted picture of the food supply. In the case of Flevoland, this problem may apply less to zoological remains, as in all the excavations examined here the bone material had been collected both by hand and by sieving. As a result, it is possible to obtain a more nuanced picture of the exploitation of animal resources. We must, however, take into account the problem of differentiated preservation. The context of the original association between animal and plant remains can no longer be identified in find assemblages that have been exposed to erosion – either chemical, biological, bacterial or mechanical. Even with the existence of apparently good preservation conditions, this is in fact always the case in Flevoland, albeit to varying degrees.

What is more, it is not always easy to answer questions about the use and specific importance of resources because of methodological variations in the presentation and analysis of data. The relative importance of a species in the food supply is generally indicated on the basis of the bone weight which – given the ratio between body weight and body size – can be regarded as a rough indicator of the meat yield. This, however, only applies to mammals and birds. The ratio does not apply to fish, so the weight of fish bones found does not indicate the meat yield from this source. The problem is that relevant details such as number and weight have not always been published, making it difficult – if not impossible – to systematically quantify the relative importance of the various groups of animals – mammals, birds and fish – and make comparisons between sites and periods.

Finally, we should briefly consider the issue of how to determine the extent to which the remains found

Positive	Negative/Neutral	Association
Animal remains	*Animal remains*	hearth
Specific selection of parts of a skeleton	(quasi) complete skeleton	pit hearth
fragmentation/crushing of bones	animal gnaw marks	refuse pit
cut marks	no sign of burning	refuse layer
signs of burning		
Plant remains	*Plant remains*	
non local plant community	local plant community	
burnt/charred	unburnt	
edible plant parts	inedible plant parts	
cultigens		

Table 4.1 Anthropogenic indicators for plant and animal remains.

constitute consumption waste or are the result of natural accumulation. After all, animal and plant remains found on an archaeological sites need not necessarily have ended up there due to human activity. Positive, negative/neutral and associative indicators (table 4.1) can be used to draw some distinctions. Positive indicators can with a high degree of certainty be related to human activity. Negative or neutral indicators suggest natural origins. Associative indicators refer to the association/relationship between the remains and other phenomena. In combination, these indicators can provide a basis for assessment, when viewed in relation to the specific find context. The degree of certainty is never absolute, however, but rather a sliding scale with no clear boundaries.

4.3 Wild and domesticated mammals as sources of food

Table 4.2 shows that a broad range of mammals are represented in the bone remains found on various sites. With the exception of the material from Dronten-N23, which can be dated to the Mesolithic, the data cover the Neolithic and the Early Bronze Age. Though bone remains (burned) were found at the Mesolithic sites of Almere-Zwaanpad, they were not from mammals. Although the Middle Bronze Age is clearly represented in the pottery from Emmeloord-J97,[144] the mixed presence of material from various periods in this context means it is not possible to say which bone remains can be attributed to the Middle Bronze Age.[145]

We cannot therefore say much about the exploitation of mammals in the Mesolithic, as it is defined in the periodisation used in this book. Out of more than 1000 fragments of bone (with a total weight of 78 g) collected from Dronten-N23 and dating to this period, only one could be identified by species: a fragment of boar/pig, probably a wild boar, given its age.[146] For the rest, only a few dozen fragments could be identified as coming from a large or medium-sized mammal. Nor is it entirely clear which phase or phases of the Mesolithic the recovered bone remains can be attributed to, since this site was in use throughout almost the entire Mesolithic. However, based on the spatial distribution of the few remains found and the presence of small clusters in relation to clearly identifiable clusters of flint, stone and fragments of charred hazelnut shell,[147] a date in the Middle Mesolithic would seem most likely. at least for the majority of the bone material collected. This date is based on the typological and technological characteristics of the flint assemblage and on ^{14}C dates.

The oldest and largest assemblage immediately following the Mesolithic is that from Hoge Vaart-A27 phase 3, which can be dated to the Early Neolithic. The spectrum of mammals in the find assemblages from the Neolithic and Early Bronze Age (table 4.2) clearly shows that large to medium-sized ungulates formed a major component of the wild mammal population, which consisted mainly of red deer and wild boar, but also included aurochs, wild horses, elk and roe deer. Other wild terrestrial mammals that may have played a role in the food supply, such as brown bears, wild cats, badgers, pine and beech martens, foxes and polecats, are considerably less common in the assemblage. Of course many animals provided other resources, such as hides and bones (see section 4.8). Semi-aquatic species such as beaver and otter are prominently represented, however, and seals are also found on an incidental basis. The latter is an aquatic species that occurs primarily in marine environments, but seals can swim upriver far inland. Small rodents and insectivores might represent 'background fauna', whose remains will probably have ended up among the other bone material without any human intervention, though we cannot rule out the possibility that people ate the larger water voles.

As well as wild mammals, domesticated animals are also represented in various find contexts (table 4.2). The oldest Neolithic context containing faunal remains, Hoge Vaart-A27 phase 3 (Early Swifterbant; see Appendix I), contains no domesticated animals at all, with the exception of the bones of a dog.[148] Nor did the Early Swifterbant context of Schokland-P14 contain any domesticated animal remains.[149] Domesticated animals, *i.e.* livestock, were present on the 300 – 350 years later sites of Swifterbant S2 and S3 (Classic Swifterbant), cattle being the most prevalent species, with domesticated pigs a close second. On other sites too, including Urk-E4 and Schokland-P14 – where there is perhaps slightly less chronological control – domesticated animals are represented that probably did

---

144 Bloo 2002.
145 Bulten, Van der Heijden & Hamburg 2002; Kerkhoven 2003
146 Van Dijk 2002, 559.

147 Wansleeben & Laan 2012a.
148 These are almost certainly the remains of just one dog (Laarman 2001, 11).
149 Gehasse 1995.

SPECIES	Hoge Vaart-A27 (phase 3)	Swifterbant-S2	Swifterbant-S3	Swifterbant-S4	Urk-E4	Schokland-P14 (layer A)	Schokland-P14 (layer B)	Schokland-P14 (layer C)	Schokkerhaven-E170	Schokland-P14 (layer D)	Schokland-P14 (layer E)	Lelystad-Kotterbos	Schokland-P14, EKW (gully)	Schokland-J78 (LNEO)	Schokland-P14, WKD-3	Schokland-J78, WKD-3	Emmeloord-J97
Terrestrial wild species																	
Alces alces	1		21				4	3	1		2		3		1		
Cervus elaphus	455	21	121	96	5	42	117	42	11		5		16		19	10	45
Capreolus capreolus	8					2	2	2					3				
Bos primigenius	20		2	1													1
Sus scrofa	35	103	47	41	1								3				53
Equus spec.	52		2						1	1	1		1		3		
Ursus arctos	4		6	1		1							2		1	3	
Felis silvestris	1		1		1			1									
Vulpes vulpes			2	1						2					1		3
Meles meles	14			1	1			1									
Martes martes/M. foina	11					1	3	1							2		
Mustela putorius			3					1			1						1
Lepus europaeus				1													
Sciurus vulgaris	4																
Subtotal	605	124	205	142	8	46	127	50	15	1	9		28		25	15	103
Aquatic wild species																	
Castor fiber	56	129	536	85	51	58	233	96	3	6	7	1	23	7	9	4	30
Lutra lutra	10	29	598	25	11	1	5	9			1					2	6
Subtotal	66	158	1134	110	62	59	238	105	3	6	8	1	23	7	9	6	36
Marine wild species																	
Phoca vitulina	13		1														
Subtotal	13		1														
Domesticated species																	
Bos taurus		12	323	163	29	27	43	24	5	7	7		31	7	41	47	81
Sus domesticus		125	34	99			3	4					1	23	26	90	53
Ovis aries/Capra hircus		5	9	29	10		2	3			9		5	3	11	4	26
Canis familiaris	11	5	57	22	5	1	2	2			2		2	1	18	21	31
									5								
Subtotal	11	147	423	313	44	28	50	33		7	18		39	34	96	162	191
Wild or domesticated species																	
Bos taurus/B. primigenius	22		1	17			6	1	2	1			1		5		
Sus domesticus/S. scrofa	806	162	2214	92	107	85	141	96	3	5	25		15				
Subtotal	828	162	2215	109	107	85	147	97	5	6	25		16		5		
Background species																	
Talpa europaea			2	1													
Crocidura russula															1		
Sorex araneus															1		
Neomys fodiens															1		
Microtus oeconomus	1		2														
Microtus spec.		1															
Arvicola terrestris	3	1					1				1			1	2	10	30
Apodemus sylvaticus															1		
Subtotal	4	2	4	1			1				1			1	6	10	30

Table 4.2: Overview of Mammal species.

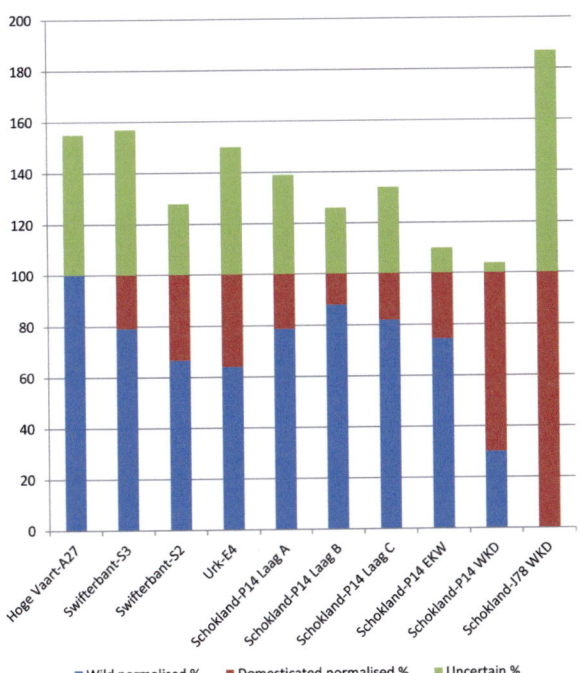
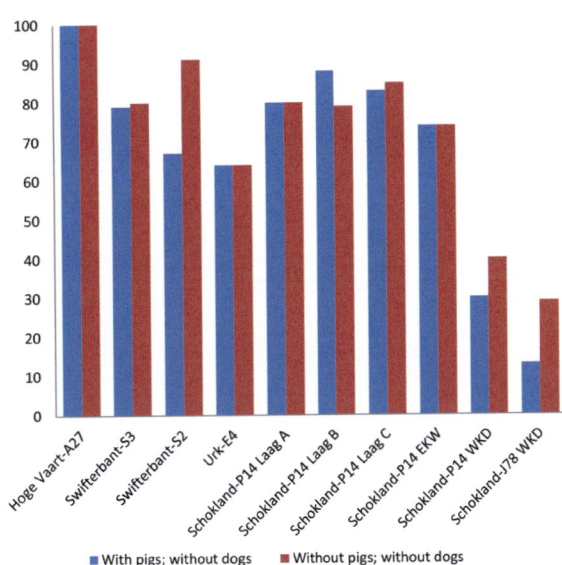

Figure 4.1: Top: relative proportion of wild and domestic mammals. The percentages for certain wild and certain domestic mammals are normalized to 100. The uncertain category indicates to what extent there is uncertainty about the actual distribution of wild and domestic mammals. Below: relative share of wild mammals including all pigs (wild, tame, uncertain) and excluding all pigs (wild, tame, uncertain). Dog is excluded from the calculation as a domestic animal.

play a role in the food supply. Generally speaking, the proportion of domesticated animals seems to be greatest in Early Bronze Age contexts.

In the debate on changes in the Neolithic food economy, great emphasis is traditionally placed on the growing importance of domesticated animals. It is therefore important to distinguish clearly (or as clearly as possible) between the domesticated animals and their wild counterparts. This is generally done on the basis of metric data. This is not always easy when it comes to Neolithic material, however, because there is a certain degree of overlap between wild and domesticated populations.[150]

Overlap in measurements is most common in pigs and wild boar, which might suggest a certain degree of interbreeding between the domesticated and wild variants.[151] The degree of (metric) overlap between pig and wild boar may not be so great in all cases, but it can make it difficult to distinguish between the two.[152] This applies not only to the Netherlands, but to other parts of Europe too.[153] The main point appears to be that, during the Neolithic, the domesticated animals had not yet changed much in comparison to their wild counterparts. This is substantiated by analysis of mitochondrial DNA, which clearly shows that

Western and Central Europe (Albarella *et al.* 2009). Measurements of skeletal elements from Neolithic contexts can be compared with these, which generally allows them to be attributed to either the wild or domesticated form (see, for example, Zeiler & Brinkhuizen 2014). Since this method has not been applied to the Flevoland data, in many cases we still find a sizeable category where it is unclear what proportion is wild and what proportion domesticated. The degree of overlap between the wild boar and domesticated pigs also varies depending on the criteria used by the researchers. This is connected not only with the variation in size between wild and domesticated pig populations in different areas, but also with more subjective factors. Sometimes, no distinction at all is drawn, and researchers maintain that this is not relevant to pigs in Neolithic contexts. The assumption is that prehistoric domestic pigs were able to forage fairly freely and that genetic mixing occurred between domesticated and wild animals, so that in a certain sense a single economically relevant population existed. Although there are arguments in support of this, we must also bear in mind that keeping domestic pigs and the decision as to whether to hunt wild pigs are based on different strategic principles, and would therefore argue that the distinction is relevant.

150 This is not however to say that there is a continuum in measurements containing no discernible interruptions. In the case of cattle and aurochs, for example, only the measurements of the largest domesticated bulls and those of the smallest auroch cows overlap (Degerbøl & Fredskild 1970).
151 Albarella *et al.* 2009.
152 For several years now, however, we have had good reference material in the form of metric data from Mesolithic wild boar from

153 Albarella *et al.* 2009.

wild boar were domesticated in different regions and at different times throughout Europe.[154]

The general overview in table 4.2 shows that the numerical dominance of wild mammal remains over domesticated animals – dogs having been excluded because of the specific social role they seem to have played – occurs in contexts associated with Swifterbant and Pre-Drouwen occupation. This dominance is not in the same proportions everywhere, however, and the absolute ratios of wild to domesticated mammals vary from one site to another (fig. 4.1). The picture is perhaps clouded by the lack of clarity regarding the category 'wild or domesticated'.[155] However, if pigs, both wild and domesticated, are excluded from the calculation of the ratio of wild to domesticated animals, this causes no shifts worthy of note in many contexts. Only Swifterbant-S2 and the Barbed Wire Culture contexts of Schokland-P14 and Schokland-J78 show a relatively stronger representation of domesticated animals: in these three particular contexts, a relatively large number of bones can be attributed with certainty to domesticated pigs.[156] It could be concluded from this that wild mammals played a significant role in food supplies throughout the Neolithic, and that the number of domesticated animals was fairly small. However, we must not forget that this is a fragmentary picture. The incidental Late Neolithic contexts (Schokland-P14, Schokland-J78) systematically show evidence for domesticated animals, albeit represented by only small quantities of material. Whatever the quantities, livestock appear to occur for the first time in Flevoland in Classic Swifterbant contexts, with an increase in the Pre-Drouwen phase. It was not until the Early Bronze Age, possibly already in the Late Neolithic, that the emphasis on livestock became much stronger, although wild mammals continued to play a role in the diet.[157] It is not easy to obtain a reliable picture of how wild and domesticated mammal populations were exploited. This is due not so much to a lack of data, but to the fact that different methods have been used for the age analysis of bones, and the data has been published in different ways, and not always in their entirety.

Nevertheless, several trends can be distilled from the data available, and it would appear that mainly mature (fully grown) animals are represented among the wild mammal populations. This applies both to furred animals and to large game. For example, the bone remains at Swifterbant-S3 suggest that otters and beavers over 2 old were selectively hunted (table 4.3); there is only one example of an animal years younger than a year old. A similar picture emerges in the material from the Swifterbant and Pre-Drouwen contexts of Schokland-P14, where the emphasis is on the 2-3 or 4 year-old age group. At Swifterbant-S4 more beavers below the age of 2 years appear to have been hunted, though fewer data are available for this site. For the Bronze Age (Barbed Wire Beaker culture), the information is limited to a single beaver aged 2-3 years recovered from Schokland-J78.

The hunting of red deer also appears to have concentrated almost exclusively on mature and adolescent animals (table 4.4), as is evidenced by the data from Hoge Vaart-A27, Swifterbant-S3 and S4, Schokland-P14 (layers B,C), Schokkerhaven-E170, Emmeloord-J97 and the Barbed Wire phase of Schokland-P14 (layer E) and Schokland-J78. The remains include bones of animals in excess of 10 years old, possibly males who would have had substantial antlers.

The picture is more diffuse when it comes to wild boar, because of the problem of distinguishing them from domesticated pigs. Given the age distribution among the other animals, it would seem likely that there was also a preference for (young) adult animals. Based on the fusion indicators for bones from Hoge Vaart-A27 phase 3, most of the wild boar appear to have been between 1 and 2.5 years old, though a considerable proportion were over 3.5 years old (table 4.5).[158] The data from this site therefore suggest that adolescent and adult individuals were selected. A similar pattern can be seen in the undetermined 'domesticated/wild pig' category at Swifterbant-S3 and S4 and Emmeloord-J97, which mainly comprises animals older than a year. At Schokland-P14 the remains of individuals less than a year old seem to occur somewhat more often, which might indicate a slightly less selective exploitation.[159]

As far as cattle are concerned, only adult and adolescent animals older than 1 or 2 years old appear to be represented at Swifterbant-S3, Emmeloord-J97 and

---

154 Larson *et al.* 2005. Attempts to obtain DNA samples from bone remains from Flevoland produced no results due to the almost complete absence of collagen.
155 A considerable quantity of bone cannot be attributed to either wild or domesticated, so it is not possible to obtain an idea of the precise ratios. If the 'wild or domesticated' group is left out, there is barely any difference between 'definitely wild' and 'definitely domesticated'. However, the problem lies in the uncertain category itself, since it is not known how great the proportions of wild and domesticated animals are within it. The uncertainty associated with this is consistent with the proportions of bone remains accounted for by this category, between 4% and 87%. These two extremes refer to the late Barbed Wire Beaker culture of Schokland-P14 and Schokland-J78.
156 Gehasse 1995; Zeiler 1997; Prummel *et al.* 2009.
157 See for example Bakels & Zeiler 2005; Zeiler 1997.

158 Laarman 2001, 20-21 see also Laarman (2001) table 10 and 11.
159 Gehasse (1995) does not always report the numbers for the different age groups.

Times of slaughter of beaver (in months) based on the degree of fusion of the epiphyses in postcranial bones, Swifterbant S3 en S4, J97, P14 layer A-C en P14 EKW (gully)
FU = fused (epiphysis) (= older than presented age)
UF = unfused (epiphysis) (= younger than presented age)

	n FU					n UF				
	S3	S4	J97	P14 A-C	P14 EKW	S3	S4	J97	P14 A-C	P14 EKW
Age										
12	-	-	-	14	-	-	-	-	-	-
24	39	7	1	96	2	-	5	2	-	-
36	-	-	-	-	-	-	-	-	46	2
48	-	-	-	2	-	-	-	-	4	2

Times of slaughter of otter (in months) based on the degree of fusion of the epiphyses in postcranial bones, Swifterbant S3 and S4
FU = fused (epiphysis) (= older than presented age)
UF = unfused (epiphysis) (= younger than presented age)

	n FU		n UF	
	S3	S4	S3	S4
Age				
12	67	2	3	-
24	44	-	4	-

Times of slaughter of otter (in months) based on eruption and wear of teeth, Swifterbant S3 and J97

	S3	J97
Age		
> 12	30	-
> 24	2	-
36	-	-
12-48	5	1
< 48	8	-

Table 4.3 Times of slaughter beaver and otter.

Schokkerhaven-E170 (table 4.6).[160] This is probably also the case at Schokland-P14.[161] The material from Swifterbant-S3, Emmeloord-J97 and Schokland-P14 does show that calves often died just after, or maybe even before, birth, which might suggest that cattle were kept at these locations at the end of the winter or in spring.

4.4 Fishing in a drowning landscape

In contrast to the evidence for mammals, we do have information about the exploitation of fish from the Mesolithic to the Early Bronze Age, albeit that the Mesolithic is represented by only one context – Almere-Zwaanpad. As can be seen in table 4.7, freshwater species account for the majority of the fish remains. These are mainly species that lived permanently in fresh water – non-migratory freshwater fish, in other words: cyprinoids such as common and white bream, perch, ruffe, pike and catfish. In comparison, the proportion of fish that spent part of their life in fresh water – migratory species – is much smaller. Migratory fish species can be divided into anadromous and catadromous species. The former swim from the sea to fresh water to spawn, while the latter make the reverse journey. Anadromous species include sea sturgeon,[162] smelt, whitefish, salmon/sea trout and the three-spined stickleback. Catadromous species include the European eel and flounder.

Although modest in number, marine fish species, sea bass and mullet, have been found in 10 of the 15 contexts (fig. 4.2). Sea bass and thinlip mullet both prefer warmer waters. Nowadays they enter Dutch coastal waters from the south in spring and towards the beginning of winter migrate via the English Channel to waters to the south of England. Mullet can swim far inland into fresh waters. Sea bass can be found in estuaries, though only in saltier water. Flatfish are also regarded here as marine species although, as already stated, the flounder is a catadromous species. It is often, however, difficult to distinguish between the remains of flounder and other flatfish that are found exclusively in salt water, such as plaice. There is a realistic chance that at least some of the flatfish remains that have not been determined by species are flounder, and could therefore have been caught in fresh water. This would thus further reduce the marine fish category. For this reason, flatfish are separated from other marine fish in figure 4.2. The only context including real marine fish is

---

160 The age data of cattle for P14, J78 and E170 were, in the absence of clearly specified data, except for one individual derived from the measured bones (Gehasse 1995, Appendix III). After all, these bones are always from mature individuals. Some data from pig / wild boar from P14 have also been calculated this way. This obviously gives a distorted picture – especially in cattle – since remains of immature animals have also been found.

161 Gehasse (1995) does not always report the numbers for the different age groups.

162 Until fairly recently, sturgeon remains from historical and prehistorical contexts were automatically attributed to the European sea sturgeon (*Acipenser sturio*). Recently, however, more and more evidence has come to light suggesting that the European sea sturgeon and the Atlantic sturgeon (*Acipenser oxyrinchus*) lived alongside each other in western Europe.

Times of slaughter red deer (in months) based on) based on the degree of fusion of the epiphyses in postcranial bones, Swifterbant S3 and J97
FU = fused (epiphysis) (= older than presented age)
UF = unfused (epiphysis) (= younger than presented age)

	n FU		n UF	
	S3	J97	S3	J97
Age				
20	2	-	-	-
32	1	-	-	-
36	-	13	-	-

Table 4.4 Times of slaughter red deer.

Times of slaughter of red deer based on eruption and wear of teeth (Swifterbant S3, J97, E170) and morphology of the antler (Swifterbant S4)

	S3	J97	E170	S4
Age				-
< 4-5 m.	-	1	-	-
> 2 y.	-	3	-	-
ca. 2,5 y.	-	-	1	-
6-7 y.	2	-	-	-
ca. 9 y.	-	1	-	-
10-11 y.	1	-	-	-
12-13 y.	-	-	-	1

Times of slaughter pig/wild boar (in months; after Habermehl 1975) based on the degree of fusion of the epiphyses in postcranial bones, Swifterbant S3 en S4, P14 layer A-E, EKW (gully) en WKD-3, J78, J97 and Hoge Vaart
FU = fused (epiphysis) (= older than presented age)
UF = unfused (epiphysis) (= younger than presented age)
Note: stage EL (= fused epiphysis) presented by FU

	n FU								n UF						
	S3	S4	P14			J78	J97	A27	S3	S4	P14			J78	J97	A27
			A-E	EKW	WKD-3						A-E	EKW	WKD-3			
Age																
12	92	4	19	1	1	-	2	4	21	2	7	-	-	-	4	-
24	100	2	11	-	1	-	4	37	53	15	7	-	-	-	3	53
24-36	2	-	-	-	-	-	-	2	4	1	-	-	-	-	-	4
36-48	4	-	-	-	-	1	1	2	6	2	4	-	-	-	-	2

Times of slaughter pig/wild boar (in months) based on eruption and wear of the teeth (excl. loose teeth), Swifterbant S3 en S4, P14 layer C-E and J97

	S3	S4	P14 C-E	J97
Leeftijd				
< 2	1	-	-	2
< 8	2	-	-	1
≥ 8	51	-	-	-
< 12	1	1	-	-
> 12	58	1	2	-
< 16	89	-	-	-
≥ 16	50	-	-	-
< 20	11	-	-	-
> 20	-	1	2	1
> 30	-	-	-	5

Table 4.5: Times of slaughter pig/wild boar.

Times of slaughter cattle (in months based on the degree of fusion of the epiphyses in postcranial bones, Swifterbant S3, P14 layer D en E, EKW (gully) and WKD-3, J78, E170 and J97
FU = fused (epiphysis) (= older than presented age)
UF = unfused (epiphysis) (= younger than presented age)
Note: stage EL (= fused epiphysis) presented by FU

	n FU							n UF						
	S3	P14			J78	E170	J97	S3	P14			J78	E170	J97
		DE	EKW	WKD-3					DE	EKW	WKD-3			
Age														
7-10	1	-	-	1	-	2	1	-	-	-	-	-	-	-
12-20	5	2	-	-	1	-	3	-	-	-	-	-	1	1
20-30	15	2	3	-	1	1	2	1	-	-	-	-	-	3
36-42	2	-	-	-	-	-	1	-	-	-	-	-	-	-
42-48	1	-	-	1	-	-	3	-	-	-	-	-	-	1

Times of slaughter cattle (in months) based on eruption and wear of the teeth (excl. loose teeth), Swifterbant S3 en S4 en J97

	S3	S4	J97
Age			
< 1	-	-	1
< 3	-	1	-
> 15-18	1	-	-
< 20-24	6	-	-
> 20-24	1	-	-
> 30	-	-	1

Table 4.6: Times of slaughter cattle.

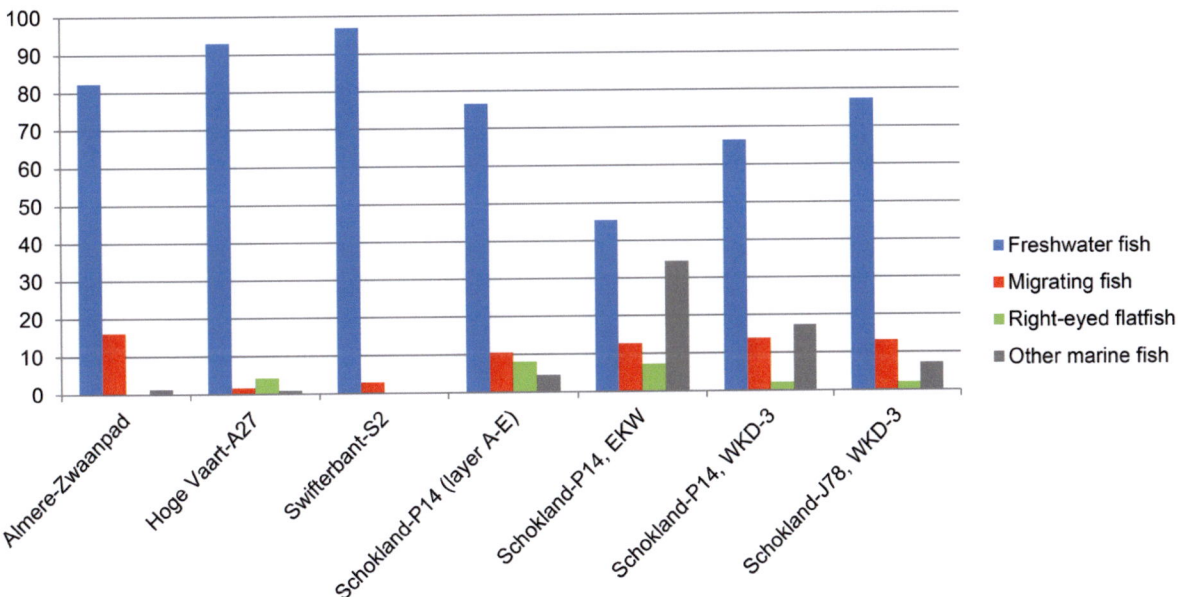

Figure 4.2: Distribution of fish species according to different aquatic environments.

SPECIES	Almere-Hout Zwaanpad	Hoge Vaart-A27 (phase 3)	Swifterbant-S3/S4	Swifterbant-S2	Urk-E4	Schokland-P14 (layer A)	Schokland-P14 (layer B)	Schokland-P14 (layer C)	Schokland-P14 (layer D)	Schokland-P14 (layer E)	Lelystad-Kotterbos	Schokland-J78 (LNEO)	Schokland-P14, EKW (geul)	Schokland-P14, WKD-3	Schokland-J78, WKD-3
Freshwater species															
Cyprinidae	39	560		52	3	11	27	37		10	1		6	239	443
Abramis brama	10		•	1							1		7	78	32
Abramis bjoerkna														26	8
Alburnus alburnus														1	
Tinca tinca			•												
Leuciscus idus	1													4	
Leuciscus leuciscus												1		4	3
Perca fluviatilis	8	103	•	31	2	1	9	7	1	5		7	1	84	329
Esox lucius	3	12	•	41	1	4	18	66	1	16	2	3	6	31	137
Gymnocephalus cernuus			•	1										3	
Siluris glanis			•	4				2		2		4	4	158	414
Rutilus rutilus			•										1	6	3
Rutilus erythropthalmus			•											5	1
Subtotal	*61*	*675*		*130*	*6*	*16*	*54*	*112*	*2*	*33*	*4*	*15*	*25*	*639*	*1370*
Anadromous/catadromous species															
Acipenser sturio			•		45										
Salmo salar			•					1					1		
Anguilla anguilla	10	29	•	4	1	4	10	11				5	124	225	
Osmerus eperlanus	1														
Platichthys flesus	1					1	1	1		1		1	1	10	9
Coregonus lavaretus		1													
Coregonus spec.	1														
Subtotal	*12*	*30*		*4*	*46*	*5*	*11*	*13*		*1*		*1*	*7*	*134*	*234*
Saltwater species															
Pleuronectidae		73				1	12	10		1			4	21	37
Mugilidae		13													
Mugil capito			•												
Dicentrarchus labrax		1													
Liza ramada	1					2	2	2		6			19	167	129
Subtotal	*1*	*87*				*3*	*14*	*12*		*7*			*23*	*188*	*166*

Table 4.7: Fish species.

Site	Technique	Date	Cultural association
Hoge Vaart-A27 phase 4	weirs, traps	4300 – 4200 cal. BC	Classic Swifterbant
Swifterbant-S5	weir?	4300 – 3950 cal. BC	Classic Swifterbant (?)
Almere Stichtsekant	weir?	2500 – 2300 cal. BC	Bell Beaker (?)
Emmeloord-J97	weirs, traps weirs, traps, hooks(?)	3400 – 2900 cal. BC 2400 – 1700 cal. BC	Funnelbeaker/Bell Beaker/Barbed Wire
Schokland-P14	hooks	Late Neolithic (?)	Bell Beaker (?)

Table 4.8 Overview of sites with evidence for fish catching. The dates provide a general range.

Hoge Vaart-A27 phase 3, where one burnt vertebra of sea bass was found.[163]

Urk-E4 and Swifterbant-S3 are the only locations where the remains of sturgeon have been found. The relative rarity of this anadromous species in the excavated assemblages is probably not a methodological distortion. This species is in fact easy to recognise, even in a highly fragmented (and/or burnt) condition, which can easily lead to numerical overestimation.[164] This might suggest that there was water in the vicinity of these locations which had an open connection to the sea during the Neolithic, allowing sturgeon to swim inland. The fact that other migratory fish remains are modest in number could be an indication that these species had only limited access to the waters of the region, or that the waters didn't provide suitable spawning grounds.[165] If fish were being caught mainly 'locally', we can conclude that the majority of the populations apparently did not migrate inland via Flevoland.

Generally speaking, the data shows that a range of fish species were caught. Insofar as we can ascertain, there does not appear to have been any overt selection in terms of size, though we cannot exclude the possibility that the larger fish represented in the bone material are associated with selection and that smaller fish are individuals that died of natural causes.[166] There is also little evidence of the exploitation of marine species. At two of the oldest sites, Hoge Vaart-A27 phase 3 and Swifterbant-S3, 'sea fish' play a very modest role (fig. 4.2), and are mainly represented by species that can also occur in fresh water, such as mullet.[167]

There are no marine fish at all at Swifterbant-S2. In the Late Neolithic context (EGK/KB) of Schokland-P14, where at first glance a relatively large proportion of 'sea fish' seems to have been found, the picture may be distorted because of the small number of remains. Thinlip mullet was found much more often in the Barbed Wire Beaker contexts of Schokland-P14 and Schokland-J78. However, as we have said, it can occur naturally far inland in fresh water.[168]

In contrast to our understanding of the hunting of mammals, we know quite a lot about the methods used to catch fish, thanks to the favourable preservation conditions at some sites. Finds indicate the use of various methods and techniques. Direct evidence of both passive and active strategies were found at four sites, in contexts that can be dated to the Neolithic and the Early Bronze Age (table 4.8).[169]

Passive catching strategies are based on the principle of 'delayed return', whereby the technique results in a certain yield at a later time. Nets, for example, are hung in the water to catch fish; the nets are then emptied at a certain moment. Another passive strategy involves fish weirs, sometimes in combination with fish traps. The weir, which may be a wattle fence structure or a dam of stacked rocks, depending on the availability of large stones, was intended to hinder the passage of the fish migrating downstream and eventually guide them into a trap.[170] In active strategies, fish are caught using lines with hooks, harpoons, eel spears or cast and drag nets. Active

---

163 The unburnt fish bones include remains of whiting and herring. These remains are however from a large concentration of unburnt fish remains which also includes perch and pike. The entire assemblage of unburnt remains, which was found on top of a sandy ridge at the interface with a loose patch/block of peat, must be regarded as a natural accumulation.
164 See also Brinkhuizen 2006.
165 This applies to a lesser extent to the most frequently encountered migratory fish species, the eel, given that it can also travel short distances over land.
166 Gehasse 1995, 99. We should however note that only mature specimens of certain species, such as mullet, swim upstream.
167 Niekus et al. (2012) report an unburnt scale of a thinlip mullet at Almere-Zwaanpad. The unburnt fish remains are said to have been found with other burnt remains. The site probably dates to around 7000 cal. BC. It is by no means certain whether unburnt fish remains, including those of thinlip mullet, can be associated with the Mesolithic remains. Around 7000 cal. BC the coast lay much further to the west and there are no indicators of the proximity of fresh water or even brackish aquatic environments in the region. Furthermore, it would be another 2000 years before the coversand ridge where the remains were found would be covered with peat. Wetter environmental conditions started to take hold much later, making it unlikely that unburnt fish remains would have been preserved for 2000 years in dry conditions.
168 Furthermore, the picture is certainly distorted, because no sieving was carried out here and mullet bones can be easily recognised (Gehasse 1995).
169 Recently new Late Neolithic fish weirs have been discovered in Almere, see Hogestijn 2019.
170 Bulten, Hamburg & Van der Heijden, 2009.

Figure 4.3: Fish hooks from Emmeloord-J97, scale 1:1 (photo: Dick Velthuizen).

techniques can also be combined with fish weirs, in which case no trap is needed.[171]

Fish hooks were found at Schokland-P14 and Emmeloord-J97 (fig. 4.3). At the latter site, the fish hooks were found in a mixed stratigraphical context in which sherds of Swifterbant, Bell Beaker, Barbed Wire Beaker and Hilversum pottery also occur. The fish hooks (which were not ^{14}C dated) cannot therefore be dated more accurately on the basis of association. Several fish hooks found at Schokland-P14 are probably Late Neolithic.[172] Relatively small bone hooks may have been used for pike or catfish. Assuming that four large antler hooks found at Emmeloord-J97, the largest complete one is over 10 cm long, can be interpreted as fish hooks, it would then be likely that they were used to catch larger fish, such as catfish. Two of them have a hole bored through them, probably to enable a line to be attached. Whilst another has no hole, this does not rule out the possibility that it was used with a line.

Evidence of the use of fish weirs and traps was found at several locations (fig. 4.4). The earliest dated evidence was found at Hoge Vaart-A27 (phase 4), where remains of three weirs with associated traps were excavated.[173] Evidence for at least ten fish weirs were found at Emmeloord-J97, as well as dozens of traps (or fragments thereof). These were divided into two main phases of activity on the basis of ^{14}C dates: Bell Beaker and Late Neolithic/Early Bronze Age.[174] Evidence of active fishing techniques was also found at this site. Strong evidence for fish weirs was found at two other locations – Swifterbant-S5 and Almere Stichtsekant. Hoge Vaart-A27 and Emmeloord-J97, in particular, have provided information on various technological details.

The final phase of activity at Hoge Vaart-A27 (phase 4) is associated with fishing in an active freshwater tidal gully. At the time when the fish weirs were in use, clay and (occasionally) thin layers of sand, were being deposited in the gully. The fish weirs probably fell into disuse when the tidal dynamics in the gully decreased around 4200 cal. BC, allowing detritus to form. This eventually led to the accretion of land as a result of peat formation.[175] The fish weirs at Emmeloord-J97 were also found in a gully that had carried flowing water under tidal influences. As the dynamics in the gully decreased and the flow of water gradually slowed and eventually became stagnant, detritus formed here too. On the basis of physical geographical research, the sequence of aquatic environments that formed in the gully can be characterised as: closed and brackish (residual tidal creek gully), open and freshwater (peat drainage channel), open and brackish (creek) and closed and brackish (residual tidal creek gully).[176] The dates attributed to the fish weirs coincide with the transitional phases, in which an open environment was transforming into a closed one.

The fish weirs were constructed using a series of upright posts made from small tree trunks driven into the clay with woven branches forming wattle fence-like screens between them. The pointed tip of the posts generally exhibit several roughly cleaved and split surfaces that may well have been created when the tree trunks were felled. Some of the posts from Hoge Vaart-A27 and the Funnelbeaker phase of Emmeloord-J97 show traces of gnawing by beavers on the pointed tip, indicating opportunistic use of the local timber supply.[177] Evidence of deliberate sharpening of posts using stone axes has only been found in weirs at Emmeloord that can be dated to the Late Neolithic and Early Bronze Age phase.[178] Some fragments of the weir screens have remained intact. It is clear from these surviving examples that separate wattle screens were positioned between the posts, the wattle sections were not woven around the posts. Another method used to fill the spaces between the posts was to attach bunches of twigs, perhaps in combination with

---

171 Von Brandt 1984.
172 Ten Anscher 2012, 458.
173 Hamburg, Hogestijn & Peeters, 1997; Hamburg *et al.*, 2001.

174 Bulten, Van der Heijden & Hamburg, 2002.
175 Gotjé, 2001; Spek, Bisdom & Van Smeerdijk, 2001b.
176 Van Zijverden 2002, 25.
177 Van Rijn & Kooistra 2001; Van Rijn 2002.
178 Bulten, Van der Heijden & Hamburg, 2002.

Figure 4.4: Emmeloord-J97: a complete fish trap in situ and the excavation of a trap. Right: "fork" (length approx. 2 m) of yew with a rectangular hole at the bifurcation (source: Bulten, Van der Heijden & Hamburg 2002).

horisontal poles or beams. Vertically-positioned bundles of bound reeds were also found at the site. These may have been used for this purpose. It is not clear how the screens of the weirs found at Hoge Vaart-A27 were constructed, as only one fragment was found.[179]

The close spacing of posts within a configuration of postholes identified as a fish weir suggests that maintenance was carried out on some of these structures. Fish weirs 4 and 10-11 at Emmerloord-J97 (see Appendix) show the clearest signs of this. It is even likely that either fish weir 10 or 11 replaced fish weir 4. The remains from other weirs – numbers 6 and 7 – comprise small clusters of regularly-spaced posts. This may be evidence for the reinforcement of the structure at the points where screens were attached. It is not unlikely that a quantity of replacement posts were kept close by, as possibly indicated by a number of horizontal posts found near one of the fish weirs at Hoge Vaart-A27.

In all probability, the fish weirs were used in combination with fish traps. The three fish weirs at Hoge Vaart-A27 phase 4 were all associated with fragments of fish traps that had disintegrated due to the effects of erosion. The weirs found at Emmeloord-J97 were also associated with fish traps. Some of these traps had been preserved in their entirety, while only fragments of others have survived.[180] The complete examples consist of a rectangular or fish-shaped chamber (or pot) with a slightly protruding mouth and a funnel-shaped entrance that was designed to prevent fish from swimming back out of the trap. The rectangular traps are longer than the fish-shaped examples, at 180 to 200 cm and 120 to 150 cm respectively. The frame of the trap was made using willow and hazel twigs bound together with twine using the *Zwirnbindung* technique. The twine was made of tree bark. Hoops in the structure maintained its three-dimensional shape. The

---

179 Hamburg *et al.* 2001.

180 Complete fish traps were found both flattened and in their original three-dimensional form. The presence of an intact three-dimensional fish trap suggests it was abandoned in a context where rapid sedimentation was occurring.

SPECIES	Hoge Vaart-A27 (phase 3)	Swifterbant-S2	Swifterbant-S3	Swifterbant-S4	Urk-E4	Schokland-P14 (layer B)	Schokland-P14 (layer C)	Schokland-P14 (layer E)	Schokland-P14, EKW (geul)	Schokland-J78 (LNEO)	Schokland-P14, WKD-3	Schokland-J78, WKD-3	Emmeloord-J97
Aquatic/moorland species													
Podiceps cristatus				1									1
Podiceps ruficollis												1	
Anas platyrhynchos	1	1	102	1	1			4	1	1	2	4	12
Anas strepera	1												
Anas penelope											1		
Anas crecca/querquedula	4									1	3		
Anas spec.			15	23						1		1	2
Aythya ferina			1										
Aythya fuligula			1										1
Aythya spec.			2								1		2
Tadorna tadorna			1										
Anatidae	229		498	1		2			1		1	2	
Anser anser	1												
Anser albifrons												1	
Anser spec.									1		3	1	
Anser/Branta spec.			1										
Merganser albellus											1		
Fulica atra	3												
Phalacrocorax carbo	1		2									1	1
Cygnus cygnus	5												
Cygnus olor	2		1						1				1
Cygnus spec.	3		4									1	
Grus grus	1		1								1		1
Pelecanus crispus	3										1		
Botaurus stellaris			1										2
Larus argentatus			1										
Charadriidae/Scolopacidae			1										
Subtotal	254	1	632	26	1	2		4	4	3	14	12	23
Birds of prey/owls													
Haliaeetus albicilla	9		6										2
Bubo bubo									1			3	
Subtotal	9		6						1			3	2
Songbirds													
Passeriformes	1										1		
Turdus spec.								1					
Garrulus glandarius			1										
Corvus corone			3										
Corvus frugilegus												3	
Corvus corax											1	5	
Subtotal	1		4					1			2	8	

Table 4.9: Bird species.

hoops found consist of either a simple fat twig, two thin branches split lengthwise and held together with twine, or a combination of a split branch and several thin twigs. Pieces of three-strand cord were found by the funnel and the top of the chamber. The cord may have been used to open and close the chamber. The thicker binders found near the funnel may have been used to attach the trap to a post, or a 'fork' that could be used to anchor it in the channel bed. A 2 metre-long, double-pronged fork made of yew, with a rectangular hole through which a cord may have been passed, was found at this site (fig. 4.4). The ^{14}C date for this object coincides with the Funnelbeaker phase of the fish weirs. The wickerwork of the fish trap is tightly woven, which might suggest that the fish traps were used mainly to catch smaller fish. This could suggest that the recovered remains of small fish do not necessarily all represent individuals that died of natural causes.

4.5 Birds in the diet

Generally speaking, bird remains are less prevalent in assemblages than those of mammals and fish. Only a handful of remains were found at most sites. In fact, at the Mesolithic sites of Dronten-N23 and Almere-Zwaanpad only a single bone fragment was found, neither of which can be identified with certainty as coming from a bird.[181] Only Swifterbant-S3 and Hoge Vaart-A27 phase 3 yielded substantial quantities of bird remains (table 4.9), although even on these locations the presence of bird remains in the assemblages appears to be limited. Bird bones represent less than 5% of the weight of mammal and bone remains in the bone assemblage from Swifterbant-S3,[182] and an even lower proportion, only 1.2%, in the assemblage from Hoge Vaart-A27 phase 3.

One common feature of all these Early Neolithic assemblages is that ducks – mainly dabbling ducks at Swifterbant-S3[183] – comprise the main group, as they do at Late Neolithic sites in the coastal region of the western Netherlands.[184] This might reflect a hunting strategy tailored to the season. Duck, like geese, are unable to fly for several weeks in late summer (July and August) during their moulting period and can be caught relatively easily in large numbers.[185] The remains of a range of other water and wetland birds have also found, both in the larger find assemblages from Hoge Vaart-A27 phase 3 and Swifterbant-S3 and in the smaller find assemblages from other sites. They include various types of geese and swans, coots, cormorants, bitterns and cranes. The most striking species is the Dalmatian pelican, remains of which were found at Hoge Vaart-A27 phase 3 and in the Barbed Wire phase of Schokland-P14.[186] Interestingly, in the Hoge Vaart-A27 phase 3 assemblage, most of the remains of water and wetland birds that cannot be identified as ducks come from the gully in the eastern section of the investigated area. Although the fact that the material is unburnt might suggest that this is a natural accumulation, the association with other remains – bones showing traces of slaughter and butchering, tools and pottery – suggests there is no reason to assume this is the case. This contrast in the distribution pattern might reflect a difference in the economic significance of different species of birds, with ducks mainly providing a source of food, whilst other water and wetland birds were being caught for other purposes (feathers?).

Next to these water and wetland birds, small numbers of remains of species from other environments have also been found, although the white-tailed eagle – present in the assemblages from Hoge Vaart-A27 phase 3 and Swifterbant-S3 – must also be regarded as a resident of wetter landscape milieus. Species of interest include the eagle-owl and the raven from the Barbed Wire Beaker phase of Schokland-P14. Remains of raven have also been found in the Barbed Wire Beaker context of Schokland-J78. Both species are relatively rare in Dutch archaeological contexts.[187] In addition, incidental remains of songbirds have been found in several contexts. Although the eagle-owl is perhaps found more commonly in wooded areas, like the raven it can also be found in open fields. Songbirds are also found in all kinds of settings. Their presence in Early Bronze Age contexts in Flevoland is not therefore unusual, despite the fact that wet conditions dominated the landscape.

4.6 Plant resources in the food economy

For a long time, only limited data were available on the exploitation of plants as a food source in the Mesolithic and Neolithic, despite the potentially favourable preservation conditions for the Late Mesolithic and Neolithic occupation periods in particular. However, important new information has become available over the past 15 years,

---

181 Van Dijk 2012; Niekus et al. 2012.
182 This refers to identified remains.
183 Dabbling ducks include mallards and teals.
184 Zeiler, 1997, 2006; Zeiler & Brinkhuizen, 2012, 2013, 2014; Zeiler & Clason, 1993. See also Chapter 6.
185 This method is still used by field biologists for researching Brent geese in Siberia. Groups of moulting geese on the water are surrounded by small boats and herded together until a net can be thrown over them (Ebbinge 2007, 135-140).

186 This species, which no longer occurs so far north, is occasionally found in prehistoric and early historic contexts; the most recent find comes from Flevoland, at a 15th-century fortress in Kuinre (Laarman, 1994).
187 The Dutch national archaeozoological database Boneinfo reports eleven eagle-owl finds, all from prehistoric or early historic contexts. There have been twelve finds of raven remains (three of which in different phases of Roman Valkenburg). The most recent comes from the 18th/19th-century site at De Hogeweide in Utrecht (Esser et al. 2010).

giving us more insight into the importance of plant foods in the diet.

4.6.1 Wild plants

The oldest assemblage for which information is available on plant material as potential food remains is that from the Mesolithic site Dronten-N23, where fragments of charred hazelnut shells have been found. The abundant occurrence of charred hazelnut shell fragments at Mesolithic sites in Northwest Europe suggests hazelnuts were an important source of food for hunter-gatherers.[188] Large concentrations of these charred shells are generally interpreted as the result of overheating while nuts were being roasted. The presence of quern-stones and hammerstones might also suggest that hazelnuts were important in the diet.[189] Nuts may have been roasted so they could be kept for longer, or to make them easier to process, as roasted nuts break more easily and are easier to grind. Roasted nuts are also believed to be more digestible due to a change in the structure of the oil, and a reduction in potentially indigestible fats.[190]

Few (no more than a few dozen), if any, remains of hazelnut shells have been found in the pit hearths at Dronten-N23.[191] This appears to be the general picture as regards Mesolithic pit hearths at other sites, too.[192] However, almost a hundred fragments were found in a single hearth at Dronten-N23. On the basis of this, it is unlikely that pit hearths should be associated with the roasting of hazelnuts. On the other hand, we cannot rule out the possibility that roasted hazelnuts – and possibly other plant foods[193] – were removed in their entirety from the hearth and, in all likelihood, consumed elsewhere. Charred remains of hazelnut shells were frequently found in the top layers of the dune at Dronten-N23, and concentrations of these shells can sometimes be identified in spatial association with flint concentrations.[194] The presence of charred hazelnut shell fragments in pit hearths could then be explained as incidental intrusion.[195] It is therefore likely that charred hazelnut remains are in fact linked to consumption near, and possibly roasting in, surface hearths. Such a situation was documented at Hoge Vaart-A27 phase 3, albeit in an early Neolithic context (fig. 4.5).[196]

In addition to hazelnuts, dogwood may also have formed part of the diet. Several charred seeds from this species were found at Dronten-N23. Although there is no direct dating evidence, they are likely to be from the Late Mesolithic, as suggested by an AMS dating of cornel (probably dogwood) charcoal.[197] This shrub may have grown in the vicinity in clearings and on the edges of woodland. The frequent occurrence of fragmented and charred dogwood seeds at other Mesolithic sites in the Netherlands and southern Scandinavia suggests that the species played a structural role in the food economy.[198] The seeds are rich in oil that is suitable for consumption, but can also be used as fuel or as an impregnating agent.

The remains of plant food sources recovered from Dronten-N23 almost certainly reflect only a very small proportion of the potentially exploited species available. Palynological data from three pit hearths suggest that fern and cattail rhizomes may also have been significant, as well as acorns. More direct evidence of a broad spectrum plant component in the diet has been found in the Early Swifterbant context of Hoge Vaart-A27 phase 3, where various charred remains from edible plants were found in association with surface hearths, alongside large quantities of burnt bone and flint knapping waste (fig. 4.5). The charred seeds of water lily were the most commonly occurring species.[199] Both the seeds – rich in protein, oil and sugars – and the starchy rhizomes of these species are edible. Charred seeds of water lily have also been found elsewhere in the Netherlands and in Denmark.[200] The charred material from Hoge Vaart-A27 phase 3 also included remains of hazelnuts and acorns, lesser celandine tubers, wild apple and raspberry. The starchy rhizomes of bulrush may also have been used as a source of food.[201]

Remains of various wild plant species that may have formed part of the diet also occur at other sites with evidence for Swifterbant and Pre-Drouwen occupation, such as Swifterbant-S3, Urk-E4 and Schokland-P14. Charred root tubers of lesser celandine, which were also found in the Late Neolithic context of Schokland-P14, may have played a substantial role as a starchy vegetable, given the plant's abundance in wet woodland and grasslands

---

188 Clarke, 1976; Zvelebil, 1994; Holst, 2010.
189 Mithen *et al.*, 2011; Holst, 2010.
190 Mason & Hather 2000.
191 Kubiak-Martens *et al.* 2012.
192 Peeters & Niekus 2005.
193 The research on Dronten-N23 focused particularly on charred remains of vegetative plant tissue (also known as archaeological parenchyma) derived from plant storage organs, such as roots, tubers and rhizomes. This did not produce any results (Kubiak-Martens *et al.* 2012).
194 Wansleeben & Laan 2012a.
195 Crombé, Groenendijk & Van Strydonk (1999) found, on the basis of ¹⁴C dates, that hazelnut fragments from heath pits at the Verrebroek-Dok site in Belgium were often older than charcoal from the same pits, and that on average the hazelnut dates are more consistent with the dates found for surface hearths.

---

196 Peeters & Hogestijn 2001; Peeters, 2007.
197 5877-5674 cal. BC (GrA-50730).
198 Regnell *et al.*, 1995; Regnell, 2012; Kubiak-Martens *et al.*, 2014. See also chapter 6.
199 Brinkkemper *et al.* 1999; Visser *et al.* 2001.
200 Kubiak-Martens 2002; Kubiak-Martens *et al.* 2014.
201 Visser *et al.* 2001.

Figure 4.5: Hoge Vaart-A27 Phase 3: cross-section of a surface hearth with a brown-orange discoloration at the base as a result of iron oxidation caused by heating. The white particles above the discoloration are burnt bone and flint. Surface hearths are generally associated with charred fragments of hazelnut shells, and charred seeds and tubers of, for example, a: galigaan (*Cladium mariscus*), b: lesser celandine (*Ranunculus ficaria*), c: cleavers (*Galium aparine*) and d: limestone (*Hippuris vulgaris*) (source: Visser *et al.* 2001).

and along riverbanks. It is one of the first rich sources of edible starchy food to become available in the spring.

Fruits and berries also appear to have been a structural part of the diet. Charred remains of wild apple were found in the Classic Swifterbant context of Swifterbant-S3 and in the Early Swifterbant context at Hoge Vaart-A27 phase 3. Charred rosehip seeds (the flesh of the rosehip is edible) and remains of hawthorn and blackberries were also found at Swifterbant-S3 and Urk-E4. Blackberry seeds were also found at Schokkerhaven-E170 and Schokland-P14 (a date more precise than Swifterbant/Pre-Drouwen cannot be given for either of these sites, however). Blackberries have not been found in the Late Neolithic context of Schokland-P14, though they do occur in the Early Bronze Age, as does hawthorn. Fruit and berries probably did not provide a substantial proportion of the calories consumed, but would have mainly been a source of sugars and vitamins.

Nuts (including hazelnuts, acorns, and water chestnut) were also a stable feature in the Neolithic and Early Bronze Age diet. Charred remains of hazelnut shells have frequently been found in the Swifterbant and Pre-Drouwen contexts of Hoge Vaart-A27 phase 3, Schokkerhaven-E170, Schokland-P14, Urk-E4 and Swifterbant-S3. The presence of waterlogged shells suggests that hazel trees grew close to the sites in question. Charred remains of acorn parenchyma suggest they continued to play a role in the diet.[202]

4.6.2 Cultivated crops

Since the discovery of a large quantity of charred cereal grains at Swifterbant-S3 in the 1970s,[203] the role of cultivated crops during the Swifterbant period and whether or not they were grown locally has been an important subject of

---

202 Waterlogged remains of water chestnut (a spine from the hard shell surrounding this starch-rich nut) were also found in the Barbed Wire Beaker context of Schokland-P14; the remains are of natural origin (Gehasse 1995, 146).
203 Van Zeist & Palfenier-Vegter 1981.

debate. It has since become clear that Swifterbant-S3 is not an isolated example. Charred cereal remains also occur at other sites from the Classic Swifterbant phase. In all cases, the remains comprise naked barley and emmer wheat. Both these species have been found at Swifterbant-S2 and S3;[204] but only naked barley was found at Swifterbant-S4.[205] In the case of Swifterbant-S2, however, it is not clear whether the remains are 'indigenous', native to the area, or had been transported there by natural processes – by water, perhaps? The concentration is low on the site and the average preservation condition poor.[206] On the other hand, the proportion of cereal remains in the charred botanical material (excl. charcoal) is high, at approx. 30% (n=110), so a secondary source is unlikely.

Charred cereal remains have also been found at Schokland-P14 (layers A-C): an internode of emmer wheat and an imprint of naked barley on potsherds from layer A, and several charred grains of emmer wheat, imprints of emmer wheat and naked barley in potsherds from layers B-C. Several grains of naked barley have been found in a posthole that has been attributed to the Swifterbant or Pre-Drouwen phase of occupation.[207] Several charred grains of naked barley and emmer wheat have also been found at Schokkerhaven-E170 and Urk-E4.[208] A grain of naked barley from the Urk-E4 provided a date that coincides with the Classic Swifterbant phase.[209] Furthermore, spikelet forks of emmer wheat and cereal pollen were also found at Schokkerhaven-E170. In this context, it is also important to note that absolutely no evidence of cereals was found at Hoge Vaart-A27 phase 3, suggesting that cereals were not yet introduced in the Early Swifterbant phase.[210]

As previously stated, evidence for the presence of cultivated crops has always raised the question of whether arable farming was practised in the Swifterbant period, or whether the inhabitants of these wetlands acquired cereals via exchange networks with groups from the sandy region, for example. The presence of chaff from naked barley, in particular, might indicate local cereal cultivation.[211] Chaff remains of emmer wheat, except for the fragments of internodes, might well constitute evidence of local cereal processing.[212] There are apparently no indications of weeds typically found in arable fields or pollen that can unequivocally be attributed to cereals.[213] However, the specific combination of weeds – pioneer communities related to the aster family which are arable weeds[214] – might constitute evidence of arable farming, while the absence of more characteristic weeds might primarily be due to the limited scale of cultivation.[215] Such characteristic plants will also not immediately occur in places where there has not previously been any arable farming. Botanical data suggest that local cultivation took place at Urk-E4, Schokkerhaven-E170 and in the Pre-Drouwen context of Schokland-P14.[216]

4.6.3 Cultivation

The archaeobotanical data from the Swifterbant period have not given rise to an unambiguous interpretation subscribed to by all researchers. However, over the past decade tillage marks have been found that show that crops were indeed grown locally.[217] Initially these shallow furrows, which suggest that soil was turned using a hoe were only identified at Swifterbant-S4. However, similar furrows were later also identified at Swifterbant-S2 and Swifterbant-S3.[218] The botanical data suggest that the cultivation of cereals was limited to riverbanks.[219] Soil micromorphological analysis at Urk-E4 suggested that an arable layer may have been present. The scratch marks from an ard were apparently identified,[220] though this interpretation is doubtful.[221]

Cereals were grown in small fields in the Classical Swifterbant context of Swifterbant-S2, S3 and S4.[222] The excavation data show that these fields can best be described as 'hoe fields', having been tilled with the aid of a hoe. The practice therefore more closely resembles horticulture than agriculture. The land was tilled by loosening and turning clods of earth (fig. 4.6), though it is not clear exactly what type of tools were used. Possibilities include implements made from the shoulder blade of an

---

204 Van Zeist & Palfenier-Vegter 1981; Prummel *et al.* 2009.
205 Raemaekers & de Roever 2020; Van Zeist & Palfenier-Vegter (1981) reported the possible presence of common wheat – a single grain – at Swifterbant-S3, though they point out that it could be a deformed grain of emmer wheat. The presence of Einkorn wheat at Urk-E4 is also debatable and possibly the result of an incorrect identification.
206 Prummel *et al.* 2009, 24.
207 Gehasse 1995; Ten Anscher 2012, 423.
208 Gehasse 1995; Vernimmen 2001.
209 4223-3967 cal. BC (GrA-16947).
210 Brinkkemper *et al.*, 1999; Visser *et al.* 2001.
211 Van Zeist & Palfenier-Vegter 1981.
212 Out 2009.
213 Out 2009, 179. Cappers & Raemakers (2008) say that cereal pollen is present in two diagrams (Lelystad and Tollebeek). According to Out (2009), there is insufficient evidence to support this and the pollen could also come from large grass varieties.
214 Ibid.
215 Schepers 2014, p 98-99.
216 Peters & Peeters 2001, p 40-42; Huisman *et al.* 2009; Weijdema *et al.* 2011; Ten Anscher 2012, 423.
217 Huisman, Jongmans & Raemaekers 2009; Huisman & Raemaekers 2014.
218 On closer inspection, the furrows were found to have been documented during the excavations in the 1970s. However, the churned layer was not recognised or interpreted as such.
219 Schepers 2014, 99.
220 Peters & Peeters 2001, p 40-42.
221 Ten Anscher 2012, 691.
222 Raemaekers & De Roever 2020: Huisman & Raemaekers 2014;Huisman, Jongmans & Raemaekers 2009.

Figure 4.6: Swifterbant-S2: during the excavations in 1964 a rumbly deposit was found, consisting of a "random mixture of the black culture layer with the gray clay of the levee" (description from the photo archive, photo: PDB Flevoland/Erfgoedpark Batavialand). The excavators at that time did not realize at that this were the oldest traces of a field in the Netherlands, and direct evidence for small-scale arable farming in the Neolithic wetlands.

Figure 4.7: Hoge Vaart-A27 phase 3: damaged T-shaped mattocks. The mattock in the middle was broken during use and newly perforated; all three specimen have heavily damaged edges due to use, scale 1:3 (source: Laarman 2001).

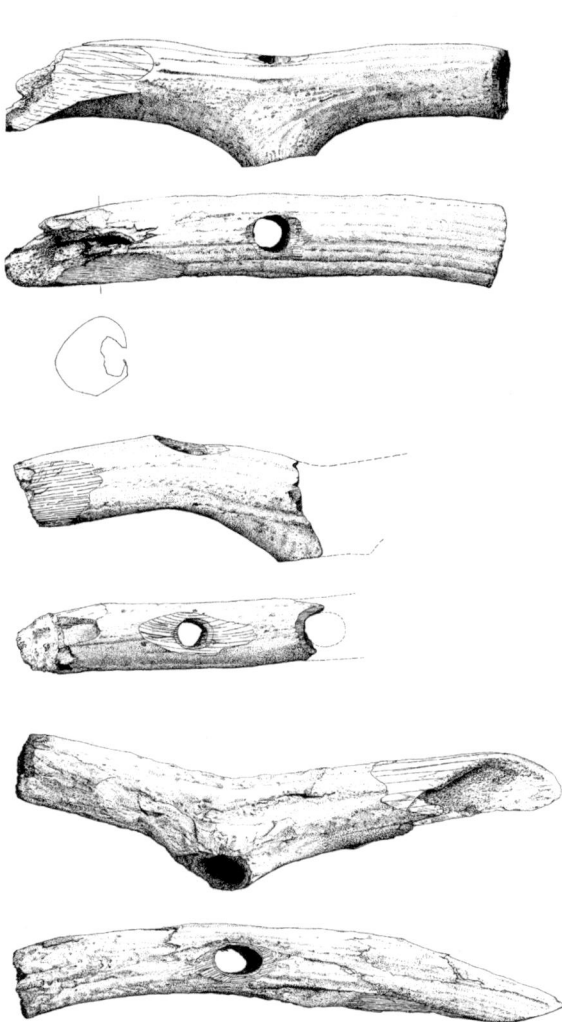

animal or a wooden blade attached to a shaft, although we have no archaeological evidence for the existence of any such tools in a Swifterbant context.

It is possible that in addition to hoe-like tools, various forms of antler axe were used in the context of crop cultivation. We must stress, however, that such tools existed long before cultivation of cereals occurred in these regions. Hoes made of elk antler and T-shaped axes of red deer antler occur early in the Mesolithic, and continue into the Early Neolithic. Twenty severely damaged T-shaped antler axes were found at Hoge Vaart-A27 phase 3, but here no evidence for crop cultivation was found (fig. 4.7). The damage consists mainly of splintering along the cutting edge and, to a lesser extent, abrasion. As a result, marks created when these tools were produced are no longer easily 'readable'. This suggests that the tools must have been under a relatively large amount of strain when they were used, but that the contact material caused only a limited amount of abrasive wear to the surface. Similar patterns of production and damage have been seen on base axes made of red deer antler from later Neolithic and Bronze Age contexts.

Use-wear analysis of base axes strongly suggests that they were used to dig out roots and tubers. Another possibility is that they were used as wedges for cleaving wood.[223] It has also been suggested that T-shaped

---

223 Peeters 1990.

axes were used to break up and loosen the soil.²²⁴ The widespread occurrence of T-shaped axes throughout large areas of Europe would seem to indicate that the activities performed with these tools formed a structural part of daily life in the Mesolithic and Early Neolithic. The breaking up and loosening of the soil to harvest roots and tubers, for example, is a digging activity, and therefore differs from tilling the soil (that is, the turning and breaking of clods of earth) for the purpose of cultivating crops, as on the fields at Swifterbant-S2, S3 and S4.

Direct evidence of arable farming in the Late Neolithic was found at Schokland-P14 in the form of an arable layer in and beneath which ard scratch marks were observed.²²⁵ The scale on which arable farming was practised in the Late Neolithic was probably greater than in the Swifterbant period, given the use of ards, which constitutes a different method of preparing the soil than that seen on the hoed fields (fig. 4.8).²²⁶ The area in which the ard marks were found has been dated on the basis of stratigraphical arguments to the Late Neolithic. An analysis of the orientation of ard marks has clearly revealed several overcutting patterns in the area, on the basis of which 14 different plough runs have been reconstructed.²²⁷ This might suggest that ploughing was repeatedly performed over a longer period, creating an arable layer 20 cm thick.²²⁸ An ard was probably used to till the soil when the land was first cleared, and when it was subsequently cleared again after lying fallow for a time.

4.7 Food preparation and consumption

The previous sections show that a wide range of food sources were exploited in the Mesolithic, Neolithic and Early Bronze Age. Some of this food would undoubtedly have been consumed without the need for any form of further preparation. Berries, fruits, nuts as well as eggs, insects and larvae, for example, could all be eaten raw. There is little archaeological evidence of these foodstuffs. The archaeological remains associated with food mainly concern the components of the diet that were processed or prepared, and involved heating. These are the processes which can be reconstructed from a technological perspective.

4.7.1 Animal foods

If we assume that burnt bone is associated with the method of food preparation, we can conclude, from the small amount of material recovered from Mesolithic

Figure 4.8 a-b (a above, b opposite page): Schokland-P14: ard marks discovered during the excavation in the 1980s. The ard marks are part of a field complex in which several phases of working have been distinguished on the basis of the direction of the ard marks (source: Ten Anscher 2012).

contexts, that both the meat from mammals (in the case of Dronten-N23) as well as fish (at Almere-Zwaanpad) was prepared using fire. The remaining bones were thrown into the fire after the meat had been consumed.²²⁹ The association between burnt bone remains and surface hearths strongly suggests that preparation took place on an open fire. It is assumed that meat – mainly larger pieces, or complete carcasses – might also have been cooked in pits, a practice for which there is, at least, ethnographic evidence.²³⁰ Heated stones were often placed in a cooking pit as a source of heat.

Such a usage has been suggested as a possible explanation for the frequent occurrence of Mesolithic pit hearths in the northern half of the Netherlands.²³¹ There is however no direct evidence for this. Remains other than charcoal – burnt bone or fragments of cooking

---

224 Smith 1989; Zvelebil 1994.
225 Ten Anscher 2012. See also section 4.3.2.
226 Ten Anscher 2012, 385-396.
227 Ibid.
228 This is a residual thickness. The thickness of the layer varies throughout the excavated area as a result of erosion.

229 The bone remains are heavily burnt (calcinated), which indicates high temperatures (>650C) (Shipman, Foster & Schoeninger 1984).
230 See for example Wandsnider (1997) for references.
231 Groenendijk 1987, 1997; Groenendijk & Smit 1990.

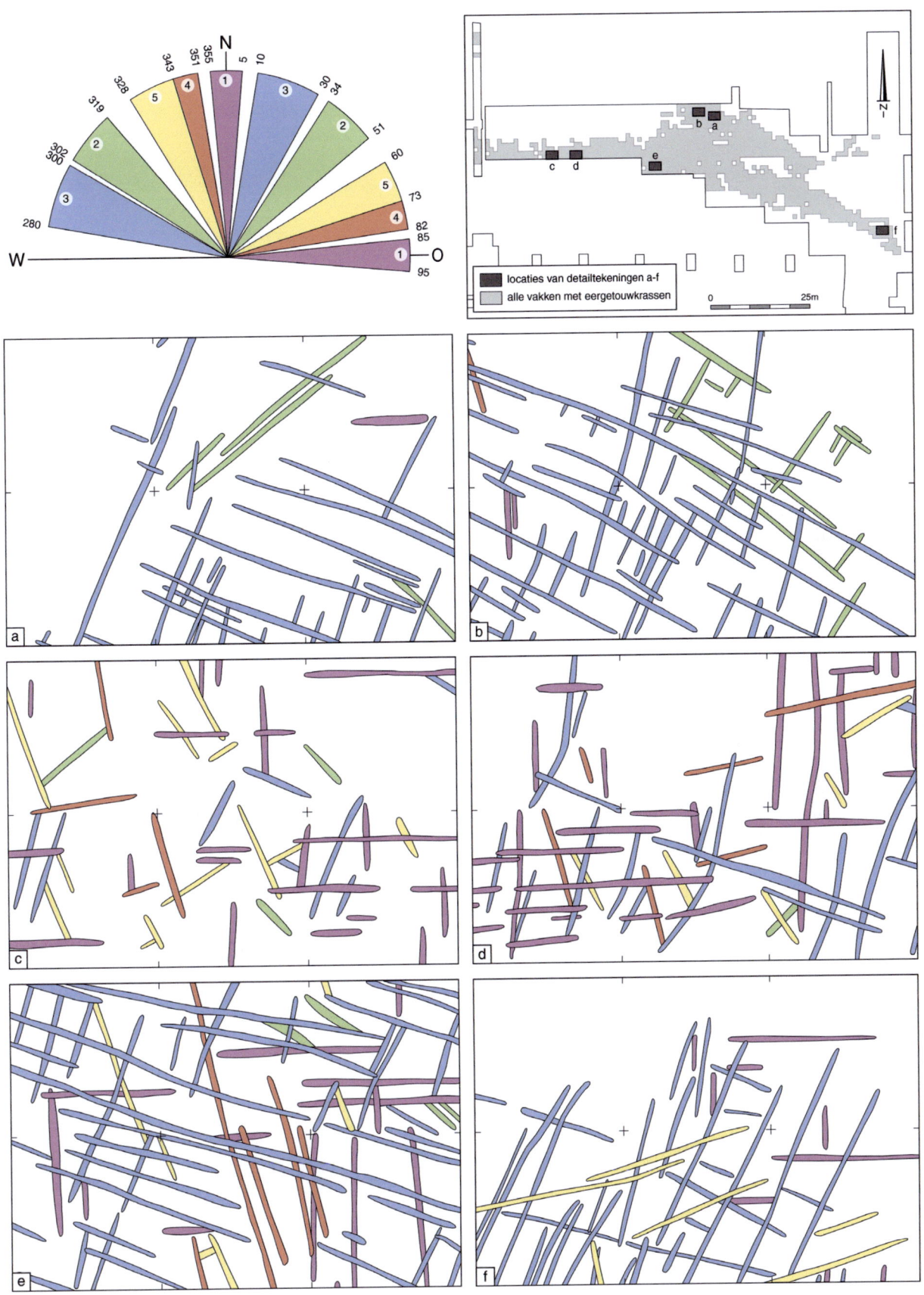

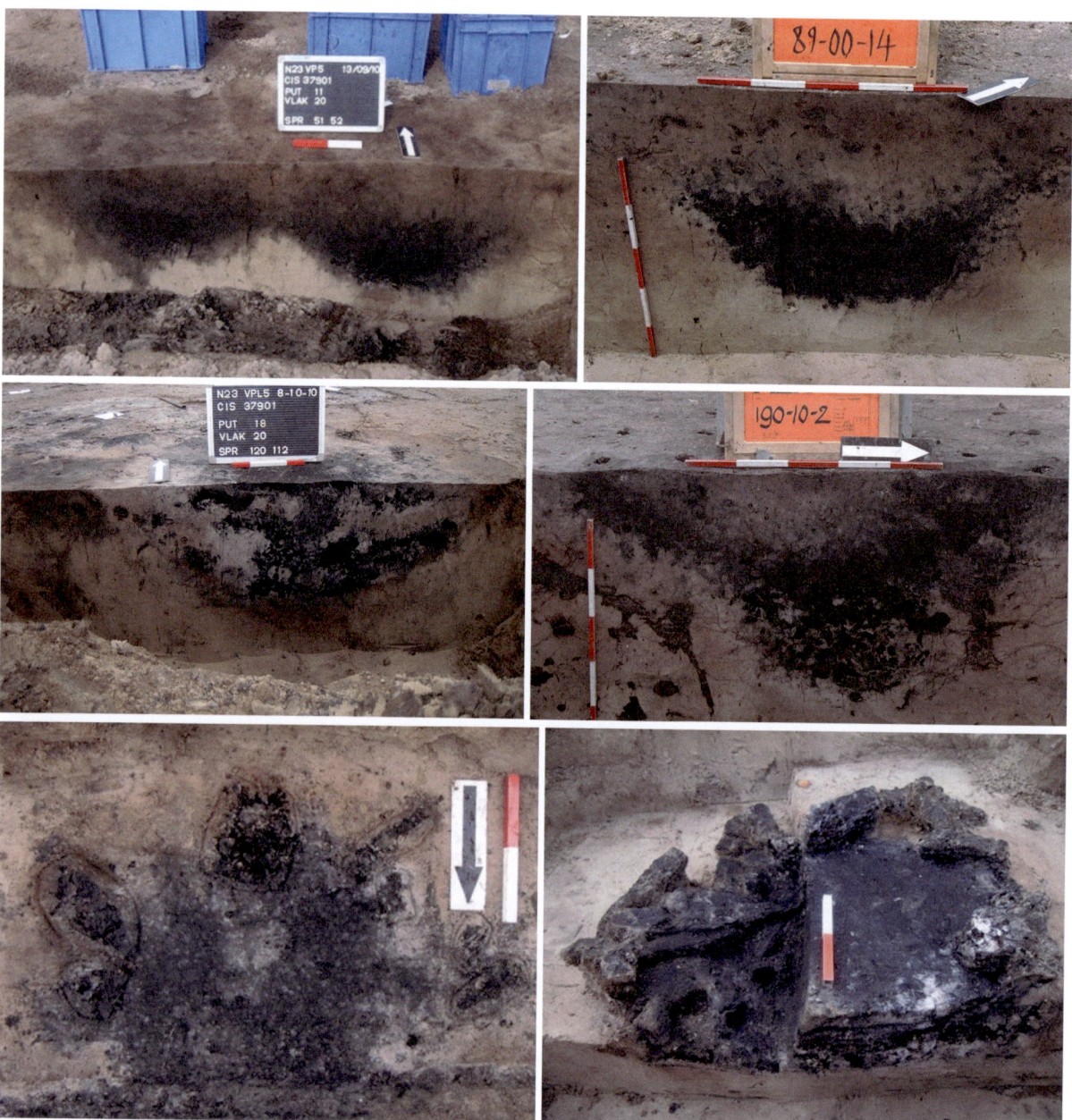

stones, for example – are rarely found in pit hearths.[232] Micromorphological analysis of thin sections from Hoge Vaart-A27 has, however, made clear that 'something' was removed from the pits after a fire had been made in them.[233] The act of removal disturbed the original fire layers, causing sand to become mixed with charcoal (fig. 4.9). Some pits were used several times, but in some large pieces of charred logs were found *in situ* in the bottom of a pit, as at Dronten-N23 and Hoge Vaart-A27.

Figure 4.9 (above and oposite page): Dronten-N23 and Almere Hoge Vaart-A27 phase 2: cross-sections of mesolithic pit hearths with different infills (left page, top and middle) and large fragments of charred wood at the bottom of pits (left page, bottom). The diffuse black colouring of the pits at the top left is most common. The pit at the top right contains more fragments of charcoal. The pits in the centre have a rumbly filling of sand mixed with charcoal. The cross-sections on the right page show pits with a flat bottom and variable filling (source: Hamburg *et al.* 2001, 2012).

---

232 For an overview of plant species identified among the charred remains, see Huisman *et al.* (2020). See also footnote 248.
233 Exaltus 2001.

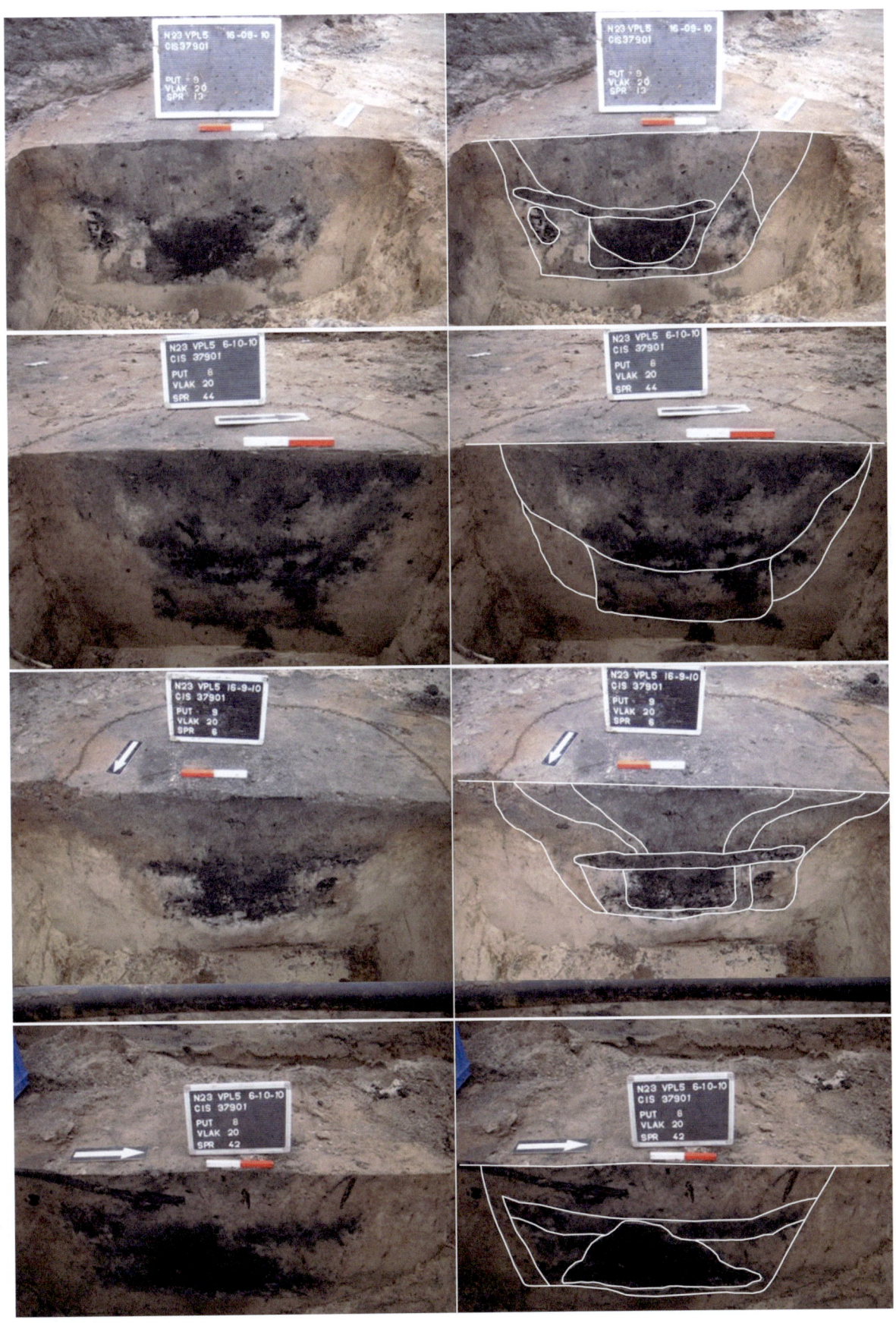

Figure 4.10: Almere Hoge Vaart-27 phase 3: heavily burnt mammal, bird and fish remains (source: Laarman 2001).

Figure 4.11: Hoge Vaart-A27 phase 3: shoulder blade (*scapula*) of presumably aurochs with a 'smoke hole' (photo: M. Dahhan, Almere). The attribution of the bone to aurochs cannot be substantiated on metric grounds; the width of the proximal part measures 62 mm.

Cracked sand grains in a thin layer under one pit hearth excavated at Hoge Vaart-A27 suggests that the fire in the hearth had been extinguished rapidly using water,[234] but this interpretation is far from certain.[235]

Preparation and consumption of meat by an open fire also took place in the Early Swifterbant context at Hoge Vaart-A27. Large quantities of burnt bone were found in and near surface hearths. Although the bone was highly fragmented, during the excavation it was possible to ascertain that larger pieces of bone were burnt *in situ*.[236] Fragmentation occurred mainly in mammal and bird bones; fish vertebrae were generally complete (fig. 4.10). The degree of combustion is the same in mammal, birds and fish remains. The degrees of combustion suggest that the bone was exposed to high temperatures, in excess of 650C. It seems highly likely that after consumption the bones were systematically discarded into the fire, where they continued to burn. A similar picture emerges for the Late Neolithic and Early Bronze Age surface hearths at Schokland-P14.[237]

Burnt bones from the hearths and unburnt bones from the gully fill at Hoge Vaart-A27, which has been correlated with phase 3, indicate that a significant proportion of the animals were consumed at the site.[238] Almost all parts of wild boar are represented, although interestingly, the remains in the gully are almost exclusively lower jaws, whilst the other body parts are represented by the burnt

---

234 Exaltus 2000, 27.
235 Personal communication Dr H. Huisman, RCE.
236 Laarman 2001, 16.

237 Ten Anscher 2012, 471-476.
238 Laarman 2001, 21-22.

material from the hearths. All body parts of red deer are also represented, though in this case mainly in the gully. Almost all the body parts of furred animals – beaver is particularly well-represented – are also present. This might mean that these animals were not cut up to be taken to other locations. This may well have happened in the case of aurochs, where the meat-bearing bones of the upper part of the front and hind legs is missing. Furthermore, the only shoulder blade found has a 'smoke hole' -in the flat part of the shoulder blade (fig. 4.11) – which could be evidence that the shoulder clod was smoked. All bones that were rich in marrow were, without exception, broken and show patterns of breakage that indicate that they were intentionally cracked, probably to extract the marrow.

In the Classic Swifterbant and Pre-Drouwen contexts of Schokland-P14 and Swifterbant-S2, S3 and S5, direct evidence for the preparation of food from animal sources is also difficult to find.[239] The same applies to the Late Neolithic and Early Bronze Age contexts at various sites. Butchering marks occur on a fairly small proportion of the bones. The heads of catfish found in the Late Neolithic context of Schokland-P14 appear to have been removed as soon as the fish were caught and left on the banks of the Overijsselse Vecht.[240] Bone remains with traces of burning are more common, which might suggest that the flesh was roasted.[241] The presence of pits with heated stones from the Early Bronze Age at Schokland-P14 suggests food, including meat, was cooked.[242] The pronounced fragmentation of larger bones might be evidence for marrow extraction.[243]

Although some meat would undoubtedly have been roasted or grilled, it could also have been cooked, for example in earthenware pots. Biochemical and microscopic analysis of encrusted residues on potsherds from Swifterbant-S3 has shown that meat and fish were prepared in the pots, sometimes in combination with vegetable/plant material.[244] Evidence for the preparation of fish has also been found in the δ15N and δ13C values measured in encrusted residues on pottery from the Swifterbant culture.[245] In addition, a reservoir effect observed in the ^{14}C dating of two Late Neolithic beaker pots from Emmeloord-J97 shows that fish was prepared in the vessels.[246]

4.7.2 Plant food sources

Like burnt bone remains, charred plant parts can indicate whether plant foods were prepared in or near a heat source. For the Mesolithic period, a connection has often been sought with pit hearths.[247] It is thought that in these pit hearths roots and tubers might have been cooked, or hazelnuts might have been roasted.[248] A lot of attention was given to this subject during the excavation at Dronten-N23. As at other Mesolithic sites, little evidence could be found to suggest that the pit hearths were used in the preparation of plant foods.[249] Charred fragments of hazelnut shells and other plant food sources are rarely found, and those that were found may have just ended up in the pits via a secondary route.[250] As stated earlier, charred hazelnut shells are generally found in a spatial association with surface hearths. It is therefore likely that charred hazelnut remains in pit hearths originally came from surface hearths, which may have been emptied into the pit. A similar pattern has been identified on Mesolithic sites outside Flevoland.[251]

Use-wear analysis of lithic artefacts from Dronten-N23 has produced evidence for the processing or grinding of hazelnuts or other nuts into meal (fig. 4.12).[252] Both base stones – querns, anvil stones – and handstones use to tap and grind were identified. The experimental grinding of hazelnuts, in particular, produced very similar use wear, so it can be assumed that a similar activity caused the use wear identified on the archaeological artefacts.[253]

The grinding of plant foods continued into the Neolithic. Use-wear analysis of lithic artefacts from Swifterbant-S2, S3 and S4 produced evidence of the processing – grinding – of plant material, although the exact nature of the plant – cereals or grasses – could not be determined (fig. 4.12).[254] As explained in section 4.6, it has been shown that cereals were cultivated in a Classical Swifterbant context for the first time, which would certainly support the assumption

---

239 Zeiler 1991, 1997; Gehasse 1995.
240 Parts of the head, in particular, were found in the gully deposits during the excavation (Gehasse 1995, 99).
241 Gehasse 1995, 98-99.
242 Ten Anscher 2012, 470-471.
243 Ten Anscher 2012, 157; See also Gehasse 1995, 48-50.
244 Raemaekers, Kubiak-Martens & Oudemans 2013.
245 Raemakers 2005.
246 The ^{14}C dates place the beaker pots, which in typological terms belong to the Bell Beaker culture, in the Funnelbeaker period. One date was taken from half a pot found over the post of a fish weir; the post itself has been dated to the period one might expect for the pot (Bulten, Van der Heijden & Hamburg 2002).

247 See for example Waugh (1916) for ethnographic examples from North America.
248 Kubiak-Martens *et al.* 2012. The anthropogenic nature of 'pit hearths' has, however, been questioned by Crombé, Langohr & Louwagie (2015). According to these authors the pits represent burnt ant nests. This hypothesis is, however, not supported by us. In our opinion, there is ample evidence for an anthropogenic nature of the pits, without excluding the possibility of an occasional pit having a natural origin. For a lengthy discussion on this topic, see Crombé, Langohr & Louwagie (2015), Peeters & Niekus (2017), Huisman *et al.* (2019), Crombé & Langohr (2020a, 2020b), and Huisman *et al.* (2020).
249 Kubiak-Martens, Langer & Kooistra 2012.
250 Hamburg *et al.* 2001; Peters & Peeters 2001; Visser *et al.* 2001; Kubiak-Martens, Langer & Kooistra 2012.
251 Crombé, Groenendijk & Van Strydonk 1999.
252 Knippenberg & Verbaas 2012.
253 Knippenberg & Verbaas 2012, 288.
254 Devriendt 2014, 94-96.

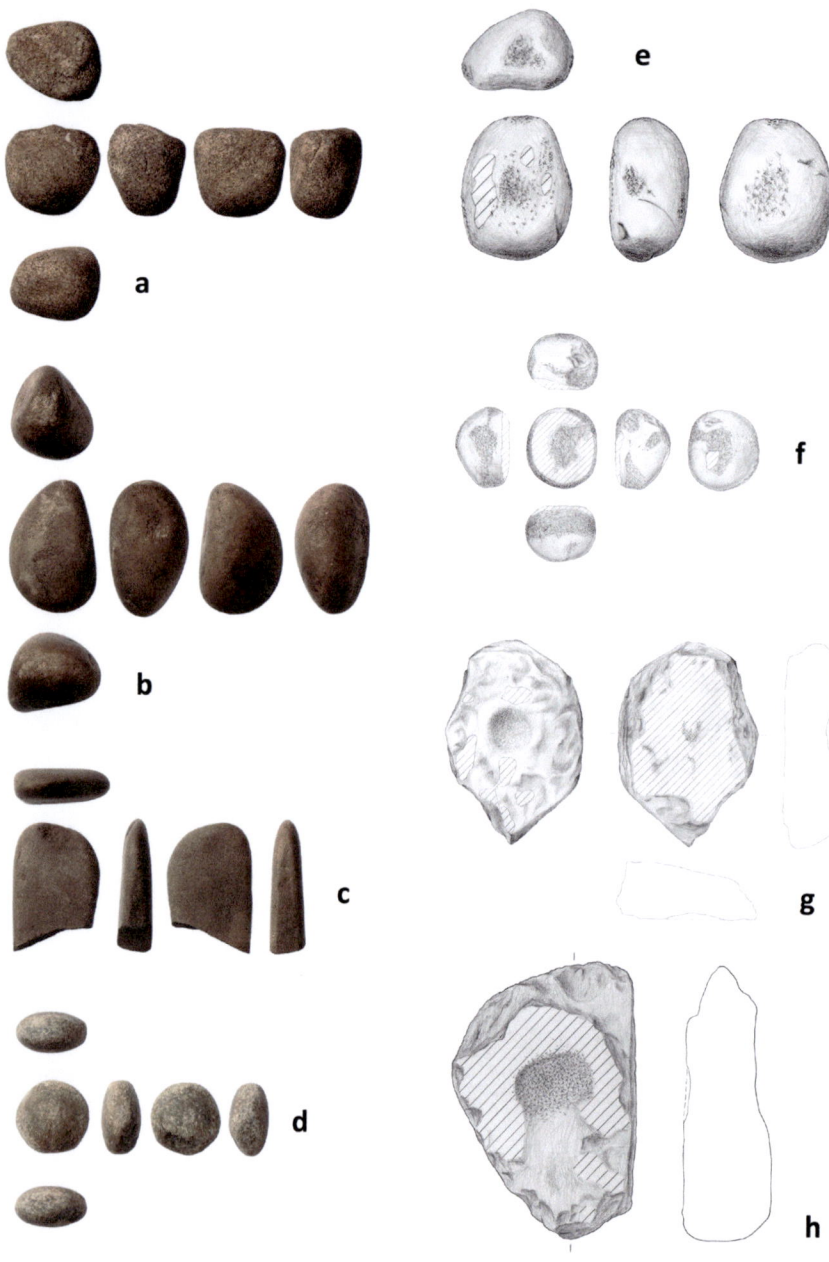

Figure 4.12: Stone artefacts from Dronten-N23 (a-d), Swifterbant S3 (e), Swifterbant S4 (f-g) and Swifterbant S61 (h); (a, c) grinding stones, (b, d) hammer stones, (e, f) hammerstone/anvil/grinding stone (polisher), (g, h) anvil/grinding stone (polisher); (a, f, g) scale 1:8; (b-e, h) scale 1:4 (source: Hamburg *et al.* 2012; Devriendt 2014).

that cereal grains may have been ground. Biochemical and microscopic analysis of encrusted residue on pottery from Swifterbant-S3 produced strong evidence that emmer wheat was processed in a foodstuff prepared in earthenware pots, probably a kind of porridge.[255] Besides emmer wheat, this food also contained animal proteins and fats. Material of animal and plant origin was also found to have been prepared in other pots, though this did not include emmer wheat (see also Chapter 5).

---

255 Raemaekers, Kubiak-Martens & Oudemans 2013.

4.7.3 Consumption

The animal and plant remains unfortunately tell us nothing about the actual consumption of food. The encrusted residues on potsherds mentioned above do however provide some information as to the nature of what was cooked or prepared in pots, assuming that this was food. Other evidence is to be found in human bone remains. The type of food that a person eats during their lifetime, be it terrestrial or aquatic, determines to a large extent which stable isotopes are incorporated into their dental and bone tissue, that is to say, in the makeup of collagen. Stable isotopes such as ^{13}C and ^{15}N, which are absorbed in the first 20 years, can be found in dental enamel. For later stages

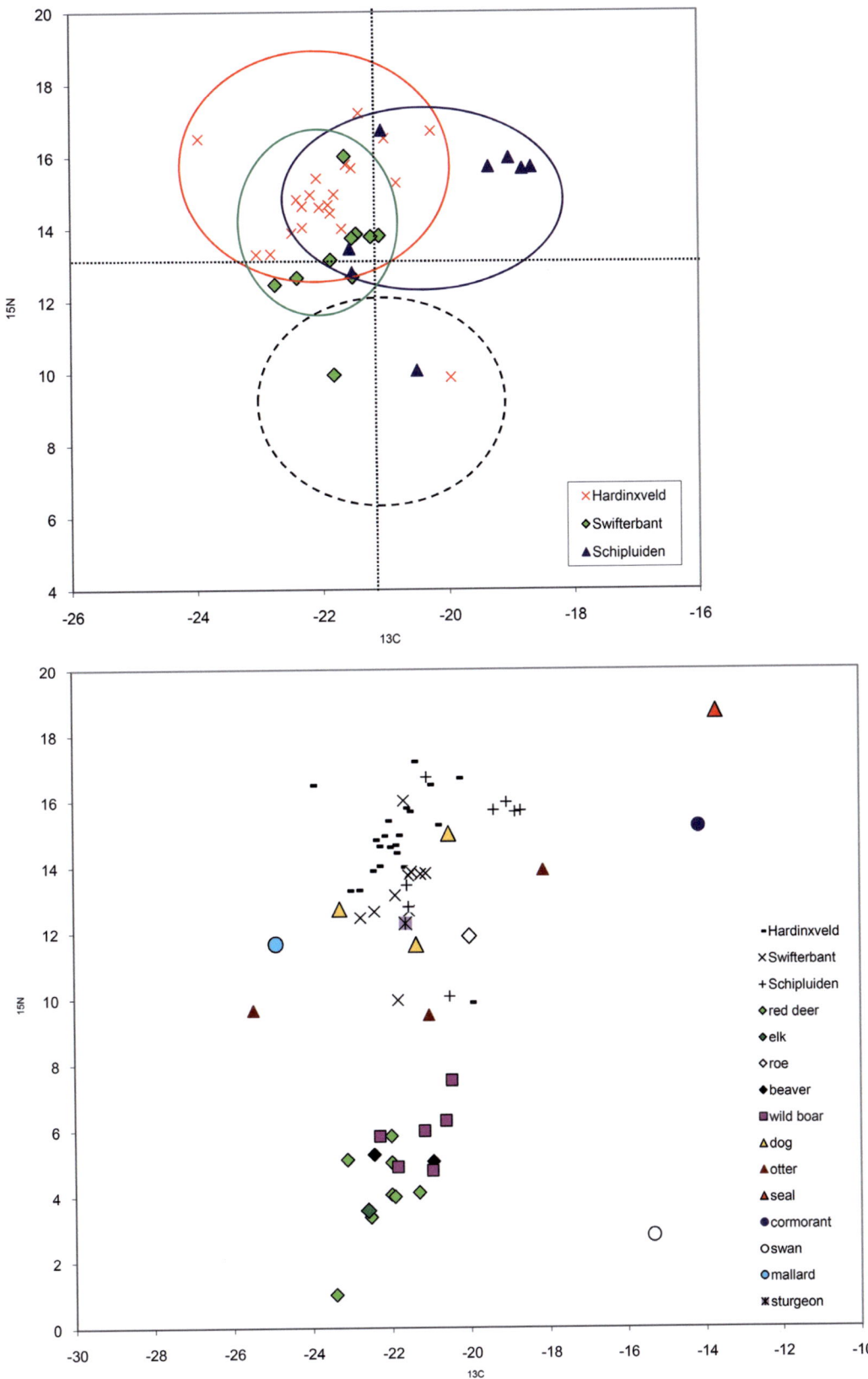

Figure 4.13: Stable Isotope ratios of human and animal remains from different neolithic sites (source: Smits & van der Plicht 2009).

of life, when no new dental enamel forms, stable isotopes from bone tissue are a source of information.

Research on human remains from the Classical Swifterbant contexts of Swifterbant-S2 and S3 has shown that the diets of these individuals consisted to a large extent of protein from terrestrial food sources (fig. 4.13).[256] Aquatic food also contributed to the stable isotope signal, albeit to a lesser extent. The fact that the landscape of the region was already dominated by wet conditions in the Classical Swifterbant does not appear to have had a huge impact on the diet, although we must of course take account of seasonal variation and the possible exploitation of an area larger than today's Flevoland (see also section 4.9.4). Furthermore, 'wet' need not necessarily imply open water. Many animals and plants from a wet, marshy habitat do not leave an aquatic isotope signal. Furthermore, cultural factors will primarily have determined people's choices regarding the food they consumed.

4.8 Resources and technology

Besides the inclusion of wild and domesticated food sources in the basic diet, all kinds of organic and inorganic resources were used as raw materials for various aspects of subsistence. Developments in the production and use of tools, dwellings, modes of transport, pottery, jewellery, clothing and footwear, to name but a few, were related to the opportunities afforded by the landscape and the choices made in a specific cultural context. The Flevoland data inform us about several of these aspects.

4.8.1 Availability of and animal resources

There is little concrete to say about the geographical origins of most of the animal and plant raw materials used in the production of utilitarian objects. Raw materials from animals that lived in the region included bone, antler and teeth.[257] The plant species used also came from vegetation that grew in the region throughout prehistory. Of course, not all animals and plants were found everywhere in the landscape, as we are not dealing here with a single biotope. Furthermore, the nature and geographical structure of the landscape changed during the Holocene. As a consequence, the availability of plant and animal resources suitable for direct exploitation was not constant neither in spatial nor temporal terms. This does not, however, mean that cultural factors, such as exchange networks for instance, did not play a role in the supply of resources.

Based on the evidence for the transformation of the palaeolandscape in Flevoland in relation to the groundwater level rise described in Chapter 3, we can draw some rough conclusions as to the likelihood that certain resources were present in the region and could have been exploited. A simple computer model shows the relative proportions accounted for by different landscape zones in relation to groundwater levels for the period 6500-2500 cal. BC (fig. 4.14).[258] The model shows that the proportion of the landscape dominated by 'dry woodland' gradually declined from the second half of the Early Atlantic in favour of 'scrub' and 'marsh woodland' and, to varying degrees, 'sedge' and 'reed zones'. The proportion of 'upland peat' zones remained relatively small over the whole period included in the model. Chapter 6 will, however, show that changes in the landscape identified on the basis of vegetation data were more complex, and that major geographical differences existed in Flevoland.

The structural groundwater level rise in the region certainly had a direct impact on the local and regional/ sub-regional availability of plant resources. Straight, voluminous trees – such as lime trees which were used for making canoes, for example – will have declined in number. By contrast, the number of trees that favoured wet conditions – such as alder and willow, which was very useful as a construction material – will have increased. The availability of reeds, used in the manufacture of containers and mats, but also as roofing material, will also have increased in the region.

This shift will have affected various groups of animals. Terrestrial mammals with a preference for drier ground will have gradually declined in number (fig. 4.14). Red deer antlers, for example, probably became increasingly scarce. The rise in the groundwater level will however, have led to an increase in the number of aquatic mammals, such as otters and beavers, particularly in the first half of the Middle Atlantic, when the landscape would have been highly differentiated. As a result, the availability of resources providing fur will have increased significantly, followed by a slight decline when environmental conditions changed due to acidification as a result of peat expansion. Acidification probably had a negative impact on the number of fish, an otter's primary source of food. Beavers would not have been affected by this, as they feed exclusively on leaves and bark.

We must however not lose sight of the fact that the use of resources need not necessarily have been dictated by those resources directly available – that is to say, those available at locations or in landscape zones in the study region. For instance, in the Early Bronze Age contexts of

---

256 Smits *et al.* 2008; Smits & Van der Plicht 2009.
257 Although horn will certainly also have been significant, no evidence of this has been found. Horn can be preserved in upland peat deposits, but none of these have been investigated in Flevoland. Examples are however known from Drenthe (Van der Sanden 2002).
258 The model combines various factors that affected the water level in the landscape: structural groundwater table rise, capillary water level rise, peat formation, clay deposits and two relatively important erosion events (Peeters 2007, 56-58).

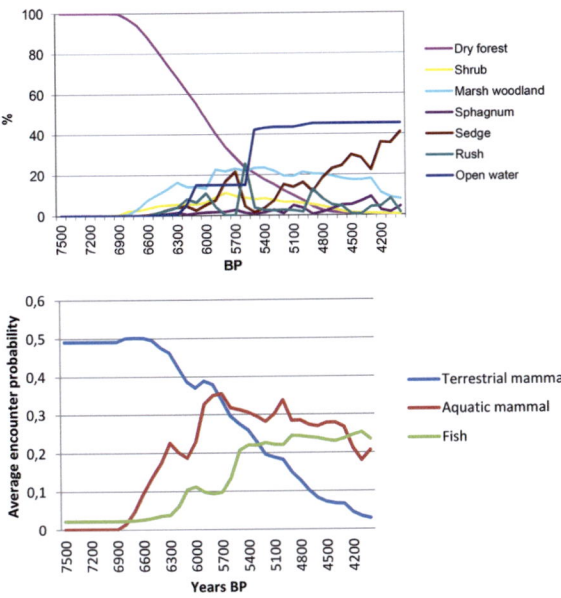

Figure 4.14: Top: graph showing the relative importance of vegetation zones (7000-4000 BP). Below: prediction of the encounter probability for several animal species in relation to the landscape model standing at the basis of the above graph (source: Peeters 2007).

Schokland-P14 and Schokland-J78, when there were many lakes and the landscape was very wet, we see more red deer than beaver. The principle of 'actualism', the use of contemporary information on the behaviour and habitat of animals to infer their behaviour and habitat in the past, is undoubtedly problematic. But even if an accurate representation of the potential presence of plants and animals in different temporal phases could be constructed on the basis of a landscape model, we would still be unable to determine where the plant and animal resources used actually came from. A considerable proportion of these resources could quite possibly have been of 'local' origin, but this is only an assumption. Stable isotopes would provide a more specific picture of the environments in which the animals lived, but such data is not available.

4.8.2 Use and selection of plant resources

Section 4.4.2 explained that fish weirs and traps were constructed using small tree trunks, branches, twigs and rope or twine made from bark fibre. Analysis of the use of wood shows that the weirs and traps were constructed using species that will have been locally available. There is no evidence to suggest that the wood came from tree stands that were managed or coppiced for this specific purpose, although evidence may suggest otherwise in relation to fish traps (see below).

Small, round tree trunks were used for the fish weir posts. The posts found from the weirs at Hoge Vaart-A27 consist mainly of alder round wood; willow and hazel were used occasionally.[259] The wattle screens between the posts were made using the branches of various species of flexible wood. In addition, some oak was also used in the structures. The fish weirs at Emmeloord-J97 were largely made of alder and birch posts, but included others made of willow, poplar, ash, hazel, oak and elm. The wood came from tree stands from various wooded zones: softwood alluvial woodland (willow and poplar) that grew beside highly dynamic and nutrient-rich rivers, hardwood alluvial woodland (elm, ash, hazel, oak) that grew higher up the riverbank, marsh woodland (willow, alder, ash) from the river basins that flooded annually, alder carrs in peat areas beyond the river basins that were subject to sporadic flooding, and birch carr from the transitional zone between peat areas under the influence of the river and more acidic peat bog. Wood from this transitional zone appears to be well-represented in the later fish weirs that date from the Late Neolithic and Early Bronze Age when birch was on average more common (fig. 4.15).

The posts from the three fish weirs at Hoge Vaart-A27 are between 4 and 8 cm in diameter. The age profile of the alder trunks used for the posts in two of the three weirs are distributed over a wide range. This suggests that the trunks came from tree stands where there was no systematic management and no fixed felling cycles (fig. 4.15). The posts in the third weir cover a narrower date range, with a clustering at around 6 to 8 years of age. This might mean that the wood comes from an alder stand that had been previously used for coppicing. At Emmeloord-J97 the diameter of the logs is generally between 6 and 10 cm, but thinner (1-5 cm) and thicker (11-15 cm) diameters were also used. The tree trunks from the various species of wood again cover a wide range in ages (fig. 4.15), suggesting that, as at Hoge Vaart A27, wood from naturally-occurring woodland resources was used. On the other hand, most of the tree trunks were from young trees (< 35 years old), which might indicate that trunks with smaller diameters were selected. Furthermore, the straightness, smoothness and the often wide annual rings around the core suggest that again, as at Hoge Vaart A27, the trunks were cut down from existing coppice.[260] There is also the possibility that such trunks had grown from the stumps of naturally fallen trees,[261] or had grown in areas where beavers were active.[262] This last option is a realistic possibility, given that several fish weirs at Hoge Vaart-A27 and Emmeloord-J97 show signs of having been gnawed. The data do not, therefore, provide any evidence that wood was obtained from systematically managed woodland.

---

259 Van Rijn & Kooistra 2001.
260 Van Rijn 2002.
261 Van Rijn & Kooistra 2001.
262 Coles 2006.

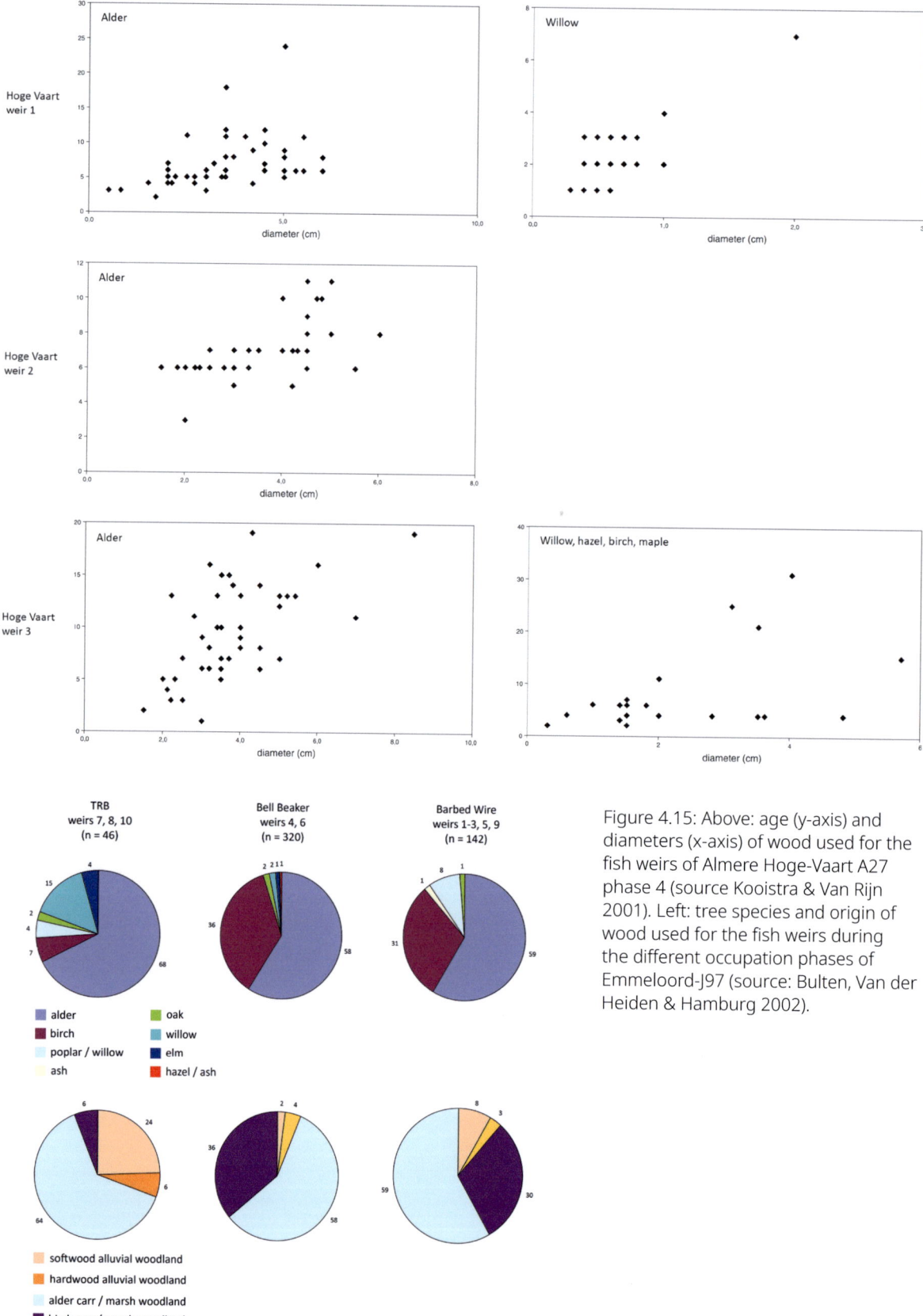

Figure 4.15: Above: age (y-axis) and diameters (x-axis) of wood used for the fish weirs of Almere Hoge-Vaart A27 phase 4 (source Kooistra & Van Rijn 2001). Left: tree species and origin of wood used for the fish weirs during the different occupation phases of Emmeloord-J97 (source: Bulten, Van der Heiden & Hamburg 2002).

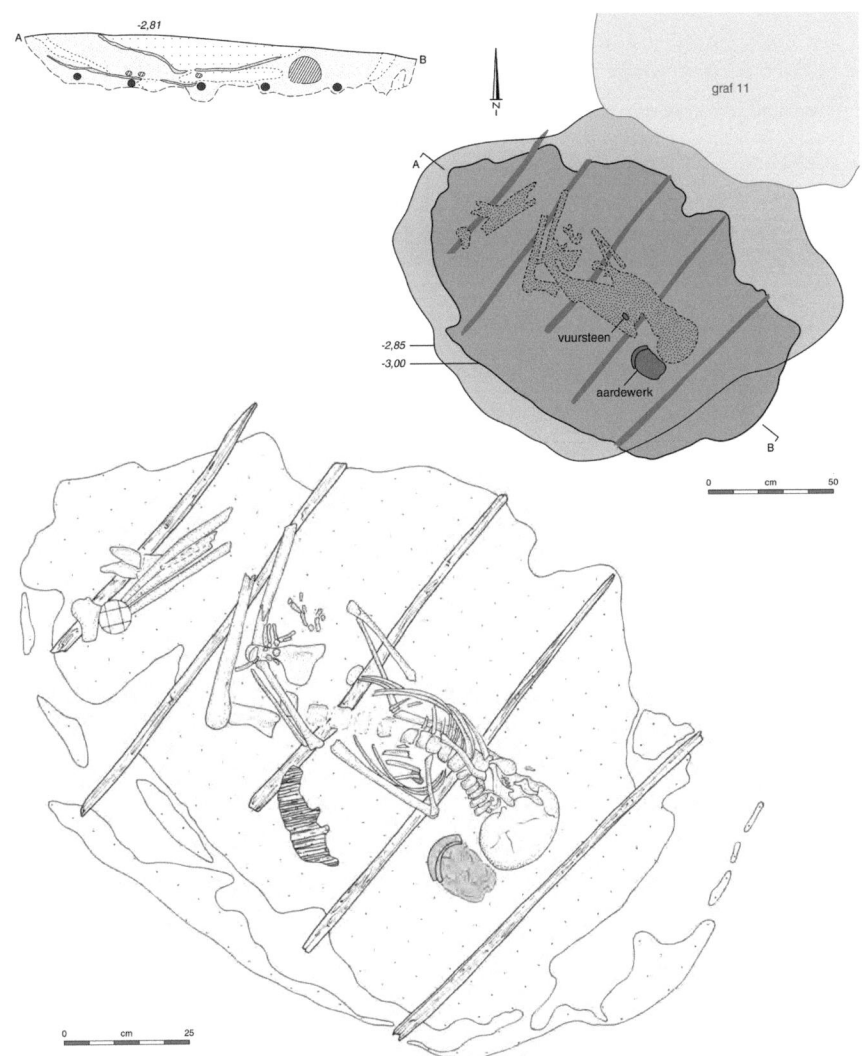

Figure 4.16: Schokland-P14: Corded Ware inhumation grave with traces of a 'coffin' and bark lining; behind the skull a complete protruded foot beaker was discovered (source: Ten Anscher 2012).

The ambiguity of the information concerning coppicing in relation to the construction of fish traps also applies to the material used to make fish traps. The traps at Hoge Vaart-A27 are made of willow shoots no more than 3 years old. The traps at Emmeloord-J97 are made of 1 and 2 year-old willow and 1 year-old hazel shoots. It is possible that the exploitation of these species did take place on a more systematic, annual basis. Like hazel, white willow and brittle willow – which occur in zone 1 – tolerate regular cutting of their shoots. This is not, however, to say that this activity signals management of the resource. The construction of fish traps, and wickerwork in general is reliant on flexible, young wood. In a landscape where beavers were active, for example, such young wood would have been available in abundance. It is certainly not unlikely that humans took advantage of this.[263]

---

263   Coles 2006.

As well as fish weirs, wooden posts of various dimensions were used in the construction of other structures. The remains of wooden stakes and posts with a diameter between approx. 5 and 10 cm were found in the Early Swifterbant context of Hoge Vaart-A27 phase 3. These were simply manufactured ash, alder and oak posts with a sharpened tip. Remains of such stakes and posts were also found at Swifterbant-S3, as well as at Schokland-P14, where they have been attributed to the Pre-Drouwen phase. The stakes and posts may have been part of light shelters or other structures such as racks and screens. Another possibility is that the stakes for instance served to secure bunches of reed used for surface elevation.

At Hoge Vaart-A27 phase 3, several heavier oak posts with a diameter of approx. 15 cm were found. These had been driven into the ground and may have been associated with a ritual context (see chapter 5). This might also apply to a 'platform' or 'jetty' made of four heavy oak tree trunks with no side branches laid out in parallel approx. 1.2 m

Figure 4.17: Almere Hoge Vaart-A27 phase 4: fragment of a paddle (photo: D. Velthuizen).

apart and at right angles to the orientation of the gully. Interestingly, the dendrochronological ages of both the heavier posts and the logs used to make the 'platform' are very close together, between c. 4650 and 4620 BC.[264] Given their location in the Late Neolithic cemetery at Schokland-P14, two Late Neolithic – Early Bell Beaker Culture – posts with diameters of approx. 25 cm, and 50 – 60 cm respectively, might be associated with graves, although it has also been suggested that they may have been 'territorial markers'.[265]

In addition, a number of Late Neolithic graves at Schokland-P14 provided evidence for the use of wood to line grave pits.[266] The bottom of grave pits was found to be lined with tree bark, as in the case of Grave 11 (fig. 4.16). This grave had also been covered over with bark. In at least one case the bottom of a grave was lined with five 3 cm-thick oak sticks which were covered with wood or bark. The walls of Grave 12 had been lined with wickerwork. There are no indications of linings in grave pits from earlier periods of the Neolithic in Flevoland.

Besides the use of wood and reeds in structures and the manufacture of objects such as fish traps and mats, other uses have also been identified at a number of sites. Although we only have preserved wooden objects from the Neolithic, use wear on flint artefacts shows that

Figure 4.18: Almere Hoge Vaart-A27 phase 3: Imprints of basketry. The photos on the right show details showing pleated structures (textile?) and very fine fiber structures; in the top right corner of the bottom photo a remnant of (un) fired clay with quartz grit is noticeable (photo: L. Klimby).

---

264   Peeters *et al.* 2001.
265   Ten Anscher 2012, 364-365.
266   Ten Anscher 2012, 331-342

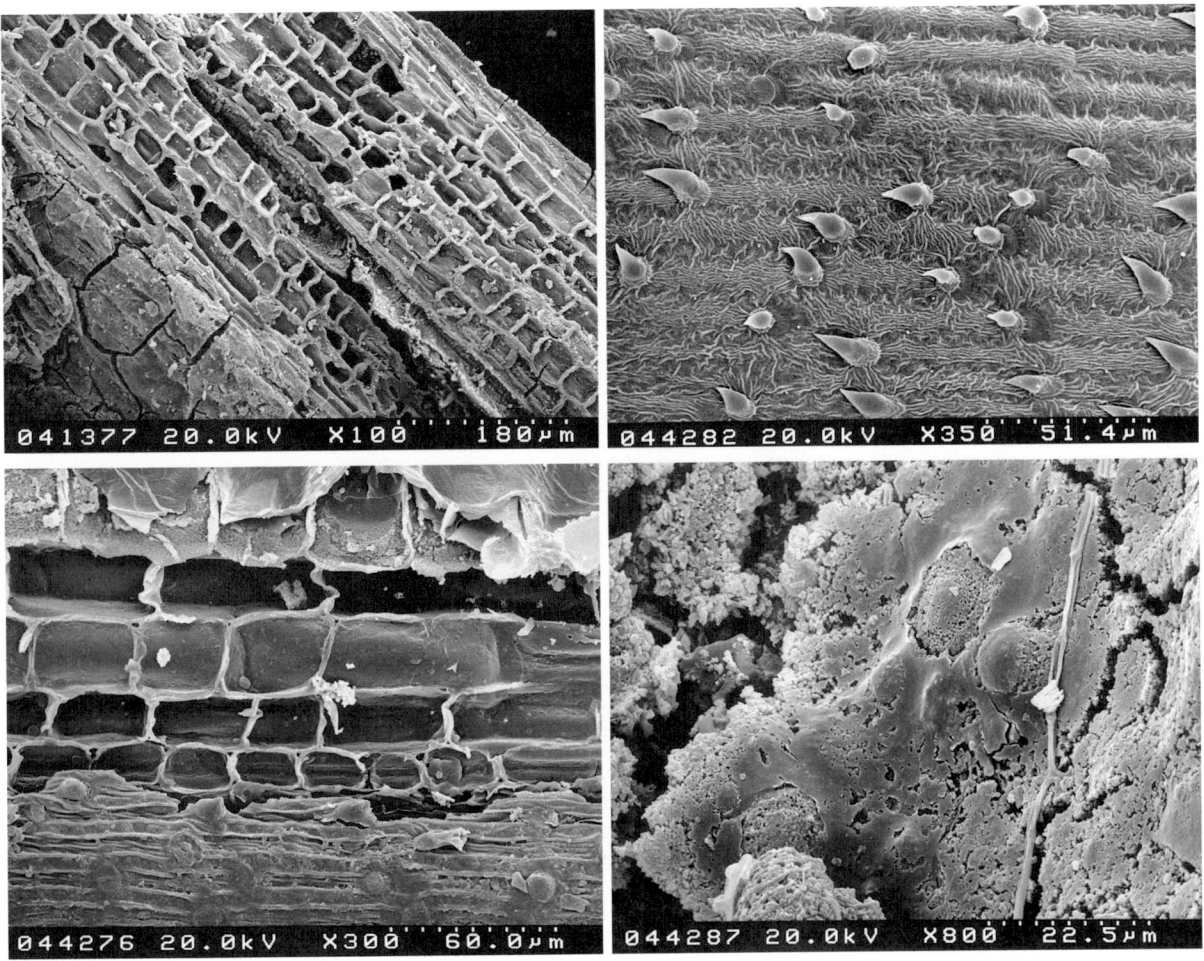

Figure 4.19: Almere Hoge Vaart-A27 phase 3: SEM photo of charred reeds from the basketry imprints (fig. 4.18). The round-oval structures in the photo at the bottom right correspond to the thorny protrusions that are present on recent grasses (source: Hamburg et al. 2001).

woodworking also took place in the Mesolithic. Analysis of flint artefacts from Dronten-N23 has shown that wood was frequently worked at this site, with activities involving wood cutting, chopping, adzing and planing, as well as bark removal and hole piercing.[267]

Wooden objects are known from various Neolithic contexts. Four fragments from paddles were found in Classic Swifterbant contexts at Swifterbant-S5 and S25, and Hoge Vaart-A27 phase 4 (fig. 4.17). The paddles were made of oak, alder and maple.[268] A two-pronged fork made of yew (discussed above) was found in the gully at Emmeloord-J97 associated with the fish weirs and traps (fig. 4.4). On the basis of a ^{14}C date, the fork van be attributed to the Pre-Drouwen phase or a slightly later Funnelbeaker phase.[269] Another object made of yew and interpreted as a 'club' was found in a Late Neolithic grave of the Single Grave Culture at Schokland-P14.[270]

One exceptional discovery that gives further insight into the use of plant materials is the series of woven matting imprints found in the Early Swifterbant context of Hoge Vaart-A27 phase 3 (fig. 4.18). The imprints were found on at least four levels in a pit (see also 4.8.6).[271] The imprints, which have remained preserved in relief, show concentric patterns of rolls varying in thickness from 1.2 to 10 mm. Using a scanning electron microscope (SEM), a sample of charred material found in one of

---

267 Siebelink et al. 2012, 253-254.
268 Out 2009, 290.
269 Bulten et al. 2002, 76-77. The AMS dating was obtained after publication of the excavation report (Bulten et al. 2002), and gave an age of 4478 ± 46 BP (UtC-12483).
270 Ten Anscher 2012, 339.
271 The investigation was performed by Dr. A. Rast-Eicher in collaboration with J. Nienker (Hamburg et al. 2001).

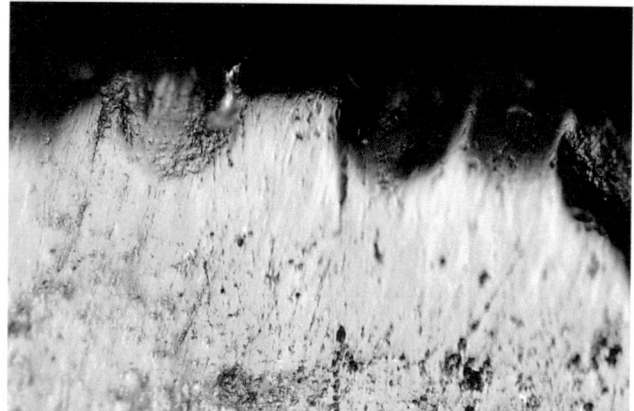
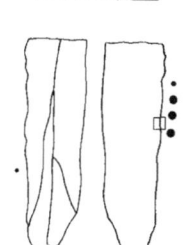
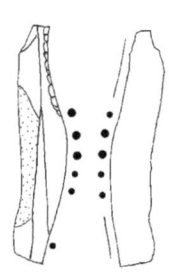

Figure 4.20: Almere Hoge Vaart-A27 phase 3: blades with use wear. This use wear is frequently discovered on blades from Mesolithic-Neolithic wetland contexts, however its origin is still unknown (source: Peeters et al. 2001).

the levels was identified as birch, either from a twig or bark. The birch may have been used to bind the rolls, which almost certainly consisted of grasses. A second sample of charred material whose orientation, like that of the birch, was associated with the roll imprints, was identified in SEM analysis as reed (fig. 4.19). Fine creases or rilling were found in the rolls. These appear to have been caused by leaves that were not removed before the mats were made. The tapering diameters of the rolls corresponds closely to that of reed stems: from 1 mm at the tip to 5 mm at the base, and to 10 mm for the leaves. Small V-shaped imprints 0.5 to 2 mm wide and 2 and 12 mm apart across the rolls probably represent the original stitching used to bind them together. In one part of the pit a much finer imprint was discovered that is probably more likely to be an imprint of a folded 'cloth' (fig. 4.18), although it was probably not made of any woven textile.

The likelihood that grasses and other plant material were commonly used in the Early Neolithic and Mesolithic to weave items such as mats and baskets is also supported by use wear found on flint blades. In the Early Swifterbant context of Hoge Vaart-A27 phase 3, large numbers of blades have been shown to have been used to flatten or squash stems (fig. 4.20).[272] Identical use wear has been found on blades from Dronten-N23,[273] Swifterbant-S2, S4 and S51[274] and on larger blades from Urk-E4.[275] Although attempts to experimentally recreate the use wear have so far been unsuccessful, it is assumed that these tools were used for craft activities for which plant resources were processed.[276]

4.8.3 The utilisation of animal resources

Direct information on the use of animal products as a raw material for tools, jewellery and clothes, for example, is only available for the Neolithic and the Early Bronze Age. There are no artefacts made of bone or antler among the

---

272 Peeters, Schreurs & Verneau 2001.
273 Siebelink et al. 2012.
274 Bienenfeld 1985, p 206-209. It should be noted that Bienenfeld (1985) associated the use wear with 'sickle polish' and harvesting. However, use-wear analysis conducted in the past few years has shown that this is incorrect (see for example Van Gijn 2010).
275 Peters & Peeters 2001. These blades are probably from the Late Mesolithic or Early Neolithic.
276 Van Gijn (2010, 66) highlights the fact that such use wear is no longer seen in the period when flint objects with sickle polish begin to appear. Further research could provide a definitive answer as to the precise function of the blades.

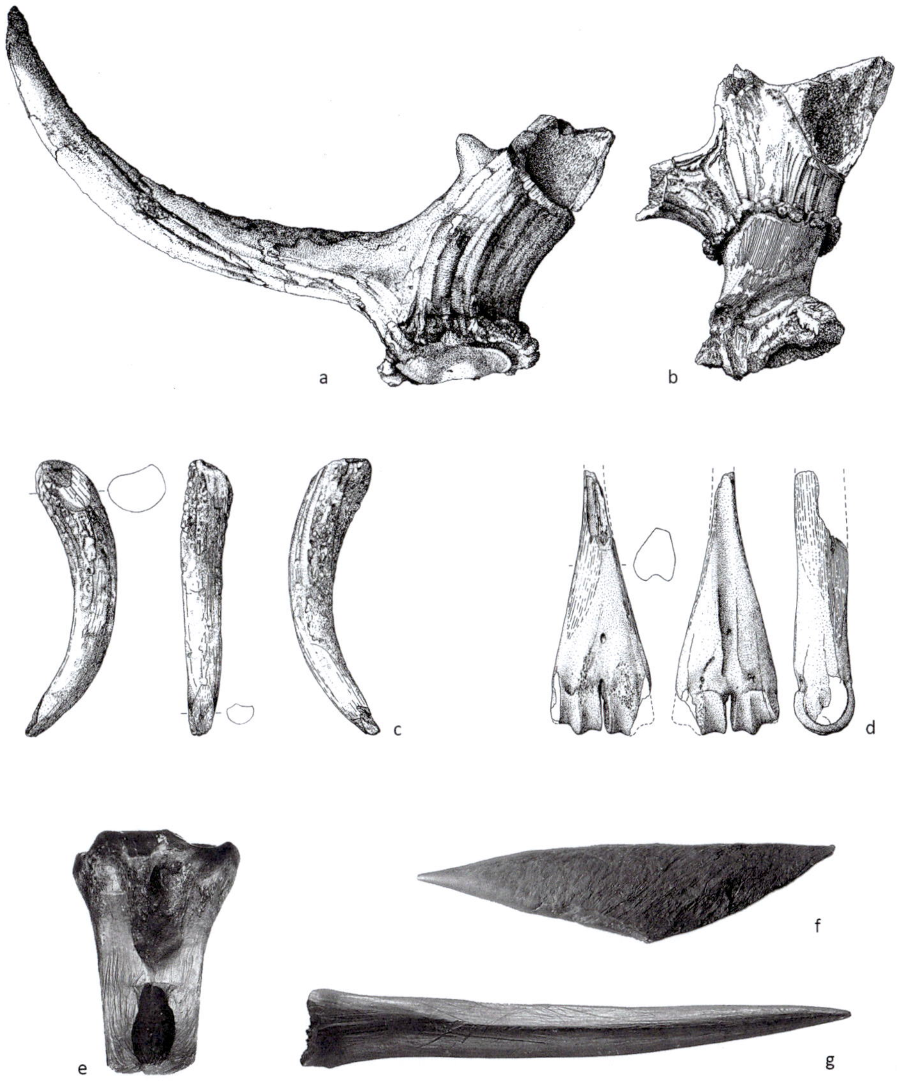

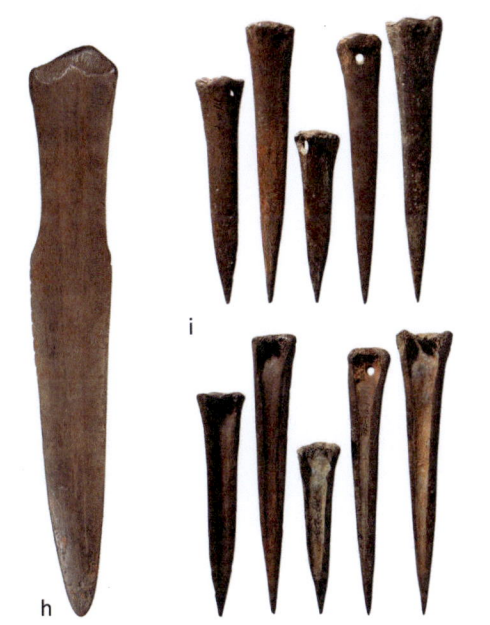

Figure 4.21: Above: artefacts of bone and antlers from Hoge Vaart-A27 phase 3 (source: Laarman 2001). (a) cut and broken off part of a shed antler; (b) cut and broken off part of a unshed antler; (c) antler with a trimmed proximal end and splintered distal end, possibly used as a punch for the production of flint blades; (d) laterally abraded and worn-through metatarsal bone of red deer; (e) socketed adze made out of an aurochs bone proximally pierced to the articular surface; (f-g) awl-like objects. (a-d) 33% actual size; (e) 50% actual size; (f-g) 65% actual size. Right: artefacts of bone from Emmeloord-J97 (source: Bulten, Van der Heijden & Hamburg). (h) dagger-shaped object with a length of approximately 16.5 cm and partially serrated edges, possibly as an imitation of a flint Scandinavian dagger; (i) awls of sheep / goat metatarsal bones. (i) 40% actual size.

few Mesolithic animal remains from Dronten-N23 and Almere-Zwaanpad.

Dronten-N23 did, however, yield indirect information as a result of use-wear analysis of flint artefacts.[277] It was concluded that the processing of hides – the scraping of dry hides and the cutting and piercing of hides – is relatively well represented in the assemblage. These might have been activities associated with the making of clothing and containers. Use wear has also been found that might be connected with the working of bone and antler. Again, this involves scraping, cutting or sawing and the piercing of material.

The remains found in the context of Hoge Vaart-A27 phase 3 provide a more specific insight into the use of animal material.[278] The bone and antler included dozens of tools and production waste. This provided evidence of the use of red deer antler – mainly naturally shed antlers – for the manufacture of mattocks, 20 heavily damaged examples of which were found in the gully (fig. 4.21). Various sawn off pieces of antler were also found. These could be associated with the production of mattocks, which were probably used to dig up roots and tubers (see 4.6.3). Other antler tools include a possible hammer made of a proximal section of antler and a possible punch made from an antler tine that may have been used to work flint, as well as other tines that have clearly been worked, but whose function is not clear.

The worked bone assemblage includes a socketed adze made from an auroch metatarsus, a sharpened rib and several awls (fig. 4.21). Two bones – an auroch metatarsus and a red deer metatarsus – were used intensively for an unknown activity which has resulted in the complete wearing away of the tools. Three sawn off horse metacarpals may be waste from the production of awls. Finally, ten front sections of wild boar mandible with the canines removed were found in the gully. These teeth may have been used as ornaments. Interestingly, the evident presence of tools and bone and antler processing waste is not reflected in the use wear on the flint material, despite a targeted search for such evidence.[279] The processing of bone and antler is barely represented as an activity in the use wear. By contrast, use-wear analysis did find that fresh hides were regularly cleaned with scrapers. This was probably more to do with the primary preparation of the hides rather than the processing of hides for clothing or other uses.

Worked bone and antler has been found in Classic Swifterbant contexts at Swifterbant-S3 and S5,[280] and a few examples have been found in the Swifterbant/Pre-Drouwen context of Schokland-P14.[281] Red deer antler was again used to manufacture mattocks from the beam of the antler. Socketed adzes were made from the radii of aurochs and cows by breaking the bone at an angle, working the cleavage plane, and piercing the proximal joint surface. Metatarsi and long bones of cattle at least, were also used to make chisels and spatulas, while awls were made of various bones – at least from horse and wild boar. Two horse incisors and a lower canine from a wild boar at Swifterbant-S3 had been pierced and can be interpreted as ornaments (fig. 4.21).[282] Pierced teeth of a pig (possible wild) and a ruminant (possibly a red deer) were found in a grave at Schokland-P14. These have also been interpreted as ornaments.[283]

Bone and antler tools and production waste have also been found in the Late Neolithic and Early Bronze Age contexts of Schokland-P14 and Emmeloord-J97.[284] Compared with the preceding periods, there seems to be a large variety amongst the tools, including many awls and chisels or spatulas made from metacarpal and metatarsal bones from red deer and other animals. Besides awls, 'needles' have also been found at Emmeloord-J97 – some pierced with an eye and some not. All are made from the metacarpals of sheep or goat. Red deer antler was again used to make mattocks and base axes. Antler was also the material used for large fish hooks with an eye through which a line would have been passed. A smaller fish hook was made of bone. A number of unique bone objects were also found at Emmeloord-J97: a possible dagger whose shape is reminiscent of Early Bronze Age flint daggers made in southern Scandinavia, a fragment of a rod with four ridges at the end (fig. 4.21). In addition, a pierced beaver tooth has been interpreted as a pendant.

Finally, it is important to consider the possibility that other parts of animals besides their bones and skins may have been used. Animal fat can, for example, be burnt as fuel. Bird feathers may also have been significant, and might have been incorporated into garments and other objects. In this connection, the presence of white-tailed eagle is important in the Swifterbant contexts of Hoge Vaart-A27 phase 3 and Swifterbant-S3 – the context of several remains at Emmeloord-J97 is unclear. In many cultures white-tailed eagles have special status, and this was probably also the case during the Mesolithic and Neolithic in Flevoland. The symbolic significance of this impressive creature is reflected, for instance, in the use of its feathers and wing bones.[285]

---

277 Siebelink *et al.* 2012.
278 Laarman 2001, 14.
279 Peeters, Schreurs & Verneau 2001.
280 Bulten & Clason 2001.
281 Gehasse 1995; Ten Anscher 2012, 420.
282 Clason & Brinkhuizen 1978.
283 Ten Anscher 2012, 360.
284 Gehasse 1995; Bulten *et al.* 2002; Ten Anscher 2012, 458.
285 Cf. Amkreutz & Corbey 2008.

	Dronten-N23	Hoge Vaart-A27 phase 1-2	Hoge Vaart-A27 phase 3	Swifterbant-S3, S4, S21-S24	Urk-E4	Schokland-P14	Emmeloord-J97
'Local' flint							
Moraine/ bryozoa flint	+++	+++	++	+++	+++	+++	+++
(Rhine/Meuse) terracce flint	o		+	?	+		+
'coastal' flint			+++				
'Exotic' flint							
Wommersom quartzite		o					
Heligoland flint (brown)							o
Baltic flint				o			
'Rijckholt'-flint				?		o	
Lousberg-flint						o	
Grand-Pressigny						o	

+++ majority
++ frequent
+ minority
o individual object
? unsure

Table 4.10: Various types of flint discovered at Flevoland sites.

4.8.4 Origin of lithic materials

We can draw more specific conclusions about the potential geographical origins of rocks and minerals. This applies in particular to the various types of flint found on the sites. We should however note that the majority of the flint nodules used come from secondary sources, *i.e.* material from a primary source – the original limestone deposit – transported to another location by natural processes (table 4.10).[286] For instance, Pleistocene Meuse alluvial gravel, which can be found far into the North Sea, is rich in varieties of flint that can be found as a primary source in southern Limburg and flint found in chalk deposits in southern Scandinavia was transported to the northern Netherlands by glaciers, ending up there as moraine debris. Although we can draw conclusions about the primary origin of many types of flint, it is often difficult to pinpoint where exactly prehistoric people might have gathered it.

As a sedimentary basin, the majority of the Netherlands serves as a receptacle for erosion and drift debris from the south, east and north. In the northwest of the country the base of the quaternary deposits (gravel, sand, clay, peat) lie over 500 m -NAP, and in Flevoland at 200 to 350 m -NAP.[287] Solid rock is found at or near the current land surface only in the eastern Netherlands (mainly sandstone) and the southeastern Netherlands (mainly limestone deposits). Flevoland therefore lies in a region where any rocks found will have been secondarily deposited by natural processes (fig. 4.22). During the Holocene period of prehistory, areas of moraine deposits (glacial till) from the penultimate ice age, the Saalian, were virtually the only places to gather rock in Flevoland. In the Noordoostpolder glacial till sediment lay fairly close to the surface, for instance at Urk, Tollebeek, Schokland and De Voorst. Further to the northeast, but outside Flevoland, large expanses of glacial till deposits could be found at or near the surface.

Ice-pushed ridges formed by glaciers can be found to the east, southeast and southwest of Flevoland. These ridges are made up of early and middle Pleistocene river deposits and moraine material. Stones could be collected on the surface of these ridges in places where erosion had occurred due to an absence of vegetation. The mainly crystalline rocks and sandstone found in the ice-pushed ridges situated to the east-southeast (on the Veluwe) were deposited by the Rhine and a river system (the Eridanos) that ceased to exist c. 500,000 years ago. The southwestern ice-pushed ridges (Utrechtse Heuvelrug) contain many rocks transported by the Meuse, including a large quantity of flint that originated in southern Limburg and neighbouring regions of Belgium.

---

286 The categories distinguished as 'local flint' primarily refer to the geological-geographic context from which a nodule is collected. The allocation depends on the weathering characteristics of the natural surface or the characteristics of the stone matrix. Most studies generally distinguish between 'northern flint' (for flint originating from boulder clay) and 'southern flint' (Rijckholt-like flint) based on the characteristics of the stone matrix. 'Coastal flint' is not systematically included as a resource category in the various flint studies. The relative importance indicated in the table is therefore based on very divergent information, whereby the absence of a category is not automatically certain.

287 De Mulder *et al.* 2003, 296.

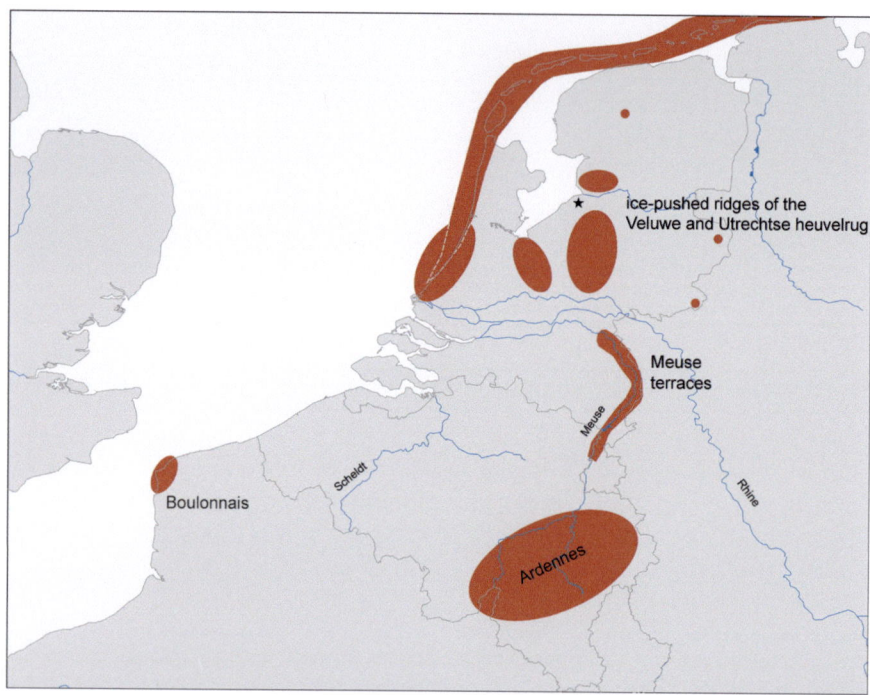

Figure 4.22: Provenance areas of amber, jet and pyrite. The star represents the Swifterbant cluster (source: Devriendt 2014).

The boulders deposited by rivers and glaciers during the Pleistocene were again subject to transportation in the Holocene, particularly the material along the coast and the banks of rivers and, later, lakes. Boulders washing up along the coast had been loosened from the seabed by the turbulence of the water, or had been scoured out of layers of glacial till in cliff faces, mainly during storms. The latter process could also occur along the banks of the large lakes that formed in the Noordoostpolder from the Subboreal onwards. It was mostly peat that was eroded in this way, although moraine material could of course be washed out near glacial till outcrops. Moraine material could also be gathered beside watercourses that bordered directly on glacial till deposits.

The majority of the flint at the sites investigated appears to come from glacial till deposits (moraine flint). The nodules are fairly small on average (fist-sized), with a highly-patinated surface consisting mainly of natural fracture planes. Fossil bryozoans occur frequently in the flint, as do cracks caused by frost. This flint is most likely to come from southern Scandinavia. Flint of southern origin – generally flint of the Rijckholt type – is relatively rare in Flevoland, and must have come from glacially-pushed Meuse terraces (terrace flint) that actually outcrop to the south of southern Flevoland, on the flanks of the Utrechtse Heuvelrug. The nodules, which are again fairly small, have a rolled and patinated surface, which again consists mainly of natural fracture planes. The flint was not, therefore, derived from its primary place of origin (southern Limburg and the neighbouring regions of Belgium). Flint that appears to have been collected on beaches (coastal flint) has also been found. These nodules generally have a heterogeneous, black, blue and grey patina combined with heavy battering marks along the edges. This material includes flint eroded out of glacial till and terrace deposits along the coast.

It is likely that, during the Mesolithic, Neolithic and Bronze Age, flint was collected at and near glacial till outcrops (moraine flint), as well as from ice-pushed ridges (terrace flint) and on beaches (coastal flint). 'Exotic' types of flint have been found at some sites. These include single or no more than a few 'isolated' artefacts made of flint types not found in 'local' deposits.

A small blade of Wommersom quartzite from Belgium and found at Hoge Vaart-A27 could be associated with a Mesolithic occupation phase (1 or 2), though this is by no means certain. Two exceptionally large flakes from a large, plate-like nodule of banded grey flint (fig. 4.23) were found in an Early Swifterbant occupation phase (phase 3) context. The origin of this flint is not entirely clear, but it could be Falster flint from the Baltic.[288] A third artefact – the middle section of a large, exceptionally regular blade, was knapped from a homogeneous, transparent, brown-grey, fine-grained flint that most likely originated in the Baltic region (fig. 4.23). Although Baltic flint is found in glacial till deposits in the Netherlands, it is unlikely that the material for these artefacts comes from this source, in view of the size of the nodules that must have been used in their production. There is no evidence for any

---

[288] The material has been examined by several specialists, but none is certain as to its origin.

Figure 4.23: Remarkable types of flint from Hoge Vaart-A27 phase 3: (a-b) large flakes of strongly banded Falster flint; (c) middle part of an exceptionally regular and large blade (residual length 48 mm) from fine-grained Scandinavian flint; (d) broken blade (residual length 50 mm) of Scandinavian flint. The 154 mm long Scandinavian dagger (e) is made of plate-shaped Heligoland flint and comes from Emmeloord-J97 (photos: Cultural Heritage Agency of the Netherlands/T. Penders; D. Velthuizen).

associated knapping debitage that would necessarily have been present had the nodules been worked on the site. Furthermore, the fragmentary blade is a product of a highly controlled blade production process. As far as we know, there is no evidence for the production of large, highly regular blades in the Netherlands, but such evidence does exist in the Baltic region.

Several pieces of flint found at Schokland-P14 originate from the Dutch-Belgian-German border area. These are flakes removed from polished axes. A fragment of a Grand-Pressigny dagger was also found in the EGK-KB context of Schokland-P14.[289] Several axe flakes and fragments of Neolithic polished axes found at sites near Swifterbant and at Urk-E4 may be of southern flint.[290] A complete dagger of brown Helgoland flint was found at Emmeloord-J97 (fig. 4.23). It can be dated on typological grounds to the Late Neolithic or Early Bronze Age.

Given the specific nature of these artefacts, it can be argued that they were brought to the sites from distant regions. These objects therefore travelled long distances, perhaps as a result of exchange between people from different regions.

Although flint is generally the most common rock found at Flevoland sites, it is certainly not the only lithic resource. Other rock – such as sandstone, quartzite, quartz, granite, diorite and amber – undeservedly receives less attention if it is not found in the form of tools. One problem may lie in the fact that it is not always possible to determine whether unworked, glacial erratics have been transported by humans. This is particularly the case at archaeological sites in locations where lots of stones occur naturally, for instance when glacial till and boulder sand are present. In Flevoland, however, this only applies to Schokland-P14. However, a convincing argument could be made that the lithic assemblage from the Swifterbant and Pre-Drouwen phases represent a selection, and are thus anthropogenic in origin.[291] This is not however true of most of the sites discussed in this book: coversand, river dune sand and river clay do not contain stones.

Other non-flint lithics found at the Mesolithic site of Dronten-N23 include sedimentary, metamorphic and

---

289 Ten Anscher 2012, 457.
290 Verneau 2001, 95; Devriendt 2014, 164-166.

291 Ten Anscher 2012, 419.

igneous rocks (table 4.10).²⁹² Sandstone and quartzitic sandstone account for the largest group (together representing approx. 65%), followed by granite and quartzite. In other contexts that can definitely be dated to the Mesolithic, sandstone and quartzitic sandstone are also the most common types. The picture is however different in Neolithic contexts. Though difficult to quantify at many sites on the basis of the data available, white quartz and granite appear to be common. Other types of rock occur less frequently. Most of the material must have come from sediments deposited by land ice, such as glacial till. Some rocks, such as phylite and trachyte, must have come from the German Eifel, while slate must come from the Ardennes. However, these rocks can also be found in the Pleistocene river deposits of the Rhine and Meuse, both of which surface in the ice-pushed ridges of the central Netherlands. There is therefore no reason to assume that rocks other than flint were brought in from far afield. In theory, all the rocks found could be sourced in the region, or just outside it.

Nor do some of the more striking 'stones', such as amber, jet and pyrite, necessarily come from distant sources.²⁹³ Although the Baltic region is known for its rich deposits of amber, in recent and early historic times this material could be collected in large quantities along the coast of the northern Netherlands, and to a lesser extent along the west coast.²⁹⁴ It also occurs in Saalian glacial deposits.²⁹⁵ There are various primary sources of jet in Northwest Europe – the southern German Alps, North Yorkshire in England – but it is also found in secondary contexts, including along the coast of the western Netherlands and Normandy.²⁹⁶ Pyrite occurs in primary contexts in the Belgian Ardennes and Normandy, and also closer to Flevoland, around Denekamp and Winterswijk in the Netherlands.²⁹⁷ Therefore, although it is possible that amber, jet and pyrite are not from Flevoland, it need not have come from far away (fig. 4.22).

4.8.5 Use of flint and other lithic material

Although flint occurs in all periods considered in this book, it was certainly not always worked in the same way. In the later Early and Middle Mesolithic the production of blades several centimetres long provided an important basis for making microlithic points (fig. 4.24). The flint used varied in quality. In the earliest complexes at Dronten-N23 segments, triangles and various other geometric forms are found, as well as sharp retouched blades of various sizes. The triangles include very small examples, only 4 to 5 mm long. Similar minuscule elements have also been found at Almere-Zwaanpad and Hoge Vaart-A27 phase 1 or 2.²⁹⁸ In principle, the blades would have been struck from a small nodule using a hammerstone. Generally, the flint core would have been reduced as much as possible. In the final stage of knapping more flakes than blades would have been removed. Flakes provided the most common basis for the production of other tools such as scrapers, borers, burins and notched tools. Unmodified flakes were also used as tools.

Late Mesolithic and Early Neolithic assemblages at Dronten-N23 and Hoge Vaart-A27 phase 3 illustrate a different flint technology (fig. 4.25). Blades on these sites are generally larger and have been knapped using indirect percussion (the punch technique). The quality of flint used appears better on average than that used in the preceding period. It would seem that production focused primarily on 2 to 3 mm thick blades with a trapezoid profile which served as a basis for trapeze points.²⁹⁹ Trapezia are the dominant point type in this period. Thicker or more irregular blades were used, sometimes in modified form, to make other tools, such as knives and borers. Flakes of various sizes were again used as a basis for objects like scrapers, borers and notched tools.³⁰⁰

Although the production of fairly large blades by means of indirect percussion definitely continued into the Classic Swifterbant period, we also begin to see the use of bipolar percussion, whereby a nodule would be reduced using a hammerstone and an anvil stone.³⁰¹ From the Middle Neolithic onwards, however, the systematic production of blades seems to come to an end and flint assemblages are dominated by flakes that have been removed from the core using a hammerstone (fig. 4.26). In this period, naturally fractured pieces of flint are also commonly used to produce tools. These tools appear to be much less standardised than those from the previous periods.

On the whole, therefore, we see a decline in the systematic production of regular blades and an increase in techniques geared to the production of flakes. The absence of clearly distinguished stratigraphical contexts at sites such as Urk-E4, Schokland-P14 and the smaller river dune sites near Swifterbant, means that we unfortunately have no clear picture of this 'transition'. Blades and blade tools were clearly in decline in the Pre-Drouwen phase at Schokland-P14.³⁰² We do not however know whether this was also the case at the other sites. It is difficult to ascertain whether the transition to the *ad hoc* use of flint

---

292 Knippenberg & Verbaas 2012.
293 Devriendt 2014, 88-90.
294 Brongers & Woltering 1978; Waterbolk & Waterbolk 1991.
295 Brongers & Woltering 1978; Waterbolk & Waterbolk 1991.
296 Huisman 1977; Muller 1987.
297 Van der Lijn 1973.

298 Peeters, Schreurs & Verneau 2001; Niekus & Smit 2006.
299 Peeters, Schreurs & Verneau 2001.
300 Borers seem to have disappeared by the Late Mesolithic.
301 Devriendt 2014.
302 Van der Kroft 2012.

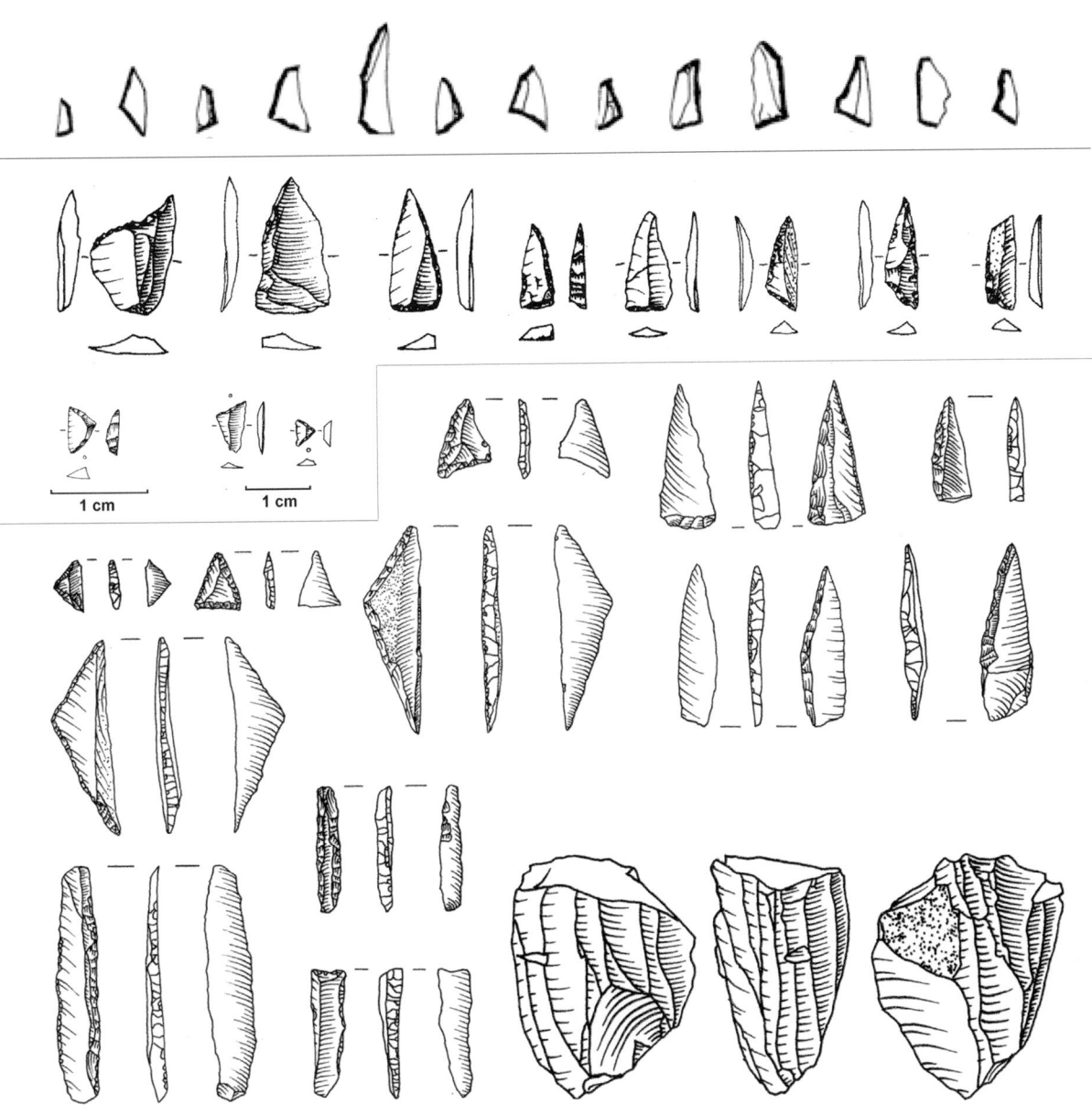

Figure 4.24: Middle mesolithic microlithic points, retouched blades and a blade core from the sites Almere-Zwaanpad (top row), Hoge Vaart-A27 (middle row) and Dronten-N23 (bottom rows). Scale 1:1, unless stated otherwise (source: Peeters *et al.* 2001; Hamburg *et al.* 2012; Niekus *et al.* 2012).

and the almost complete disappearance of the blade-production technology marks a sharp temporal boundary, or rather reflects a gradual process. The picture at Hoge Vaart-A27 phase 3 as well as the sites at Swifterbant and Schokland-P14 do suggest a gradual shift.[303] The absence of 'pure' find assemblages from the Middle and Late Neolithic and the fact that find assemblages from the Early Bronze Age (Schokland-P14) have not yet been studied also makes it unclear how flint technology continued to develop.[304]

Other types of lithics found at the different sites were used in a great variety of applications. The majority of non-flint lithics found at the Mesolithic site of Dronten-N23 consist of small and large splinters and lumps.[305] Seeing as fragmentation can be caused by heating, it is possible that these may be examples of cooking stones that

---

303 The shift from technology focused on blades to the ad hoc use of flint has also been observed elsewhere in the Netherlands and in other countries.

304 Verneau 2002; Van der Kroft 2012.
305 Knippenberg & Verbaas 2012.

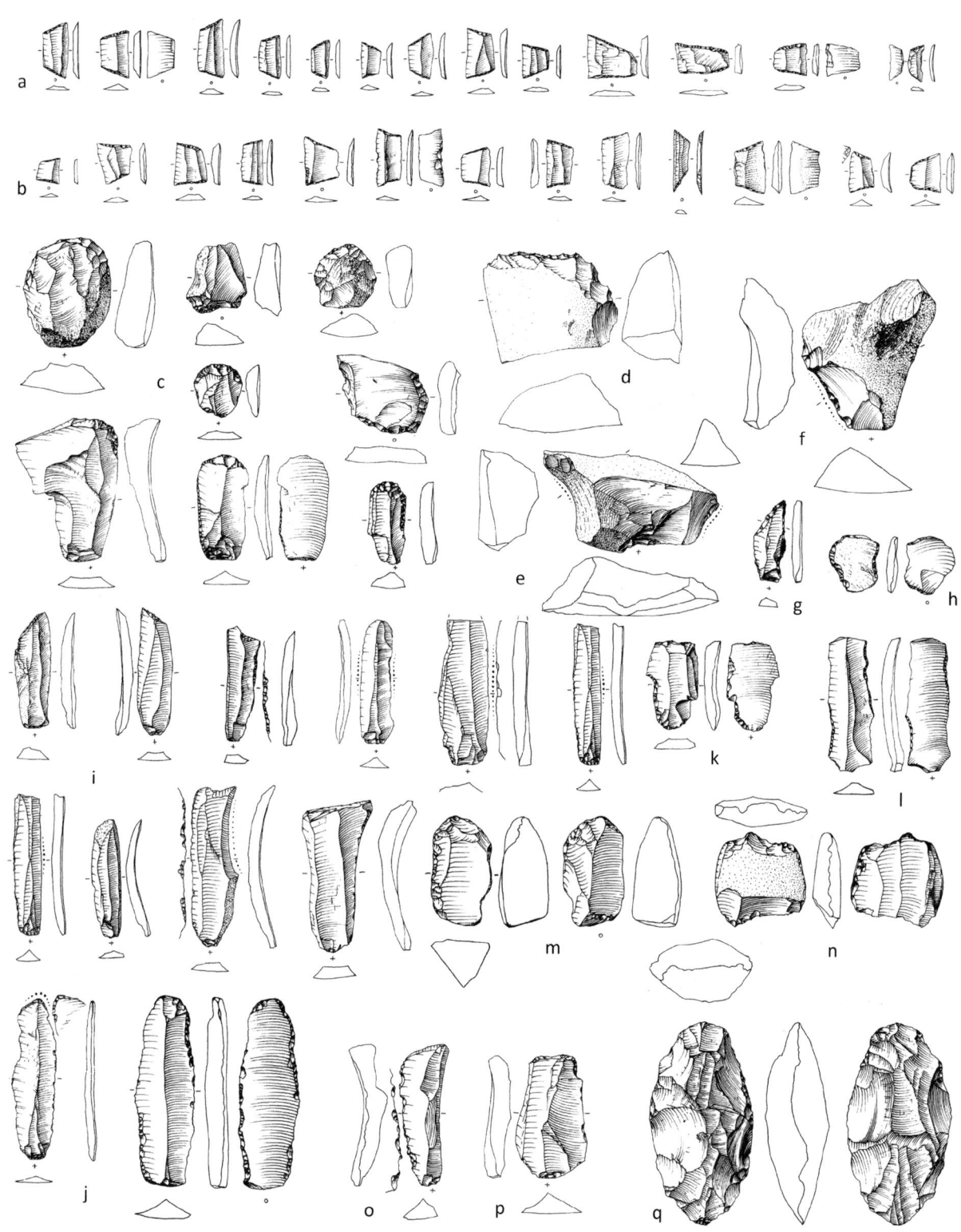

Figure 4.25: Almere Hoge Vaart-A27 phase 3: overview of flint artefacts (a-b) trapezoidal points; (c) scrapers; (d) denticulated artefact; (e-f) flakes used for woodworking; (g) borer; (h) retouched flake; (i) used blades; (j) blades with strong rounding, possibly used as strike-a-lights; (k) staked, truncated blade; (l) notched blade / Montbani blade; (m-n) strike-a-lights; (o) serrated blade; (p) truncated blade; (q) core axe. Scale 1:2 (source: Peeters *et al.* 2001).

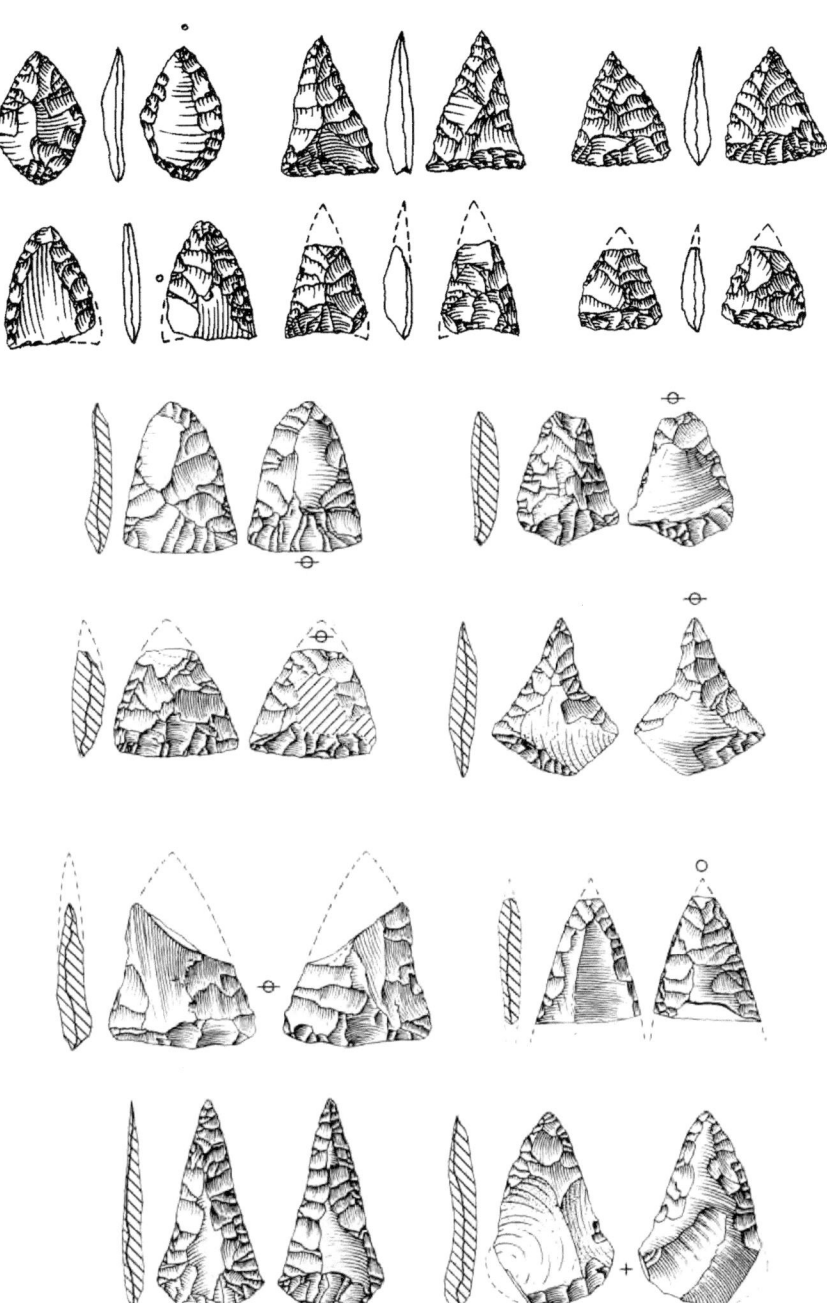

Figure 4.26: Points with surface retouche, likely from the Pre-Drouwen phase of Urk-E4 (top two rows) and Schokland P-14 (lower four rows). Scale 1:1 (source: Peters & Peeters 2001; Van der Kroft 2012).

disintegrated due to the effects of thermal shock. Smaller splinters can also be created when other materials, such as flint, are worked, whilst flakes show that non-flint lithics were also worked. Tools have been recognised among the material. Use-wear analysis suggests they may have had a range of functions. These include hammerstones and grinding stones (fig. 4.12) which were used not only for working flint and other stone, but also for crushing, burnishing and polishing both soft material – plant material, soft rocks – and hard material, such as hard rock and bone.[306] A sandstone macehead (*Geröllkeul*) from Swifterbant-S21-S24, found broken in a heath pit, dates to the Late Mesolithic (fig. 4.27).[307] A small quantity of sandstone, quartzitic sandstone and quartzite from Hoge Vaart-A27 is probably from the Mesolithic, and includes distinct flakes and an 'arrow shaft polisher' (fig. 4.27).[308]

---

306 Knippenberg & Verbaas 2012.
307 Drenth & Niekus 2009a/b.
308 Peeters 2011.

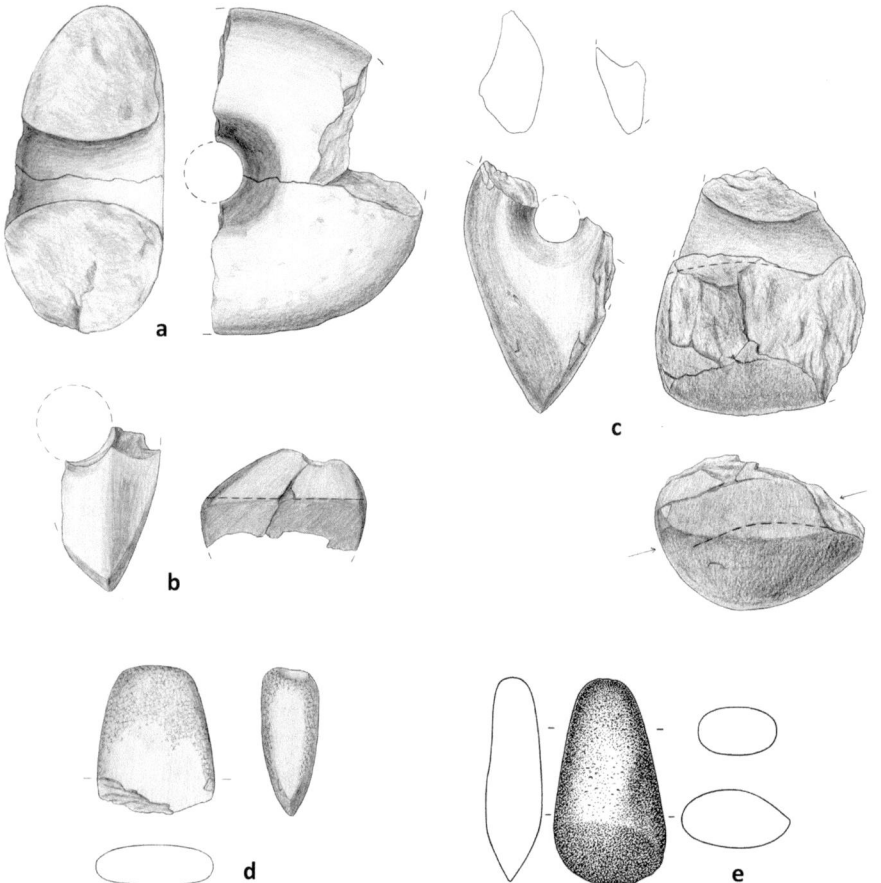

Figure 4.27: Stone tools Swifterbant S21-24 (a) and Swifterbant S3 (b-e). (a) Mace-head; (b-c) shaft-hole axe; (d-e) oval axes. Scale 1:2 (source: Devriendt 2014).

A relatively large quantity of non-flint lithic material, mainly fragmented white quartz and granite was also found in the context of Hoge Vaart-A27 phase 3. This material was probably primarily used for tempering clay (see section 4.8.6). A small number of fire-fractured cobbles of quartzitic sandstone may have been associated with cooking. A broader range of applications have been identified in the Classic Swifterbant contexts of Swifterbant-S2, S3, S4, S21-24 and others.[309] Besides hammerstones and grinding stones, anvil stones, querns, axes and adzes were also found on these sites (fig. 4.27). The axes and adzes are made of amphibolite and diabase. A similar range, in functional and typological terms, was found in the Swifterbant and Pre-Drouwen contexts of Schokland-P14.[310] The axes from this site (*Felsoval-Beile*) are made of quartzite and possibly biotite.

As well as tools, stone would also be used to make ornaments (fig. 4.28). Simple, pierced pendants and flat beads of sandstone, quartzitic sandstone, quartzite and shale have been found in Swifterbant culture graves. Most of the pendants and beads are made of amber, with the occasional piece of jet.[311]

4.8.6 Pottery production

The earliest pottery found in the region comes from Hoge Vaart-A27 phase 3, and has been assigned to an early phase of the Swifterbant culture.[312] From this phase on, pottery is a permanent feature of the material culture. Clear technological trends can be identified in the diachronous shifts in the type of temper used, attachment techniques and average wall thickness.[313]

---

309 Devriendt 2014.
310 Ten Anscher 2012, p 419-420.
311 Devriendt 2014.
312 Haanen & Hogestijn, 2001; Peeters, 2007; Peeters, 2010.
313 Several aspects generally regarded as indicative of the firing process have been disregarded here. For instance, the colour of sherds seems to be highly influenced by the lithological context: pottery from sandy layers, of whatever date, is largely light-coloured (yellow-brown to orange-brown), whereas pottery from peat or clay is generally darker in colour (dark olive-green/-brown to dark grey-black). The hardness of the sherds is apparently heavily influenced by local hydrological conditions at the find spots; after years in the repository they are now considerably harder (see Ten Anscher 2012, 40).

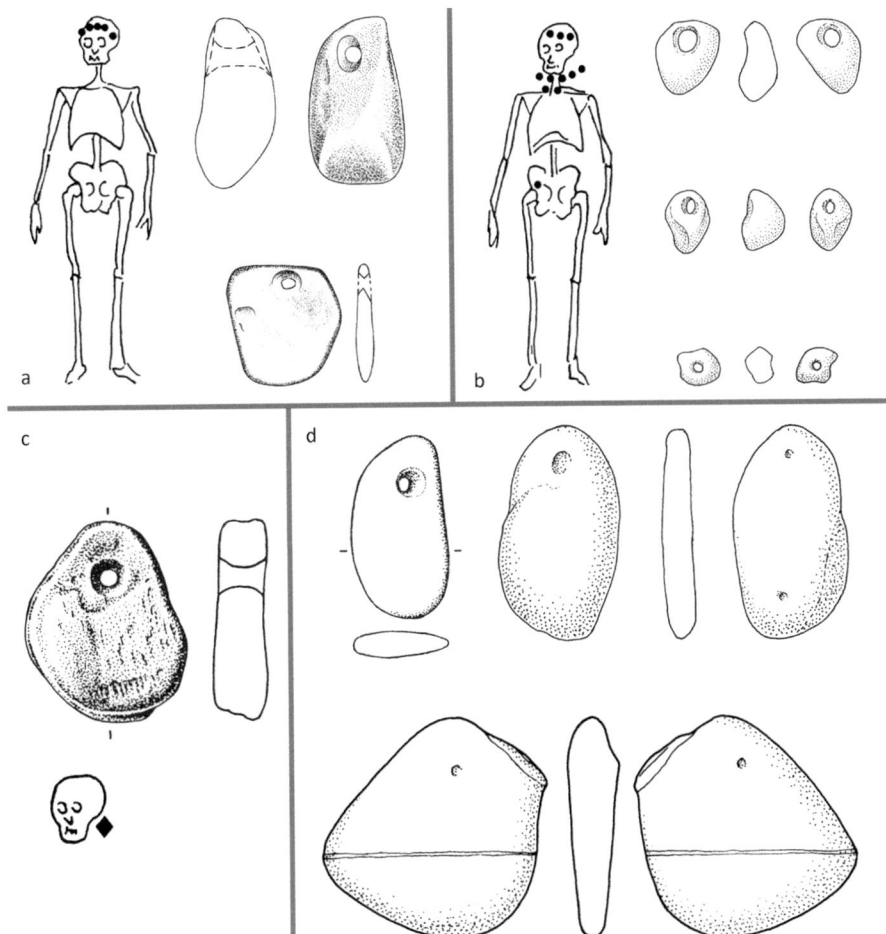

Figure 4.28: Amber ornaments of Swifterbant S2 (a-b), jet ornaments of Swifterbant S21-S24 (c) (quartzitic) sandstone ornament (d). The ornaments of Swifterbant S3 show perforation attempts and the horizontal lines are indications of a quartz vein, scale 1:1 (source: Devriendt 2014).

Zuidelijk Flevoland provides the type site for the early phase of the Swifterbant culture, with Hoge Vaart-A27 phase 3.[314] Oostelijk Flevoland is important for the Classical Swifterbant culture, with Swifterbant-S2 and S3.[315] Mineral tempering (mainly crushed quartz, but also crushed granite) is typical of the oldest pottery. In the Classical phase, on the other hand, the preference was for organic, plant tempering, frequently mixed with stone grit. The pots are made by hand, using coils of rolled clay (fig. 4.29), with mainly U-joins (clay rolls pressed more or less directly on top of each other), though Hb-joins (oblique attachment) gradually become more common. Pottery with Hb-joins often exhibits has a reversal in the direction of the coil application at the largest belly circumference or at the shoulder of the vessel, the lowest part of the seam running over from the inside to the outside. In addition to a coiled structure, point and knob bases are also made from a ball of clay, sometimes with a disc of clay on top. Knobs and eyelets (Ösen) seem to have been attached by simply pressing into the wall. The wall thickness of vessels increases over time, from an average 8 mm to 9-10 mm. Nothing can be said about the finishing of the walls in the earliest phase. Most of the later pots appear to have been smoothed with a wet finger. Some have been polished using a hard object. Roughening of the surface occurs towards the end of the Classical phase.

In the Classical Swifterbant phase there is a correlation between tempering and wall thickness. Pottery with only organic tempering is generally thicker-walled than pottery with a combination of organic and mineral tempering, while pottery with only stone-grit tempering has the thinnest walls.[316] It would appear that the proportion of thin-walled pottery (of good quality) tempered with a combination of organic material and stone grit gradually increased at Swifterbant-S3, and that such pots were used for the preparation of cereal porridge.[317]

The latest dated pottery comes from the Noordoostpolder; very little has been found in Oosterlijk and Zuiderlijk Flevoland. The type site for the latest phase

314 Haanen & Hogestijn 2001; see also Peeters 2010.
315 De Roever 1979; 2004; Raemaekers 1999, 28-35.
316 De Roever 2004, 52; Raemakers 2015.
317 Raemakers 2015.

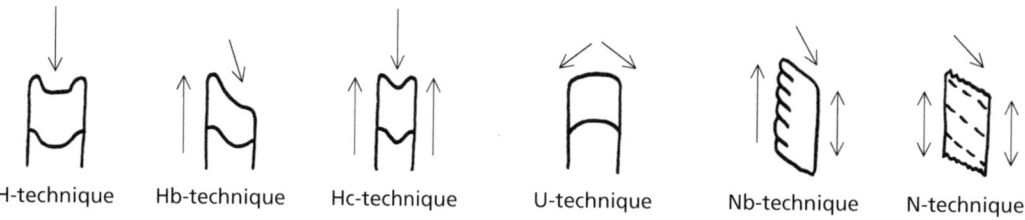

Figure 4.29: Coil built-up of Ertebølle pottery. In Swifterbant pottery generally the U and Hb technique is observed (source: Raemaekers 2011).

of the Swifterbant culture is Schokland-P14, with layer B. This phase can also be regarded as the transitional phase to the Funnelbeaker culture.[318] The technological trends visible at Swifterbant-S3 continue here. The pottery from layer B has mainly organic tempering with a mineral additive (generally crushed granite). Hb-joins dominate. Flat bases are also found in this phase, made from a single or double disc of clay. In a few cases, a deep incision was made in the pot wall, thereby increasing the adhesion plane for knobs (and possibly also eyelets, Ösen). The vessel wall thickness is generally around 7mm, slightly thinner than in the previous period. Polished walls appear to be more common than smoothed walls. Roughening of the surface remains relatively rare. As in the Classical Swifterbant phase, pottery tempered with stone grit has, on average, thinner walls than those vessels with a organic tempering component. Thick-walled pottery is also more likely to have U-joins.

With its C and D layers, Schokland-P14 is also the type site for the Pre-Drouwen phase of the Western Group of the Funnelbeaker culture.[319] Only a small proportion of the Pre-Drouwen pottery – the 'definite' Pre-Drouwen pottery – can be properly distinguished on typological grounds from the late Swifterbant tradition from which it developed. In technological terms, however, the definite and other Pre-Drouwen pottery does not differ from that of the Late Swifterbant tradition, though the trends identified above do continue. Pure mineral tempering (mainly granite) gradually becomes more important than tempering with an organic component. Hb-joins dominate; U-joins are now rare. The average wall thickness is 7-9 mm. Polished walls appear to become more common than smoothed walls. Surface roughening probably ceased entirely in the course of the Pre-Drouwen phase. A collar (from a collared flask) seems to have been attached as an applique. Ceramic baking plates appear to be made from a single disc of clay.

Schokland-P14 remains by far the most important pottery site for the later Neolithic periods (the 'classic' Funnelbeaker culture with phases 1-7, the Single Grave culture – particularly the latest phase – and the Bell Beaker culture) and the Early Bronze Age (Barbed Wire Beaker culture).[320]

The 'classic' Funnelbeaker pottery is tempered almost exclusively with stone grit (crushed granite; the thicker the walls the larger and greater in number the tempering particles) and has an average wall thickness of 5-7 mm. From this point on only oblique Hb-joins occur, in so far as any can be identified. The pottery was probably mostly polished. Handles and grips have a plug inserted into the wall of the pot, increasing the adhesion surface. Feet and mouldings were applied as appliques.

The Late Neolithic beakers (fine ware) are, where the tempering is visible, mainly tempered with sand. Late Bell Beakers are generally tempered with granite grit, mixed with sand and grog. The wall thickness is around 5-6 mm. The coarse ware, with average wall thicknesses around 8 mm, was initially tempered mainly with sand or granite grit; later (in pot beakers), granite grit becomes more important. Thicker sherds generally have more, and coarser, tempering (granite rather than sand). The pottery was probably mainly polished, though smoothed walls also seem to occur.

The earliest Barbed Wire Beakers are similar to the late Bell Beakers in terms of their tempering. The earliest beakers are less robust and have finer tempering than the more recent, and are also a fraction thinner on average, at around 5 mm rather than around 6 mm. This fine ware, however, disappears fairly quickly from the repertoire. The average wall thickness increases to around 9 mm, and the late Barbed Wire vessels are tempered with more abundant, larger particles (mainly crushed granite). While the earliest Barbed Wire vessels are often polished, most of the later examples have smoothed surfaces.

The trend towards coarser ware identified above, which becomes noticeable during the Late Bell Beaker culture, manifests itself gradually more clearly during the Early Bronze Age and continues into the Middle Bronze Age. However, as a result of erosion, hardly any Middle Bronze Age pottery has been found in the Noordoostpolder. The technological developments in pottery identified in

---

318 Ten Anscher 2012; 2015.
319 Ten Anscher 2012, 114-123; Ten Anscher 2015.

320 Ten Anscher 2012, 159-164; 181-183; 233-234; 261-262.

the Middle Neolithic and later do not appear to be specific to the region, but are consistent with developments in neighbouring regions.

Evidence of local pottery production has been found at several sites. At Hoge Vaart-A27 lumps of worked clay and slabs of apparently unfired clay were found within a concentrated zone of the site, in association with a considerable quantity of broken and crushed quartz (fig. 4.30). White quartz grit was frequently used as tempering in the pottery from this site. A dense concentration of poorly-fired pottery sherds were found within the same zone. These sherds appeared to cover a pit in which imprints from woven reed mats were found (see 4.8.2).[321] The mats may have been used as a surface for the preparation of clay – to mix the clay with stone grit tempering – as part of the pottery production process.[322] The clay may have come from the gully that had formed in the low-lying landscape beside the sandy ridge before the start of this phase of occupation on the site.[323]

Other evidence for local pottery production has been found at Schokland-P14 and Swifterbant-S3. This includes two small rolls of tempered clay, which were probably accidentally fired, and a lump of 'squeezed', fired clay.[324] Analysis of the diatomes in pottery sherds as well as the locally-occurring clay does suggests that the clay used was sourced locally.[325] The evidence from Swifterbant-S3 comes from x-ray diffraction analysis and the diatome content of several sherds, the results of which were compared with those from several clay samples.[326] Here, too, local clay appears to have been used as a raw material. The absence of diatomes in several sherds shows, however, that non-local clay must also have been used. It probably came from neighbouring sedimentation areas situated further inland, where freshwater clays are typified by the absence of diatomes.[327]

4.8.7 Wood tar production?

Evidence of Mesolithic activity in Flevoland is based to a large extent on the presence of large numbers of pit hearths. As has been stated, these pits have mainly yielded charcoal and charred organic material, for instance at Dronten-

Figure 4.30: Hoge Vaart-A27 phase 3: concentration of poorly fired pottery, unfired clay and chunks of broken quartz. The whole was lifted as a block for research in the laboratory under controlled conditions. Impressions of reed mats were discovered (see figure 4.18), which were repeatedly placed at several levels in a shallow pit. This has been interpreted as an activity zone for the production of pottery (source: Hamburg et al. 2001; Peeters 2011).

N23 and Hanzelijn-Drontermeer. This charred organic material, some of it accreted to charcoal fragments, has a homogeneously glassy or tar-like appearance (fig. 4.31) and it has been suggested that it may be associated with the production of tar from pine wood.[328] Archaeological finds from Mesolithic and Neolithic sites in northwest Europe show that wood tar was used as an adhesive – for hafting flint tools – and may also have had a 'medicinal' use, as suggested by the occurrence of lumps of masticated wood tar.[329] However, little is known about the production of wood tar in prehistory.

In order to investigate the wood tar production hypothesis further, several samples of this charred glossy material from Dronten-N23 and Hanzelijn-Drontermeer underwent a physico-chemical analysis and were also examined under a scanning electron microscope (SEM).

---

321 Hamburg et al. 2001.
322 Clay particles left on the mats may eventually have caused the patterns to become fixed; the matting itself has disappeared.
323 Geochemical analysis of samples of the lumps of clay found, the potsherds and several layers of clay in the gully has found similarities in the composition of main and trace elements (Jansen & Peeters 2001). It is however difficult to interpret the data because many processes – including recent seepage – could have impacted on the current chemical composition of the clay, which need not necessarily be the same as the original composition.
324 Ten Anscher 2012, 90-91.
325 Jansma 1990; Ten Anscher 2012, 90.
326 De Roever 2004, 120.
327 De Roever 2004.
328 Kubiak-Martens et al. 2011.
329 Aveling & Heron 1998; Baumgartner et al. 2012; Bokelmann 1994; Larsson 1983.

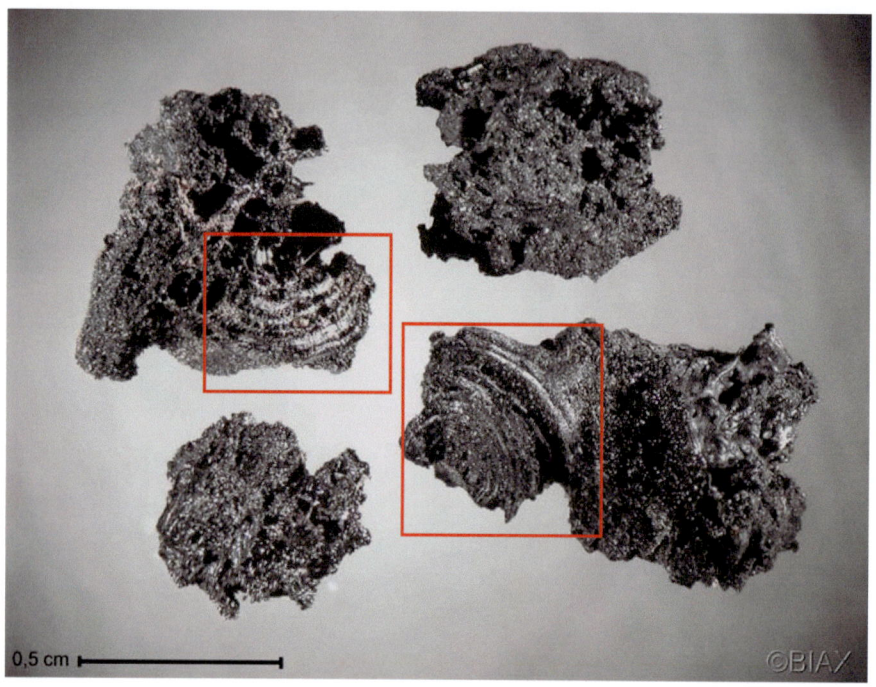

Figure 4.31: Dronten-N23 vitrified charcoal showing the charred remains of the original structure of the wood (source: Hamburg *et al.* 2012).

Samples from the bottom of pit hearths were also subject to a physical and chemical analysis in order to obtain a better understanding of the processes that took place in these pits.[330] The SEM analyses showed that this homogeneous glossy or tar-like material had a porous structure, suggesting that it was originally at least semi-liquid. It could be established that the liquefaction of the wood began in the thin-walled tissues: the spring wood, the radial wood and the resin channels (fig. 4.31). Eventually a liquid mass formed that charred as the temperature rose.[331]

The physico-chemical analysis of soil samples from the sections of four Late Mesolithic pit hearth provided evidence of the destructive distillation of wood in anoxic or oxygen-poor conditions. This process resulted in the formation of wood tar and charcoal. It was established that the degree of thermal degradation in the pits would have increased with depth, suggesting that the highest temperatures were achieved at the bottom, where the wood tar would have overheated. In all the soil samples analysed from Dronten-N23 the organic fraction was found to have been converted to solid carbon at temperatures ranging from 200 to 500C. Extensive carbonisation occurred within this temperature range, and the tar overheated at temperatures in excess of 400 ºC.[332]

This raises the question of which types of wood might have been used to produce tar. Coniferous wood and the bark of birch are well-known raw materials for vegetable tar. The determination of the type of wood used to produce the heavily charred, thermally-degraded remains from the Flevoland sites was generally no longer possible. Retene – and derivatives of retene – were found in a Middle Mesolithic sample from Dronten-N23 that underwent physical and chemical analysis. Retene is a biomarker for the thermal decomposition of pine during dry distillation (pyrolysis). One of the Late Mesolithic pits at Hanzelijn-Drontermeer provided evidence of tar formation from birch bark.[333] In Middle and Late Mesolithic pits dating to the Atlantic period, where charcoal from deciduous trees dominates, tar-like material was only found in pits where charcoal from coniferous trees was also present. If the tar was not created accidentally, pine and birch could have been used as the main raw materials. Other types of wood identified in the charcoal from pit hearths – mainly oak in the Atlantic -are more likely to have been used as fuel. Both pine and birch grew in abundance in Flevoland throughout

---

330 Kooistra *et al.* 2009; Kubiak-Martens *et al.* 2012.
331 Kooistra *et al.* 2009.

332 It is important not to confuse the tar-like material found in the pit hearths with vitrified wood. Vitrification of wood occurs at high temperatures, and involves the variable fusion of anatomical components in the wood itself, leading to homogenisation of the structure (Marguerie & Hunot 2006).
333 Kooistra *et al.* 2009. Kubiak-Martens *et al.* 2012.

the Mesolithic, though pine gradually disappeared from the landscape during the Atlantic.[334]

Whilst the data suggest that wood tar may have been produced in pit hearths during the Mesolithic, we still cannot rule out the possibility that it was created unintentionally during a heating process *similar* to that used during tar production.[335] On the other hand, it is fairly unlikely that wood tar produced accidentally as a by-product of another process would then have been deliberately 'harvested'.

4.9 Habitation patterns

So far, the focus has been mainly on the nature and interpretation of material remains. This section looks at the diversity in the use of certain locations based on the spatial structure and organisation of habitation areas, functional aspects, duration of use and position in the landscape. Any evidence of geographical connections between locations and transport will also be explored. It is important to bear in mind that the sites have not all been investigated with equal intensity, that the number of 'windows of observation' is limited and that the preservation conditions are variable. This necessarily has implications for our understanding of various aspects of occupation. Table 4.11 lists the features and other archeological remains found at the most important sites, distinguishing as much as possible between different occupation phases or contexts. Functional variation is considered below on the basis of the features listed in the table.

4.9.1 Mesolithic

Many Mesolithic sites are situated on river dunes and coversand ridges that had formed during the Late Glacial and possibly the Early Holocene. As with the glacial till outcrops, these landscape features represented the highest elements in the former Mesolithic landscape. These landscape features now lie relatively close to the surface in several parts of Flevoland, where they are reasonably accessible for investigation. As a result, the archaeological information available on the Mesolithic occupation of the area refers mainly to the exploitation of these landscape features, since the deeper landscape zones have seen only very limited investigation (see chapter 2).

The earliest evidence of the presence of Mesolithic hunter-gatherers in the region comes from Dronten-N23, where charred hazelnut remains have been dated to the second half of the Preboreal.[336] Mesolithic activity at this site continued into the second half of the Middle Atlantic. The picture from this and other sites where smaller and large-scale investigations have taken place, is of Mesolithic contexts consisting essentially of scatters of flint and charcoal. There are also large numbers of pit hearths. It is important to note that features like pit hearths rarely occur after c. 5000 cal. BC – the temporal boundary between the Mesolithic and Neolithic.[337]

^{14}C dates show that the Mesolithic sites are the cumulative result, or palimpsests, of various phases of occupation that may have been separated by long periods of time in between (Appendix I).[338] The identification of concentrations of occupation waste within a finds scatter is no simple matter, as we could be dealing with the residue of a single phase of use or, alternatively, an aggregate of material that has become mixed after several phases of use. It is therefore virtually impossible to recognise the original spatial structure of an area, in terms of activity zones for example, that would have been in use at any one time. It has however become possible, through careful examination of the zones where relatively little accumulation of material has taken place, to gain a better understanding of the nature of the activities that caused the formation of these complex 'aggregates' zones.[339]

There is evidence of activity at the Dronten-N23 site throughout most of the Mesolithic. Over time, this activity created a sizeable palimpsest of lithic material, charcoal and pit hearths.[340] Flint clusters documented during the excavations do however often appear to be associated with charred hazelnut remains, and occasionally a few tools made from other lithics. Many of the flint clusters at the sites were situated outside the zones containing pit hearths, so that, regarding these two aspects, there is some spatial differentiation. Although this might suggest the existence of different activity zones, there are no indications for the more or less simultaneous use of these two zones. Some arguments have been put forward in for this.[341] For example, the oldest flint clusters predate the oldest pit hearths, while the youngest pit hearths are younger than the youngest flint clusters. Furthermore, the number of pit hearths increases in the Late Mesolithic,

---

334 See chapter 6 for more information on changes in the vegetation.
335 We should note that not all of the phenomena labelled 'pit hearths' or 'charcoal pits' by archaeologists will have had the same function.
336 Other early dates – in the first half of the Boreal – have come from Almere-Zenit (charred hazelnut remains) and Zeewolde-OZ50 (charcoal). In contrast to Dronten-N23, the dates from these two sites are from borehole samples and an incidental observation of an 'isolated' pit hearth respectively. As a result, nothing further is known about the context from which the dated samples were taken.
337 Peeters 2009; Raemaekers & Niekus 2009.
338 Bailey 2007.
339 Peeters 2007, 2010; Wansleeben & Laan 2012a.
340 The highly-fragmented nature of the burnt bone makes it almost impossible to determine species (Van Dijk 2012, 561).
341 Müller *et al.*, 2012, 397.

Site/context	Artefact scatter	Refuse layer	Pit hearth	Surface hearth	Water pit/pit	Ditch	Stake row/stake cluster	Fence	Houseplan	Agricultural fields	Fish weir/fish trap	Wild mammals	Birds	Fish	Livestock	Cereals	Small spectrum	Broad spectrum	Hunting	Catching of fish	agriculture/animal husbandry	Domestic activities	Grave/graves	Depositions
Dronten-N23 Middle Mesolithic	V		V	V														V			V			V
Almere-Zwaanpad Mesolithic	V		V											V			V	V			V			
Hoge Vaart-A27 Mesolithic (phase 1)	o		V																?		?			
Swifterbant riverdunes Mesolithic	V		V																V		?			
Urk-E4 Mesolithic	o		V																V		?			
Hanzelijn-Drontermeer Mesolithic			V																					
Dronten-N23 Late Mesolithic/Early-SW	o		V																V		?		V	
Hoge Vaart-A27 Late Mesolithic (phase 2)	o		V																?					
Hoge Vaart-A27 Early SW (phase 3)	V	V		V	V		V					V	V	V				V	V			V		V
Swifterbant-S2 Classic SW	V	V							V							V					V	V	V	
Swifterbant-S3 Classic SW	V	V						V	V							V					V	V	V	
Swifterbant-S4 Classic SW	V	V							V							V					V	V	V	
Swifterbant-S5 Classic SW	V	V									V								V					
Hoge Vaart-A27 Classic SW (phase 4)									V	V									V					
Emmeloord-J97 SW		o																				?		
Urk-E4 SW	V	V							?		V	V	V	V	V			?		?	V		V	V
Schokkerhaven-E170 SW		V																				V		
Swifterbant-S25 SW/Pre-Drouwen		V																						
Schokland P14 SW/Pre-Drouwen	V	V			V		V		V			V	V	V	V							V		
Schokland P14 TRB	V	V			V																	V	V	V
Schokkerhaven-E170 TRB		V					V															V		
Emmeloord-J97 TRB		o								V							V			V		?		
Schokland P14 EGK/KB		V	V	V	V	V			V			V	o	?	V	V			V		V	V	V	
Almere-Stichtsekant KB?										V									V					
Schokland P14 WKD		V	V	V								V	o	V	V	V						V		
Emmeloord-J97 WKD		o									V								V		V	V		

- diffuse finds layer (without clear concentrations)
- Isolated, functional phenomenon
- Spatially structured unit
- Refuse categories
- Functional-technologic spectrum
- Ritual-ideologic phenomena

Table 4.11: Features and archaeological remains discovered at several sites in Flevoland.

whilst the number of flint clusters decline. Finally, pit hearths are a continually-occurring phenomenon, though there are several short hiatuses around the time that dates from hazelnut associated with the flint begin to occur again. Such discrepancies have also been identified at other sites outside the region (see chapter 7). There is, therefore, evidence for a partial overlap in the date range of flint clusters and pit hearths, but the activities associated with these phenomena may not have been carried out simultaneously.

As we have said, the flint clusters were associated with charred hazelnut remains and, in some cases, with tools made of other lithics and a small amount of burnt bone (fig. 4.33). However, the clusters vary in size, and

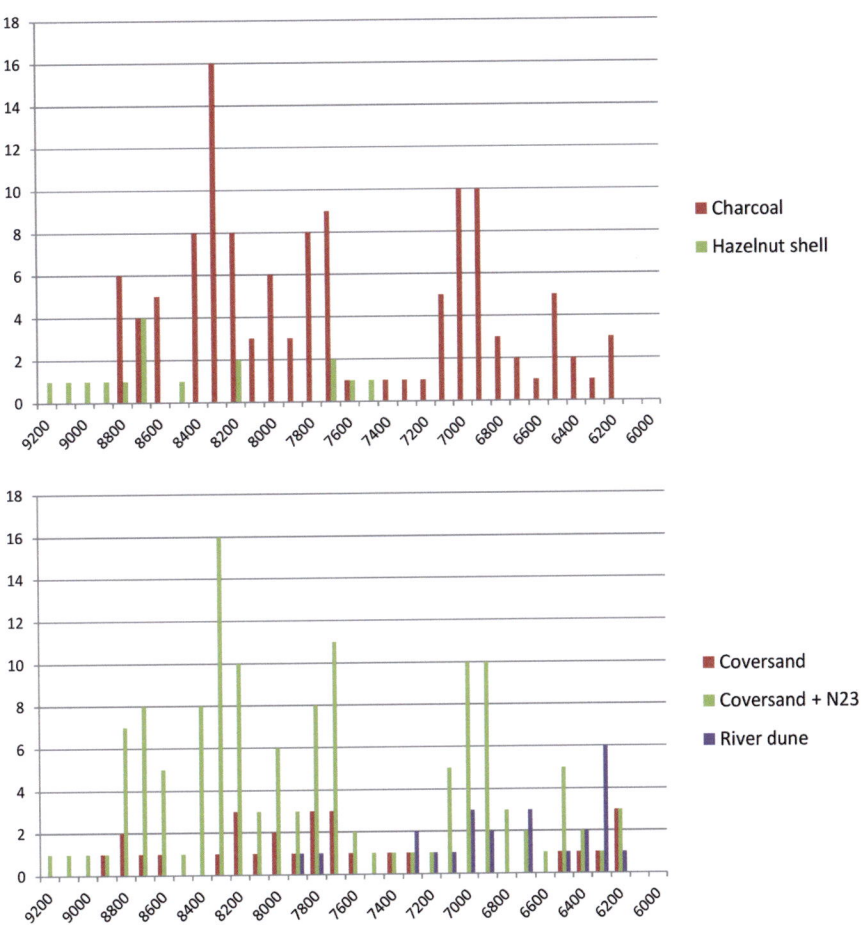

Figure 4.32: Top: number of ^{14}C dates based on charcoal and charred hazelnut fragments; the charcoal dates usually come from pit hearths. Bottom: number of ^{14}C dates related to landform; an important number of dates of the coversand locations come from Dronten-N23. Remarkably, the dates of mesolithic habitation on river dunes are younger than 8000 BP. Because these dates relate to pit-hearths, it cannot be ruled out that this observation is solely a reflection of activities on river dunes during which pit-hearths were created. It is quite possible that river dunes were previously used in other functional contexts, which, however, are not reflected in the ^{14}C dating.

sometimes appear to 'merge', producing an amorphous pattern of variable densities of material. Larger, relatively high-density clusters contain smaller concentrations with relatively more visible material. Such patterns can emerge due to the random aggregation and accumulation of material that repeatedly lands on the surface every time the location is used or some other event occurs.[342] The ongoing process of accumulation results in more or less repeating patterns.[343] Importantly, understanding this process can help to form an idea on the systematic nature of the underlying activities.[344]

An analysis of the flint material within the extensive assemblage at Dronten-N23 has led to the identification of at least two spatially-distinct zones each with a different functional basis. The tool assemblage in the first context comprises approximately 50% microlithic points and small blades with steep retouching, while the rest is made up of scrapers and retouched flakes (concentrations 1, 2 and 5). The second context is dominated by microlithic points (80-85%), while scrapers are relatively rare (concentrations 3 and 9). It is possible that the second context represents a more specialist use of the location, perhaps related to hunting, whilst the first context appears to be related to carrying out more domestic activities.[345] It is important to note, however, that use-wear analysis has made it clear that the typological categories are functionally heterogeneous. The microliths which have traditionally been regarded as

---

342 Wansleeben & Laan, 2012a, 109-111.
343 Peeters, 2007; 2010; Wansleeben & Laan, 2012a.
344 Peeters 2007, 2010.

345 Niekus, Knippenberg & Devriendt, 2012, 240.

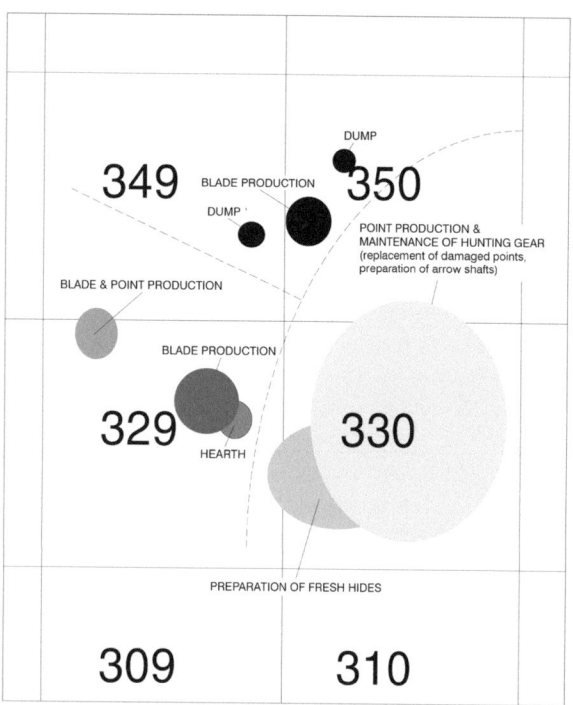

Figure 4.33: Almere Hoge Vaart-A27 phase 3: interpretation of activity areas in the northern concentration (source: Peeters 2007).

projectile points now turn out to have been used for all manner of purposes. The sharp retouched blades would have been used for cutting and piercing, for example.[346] Unmodified flakes and blades were also frequently used as tools for many different purposes. Only scrapers appear to be reasonably function-specific, and are linked mainly to the working of hides. The typological variability in an assemblage does not therefore appear to represent the functional variability, which means that our hypotheses on how locations in the landscape were used are almost certainly too simplistic.[347]

The partially-investigated Middle Mesolithic site at Almere-Zwaanpad shows similarities with the microlith-dominated context of Dronten-N23. At Almere-Zwaanpad almost 90% of the modified artefacts can be classed as microliths. In addition, the processing waste on the site contained many microburins that indicate microlith production. As such, it is likely that this site was connected with hunting activities. As no use-wear analysis was carried out on the artefacts, it is not clear to what extent the problem of functional variation also applies here. The flint distribution at Almere-Zwaanpad appears to be associated with a surface hearth, close to which charred remains of hazelnut shells and fish were also found.

Given the small size of the area investigated, it is also possible that the relative predominance of microliths and associated production waste represents only part of the activities carried out in the site.

As indicated above, pit hearths are present in abundance in Mesolithic contexts – Hoge Vaart-A27 phases 1 and 2, Urk-E4, Swifterbant river dunes – though it is generally difficult to determine a link between these phenomena and scatters of mobilia. Excavated pit hearths generally contain little material besides charcoal and other charred plant remains. Fragments of hazelnut shells and inorganic material, such as flint, are relatively rare. This might suggest that pit hearths were not used near to other activities practised at the site.[348] It is, however, also possible that pit hearths were created and used in a context that was spatially distinct from other activities. As has previously been stated, the latter is a more likely scenario at Dronten-N23. At Hanzelijn-Drontermeer, apart from 38 pit hearths, no other finds were recovered except for two pieces of flint.[349] This suggests that the activities associated with pit hearths were not necessarily related to settlement sites.

We can conclude on the basis of the data available that the Mesolithic sites on river dunes and coversand ridges were made up of temporally distinct find assemblages that can be linked to various behavioural contexts.

Distributions of charred hazelnut shells, bone – if preserved – and flint with a relatively narrow spectrum of morphological and functional categories associated with surface hearths. It appears that this type of context is likely to be associated with the short-lived specialised use of sites, though use-wear analysis is needed to show whether the diversity of functions is as narrow as the typological variability suggests (Almere-Zwaanpad, Dronten-N23 concentrations 3 and 9).

Distributions of charred hazelnut shells, bone – if preserved – and flint with a relatively broad spectrum of morphological and functional categories associated with surface hearths. This type of context is associated with the more or less short-lived use of sites in a domestic setting (Dronten-N23 concentrations 1, 2 and 5). The functional diversity is probably greater than the typological variation in flint and other lithic tools suggests.

Pit hearths that occur in isolation or in clusters and are (probably) not associated with activities for which flint was used, or in which animal remains were deposited or simply discarded in the fire. It is not clear what functional diversity exists within the 'pit hearths' category. Some of them may have been used to produce wood tar, and

---

346 Siebelink *et al.*, 2012, 266.
347 Cf. Siebelink *et al.*, 2012, 259.

348 Cf. Groenendijk 1989.
349 Prangsma 2009. It is however possible that there was originally more flint there which has disappeared due to erosion.

possibly also for preparing plant-based food, although no direct evidence of this has been found to date.

Since only sites on river dunes and coversand ridges have been excavated so far,[350] it is not possible to conclude whether the nature of the Mesolithic activities within these landscape units is also representative of the plains and gently undulating parts of this coversand landscape. On the basis of ^{14}C dates for archaeological material – charcoal from pit hearths and charred hazelnut shells – there does seem to be an emerging difference in terms of when certain activities first started, on the coversand ridges on the one hand and on river dunes on the other (fig. 4.32). The earliest datings – c. 8400 cal. BC – fall in the second half of the Preboreal, and are related to charred hazelnut shells from coversand sites. These dates should probably be linked to surface hearths. The earliest dates for pit hearths on coversand ridges fall in the first half of the Boreal. The earliest dates for river dune sites – c. 7300 cal. BC – are based on charcoal from pit hearths, and these fall in the second half of the Boreal. A considerable proportion of the dates from coversand sites come from Dronten-N23. Even if data from this site are excluded, the river dune sites still appear to begin later than those sites on coversand ridges. This may be connected with the formation of the river dunes themselves. These geomorphological units were formed not only in the Late Glacial but also in the Early Holocene.[351] The relatively late 'start date' for activities on river dunes in Flevoland may be an indication that these dunes formed later.[352]

4.9.2 Swifterbant, Pre-Drouwen and Funnel Beaker occupation

River dunes and coversand ridges that evidently played a role in the exploitation of the landscape were also important in the context of Swifterbant occupation. In contrast to the Mesolithic, information on the exploitation of other landscape zones – the river flood plain, lowland peat bogs and glacial till outcrops – is available for the Swifterbant period. Thanks partly to the generally better preservation conditions in which remains of Swifterbant occupation have been found, we have a more differentiated impression of the nature of the occupation. In chronological terms, however, our picture of the Swifterbant period is still unbalanced due to the underrepresentation of sites from the early phase.

Early Swifterbant

The earliest well-documented Swifterbant context is Hoge Vaart-A27 phase 3, situated in the catchment basin of the river Eem.[353] The coversand ridge, that was still used at the end of the Mesolithic for activities associated with pit hearths, was used again after a brief interruption between c. 5000 and 4900 cal. BC for an entirely different functional constellation (see Chapter 3). The coversand ridge was a relatively dry headland within what was by now a wetland environment. A more or less isolated concentration of flint was found in a slightly lower-lying part of the sandy ridge, in association with a surface hearth (fig. 4.33). The analysis of the flint artefacts has shown that blades were knapped near the hearth as a basis for the production of trapeze points.[354] Those points were also produced here. In another zone close to the hearth, the assemblage consisted of mostly tools, mostly damaged trapezia, scrapers and used blades. Use-wear analysis has shown that some points were damaged due to their use as projectiles.[355] Scrapers were used exclusively to clean fresh hides, while blades were also used to work soft plant material. Several small fragments of pottery and some burnt bone were also found in the same 100 m² area. Charred fragments of hazelnut shells were also found in the hearth. These fragments have been dated to between 4790 and 4550 cal. BC.[356]

A large concentration (approx. 50 x 15 m) of occupation remains was found on the higher part of the sandy ridge, within which 120 surface hearths have been documented in association with large quantities of flint, burnt bone, and limited quantities of potsherds. Crushed granite, quartz and sandstone boulders were also found. Interestingly, the composition of the flint assemblage in this larger concentration is similar to that of the smaller concentration described above. An in-depth analysis of spatial patterns has produced convincing evidence that the large concentration essentially consists of an aggregate of occupation remains that were repeatedly abandoned in a functional context similar to that described above for the more isolated smaller concentration.[357] In this context, the maintenance of the hunting inventory and the primary processing of hides played an important role. If every 'instance of use' was associated with a single hearth, the

---

350 Mesolithic flint in the form of microlithic points has also been found at Schokland-P14 (Van der Kroft 2012). Trapezia have also been collected from this find spot, as well as regular blades that appear to have been produced using the punch technique. This material can be dated to the Late Mesolithic, and also to the Swifterbant period. Since little contextual information has been published on the Mesolithic, we cannot draw any further conclusions as to the nature of this phase of occupation. Ten Anscher (2012) looks in detail at the nature of the Swifterbant period.
351 Peeters *et al.* 2015, 303.
352 This requires investigation, however.
353 Although several ^{14}C dates and some of the flint suggest that activity occurred at Schokland-P14 in the early Swifterbant phase, there are no reliable contexts on which to base any conclusions about the nature of this occupation (cf. Ten Anscher 2012, 425).
354 Peeters & Hogestijn, 2001; Peeters, 2007.
355 Peeters, Schreurs & Verneau, 2001.
356 GrA-21376: 5820 ± 50 BP
357 Peeters & Hogestijn 2001; Peeters, 2007, 2010 (CAA paper); Merlo, 2010.

large concentration represents some 120 'instances of use' between c. 4900 and 4550 cal. BC.

Although the maintenance of hunting gear and the primary processing of fresh hides are the dominant functions represented in the flint component from Hoge Vaart-A27 phase 3, the site was also used in another context. It was suggested in section 4.8.2 that there is strong evidence for the local production of pottery. The poor quality of the fired ware and the fact that the pots had often broken on the spot (fig. 4.34) suggests that the pottery was produced for short-term local use.[358] It is, however, not known what the pots were used for, although $\delta^{13}C$ and $\delta^{15}N$ values in encrusted residues suggest they may have been used to prepare aquatic foodstuffs, amongst other things.[359] The presence of mattocks could be associated with the loosening and extraction of the starch-rich rhizomes of aquatic and marsh plants in the adjacent low-lying landscape hollow. In addition, a pit that had gradually filled with sandy and peaty material has been interpreted as a natural well.[360] Activities of a ritual nature on and around the sandy ridge, which was further 'engulfed' by the expanding marshland during phase 3, include the deposition of flint artefacts (see Chapter 5).

Although evidence of local production and use of pottery, and of ritual activity, could be interpreted as an indication of long-term habitation at this site, no clear traces of shelters have been found. It was not possible to identify any configurations within the horizontal distribution of post and stakehole features that could be interpreted as a dwelling structure.[361] Very few heavier posts with a diameter of more than 10 cm have been found, whilst elsewhere wood from such posts often remains preserved. Two oak posts were found on the banks of the gully, close to two auroch skulls. Another heavy post that had been driven through the peat into the sandy substrate was found close to one of the flint deposits. Dendrochronological analysis placed the death of the tree at c. 4646 BC, which is close to the age of the reed peat through which the stake was driven.[362]

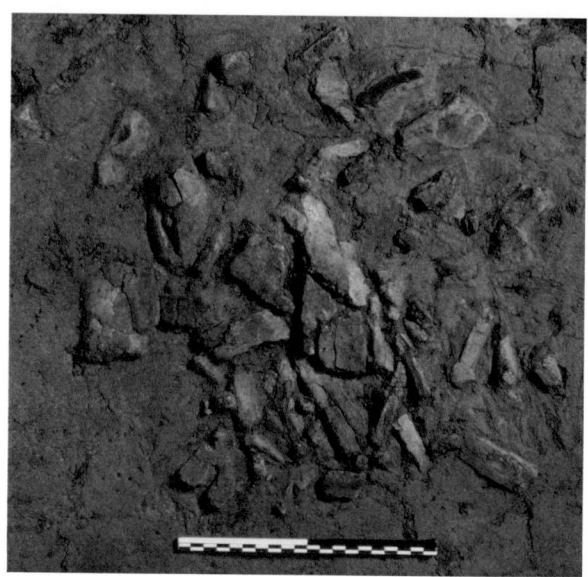

Figure 4.34: Almere Hoge Vaart-A27 phase 3: a broken vessel (Early Swifterbant).

The data available strongly suggest that use of the site during the third phase was characterised by a context in which activities focused on short-term use. Some of these activities were certainly related to hunting, but it is also likely that food was prepared and consumed at the site. Plant-based food would probably have been gathered in the immediate surroundings. The evidence of ritual activity indicates that the Hoge Vaart-A27 site was not exclusively of economic significance.

Classical Swifterbant, Pre-Drouwen and Funnel Beaker

In the catchment basins of the prehistoric Overijsselse Vecht and Hunnepe rivers occupation remains from the Classical Swifterbant phase are located on the 'levees' in the flood plain, on river dunes and on the glacial till outcrop at Schokland. These last two contexts also contain occupation remains from the Pre-Drouwen and Funnel Beaker phases. Interpreting the occupation remains on the river dunes and till outcrop is no simple matter because of the presence of earlier and also often later occupation remains. Furthermore, the highest parts of almost all river dunes and sandy ridges with till deposits located in the subsoil have been eroded. This is less of a problem at the 'levee sites' which, given their specific landscape context, were used for only a relatively short time (see also section 4.4.5).

As reported in the previous section, no structures from the Early Swifterbant phase have been found that could be interpreted as the ground plans of houses or lighter forms of shelter. This is not to say that there were no houses or other structures, as we do know of one well-documented context from this early phase. The picture is different,

---

358 Peeters 2010 (ASLU).
359 Ref. Crombé/Van Strydonck. It should be noted that the analysis of several sherds has shown that cracks and pits on the inside of the pot appear to have been filled with a black, highly glossy graphite-rich substance, which might be associated with tar (Jansen & Peeters 2001, 45). Further investigation is required to determine the function of the pottery.
360 Hamburg et al. 2001, 14.
361 Hamburg et al., 2001.
362 Two ^{14}C dates for the peat give its age as between c. 465 and 4330 cal. BC (Peeters et al. 2001, 8). The dated peat samples were however taken from a point higher than the peat around the dated oak post. Based on the depth of the sand around the post and the dating of other peat samples, it is likely that the peat started to grow around 4700 cal. BC.

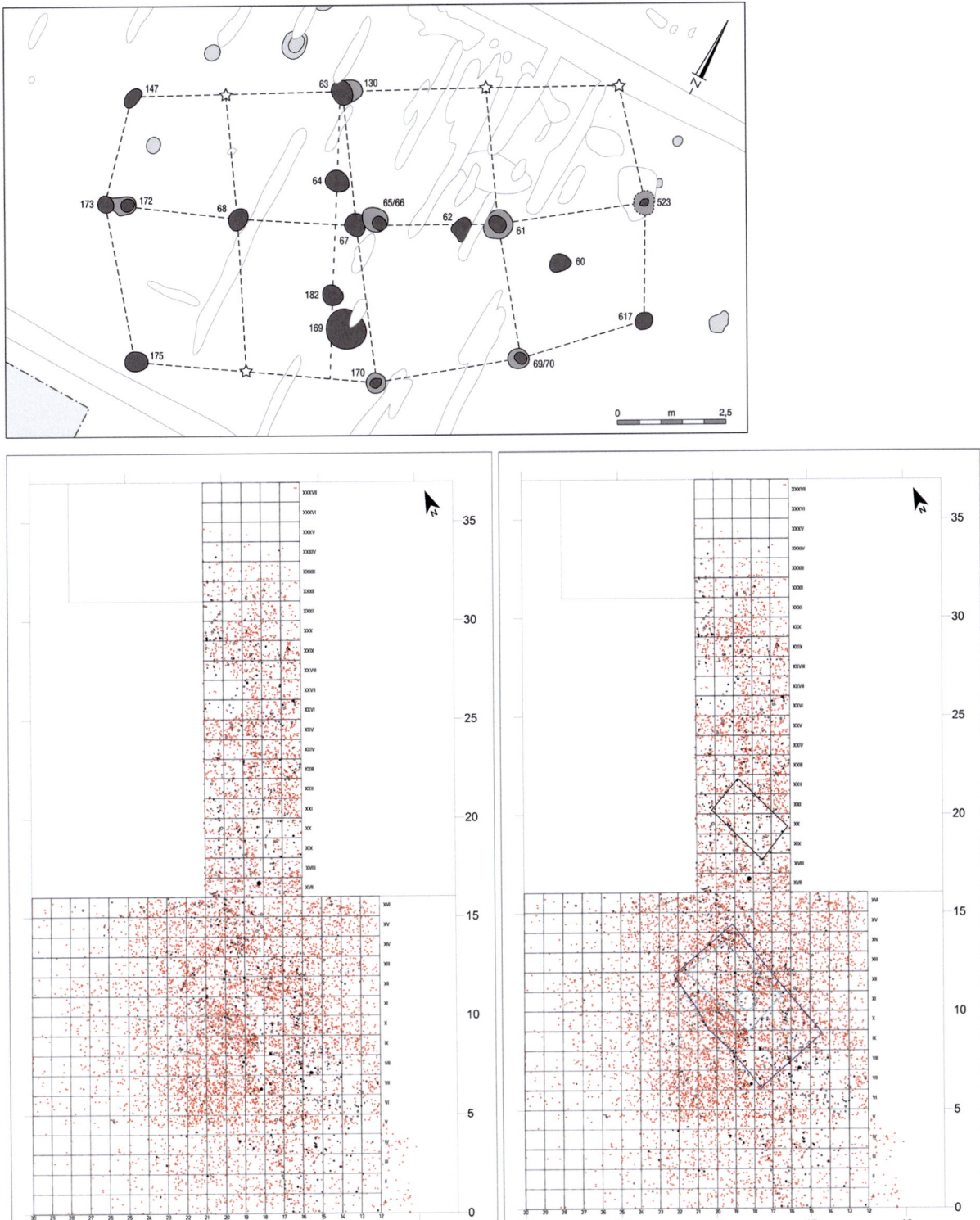

Figure 4.35: Top: Schokland-P14 houseplan (Pre-Drouwen phase) (Ten Anscher 2012). Bottom: distribution of flint (red) and postholes (black) and postulated houseplan Swifterbant-S3 (source: Devriendt 2014).

however, when it comes to the Classical Swifterbant and Pre-Drouwen occupation. At Schokland-P14 four structures have been identified that have been interpreted as house plans and assigned to the Pre-Drouwen phase, although a Swifterbant date cannot be ruled out (fig. 4.35).[363] Two structures are more or less complete; the two other structures, whilst incomplete, show many similarities in layout, however. The two-aisled structures were some 12-13 metres long and 6 metres wide, with sunken central posts approx. 25-30 cm in diameter. The ground area was slightly trapezoidal in shape and the walls of each structure would have bowed outwards slightly. These fairly sturdy buildings are regarded as permanent structures, though it is not clear whether they existed simultaneously.[364]

One or more lighter structures may have stood at the riverbank site of Swifterbant-S3 (Classical Swifterbant phase). An 8 x 4.5 metre rectangular structure containing a surface hearth can be distinguished on the basis of linear configurations in the distribution of posts, stakes and material remains (pottery, flint, stone) (fig. 4.35).[365] The overall shape and slightly bowed walls are similar to the houses at Schokland-P14.[366] The hearth had been remade several times on the same spot, indicating that it was used repeatedly and that the structure was maintained.[367] The close proximity of several linear post and stake configurations and patterns in the distribution of waste might also suggest that the structure was rebuilt.[368] A smaller structure may be present several metres from this one, though the evidence for this is less convincing.

If we assume that the heavy structures at Schokland-P14 are dwellings then, along with evidence of arable fields, these may constitute an indication of a settlement of a more permanent character at this relatively high-lying site beside the prehistoric river Overijsselse Vecht.[369] Although there is no direct evidence, the extensive river dune at Schokkerhaven-E170 may also have had a more settlement like function. A palisade may have been present here during the Funnel Beaker phase, which might suggest there was an enclosed settlement at the site.[370] There is no evidence of any such structure in the Funnel Beaker period at Schokland-P14.[371]

Although there is evidence of a lighter (possibly dwelling) structure at Swifterbant-S3, the general picture of occupation at the riverbank sites suggest repeated short-term use. The research carried out over the past few decades has also shown that any interpretation of how these sites were used is anything but straightforward. A varied pattern of functions has emerged which, during the relatively short periods of use of the locations, changed over time.

It now appears that several levee sites were initially used as 'dwelling areas'. For this purpose the sandy clay substrate was covered with plant material (reeds).[372] After what was probably a short period of activity, on locations such as Swifterbant-S2, S3 and S4, the previously applied layers of plant material were subsequently dug into the ground and the locations were given over to growing crops (figure 4.36). Frequent flooding from the neighbouring gully deposited layers of clay on the fields, making them unworkable. Layers of plant material were then once more laid over the ground. The nature of the activities that subsequently took place at these sites appears to be varied, though proper interpretation is not always possible. Furthermore, human graves have been found at a number of levee sites – Swifterbant-S2, S4, S11 – suggesting these sites had some ritual or ideological significance.

Interestingly, human graves that can be linked to the Classical Swifterbant phase are also present at the river dune sites of Swifterbant-S21 and S22-S23, though there is no evidence of simultaneous occupation there (see also chapter 5). A similar situation occurs at Dronten-N23, where an isolated grave was discovered.[373] These are river dunes with evidence for occupation activity during the 'dry' Mesolithic, but which appear to have acquired another function and significance in the 'wet' Swifterbant phase. Burials also took place in higher parts of the landscape where there were still occupied settlement sites – Urk-E4, Schokland-P14.

Our overall picture of Pre-Drouwen and Funnelbeaker occupation in the Flevoland part of the Overijsselse Vecht and Hunnepe river basin is very unclear. Landscape features that remained accessible for a fairly long time thanks to their elevation remained in use as the water table rose in the surrounding landscape. The river dune at Urk-E4 became overgrown with peat at the start of the Funnelbeaker period. Only at Schokkerhaven-E170 and Schokland-P14 were the remains of settlement activity found dating to the Funnelbeaker period. The remains from Schokkerhaven-E170 include the possible palisade (mentioned above) A grave at Schokland-P14 probably also dates from this period, as well as the probable ritual deposit of a small wooden bowl.[374] Some Funnelbeaker pottery has also been found at Emmeloord-J89, and one ^{14}C date from Emmeloord-J78 fits in the period. On the whole, however, only a very limited amount of diagnostic

---

363   Ten Anscher 2012, 375-382, 426, 428.
364   Ten Anscher 2012, 376.
365   De Roever, 2004; Devriendt, 2014.
366   Ten Anscher 2012, 383.
367   De Roever 2004, 100.
368   Devriendt, 2014.
369   Ten Anscher 2012, p 427-428.
370   Hogestijn 1990.
371   Ten Anscher 2012, 525.

372   Huisman & Raemaekers 2014.
373   Hamburg et al. 2012.
374   Ten Anscher 2012, 433.

Figure 4.36: Swifterbant S2: hoe-field in horizontal view (A) and cross-sections (B-C). C is a detail from B clearly showing material from the dark cultural layer embedded in clayey levee deposits (source: Huisman & Raemaekers 2014).

Funnelbeaker material has been found. Several fish weirs and associated traps at Emmeloord-J97 that have been dated to the Funnelbeaker period do suggest that the river catchment area was a part of the regularly exploited landscape. Fish weirs are, after all, 'permanent' structures that require maintenance.

The prominent occurrence of multi-purpose sites from the Classical Swifterbant Culture and the subsequent Pre-Drouwen and Funnelbeaker occupation in the Overijsselse Vecht and Hunnepe river basin contrasts sharply with the Eem river basin. Apart from the fish weirs found at Hoge Vaart-A27 phase 4, which are contemporaneous with the levee sites at Swifterbant, no further remains of activity have been found for this period in this region. Only a pot found with a stone axe on the southern edge of Flevoland[375] has been dated to the end of the Pre-Drouwen phase,[376] although it is not clear whether this date is the result of a reservoir

---

375 Vlierman 1985.
376 Peeters 2007; see also Ten Anscher 2012, 126, note 65.

effect.³⁷⁷ This 'blank spot' on the map could be the result of the limited attention given to stratigraphically younger palaeolandscape units, or the strong focus that has been placed on investigating the coversand landscape. On the other hand, it is equally likely that the picture is connected with palaeogeographical developments within this part of the landscape (see Chapter 6 for further discussion).

4.9.3 Late Neolithic and Early Bronze Age

Outside the Noordoostpolder, no specific evidence for the Late Neolithic (Single Grave culture, Bell Beaker culture) and the Early Bronze Age (Barbed Wire Beaker culture) has been found in Flevoland. This is as a result of the erosion of the peat landscape (see also chapter 3).³⁷⁸ The incidental presence of Bell Beaker sherds at Swifterbant-S2 suggests that activities also extended as far as this area.³⁷⁹ A ¹⁴C date for a fish weir near Almere is also contemporaneous with the Bell Beaker Culture, and provides evidence of fishing in the Eem river basin.³⁸⁰ The most informative sites are located in the Noordoostpolder. Schokland-P14 and Emmeloord-J97 are the most prominent, though the picture presented by the latter is fragmented. Schokland-P14 illustrates well how the function of the excavated site changed. The rising water levels lead to a decrease in the amount of available land and the function of the excavated terrain changed.

There was probably a settlement, possibly permanently occupied, from the Late Single Grave culture and Early and Late Bell Beaker culture on the sandy ridge on which Schokland-P14 is located.³⁸¹ Evidence for this settlement consists of pottery interpreted as cooking pots and storage jars, a cemetery and a field system that may have been related to a ditch system that is assumed to have bordered on a trackway.³⁸² This trackway may have served as a cattle path along which animals would have been driven from the meadow to a watering hole or ford, as suggested by the presence of cattle hoof impressions in and between the ditches and on the banks of the former Overijsselse Vecht. No evidence of dwelling structures was found in the excavated area, though they probably lie on a higher part of the sandy ridge.³⁸³ From this perspective, the investigated part of the site can be interpreted as the peripheral zone of a much larger occupied area,³⁸⁴ within which people lived, prepared and consumed food, grew cereal crops and kept livestock, produced and used pottery and buried their dead.

The rising water levels would, however, have changed the usage of the investigated area.³⁸⁵ Growing crops must have become increasingly problematic, and fields would have had to be created on higher parts of the ridge. The cemetery also appears to have been abandoned. The ditch system may still have functioned (at least partly) as a drainage facility, while the levee zone was reinforced with boulders near the start of the possible trackway. In somewhat drier periods hearths were laid out along the levees, at which locations animal and plant-based food was processed and consumed. These activities may have been associated with seasonal pasturing close to the Overijsselse Vecht, for example, and the watering of livestock on the banks of the river.

A series of fish weirs and traps at Emmeloord-J97 represents a specific economic activity associated with the Bell Beaker occupation of the prehistoric Vecht river basin. Some of the bone and antler tools – including fishing hooks (see section 4.8.3) – should perhaps be assigned to this period. Fragments of bell beakers and pot beakers show, however, that here, too, activities associated with a settlement took place.³⁸⁶ The problematic context in which these finds were made, however, makes it impossible to obtain a clearer idea of the nature of the occupation. Nevertheless, a complete dagger made of Helgoland flint and of the Scandinavian type – an item imported from northern Germany or Denmark -, and part of a pot beaker put over the post of a fish weir, might also suggest ritual activity.

The occupation at both Schokland-P14 and Emmeloord-J97 continued into the Early Bronze Age, with the Barbed Wire Beaker culture. The picture does not become any clearer, again as a result of the erosion of the landscape. Actual activity remains that can provide any information about the nature of the occupation – other than potsherds or other objects -have been found only at Schokland-P14.³⁸⁷ Occupation along the levees of the Vecht was probably short-lived, and consisted of activities involving the hearths and cooking pits. The locations were used for the preparation and consumption of food, as well as the processing of raw materials such as bone, hides and flint. The riverbank vegetation was regularly burnt down to provide access. Human footprints found in the riverbank zone clearly show that both adults and children were present (fig. 4.37). Cow hoofprints also show the presence of cattle. Evidence for Early Bronze Age activity was also found further from the gully. It is likely that seasonal pasturing of livestock occurred in higher parts of

---

377 Raemaekers 2005.
378 Raemaekers & Hogestijn 2008.
379 Raemaekers & Hogestijn 2008
380 Van de Geer 2013, 25-27; Hogestijn 2019.
381 Ten Anscher 2012, 461-462.
382 Ten Anscher 2012, 403.
383 Ten Anscher 2012, 462.
384 Ten Anscher 2012, 463.

---

385 Ten Anscher 2012, p 462-463.
386 Bloo 2002.
387 Ten Anscher 2012, p 465-486.

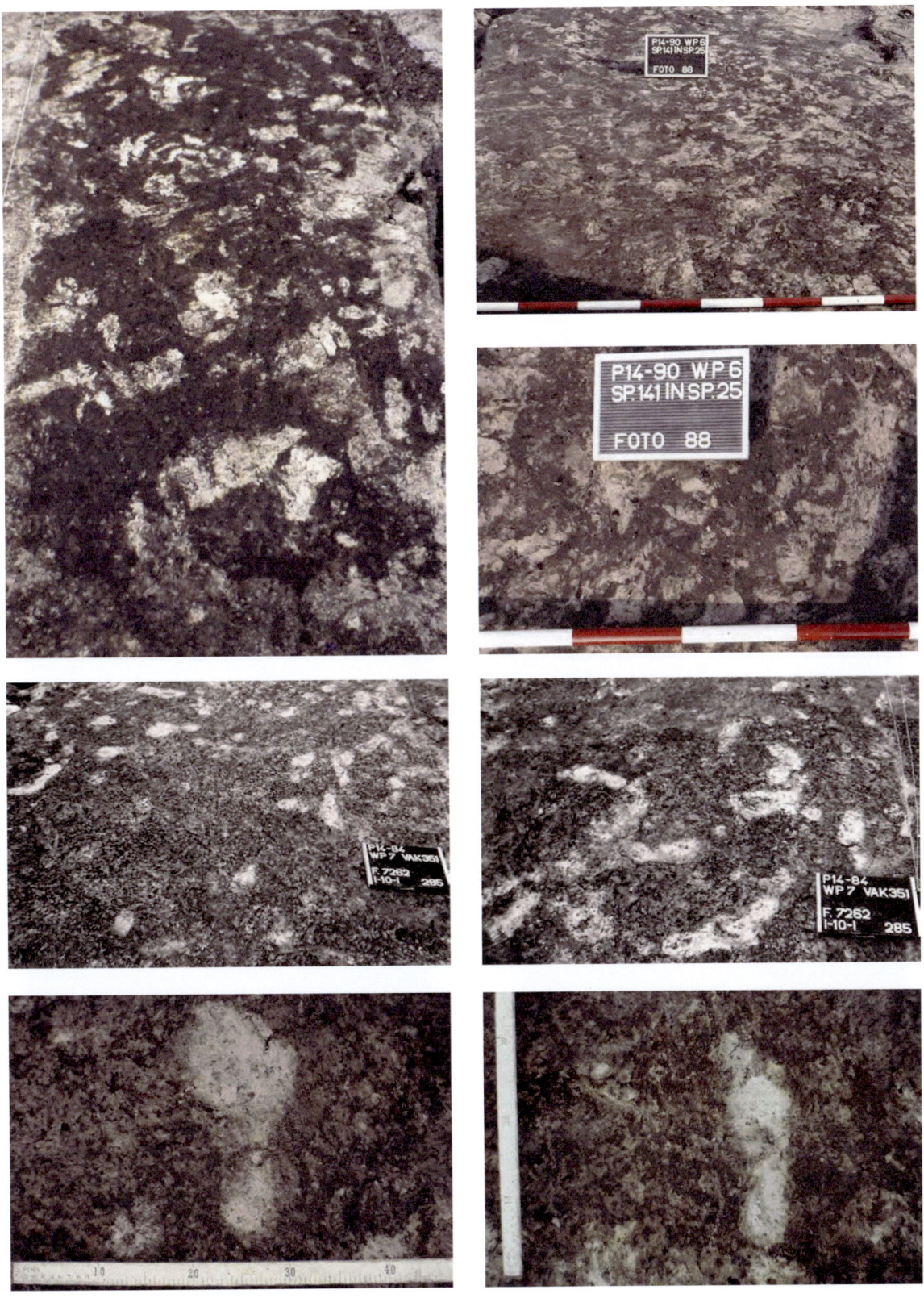

Figure 4.37: Schokland-P14: cattle-hoof and human foot imprints (Barbed Wire phase) (source: Ten Anscher 2012).

the landscape, and that animals were accompanied to the river to drink.[388]

4.9.4 People on the move

Determining the use of specific sites in the landscape cannot be seen separately from the movement of individuals and the transportation of materials. The connections (travelling routes) between sites are at least as important as the sites themselves. These connections played an important role in the spread of information, the supply of all kinds of resources, and contact with other, both known and unknown, groups.[389] Direct archaeological evidence for such routes is generally difficult to find. Vessels such as canoes and boats are found only in favourable preservation conditions, and only provide evidence of transport by water. Evidence of terrestrial routes is even more difficult to find. Other indirect evidence for travel can be found in 'exotic' raw materials or objects, or in the isotopes in bone remains.

It is only logical that people would have moved around between the locations they used. Signs of this lie in evidence of treading and trampling, as found in various contexts at Schokland-P14.[390] The questions, however, are not only how much did people move between sites, but how did they move? In a landscape growing gradually wetter, waterways would certainly have played a key role. The network of larger and smaller watercourses would have assumed a different character and pattern over the course of time, with larger bodies of water – lakes – forming as a result of the structural rise in the water level (see also Chapter 6). No canoes or paddles from the Mesolithic have been found in the region, although it is highly probable that this is at least partly a result of preservation conditions. By contrast, fragments of four paddles (fig. 4.17) have been found in the Classical Swifterbant contexts of Hoge Vaart-A27 phase 4 and Swifterbant-S5 and S25. All of them were found in gullies and provide direct evidence for transportation by water.[391]

As the marshy, waterlogged areas expanded, it would have become increasingly difficult to move around via terrestrial routes, although this did not mean travel was by then exclusively by water. Measures would certainly have been taken to provide access to areas by laying tracks using branches, logs or planks. Although no evidence of this has been found in Flevoland, we do know that this occurred in other marshy areas of the Netherlands after the Funnelbeaker period. One key factor in the accessibility and traversability of marshlands would have been the type of vegetation growing there. Dense carrs are for example difficult to access and cross. In such an environment, connections between sites and the transportation of people and materials would have necessarily been mainly by water. Very wet sedge peat bogs can also present an obstacle. However, as on upland peat bogs, laying trackways can have provided a solution in these environments.

Besides the physical accessibility of landscape zones, the distance to be covered would also have helped determine the choice of transport. An extensive network of waterways could facilitate connections between areas located far apart. In the study region, the main waterways were the Overijsselse Vecht, Tjonger (or Kuinder), the Hunnepe and the Eem rivers. These could provide access to several areas. That is not, however, to say that people travelled long distances only by water. In wet landscapes with extensive river systems, people can also travel overland, carrying a canoe from one river system to another, for example.[392]

We can gain some insight into how much an individual moved about during their lifetime from a study of the stable isotopes in tooth enamel and bone. Every individual builds up a 'signal' that reflects the geological conditions in their area of origin. This is determined by the geographical origin of the food and water they consume. By analysing stable isotopes in human dental enamel dating from the Classical Swifterbant phase, archaeologists have been able to determine that the populations investigated at Swifterbant-S2 and S3 were homogeneous and had local origins (fig. 4.13).[393] Since the measurements were taken from enamel, we have to conclude that all these individuals spent their early years in the region. Only one individual from Swifterbant-S2 had a signal that did not conform to the geology of the region, which indicates that this individual grew up elsewhere, or at the very least ate a different diet.

This picture of a homogeneous local population would seem to be in line with the general picture of the use of local resources for tools and ornaments, for example (see sections 4.8.4 and 4.8.5). Raw materials, such as different types of flint, mostly originated in the region, or were at least available there. Only a single object appears to have been made of 'exotic' material, which might indicate contacts with areas further away. How large the exploited 'region' actually was is difficult to say. The borders of present-day Flevoland are of course an arbitrary boundary. Isotope analysis has also shown that the diet of the Swifterbant population included a large amount of terrestrial and less aquatic food, even though conditions in the landscape were

---

388 Ten Anscher 2012, 484.
389 See various contributions in Whallon, Lovis & Hitchcock 2011.
390 Ten Anscher 2012, 477-478.
391 Casparie & De Roever 1992; Peeters 2007; Raemaekers *et al.* 2013/2014.

392 Lovis & Donahue 2011.
393 Smits *et al.* 2008; Smits & Van der Plicht 2009.

dominated by water (see section 4.7.3).[394] This might be an indication that individuals moved (perhaps seasonally) around a region much larger than the region covered by the province of Flevoland today (see also Chapter 7). Other factors are also important to consider: cultural factors: taboos, preferences, traditions. The availability of certain food sources in a region does not necessarily mean that all these sources would have formed a part of the diet.

4.10 Conclusions

The discussion above makes it clear that the archaeological record of Flevoland represents a range of activities associated with daily life. Although the data varies sharply in terms of both quantity and quality and in temporal and spatial terms, particularly as a result of variable taphonomic processes, it shows that exploitation of the landscape during the Mesolithic and Neolithic was actually based to a substantial extent on the resources occurring naturally in the region, both food and non-food resources.

The earliest evidence that livestock and cultivated crops formed part of the food economy is found in Classical Swifterbant contexts. Later in the Neolithic, food production appears to play a more prominent role, though hunting and fishing nevertheless remain significant. It is not, however, clear to what extent this slight shift towards food production influenced the settlement pattern, for example. Whether the Pre-Drouwen 'settlement structures' at Schokland-P14 are a reflection of more permanent occupation on the site remains unclear. Nor do we have a good idea of subsistence patterns in the Mesolithic, mainly because of a lack of well-preserved food remains from this period. Although Hoge Vaart-A27 phase 3 can be interpreted as a hunter-gatherer context (ceramic Mesolithic), this does not mean that it is representative of the entire Mesolithic, or even of the Late Mesolithic. Indeed, this seems unlikely.

As well as evidence of the incorporation of small-scale food production into the extended broad spectrum economy, other evidence exists for the exploitation of resources. The production of wood tar in pit hearths – if this did indeed take place – during the Mesolithic was a structural aspect of the technology. During the transition from the Mesolithic to the Neolithic, as the pit hearths disappeared, the production of pottery appeared to gain a foothold, although the background to this is largely unknown. Many aspects indicate both continuity and discontinuity in the use of the landscape This was an environment which, as will be explained in Chapter 6, was undergoing dramatic change. These changes in the landscape appear to have had some influence on the potential for exploitation. Of course the current arbitrary boundaries of the region did not exist in prehistory, and the region formed part of an extensive landscape in which rivers and pathways provided socially significant links. There is evidence of relations with regions further afield, albeit scarce. The extent to which the landscape had a sacred or ritual significance is explored in the following chapter.

---

394 Smits *et al.* 2008; Smits & Van der Plicht 2009.

Chapter 5

People, ritual and meaning

D.C.M. Raemaekers

5.1 Introduction

In 1984, Richard Bradley suggested that "in the literature as a whole successful farmers have social relations with one another, whilst hunter-gatherers have ecological relationships with hazelnuts".[395] With this proposition as a background, it could be argued that archaeological research in Flevoland is dominated by an ecological-deterministic and cultural historical paradigm. The many ecological sources of information available have led to a great deal of attention being paid to landscape, vegetation, existence and exploitation, while the research into material culture focused on the development of a chronological framework. These are all aspects that have also been discussed in other chapters in this book. This could give the impression that the prehistoric inhabitants of Flevoland were indeed exponents of the hunter-gatherer theory as outlined by Bradley.

As has been discussed in passing in the previous chapter, it is evident that for the prehistoric inhabitants of the region, which now comprises the three polders of Flevoland, more was important to them than just how to find the daily food supply. The archaeological record itself provides many starting points for a perceptive design of the cultural world by means of burial ritual, different treatments of human skeletal material, depositions of material culture and the conceptual connection between material culture and meanings (materiality). These aspects of Flevoland archaeology are central to this chapter.

5.2 Burial practice

5.2.1 Introduction

When an individual dies, there is a wide spectrum of actions that could be performed on or with the body, each with its own archaeological impact..[396] Practices that are most in line with modern rituals are those of interment (inhumation) and cremation. Evidence for these rituals after death in prehistoric Flevoland are discussed below. Given that, by definition, there is a grave in the case of inhumations and an archaeologically recognisable context by cremations, the term burial practice is used here. The archaeological data makes it clear that other rituals associated with the deceased must be taken into account: loose disarticulated bones have been found at many sites. These may originate from disturbed graves, from deliberately opened graves, or may be the result of excarnation practices, where the human body is de-fleshed by exposing it to the elements (and to animals). In addition, apparently selected, dislocated skeletal bones have been found in graves.

Thanks to the relatively good conservation conditions in Flevoland, a large number of prehistoric burials have been discovered (table 5.1). Nevertheless, the common thread

---

395   Bradley 1984, 11.
396   See, for example, Meyer-Orlac 1982, 139.

in publications dealing with the bone material is that the preservation of skeletal remains is poor.[397]. This apparent contradiction is the result of the specific genesis of the Flevoland landscape (see Chapter 3): no sedimentation took place until the very moment of inundation of the landscape, which means that sometimes hundreds of years of erosion and degradation may already have taken place before the preserving effect of covering sedimentary layers could halt the process.

The poor conservation of human bone material has various consequences for our dataset. Firstly, attempts on various sites to use ^{14}C dating methods on skeletal remains has not met with success. The amount of remaining collagen in the bones turned out to be to low to give a reliable age determination. Secondly, poor conservation leads to interpretation problems. For example, six "possible graves" have been recorded on the site of Schokland-P14.[398] These "possible graves" are features that resemble known graves in shape and size and from which, in some cases, the bones recovered were so badly degraded that it could not be established with certainty that they were actually human. The interpretation of these features as graves is therefore uncertain. However, the same conservation problems have affected the known graves. In these contexts, not only more or less complete skeletons have been found but, in a number of cases, only one or a few bones of an individual (for example, Swifterbant-S2-VIII-2 and Urk-E4-7-III). These are generally more robust skeletal parts, such as skulls and long bones. It is therefore sometimes extremely difficult to determine whether these are the remains of an extremely poorly preserved, but originally complete skeleton, or evidence for a *pars pro toto* burial, in which only a few selected, dislocated skeletal parts were buried. This latter phenomenon is possibly witnessed by the discovery of a skull at Swifterbant-S22, for example, in a 'grave' that was hardly bigger that the skull buried in it (Swifterbant-S22-I). Finally, poor conservation may lead to a limited number of grave goods: there are many conceivable organic grave goods that would certainly not have been preserved.

Where ^{14}C dates are missing, the Flevoland graves have been dated by association with the geological layers in which they have been found (Dronten-N23, Swifterbant-S2 and Swifterbant-S4, Urk-E4). The grave in Dronten-N23 cuts through a ^{14}C dated pit hearth (charcoal; GrA 6455 ± 40 BP: 5484 – 5340 cal BC (2σ)). The stratigraphical relationship indicates that the grave is, in any case, later that this date (*terminus post quem*). The latest possible date for the burial is based on a dendrochronological date from an oak tree. The good conservation of the oak led excavators to assume that the tree had been covered relatively quickly by sediments after falling. The dendrochronological date is 4799 cal BC.[399] This has led to a dating of the Dronten-N23 grave to between 5400-4800 cal BC (see table 5.1).[400] There are six ^{14}C dates available for finds from the cultural layer in Swifterbant-S2 and Swifterbant-S4.[401] Based on these dates, both sites can be placed in the period 4300-4000 cal BC. It is thereby assumed that the general dating of the graves also falls within this range (see below). The graves from Urk-E4 have been dated using a combination of three arguments. The strongest argument is the absence of any Funnelbeaker culture finds with the characteristic deep-grooved, or stab-and-drag decoration (*Tiefstich*), as well as three 14C dates from the peat layer that suggest that the top of the dune was covered at the latest in 3400 cal BC. This provides a *terminus ante quem* date for the burials. The second argument is the fact that skeletal material has been preserved. This in itself suggests that the graves should not be dated thousands of years before the formation of the covering peat layer. Thirdly, the pottery from the site appears to date from the period 4200-3400 cal. BC. It is therefore possible that, in this period, not only occupation but also burial took place.[402] At the site of Schokland-P14, two grave groups can be differentiated on the basis of ^{14}C dates. The undated graves have been assigned to these groups depending on the differentiation in preservation exhibited by the dated graves.[403]

It is important to note that all the graves have been discovered at sites that were also settlements. As a result, the possible contemporaneity of burial and settlement evidence has been the subject of discussion for Swifterbant-S2. Contemporaneity can be argued due to the fact that the graves were found in the cultural layer and the grave fill could not be distinguished from this. Moreover, there was no clay deposit recorded that stratigraphically separated the graves from the cultural layer and during the excavation no grave cuts could be identified on the surface.[404] There are also arguments that suggest that the cemetery may have been older than the settlement. The spatial distribution of the flint[405] and pottery[406] make it clear that the settlement activities took no account of the prior existence of any graves: the spread

---

397 Ten Anscher 2012, 313 (Schokland-P14); d'Hollosy & Baetsen 2001, 59 (Urk-E4); Baetsen & Kootker 2011, 147 (Dronten-N23); Constandse-Westermann & Meiklejohn 1979, 254 (various sites in Swifterbant).
398 Ten Anscher 2012, 344-348.
399 DRT00050 (Dendro-code RING).
400 Baetsen & Kootker 2011, 153.
401 Overviews of the dating are published by Lanting & Van der Plicht (1999/2000, 59); De Roever (2004, table 2); Peeters (2006, appendix) and Devriendt (2013, table 2.4).
402 Peters & Peeters 2001, 122.
403 Cf. Ten Anscher 201, 352-357.
404 De Roever 2004, 25
405 Deckers, 1979 154.
406 De Roever 2004, 25 and appendix 9.

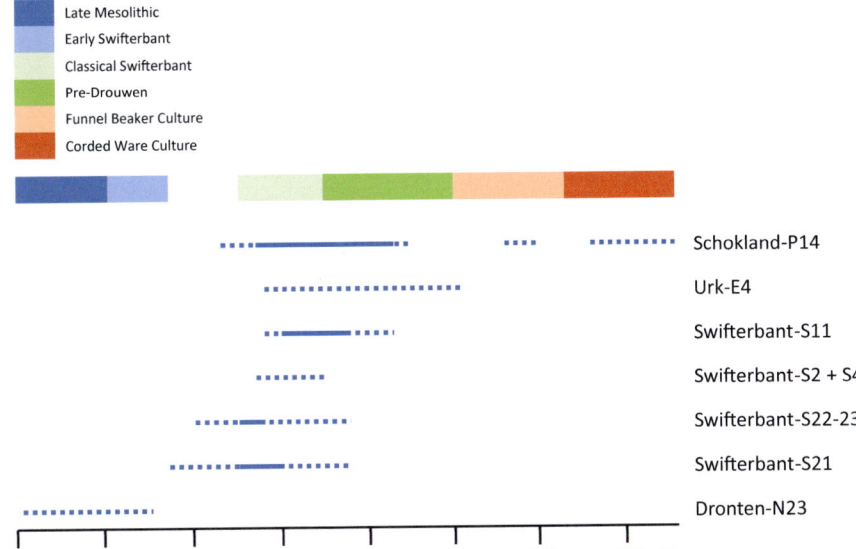

Figure 5.1: Schematic range of dates of prehistoric graves in Flevoland. Presumed ranges are dotted.

of the finds continues over the area of the cemetery. De Roever argues convincingly that the site was first used for burial and that settlement activities took place later on the same location.[407]

Although the contemporaneity of cemetery and settlement cannot be determined with any precision, the conclusion could at least be drawn that the same type of landscape locations were chosen for cemeteries and settlements: sand dune ridges (Dronten-N23, Swifternant-S11, Swifterbant-S21-23; Urk-E4, Schokland-P14) and levees (Swifterbant-S2 and Swifterbant-S4). To adopt this as a general conclusion is premature, however, since research in Flevoland to date has, without exception, focussed on excavating the settlements. Until excavations are carried out on sites without settlement remains, it will continue to be unclear whether the same landscape locations were indeed chosen for both types of sites. In this respect we must keep in mind that the current picture is distorted.

The archaeological record for Flevoland has yielded inhumation graves on seven different locations (table 5.1). Three different periods can be distinguished based on the dates attributed to these graves (fig. 5.1). The burial practice in each of these periods is discussed below.

5.2.2 Late Mesolithic and Early Swifterbant

The first period in which inhumations are present in Flevoland is the Late Mesolithic. The number of excavations with finds assemblages from this period is limited. As a consequence, the number of graves is also limited.

The excavation at Dronten-N23 revealed an oval pit measuring c. 2 x 1 m which contained some bone remains. Given the size and dimensions of the pit, the suspicion arose during the excavation that this may in fact be a grave. The pit was therefore lifted as a block so that a more controlled excavation of its contents could take place away from the site. Physical-anthropological analysis indicated that the excavated skeletal remains (largely only present in silhouette) belonged to an adult female. The degree of tooth wear (attrition) suggests an estimated age at death of between 35-45 years old. The woman had been positioned laid out on her back (fig. 5.2). No grave goods were found during the excavation. Dating evidence for the grave is indirect (see above). The date of the burial was determined by the fact that the grave had cut through some older features that had been ^{14}C dated (giving a *terminus post quem*). The latest possible date for the burial is established by the ^{14}C date from the oak tree (see above) which is assumed to have died just before the sand ridge became overgrown with peat, thus bringing the exploitation possibilities here to an end (*terminus ante quem*). On this basis, the skeletal remains can be dated to the period 5400-4800 cal. BC.[408] This places the burial in the Late Mesolithic -Early Swifterbant phase.

An unexpected clue for a prehistorical burial practice in Flevoland was recovered during a borehole survey in Almere-Europakwartier (site 7). A section of red-coloured powder was recorded in a core taken from a depth of c. 3.5 m below current ground level on a small, covered sand ridge. Once this powder had been identified as red ochre, the idea arose that the core section may have come

---

407 De Roever 2004, 25.

408 Baetsen & Kootker 2011.

Nr	Phase	Location	Burial	Orientation	Position	Grave gifts	Sex	Age	Date range (calBC)	GrA-nr	Result
1	Late Mesolithic/Early Swifterbant	Dronten-N23	Dronten-N23-1	S-N	supine	no	female	35-45 yrs	5400-4800		
2	Classical Swifterbant/Pre-Drouwen	Swifterbant-S21	Swifterbant-S21-744	SW-NE				20-35 yrs	4680-4340	39709	5640 ± 70
3	Classical Swifterbant/Pre-Drouwen	Swifterbant-S21	Swifterbant-S21-XI	S-N	supine		female (1)	20-55 yrs	4450-4260	38134	5490 ± 35
4	Classical Swifterbant/Pre-Drouwen	Swifterbant-S21	Swifterbant-S21-485	SW-NE				20-35 yrs	4360-4040	39708	5400 ± 70
5	Classical Swifterbant/Pre-Drouwen	Swifterbant-S21	Swifterbant-S21-IV	S-N	supine		female (2)	20-55 yrs	4350-4180	33541	5425 ± 35
6	Classical Swifterbant/Pre-Drouwen	Swifterbant-S21	Swifterbant-S21-III	NW-SE	supine		male (3)	20-35 yrs	4230-3950	38133	5200 ± 35
7	Classical Swifterbant/Pre-Drouwen	Swifterbant-S21	Swifterbant-S21-V	WNW-ESE				adult	4600-4000		
8	Classical Swifterbant/Pre-Drouwen	Swifterbant-S22-23	Swifterbant-S22-VII	SW-NE	supine		female (2)	20-35 yrs	4550-4360	33542	5650 ± 35
9	Classical Swifterbant/Pre-Drouwen	Swifterbant-S22-23	Swifterbant-S22-VIII	SW-NE	supine	two flint blades?	male (2)	35-55 yrs	4440-4260	38135	5480 ± 30
10	Classical Swifterbant/Pre-Drouwen	Swifterbant-S22-23	Swifterbant-S22-II	SW-NE	supine		female (3)	adult	4500-4170	39712	5500 ± 70
12	Classical Swifterbant/Pre-Drouwen	Swifterbant-S22-23	Swifterbant-S22-VI	WSW-ENE	supine		female (1)	35-55 yrs	4350-4070	42739	5400 ± 40
11	Classical Swifterbant/Pre-Drouwen	Swifterbant-S22-23	Swifterbant-S22-IX	WSW-ENE	supine		male (3)	35-55 yrs	4340-4080	38139	5400 ± 30
13	Classical Swifterbant/Pre-Drouwen	Swifterbant-S22-23	Swifterbant-S23-XII	E-W				adult	4330-4066	38140	5370 ± 30
14	Classical Swifterbant/Pre-Drouwen	Swifterbant-S22-23	Swifterbant-S22-I		skull burial	jet bead	female (3)	35+ yrs	4330-3970	39711	5295 ± 70
15	Classical Swifterbant/Pre-Drouwen	Swifterbant-S2	Swifterbant-S2-I	NNW-SSE	supine		female (1)	adult	4300-4000		
16	Classical Swifterbant/Pre-Drouwen	Swifterbant-S2	Swifterbant-S2-II	NNW-SSE	supine		female (2)	35+ yrs	4300-4000		
17	Classical Swifterbant/Pre-Drouwen	Swifterbant-S2	Swifterbant-S2-III	NNW-SSE	supine		male (2)	20-35 yrs	4300-4000		
18	Classical Swifterbant/Pre-Drouwen	Swifterbant-S2	Swifterbant-S2-IV	NNW-SSE	supine		male (1)	20-55 yrs	4300-4000		
19	Classical Swifterbant/Pre-Drouwen	Swifterbant-S2	Swifterbant-S2-V	NNW-SSE	supine	eight amber beads	female (2)	adult	4300-4000		
20	Classical Swifterbant/Pre-Drouwen	Swifterbant-S2	Swifterbant-S2-VI	NNW-SSE	supine		male (2)	20-35 yrs	4300-4000		
21	Classical Swifterbant/Pre-Drouwen	Swifterbant-S2	Swifterbant-S2-VII	NNW-SSE	supine		female (3)	adult	4300-4000		
22	Classical Swifterbant/Pre-Drouwen	Swifterbant-S2	Swifterbant-S2-VIII-1	NNW-SSE	supine			3.5-4 yrs	4300-4000		
23	Classical Swifterbant/Pre-Drouwen	Swifterbant-S2	Swifterbant-S2-VIII-2		three rib fragments			adult	4300-4000		
24	Classical Swifterbant/Pre-Drouwen	Swifterbant-S2	Swifterbant-S2-IX	NNW-SSE	supine	five amber beads, one stone bead, perforated boar tusk	male (2)	20-55 yrs	4300-4000		
25	Classical Swifterbant/Pre-Drouwen	Swifterbant-S4	Swifterbant-S4-I	S-N	supine	one amber bead		7 yrs	4300-4000		
26	Classical Swifterbant/Pre-Drouwen	Swifterbant-S11	Swifterbant-S11-42	N-S				20-35 yrs	4230-3970	39707	5170 ± 70
27	Classical Swifterbant/Pre-Drouwen	Swifterbant-S11	Swifterbant-S11-I	WNW-ESE	supine		male (2)	20-55 yrs	4230-3790	38131	5255 ± 35
28	Classical Swifterbant/Pre-Drouwen	Urk-E4	Urk-E4-1	N-S	crouched		male?	25-35 yrs	4300-3400		
29	Classical Swifterbant/Pre-Drouwen	Urk-E4	Urk-E4-3	WNW-ESE	supine	five amber beads		25-35 yrs	4300-3400		
30	Classical Swifterbant/Pre-Drouwen	Urk-E4	Urk-E4-4	WNW-ESE	supine		female?	40-45 yrs	4300-3400		
31	Classical Swifterbant/Pre-Drouwen	Urk-E4	Urk-E4-5	WNW-ESE	supine			21-25 yrs	4300-3400		
32	Classical Swifterbant/Pre-Drouwen	Urk-E4	Urk-E4-7-I	NNW-SSE	supine			25-35 yrs	4300-3400		

Nr	Phase	Location	Burial	Orientation	Position	Grave gifts	Sex	Age	Date range (calBC)	GrA-nr	Result
33	Classical Swifterbant/Pre-Drouwen	Urk-E4	Urk-E4-7-II	**NNW**-SSE				adolescent or adult	4300-3400		
34	Classical Swifterbant/Pre-Drouwen	Urk-E4	Urk-E4-7-III	**NNW**-SSE	upper arm		female?	9-14 yrs	4300-3400		
35	Classical Swifterbant/Pre-Drouwen	Urk-E4	Urk-E4-8	**NNW**-SSE	supine	no		33-45 yrs	4300-3400		
36	Classical Swifterbant/Pre-Drouwen	Schokland-P14	Schokland-P14-3		skull burial(s)	no		7-18 yrs (plus 5-18 yrs?)	4450-3970	12612	5380 ± 120
37	Classical Swifterbant/Pre-Drouwen	Schokland-P14	Schokland-P14-4-1	**WNW**-ESE	crouched, on left side	no	female?	18-25 yrs	3930-3650	15426	4970 ± 40
38	Classical Swifterbant/Pre-Drouwen	Schokland-P14	Schokland-P14-4-2	**WNW**-ESE	crouched, on right side	antler fragment?		18-25 yrs	3950-3710	15427	5030 ± 40
39	Classical Swifterbant/Pre-Drouwen	Schokland-P14	Schokland-P14-4-3	**ESE**-WNW	supine or crouched on right side	Bone beads		18-35 yrs	4400-3700		
40	Classical Swifterbant/Pre-Drouwen	Schokland-P14	Schokland-P14-4-4		partial burial of teeth			12-18 yrs	4400-3700		
41	Classical Swifterbant/Pre-Drouwen	Schokland-P14	Schokland-P14-4-5		partial burial of teeth			2-9 yrs	4400-3700		
42	Classical Swifterbant/Pre-Drouwen	Schokland-P14	Schokland-P14-4-6		partial burial of teeth and jaw fragment			18-25 yrs	4330-3990	16188	5330 ± 80
43	Classical Swifterbant/Pre-Drouwen	Schokland-P14	Schokland-P14-5	**ESE**-WNW	?	no		6-10 yrs	4230-3810	16186	5200 ± 60
44	Classical Swifterbant/Pre-Drouwen	Schokland-P14	Schokland-P14-1	**WNW**-ESE	supine	no		young adult?	4400-3700		
45	Classical Swifterbant/Pre-Drouwen	Schokland-P14	Schokland-P14-2	**WNW**-ESE	supine	no		18-25 yrs	4400-3700		
46	Classical Swifterbant/Pre-Drouwen	Schokland-P14	Schokland-P14-6	**WNW**-ESE	?	no		35-45 yrs	4400-3700		
47	Classical Swifterbant/Pre-Drouwen	Schokland-P14	Schokland-P14-7	**SE**-NW	?	no		12-18 yrs	4400-3700		
48	Classical Swifterbant/Pre-Drouwen	Schokland-P14	Schokland-P14-8	**E**-W	supine?	no	male?	below 50 yrs	4400-3700		
49	Uncertain	Schokland-P14	Schokland-P14-9	**NW**-SE	crouched, on left side		male	25-35 yrs	3200-3000?		
50	Late Neolithic	Schokland-P14	Schokland-P14-10	**SE**-NW	crouched, on right side	SGC beaker type 1d	female	25-35 yrs	2800-2400		
51	Late Neolithic	Schokland-P14	Schokland-P14-11	**W**-E	crouched, on right side	no	male	25-35 yrs	2800-2400		
52	Late Neolithic	Schokland-P14	Schokland-P14-12	**W**-E	crouched, on right side	wooden club	male	40-45 yrs	2800-2400		
53	Late Neolithic	Schokland-P14	Schokland-P14-13	**W**-E	crouched, on right side	no	male	30-40 yrs	4400-3700		
54	Late Neolithic	Schokland-P14	Schokland-P14-14	**SW**-NE	crouched, on right side	six flint tools	male	35-45 yrs	2800-2400		
55	Late Neolithic	Schokland-P14	Schokland-P14-15	?	?				2800-2400		

Table 5.1: Overview of graves and other human remains. Orientation in bold refers to the location of the head.

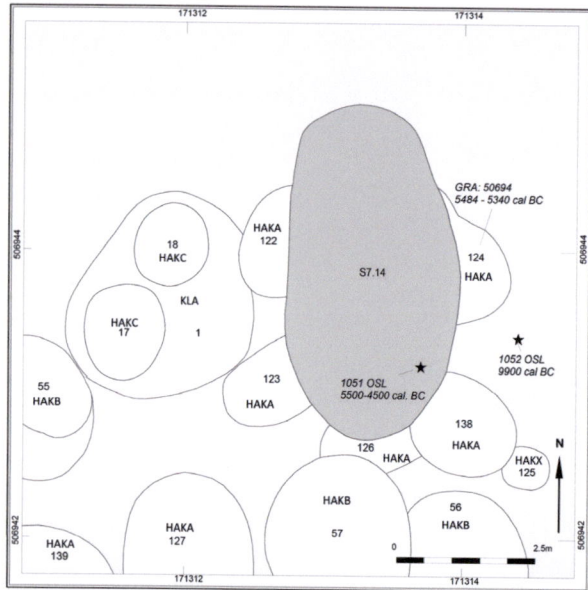

Figure 5.2: Dronten-N23: grave with poorly preserved remains of a female individual. The burial pit crosscuts mesolithic pit-hearths, providing a maximum age for the burial (source: Baetsen & Kootker 2011).

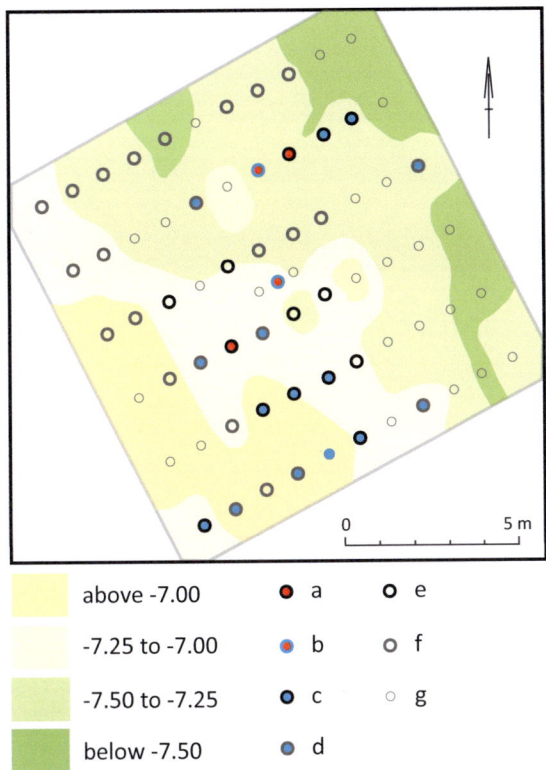

Figure 5.3: Almere-Poort: possible Mesolithic burial field, projected on a DEM of the Pleistocene surface. Circles: boreholes, (a) ochre and a large amounts of charcoal, (b) ochre and a 'mottled' layer, (c) 'mottled' layer with large amounts of charcoal, (d) mottled layer and some amounts of charcoal, (e) large amounts of charcoal, (f) some amounts of charcoal, (h) no archaeological indicators (source: Raemaekers, Borsboom & Müller 2003)

from a Mesolithic grave. The relative dating of the grave is linked to the depth from which the core section was taken on the sand ridge. A later date would be impossible because the dune had already been covered by a sediment layer around 5100 cal BC. The interpretation of this feature as a grave is based on recorded occurrences of ochre in various Mesolithic graves throughout Europe.[409] Additional boreholes were made on the site using a much closer-spaced survey grid. A further three core sections were found to contain red ochre. Another 19 cores showed evidence for a disturbance of the regular stratigraphy. It is possible that these cores record the presence of features (graves?) in the substrate, and that they provide evidence for the presence of a Mesolithic cemetery. Since no subsequent excavation took place on the site, this hypothesis cannot be verified (fig. 5.3).[410] No ^{14}C dating was carried out to determine the age of the possible graves, but a *terminus ante quem* based on the depth of these features is estimated at 5400 cal. BC.

At A27- Hoge Vaart (phase 3), some fragments of burnt human skeletal remains were identified mixed into the extensive spread of archaeological material across the site. The remains include the odontoid process of the axis, a toe phalanx, a fragment of pelvic bone, and a fragment of a jawbone dating to the Early Swifterbant phase.[411] The jaw fragment is from a child, the rest of the bones from adults. It is not clear whether these remains originate from disturbed cremations, or are the result of other types of ritual behaviour. This latter interpretation might be supported by other significant finds recovered close to the cremated remains: a fragment of an arrow shaft sharpener, a lump of red ochre and an assemblage of bone remains from furred animals.[412] A third possibility is that the skeletal remains circulated within society[413] and were only later burned, together with the animal bones.

The interpretation of the burnt human remains from A27-Hoge Vaart as representing deliberately cremated remains is not supported by any parallels found within Flevoland, but outside the region there are clear indications for cremations in the Late Mesolithic. At Oirschot (Noord-Brabant), for instance, site documentation records a pit found in the middle of a flint scatter, containing the cremated remains of an individual, a child between 10-13 years old.[414] These remains have been ^{14}C dated to the period 7515-7196 cal. BC.[415] Three pits containing cremated remains were discovered in Rotterdam (Zuid-Holland).[416] A ^{14}C date for these remains places the cremations in the period 7304-7047 cal. BC.[417] On a site in Dalfsen (Overijssel), cremated remains were recovered from the fill of one of more than twenty pit hearths. In all probability these represent the remains of an adult female, possibly including remains from a second individual.[418] A ^{14}C date does not seem to be very reliable,[419] so the dating of the cremated remains is entirely based on their spatial association with the pit hearths. These sites indicate that both inhumation and cremation were both common burial practices in the Late Mesolithic, and therefore also in Flevoland.

---

409 Newel, Constanse-Westermann & Meiklejohn (1979) present an overview of European Mesolithic graves. Red ochre is found in nine of the graves in the catalogue.
410 Raemaekers, Borsboom & Müller 2003.
411 Laarman 2001, 11.
412 Peeters 2007, 199.
413 See 5.3.
414 Arts & Hoogland 1987, 177-179.
415 Lanting 2001. GrA-13390: 8320 ±40 B
416 Drenth & Niekus 2009a, 760.
417 Drenth & Niekus 2009b, 92. This refers to the dating GrA 43443: 8135 ± 45 BP
418 Verlinde 1974, 116.
419 Lanting & Van der Plicht 1997/1998, 142.

5.2.3 Classical Swifterbant and Pre-Drouwen

We have relatively good evidence pertaining to the burial practice from this period because of the large number of graves excavated at seven sites. Inhumations are generally the norm in this period, with cemeteries containing up to 16 individuals.

Burial sites and dating

The burial field at Swifterbant-S2 (fig. 5.4) was discovered during the first excavations at Swifterbant carried out by the *Rijksdienst voor de IJsselmeerpolders* in 1964. Documentation shows the presence of ten individuals buried in nine graves. It is striking that the graves all have approximately the same orientation and do not as a rule overlap. An important exception is formed by graves V and VI. These either represent a double grave, or the individual in grave V was buried at a later date, on almost the same location as grave VI. The best-known grave is Swifterbant-S2-IX (fig. 5.5). This is the grave of an adult male in which various objects were found that have been interpreted as personal belongings. These include six amber beads and hangers and a pierced boar's tooth. This individual is often referred to as 'the headman of Swifterbant', which is a particularly far-reaching interpretation for such a relatively unspectacular burial. The female burial Swifterbant-S2-V has more grave goods (eleven amber beads and hangers), but is never referred to as being of 'the head woman of Swifterbant'.[420] The Swifterbant-S2 cemetery has been dated on the basis of available ^{14}C dates from the finds combined with the geological stratigraphy.

The excavations at the site of Swiferbant-S4 uncovered the grave of a child (fig. 5.6). This is the only site on which a single grave has been recorded. The grave has been dated in the same way as the cemetery at Swifterbant-S2. In the knowledge that the distance between individual graves in previously excavated cemeteries had generally been limited to 1 – 2 m, a zone of 5 m was subsequently stripped around the grave in the expectation of uncovering more graves. This proved not to be the case. It is not clear why this apparently isolated grave did not form part of a larger cemetery, like all other graves of this period.

The cemeteries of Swifterbant-S21 (fig. 5.7) and Swifterbant S22/S23 (fig. 5.8) are situated on the same river dune, Swifterbant-S21 on the north side, the two excavation areas of Swifterbant-S22 and Swifterbant-S23 on the west side. The graves at Swifterbant-S22 and Swifterbant-S23 are counted as one cemetery since the single grave at Swifterbant-S23 lies only a short distance from, and directly in line with, the graves at Swifterbant-S22. All but one of the graves at Swifterbant-S21 have been ^{14}C

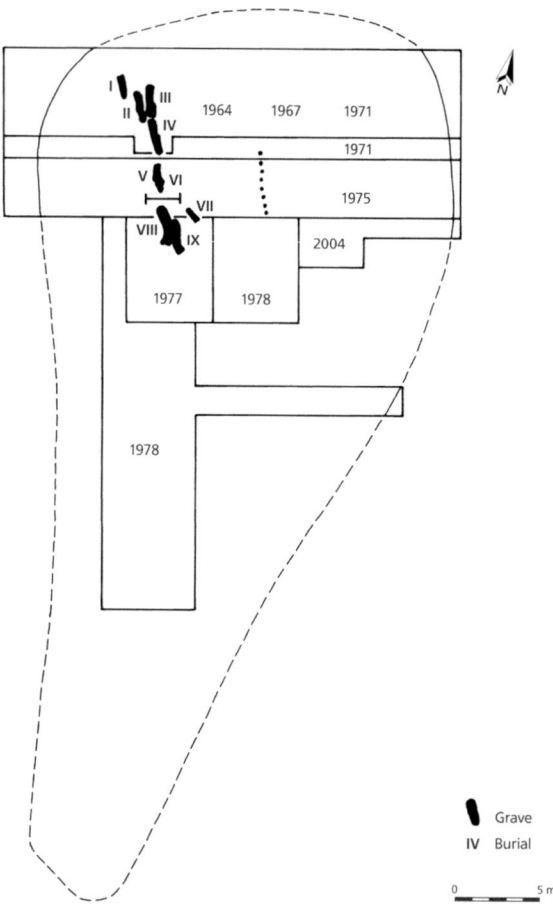

Figure 5.4: Swifterbant-S2: burial field (Classical Swifterbant) (source: Raemaekers, Molthof & Smits 2009).

dated. It is interesting to note that the Swifterbant-S22/S23 graves all form a line across the contours of the dune: the line of graves stretches from grave VII on the flank to grave XII on the highest part of the dune. This implies that when deciding the location for new graves in the group, the existing graves would still have been visible (just as at Swifterbant-S2). The cemetery was in use over a maximum period of 580 years. The spatial structure of the cemetery at Swifterbant-S21 is less clear. The cemetery seems to have been divided into a few subgroups with a period of use spanning 110-730 years. A comparison of the two cemeteries makes it clear that the period of use does not determine whether the spatial structure of the cemetery is recognisable to us. It is therefore unclear how the differences in layout can be explained.

The actual position of the two graves at Swifterbant-S11 is difficult to determine as the excavation plan and the

---

420   Devriendt 2008, 386-387.

Figure 5.5: The 'headman' of Swifterbant (Swifterbant-S2 grave IX). On the skull amber pendants were found (© University of Groningen, Groningen Institute of Archaeology).

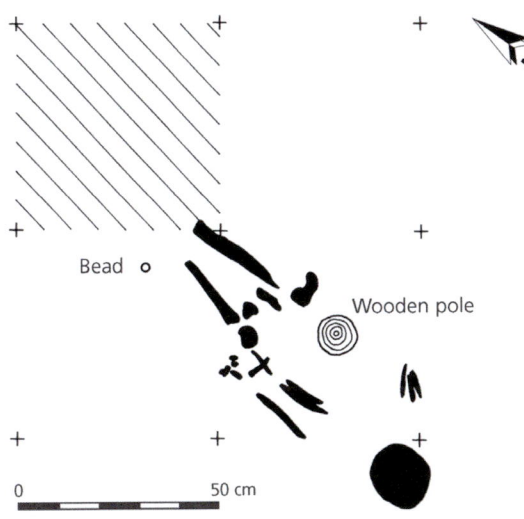

Figure 5.6: Swifterbant-S4: burial (Classical Swifterbant) (source: Raemaekers, Molthof & Smits 2009).

location of the graves within it are not known.[421] The finds themselves at least endorse the importance of the dunes as a location choice for cemeteries. Both graves are ^{14}C dated.

The cemetery at Urk-E4 (fig. 5.9) is particularly badly preserved (fig. 5.10). This is undoubtedly a consequence of the site being located high above the then groundwater level. It also means that the ^{14}C dates from skeletal remains are unreliable. The dates in table 5.1 are those of the archaeological finds from the site (pottery and flint). In addition, the absence of pottery with stab-and-drag decoration is an indication that there was no activity on the site after 3400 cal. BC.

The last cemetery is that of Schokland-P14 (fig. 5.11). On this site inhumation graves dating to the Classical Swifterbant/pre-Drouwen period lay close to others dating to the Late Neolithic. Many graves have been ^{14}C dated, but the dates of these are mostly unreliable. The bones that were very weathered had a too low collagen yield (about 2%) and were therefore discarded as unreliable.[422] The reasonably preserved bones on the other hand, had a collagen yield of approximately 20%,[423] but the $δ^{13}$C values were so negative that the contamination of the samples with humus must be assumed.[424] This made the results from these samples equally unreliable. The dates listed in table 5.1 are considered reliable by the researchers involved.[425] These dates have not been obtained from the collagen fraction, but rather from the carbonate fraction of the skeletal material. The other graves were added to the two grave groups on the basis of cultural characteristics and the degree of preservation.[426] A total of eight graves containing a minimum of 13 individuals can be attributed to the Classical Swifterbant/pre-Drouwen period. The very poor level of preservation is striking (fig. 5.12).

The occurrence of cemeteries suggests that certain locations in the landscape were selected as burial sites and retained this function over a long period of time. The available ^{14}C dates indicate that the excavated cemeteries

---

421 According to the excavator, the two graves were situated in the middle of the excavation (personal comment R. Whallon 2007).

422 Ten Anscher 2012, 353.
423 Ten Anscher 2012, 353.
424 Lanting & Van der Plicht 1999/2000, 59. This conclusion has been adopted by Ten Anscher (2012, 354).
425 Ten Anscher 2012, 353; Lanting & Van der Plicht 1999/2000, 59.
426 Ten Anscher 2012, 352-356; Lanting & Van der Plicht 1999/2000, 59, 77.

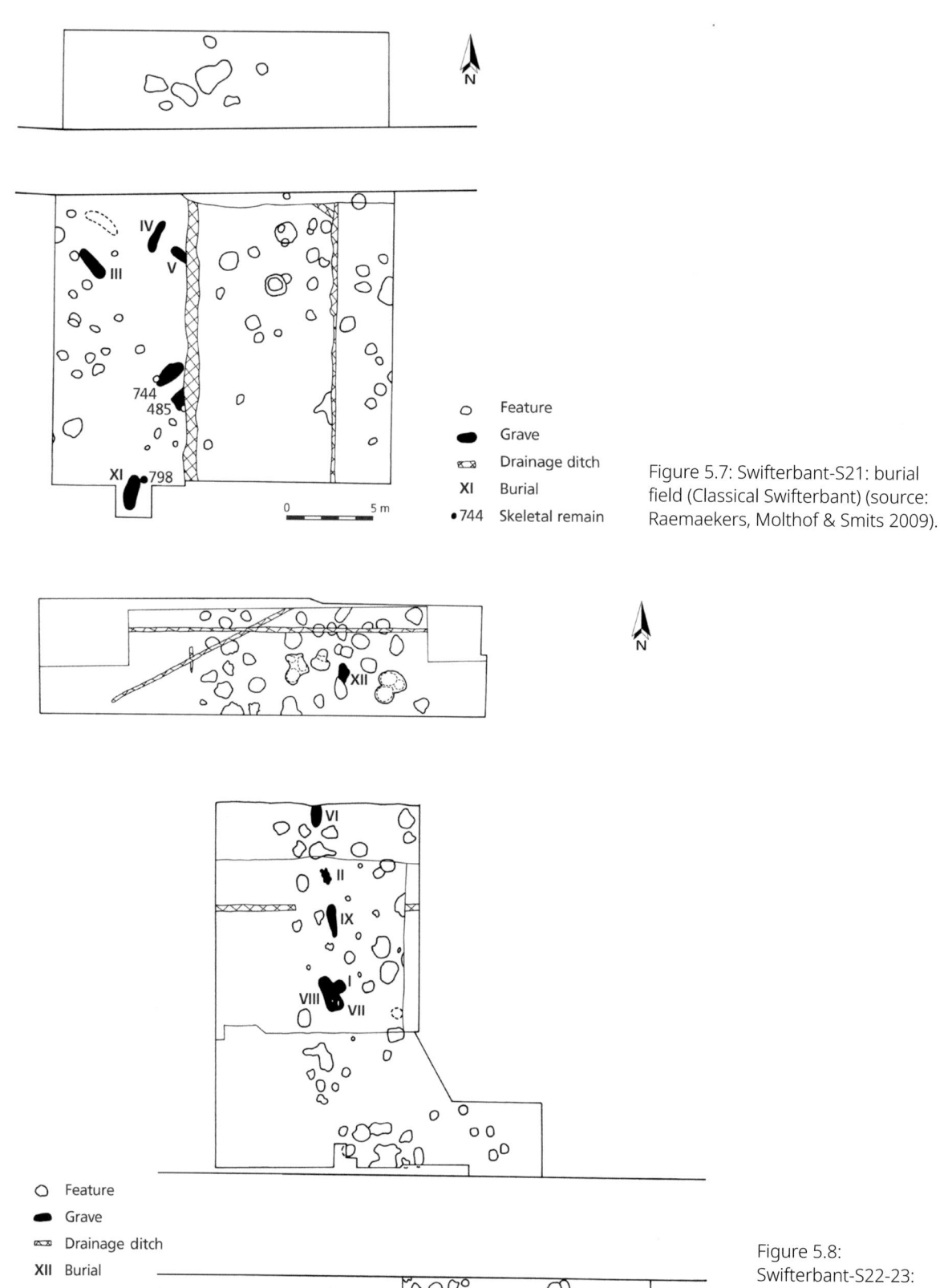

Figure 5.7: Swifterbant-S21: burial field (Classical Swifterbant) (source: Raemaekers, Molthof & Smits 2009).

Figure 5.8: Swifterbant-S22-23: burial field (Classical Swifterbant) (source: Raemaekers, Molthof & Smits 2009).

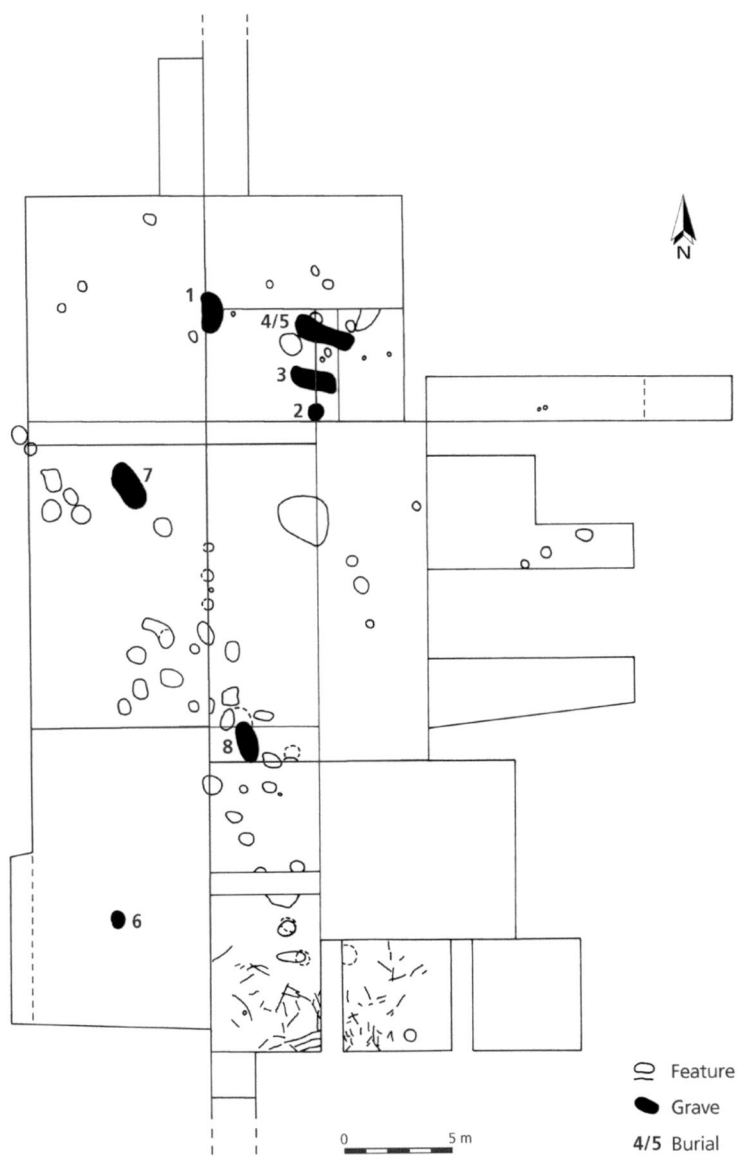

Figure 5.9: Urk-E4: burial field (Classic Swifterbant/PreDrouwen) (source: Raemaekers, Molthof & Smits 2009).

were in use over a period of several centuries (see fig. 5.1). Table 5.1 makes clear that there were no general rules regarding the orientation of graves during this period. It is, however, interesting to note that where the position of the head is known in a grave, that this generally has an orientation between north-north-west and south-west (this accounts for 36 of the 41 graves where an orientation could be determined). For the different cemeteries, the range of variables within the graves is often more limited (Swifterbant-S2; Swifterbant-S22/S23). The cemeteries of Urk-E4 and Swifterbant-S21 do, however, show more variation. One hypothesis for this could be the longer period of use of these cemeteries.[427] That said, the ^{14}C dates

from Swifterbant-S21 suggest a period of use similar in length to Swifterbant-S22/S23 and similar to the assumed period of use of Swifterbant-S2.

Body positioning

The majority of bodies were buried laid out extended on their back. Others are found laid on their side with their legs flexed and raised. The available ^{14}C dates suggest that there is a recognisable chronological development in this. There are ^{14}C dates for ten individuals for whom the positioning of the body in the grave is recorded. Six of these are laid out on their back and can be reliably dated to before 4000 cal. BC; for two others, a later date is equally possible. The two individuals laid out on their side with flexed legs were both buried after 4000 cal. BC. The available data from Flevoland makes two options

---

427 Cf. Raemaekers, Molthof & Smits 2009, 491.

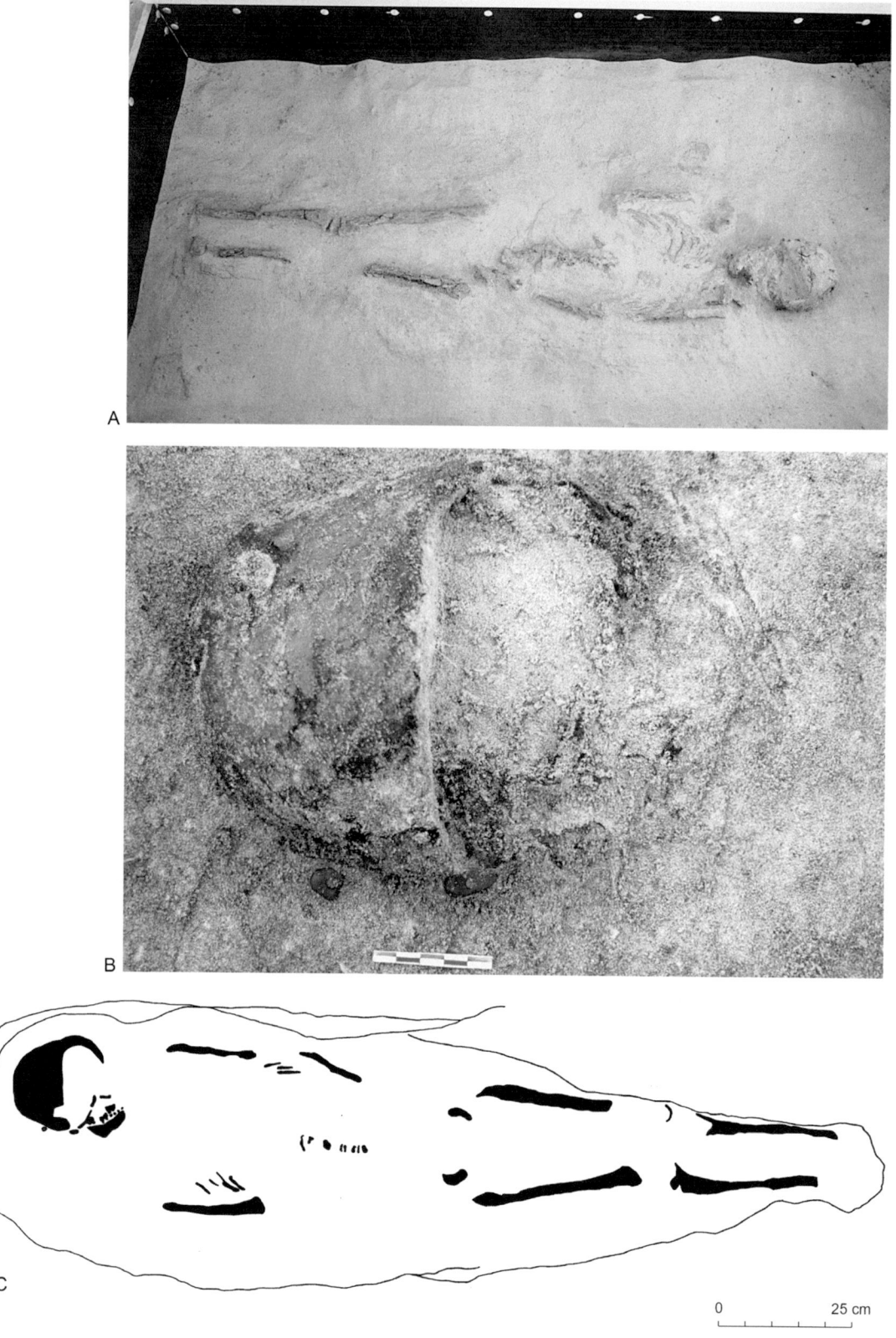

Figure 5.10: Urk-E4: burial 3 showing the poorly preserved skeletal remains (source: d'Hollosy & Baetsen 2001).

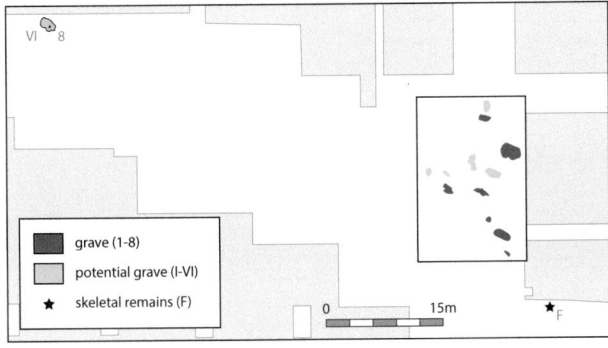
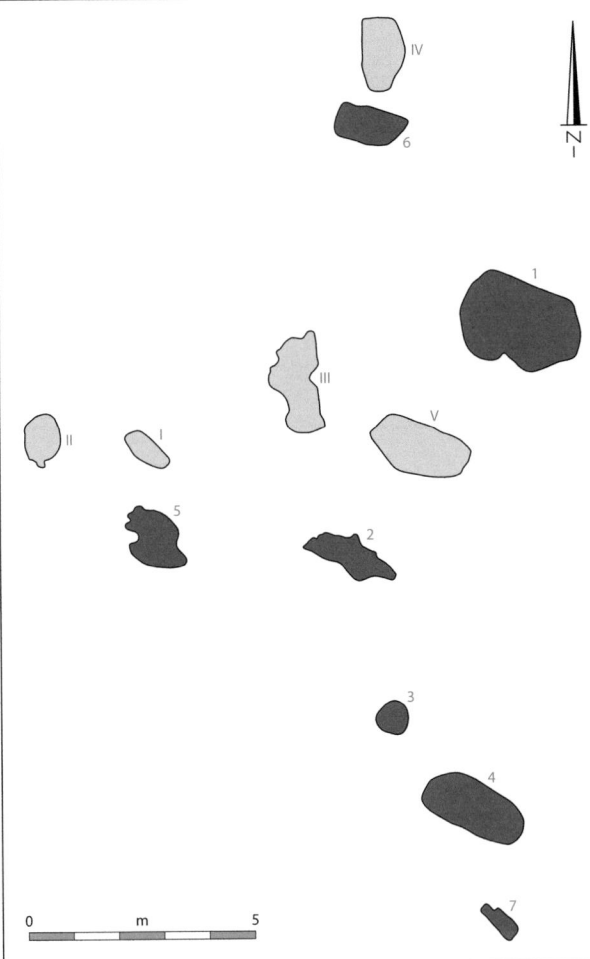

Figure 5.11: Schokland-P14: burial field (Classical Swifterbant/PreDrouwen) (adapted from Ten Anscher 2012).

possible. Firstly, the tradition of extended supine graves may well have fallen out of fashion around 4000 cal. BC (the graves from Swifterbant-S21-III and S11-I are dated before 4000 cal. BC). Secondly, it is possible that both burial positions were common in the centuries after 4000 cal. BC (the graves from Swifterbant-S21-III and S11-I are dated after 4000 cal. BC).

Grave goods

Grave goods seem to be scarce. What we do have is jewellery in six graves, an antler fragment in one grave and possibly two flint tools in another. Bearing in mind the poor preservation conditions, this assemblage could be the proverbial tip of the iceberg. In more or less contemporary Late Mesolithic Denmark, we have the site of Vedbæk-Bogebakken. Excavations discovered a double grave containing the body of a woman of around 18 years old and the body of a new-born. A total of 190 pierced teeth from red deer and wild boar and an unknown number of pierced snail shells were found at the woman's head. Several rows of the same pierced snail shells were found near her pelvis, plus a row of around 50 pierced teeth from red deer, but also teeth from seal and moose. The new-born was buried right next to the woman and lay on the wing of a swan with a flint blade on its pelvis.[428] If such a similar burial had ever taken place at Schokland-P14 for instance, then it is possible that due to the poor preservation conditions, only the skull of the woman and the flint blade would have survived. We therefore have to be cautious when interpreting the significance of the grave goods we do find.

The inclusion of jewellery in a grave does not seem to be related to either gender or age, given its presence in the graves of men, women and children. The antler fragment was found in a grave at Schokland-P14 with the remains of several individuals (see below), so it is difficult to see if it was associated with individual 2 or 3. The fragment is assumed to be the remains of a mattock or pick that had been used to reopen the grave at some point and should therefore not be seen as a grave good.[429] The flint tools from Swifterbant-S22-VIII have been described in the past as transverse arrowheads,[430] but were unfortunately not illustrated. One of these tools has recently been found again and turned out to be a blade.[431] It is very likely that the second tool was also previously wrongly identified. Whilst it is possible that the blades were included in the

---

[428] Albrechtsen & Brinch Petersen 1976, p 8-9 and figs 9 and 10. It should be noted that a somewhat similar grave has been found recently in a Swifterbant context at the site of Nieuwegein, near Utrecht (Netherlands); these finds have yet to be published.
[429] Ten Anscher 2012: 320-321.
[430] De Roever 1976: 217-219; Meiklejohn & Constandse-Westermann 1978, 58.
[431] Raemaekers, Molthof & Smits 2009, fig. 5a.

Figure 5.12: Schokland-P14: burial 9 (source: Ten Anscher 2012).

grave as grave gifts,[432] it is equally possible that both flint artefacts were not actually deliberate gifts, but were settlement debris, accidentally included in the earth used to back-fill the graves.[433]

Gender and age

The gender of 30 adults has been established, with varying degrees of certainty: 15 male and 15 female (Table 5.1). This balance suggests that gender played no role in deciding whether the deceased should be buried.

The age at death is difficult to summarise due to the different classifications used. Leaving all the uncertainties surrounding gender determination aside, we can identify two adult females aged up to 35 and 5 females aged between 35 – 55. The balance is slightly different for the adult males, with four aged up to 35 and another two aged between 35 -55. It is not clear whether these differences are the result of differences in life expectancy, gender related selection criteria or are simply due to the limited size of the dataset.

There are a total of 8 or 9 children's graves (depending on the number of individuals at Schokland-P14-3). The age of the children at death varies from toddler (Swifterbant-S2-VIII-1) to – especially – teenagers. Various conclusions can be drawn from this. In the first instance, the group for the most part, represents children who had survived the years in which the risk of death was highest (up to about five years). It may mean that these children were therefore regarded as individuals in this society and were no longer simply interpreted as 'children'. Possibly they thereby also met the conditions to be buried. Younger deceased children would possibly have been treated in a way that left no archaeological remains at the excavated sites. Bones of younger children are smaller and more fragile than those of older individuals. This means that the chances of preservation are not as high. As a result, young children will, by definition, be under-represented. The dataset lacks graves of young children so structurally that a cultural interpretation (different burial rituals) is plausible.

In addition, different characteristics come together in two subgroups of children's graves. The children's graves at Swifterbant-S2-VIII-1 and Urk-E4 7-III form the first subgroup. On both sites the orientation of the graves corresponds with the dominant orientation within the cemeteries. In addition, the grave also contained the remains of a second, adult, individual. In Swifterbant-S2-VII-1, 3 rib fragments have been added to the child's grave; in Urk 7-III, due to poor conditions of preservation it is uncertain whether the grave includes a selection of loose bones from an adult, or a very poorly preserved complete skeleton. The second subgroup comprises the graves Schokland-P14-5 and Schokland-P14-7.

---
432 Meiklejohn & Constandse-Westerman 1978, 58.
433 Devriendt 2013, 151.

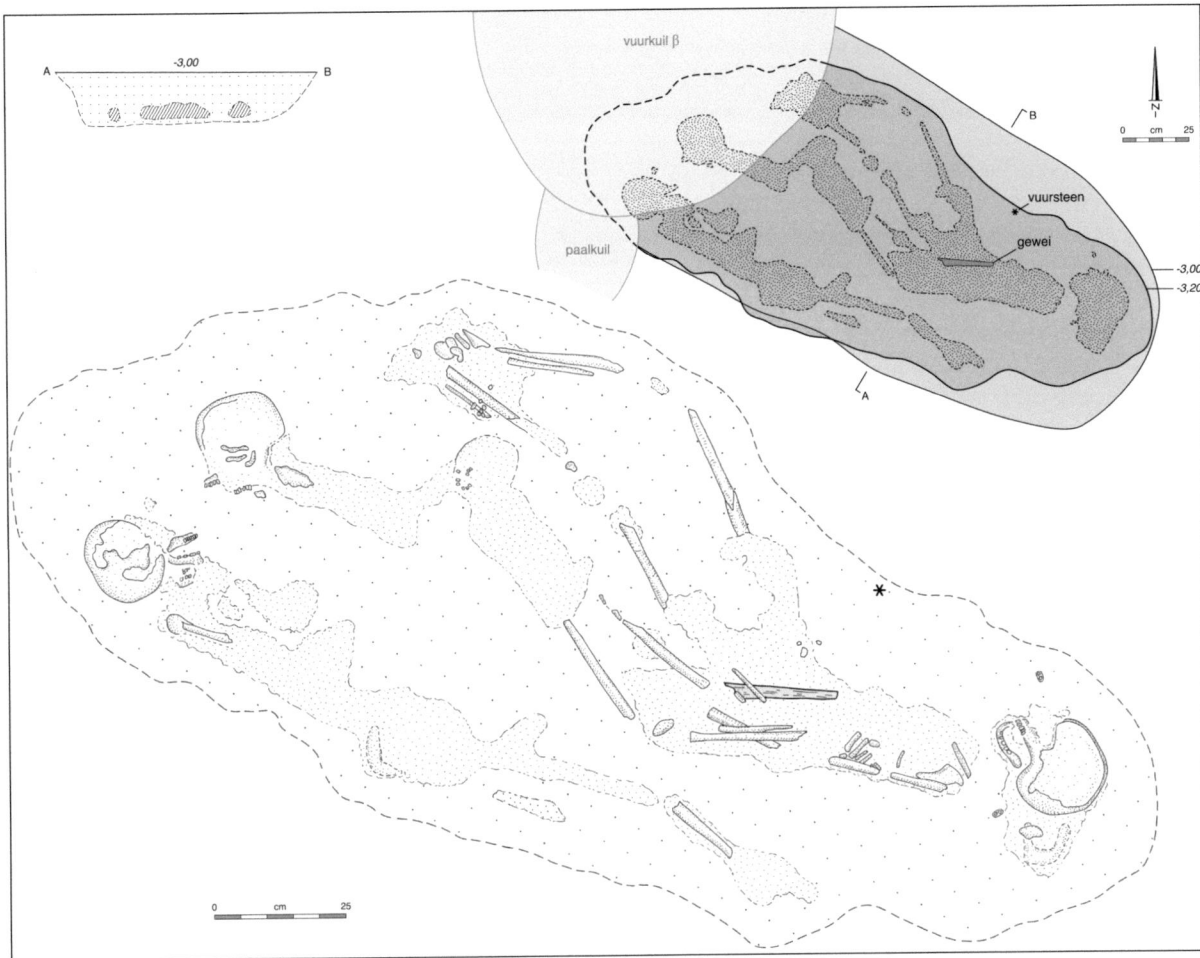

Figure 5.13: Schokland-P14: burial 4 in which a minimum of six individuals were buried (Classical Swifterbant/PreDrouwen) (source: Ten Anscher 2012).

The orientation of these graves is diametrical to the dominant orientation in the rest of the cemetery. There were also no bones from other individuals in the graves. The isolated child's grave Swifterbant-S4-I cannot be associated with either of these groups, neither can the skull burial Schokland-P14-3. The two poorly-preserved child graves of Schokland-P14-4 and Schokland-P14-6 are also not easy to interpret. The variety in the burial rituals witnessed for children is difficult to explain. Possible explanations include local differences, differences attributed to the gender and status of the deceased child (not recognisable in the grave goods), differences in the cause of death or even differences between the ages of the graves (not identifiable in the available ^{14}C dates).

Number of individuals

A characteristic of graves from this period are the occurrence of multiple burials. These include graves containing an adult and child as well as graves containing two or more adults. The poor state of preservation means its almost impossible to ascertain if we are dealing with the burial of several individuals at the same moment, or whether the grave had been repeated opened so that later interments could be added to the primary grave. The most exceptional multiple grave is Schokland-P14-4 (fig. 5.13).[434] The grave contained the remains of at least six individuals. Three of these individuals were found almost completely intact; the other individuals were represented by only a few teeth or jaw fragments. The poor preservation and the incompleteness of the majority of individuals are arguments for not dismissing the idea of higher numbers of burials being originally present in the cemeteries. The combination of a few partially preserved individuals with much more completely preserved individuals may be grounds to suggest that the grave was repeatedly reopened. The incomplete individuals should then be

---

434 Ten Anscher 2012, 319-325.

seen as the earliest interments, whilst the more complete individuals could represent the last phase in the use of the grave. This interpretation should, in theory, be able to be substantiated on the basis of the chronology extrapolated from the available ¹⁴C dates. Since, with the exception of the information included in table 5.1, these dates are unreliable, this argument will not be expanded on here. Nevertheless, the grave of Schokland-P14-4 remains the most notable example of the practice of adding new inhumations to existing graves.

The dating of grave Schokland-P14-9 is uncertain because of the absence of any direct dating evidence. The disturbed grave contained the remains of an adult male buried with the legs bent in a crouched position, laid on his left side.[435] A large number of sherds from a Funnelbeaker bowl were recovered some distance away. This bowl may have been buried as a grave good in grave 9.[436] If this was the case, then grave 9 represents a single use of the grave plot, and can be dated between the two main phases of the cemetery. Grave 9 is certainly older than the Late Neolithic graves. Firstly, grave 9 is stratigraphically cut through by the Late Neolithic grave Schokland-P14-14. Secondly, the burial position of the body in grave 9 deviates from the burial rites documented in the Late Neolithic graves.[437] A much earlier date for grave 9 cannot be ruled out, as it is unclear exactly where the Funnelbeaker dish was found, which brings into question its association with the grave and its interpretation as belonging to any grave goods. This assessment removes the only argument that would contradict the dating of grave 9 as part of the Swifterbant/Pre-Drouwen cemetery. A dating of grave 9 to this earlier phase can be argued on the basis of the similarities of positioning and orientation between grave 9 and the Swifterband/Pre-Drouwen cemetery.

5.2.4 Late Neolithic

At Schokland-P14, a small Late Neolithic cemetery was laid out on the same site as the Swifterbant/pre-Drouwen cemetery.[438] The new cemetery consisted of six graves (fig. 5.14). Five of these graves were relatively well preserved, so that evidence remained to show that the deceased had been buried in a coffin or burial chamber constructed of bark and wood. (fig. 5.15). The bodies had all been placed lying on their right sides with legs bent in a crouched position. One of the graves was a female, buried with a beaker of the Single Grave Culture. The other four burials were male. Two of these contained grave goods: a wooden club made of yew and six flint knives. The

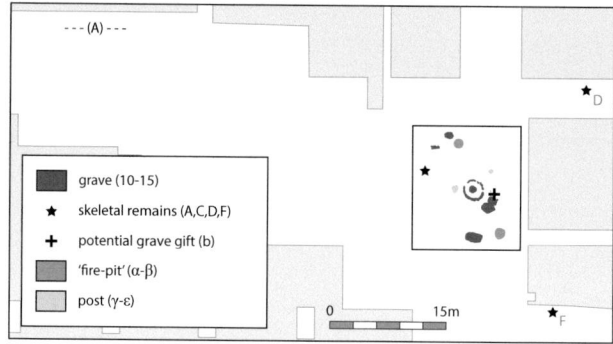
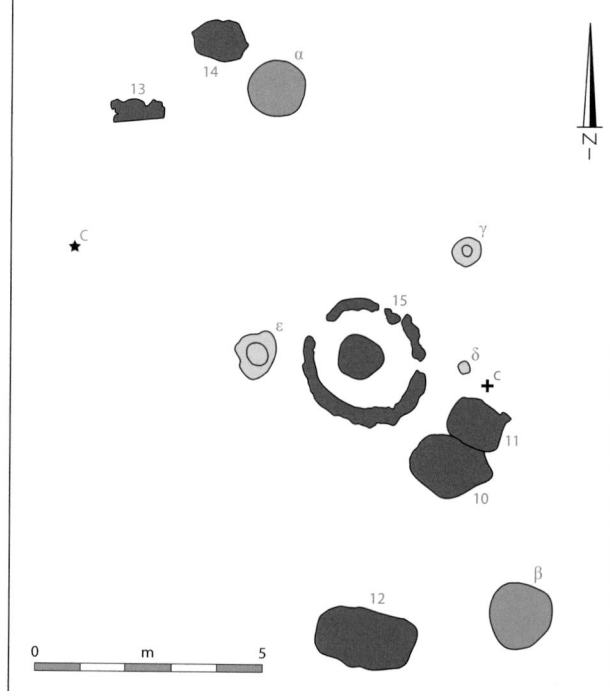

Figure 5.14: Schokland-P14: Late Neolithic burial field (source: Ten Anscher 2012).

homogeneity of grave construction and body positioning within the group is striking.

The sixth grave is a burial surrounded by a circular ditch. There were wooden posts in the ditch. A round pit with a diameter of 90 – 100 cm was discovered in the centre of this circular ditch.[439] In size, the pit would have been big enough to contain the crouched burial of an adult, but could also have been a child's grave. No bone remains were found in the pit. Possibly due to the effects of erosion, there is no visible evidence for any burial mound under which the body may have been buried.[440] A flint

---

435 Ten Anscher 2012, 329-331.
436 Ten Anscher 2012, 349. The distance between grave 9 and the bowl is not documented.
437 Ten Anscher 2012, 355-356.
438 Ten Anscher 2012, 331-344.

439 Ten Anscher 2012, 344.
440 Ten Anscher 2012, 344.

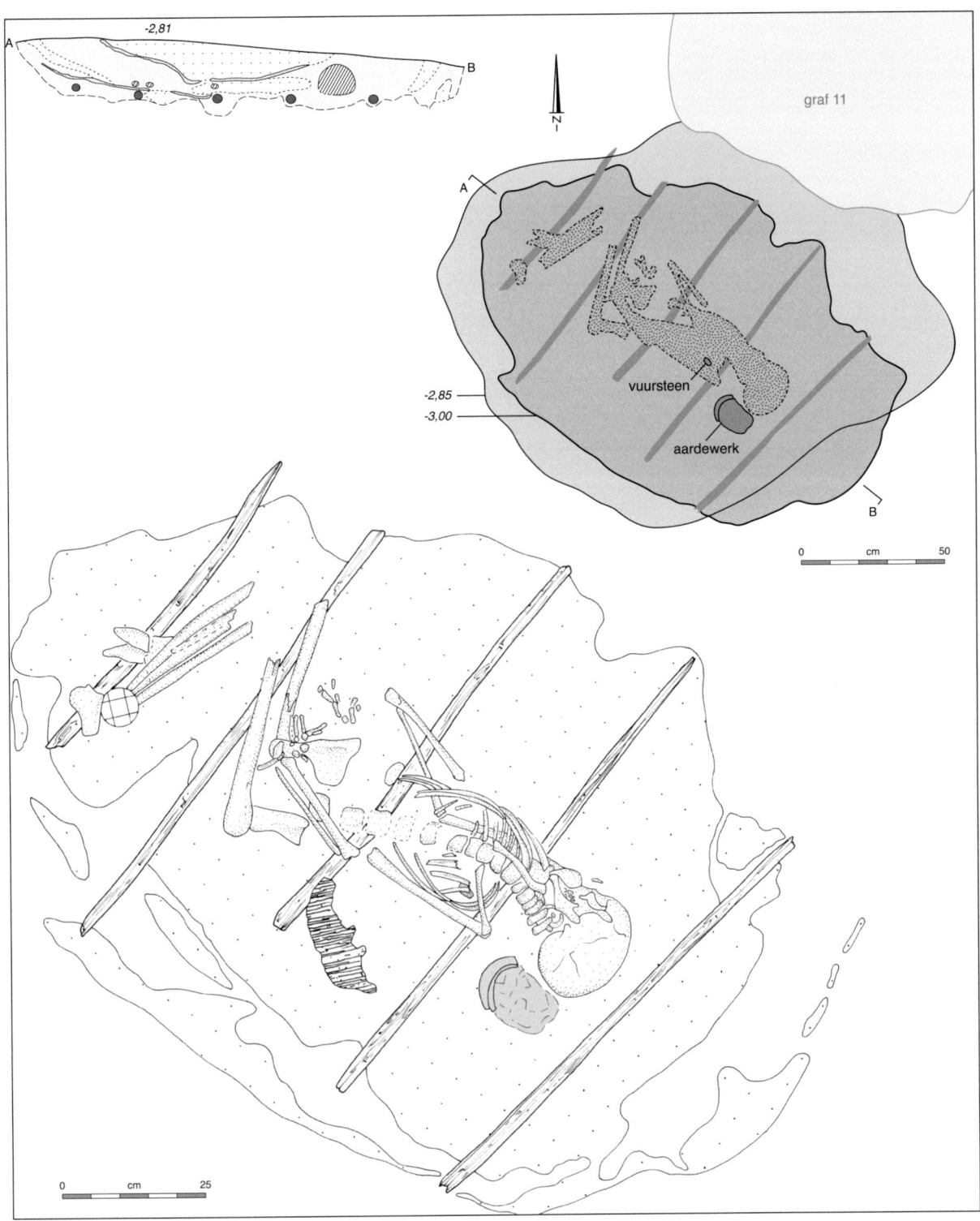

Figure 5.15: Schokland-P14 grave 10 (Corded Ware); remarkable is the presence of a wood and bark coffin/chamber (source: Ten Anscher 2012).

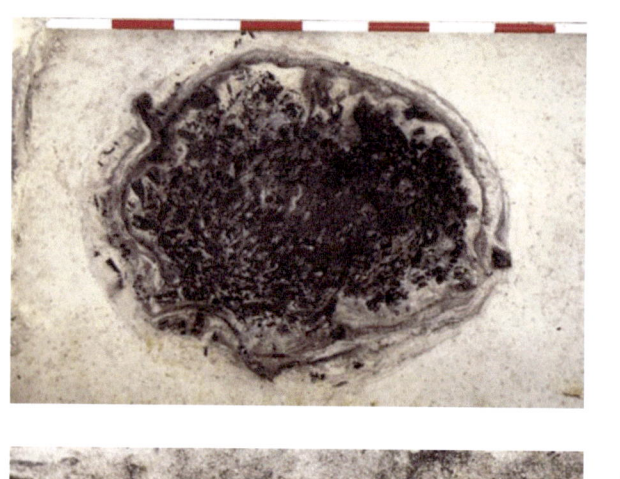

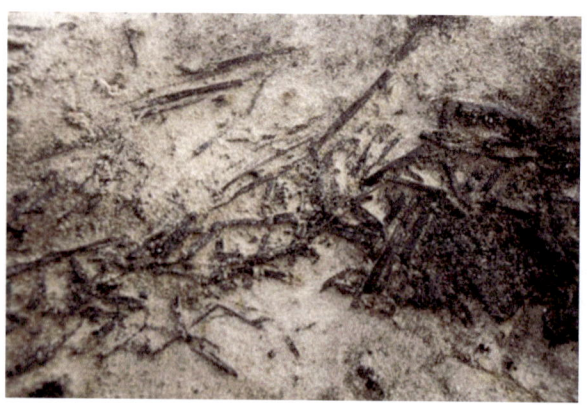

Figure 5.16: Schokland-P14: 'Fire-pit' associated with the Late Neolithic grave field (source: Ten Anscher 2012).

dagger was found a distance of *c.* 60 cm away. This may have belonged to the grave Schokland-P14-15.[441]

Close to the grave, two features were excavated that have been interpreted as pit hearths: pits with a diameter of 130-140 cm, an original depth of *c.* 50-80 cm and containing a large quantity of charcoal (fig 5.16).[442] These pits may well have had a function in the rituals performed at the burial site.[443] Three wooden posts have also been interpreted as belonging to the Late Neolithic cemetery. Although a Late Neolithic date remains plausible, an association with the cemetery cannot be established beyond doubt, especially considering that the Schokland-P14 site was also in use as a settlement in this period.[444]

5.2.5 Conclusions

This section presents the evidence for prehistoric burial ritual in the province of Flevoland. The evidence for the existence of cremations is limited to (possible) burnt human remains recovered from A27-Hoge Vaart, so the conclusions focus on the evidence for inhumations. In the province of Flevoland, inhumation graves are known from a period that covers several millennia (table 5.1). Using the data, a general picture can be sketched in which similarities and changes throughout the period form the common thread.

The most important similarity is that, without exception, the graves have been discovered at sites that were also in use as settlements. It has already been stated in the introduction that this is primarily the result of the fact that these settlement locations were the reason for research in the first place. The chronological relationship between site and grave is, generally, difficult to determine (Swifterbant-S2, Swifterbant-S3, Swifterbant-S4, S11, Urk-E4 and Schokland-P14). At Dronten-N23 the evidence suggests that hunter-gatherers continued to return to the settlement site over the course of millennia. For Dronten, the grave is interpreted as the very last activity that took place.[445] A similar conclusion could be drawn for the dune on which Swifterbant-S21 and Swifterbant-S22/S23 lie. The flint[446], the ^{14}C dates from charcoal[447] and the pit

---

441 Ten Anscher 2012, 349.
442 Ten Anscher 2012, 349-351.
443 Ten Anscher 2012, 371.
444 Contra Ten Anscher 2012, 351-252.

445 Müller *et al.* 2011: 401.
446 Devriendt 2013: 146-151; Price 1981.
447 Lanting & Van der Plicht 1997/1998: 145.

hearths[448] all make clear that Mesolithic settlement on the dune where the two cemeteries were found dates to the period 6700-5000 cal. BC.[449] The graves date several centuries later (4600-4000 cal. BC). Evidence for settlement during the period of cemetery use is very limited. Typical Neolithic lithic artefacts such as quern-stones have not been found.[450] Excavations on Swifterbant-S25, in the flank zone of the dune, show that the function of the site differs greatly to other contemporary levee sites. The use of flint on the site focuses more on consumption than on production (less debitage and cores) and the small amount of pottery found seems to represent a thinner-walled subset of the pottery more typical on settlement sites. The small amount of pottery present could be interpreted as being part of a death ritual.[451] This would mean that the dune on which Swifterbant-S21 and Swifterbant-S22-S23 lie in the period 4600-4000 cal. BC should primarily be interpreted as a cemetery. Both locations illustrate that we have to be cautious in concluding that the graves were dug in areas of inhabited settlements. In both instances, the same location was first used for settlement and then later re-used as a cemetery. It is unclear whether such an interpretation of changing use can be applied to other locations were graves have been found, for instance Urk-E4.

The dataset from Flevoland makes it clear that inhumations were common during the Late Mesolithic – Late Neolithic. The positioning of the body in the grave shows little variation, with the exception of a change around 4000 cal. BC (zie below). The different periods display no variation in the age at death. The underrepresentation of graves of young children is also a characteristic of the whole dataset. Grave goods are scarce throughout the whole period. In the Late Neolithic cemetery at Schokland-P14 grave goods are found in most of the graves, but are still limited in number.

Any analysis of changes in the burial ritual is limited to a comparison on the period Classical Swifterbant/Pre-Drouwen and the Late Neolithic. It appears that until around 4000 cal. BC, the deceased were only buried in an extended position on their bags. After 4000 cal. BC crouched burials, with raised legs begin to appear. The limited number of reliable ^{14}C dates do not allow conclusions that could confirm either the replacement of the first tradition with the second, or that both burial traditions existed side by side in the pre-Drouwen period. Taking a wider geographical area into consideration, the sites of Schipluiden and Ypenburg can be used as a mirror. Both sites belong to the Hazendonk-group and date to the first half of the fourth millennium cal. BC which makes them contemporary with the pre-Drouwen period. Both sites are settlements where graves have also been found.[452] Schipluiden has seven graves with the same number of burials[453]; Ypenburg has 32 graves containing 42 burials.[454] Both sites have examples of burials in an extending position on the back (n=3) as well as crouched burials, with raised legs (n=30). The comparison supports the second hypothesis put forward, that in Flevoland both burial traditions probably existed side by side after 4000 cal. BC.

The Late Neolithic cemetery of Schokland-P14 forms a great contrast to the preceding period. Here evidence shows a standard crouched burial tradition, with the body being positioned on its right side. Another important difference is the evidence for coffins or grave chambers from wood and bark in the Late Neolithic cemetery, as well as the appearance of whole pots as grave goods. The burial mound surrounded by a ring of poles is a new addition to the Flevoland dataset. All these aspects connect seamlessly with the Corded Ward burial mounds that we find in large parts of northern Europe.[455]

5.3 Other cultural practices with human bones

The surviving burial ritual as described above is certainly not the only evidence that exists for the way in which the deceased were treated in the past. At various cemetery sites evidence has been found for graves containing selected skeletal parts and for individual disarticulated bones being found outside a grave context (Swifterbant-S2, Swifterbant-S4, Swifterbant-S21, Urk-E4). In addition, loose, disarticulated human skeletal parts have been found on sites where there is no evidence for graves (Hoge Vaart-A27; Swifterbant-S3). Evidence has been found in the Late Mesolithic and in the Classical Swifterbant/pre-Drouwen period. It is important to reiterate here that our field of vision is limited to settlement contexts: the manipulation or use of human skeletal material is visible only if its manifestation falls within this specific field of vision.

Skeletal remains that are found outside a burial context cannot be interpreted without some degree of ambiguity. There are a number of finds in Flevoland for which it is not clear whether they come from a disturbed grave or whether they represent a skull burial or another sort of manipulation of human skeletal material. The following finds are not taken account in this discussion because of the lack of provenance data.

---

448 Price 1981: fig. 4.
449 Geuverink, Raemaekers & Devriendt 2009: fig. 2.
450 Devriendt 2013: tabel 4.4.
451 Raemaekers et al. 2014.

452 See also Louwe Kooijmans 2009.
453 Smits & Louwe Kooijmans 2006.
454 Baetsen 2008.
455 E.g. Drenth & Lohof 2005 (the Netherlands); Hübner 2011 (Denmark); Strahl 1990 (North West Germany).

Nr	Site	Skeletal element	Sex	Age	Date range (calBC)	Remark
1	Hoge Vaart-A27	lower jaw fragment		child	4950-4460	
2	Hoge Vaart-A27	skull fragment		adult	4950-4460	
3	Hoge Vaart-A27	axis fragment		adult	4950-4460	
4	Hoge Vaart-A27	phalanx		adult	4950-4460	
5	Hoge Vaart-A27	pelvis fragment		adult	4950-4460	
6	Swifterbant-S21	skull		young adult	4600-4000	
7	Swifterbant-S21	mandibula		young adult	4600-4000	
8	Swifterbant-S21	skull	male (3)	adult	4240-4040	GrA 38138: 5305 ± 30 BP
9	Swifterbant-S2	humerus			4300-4000	Belongs to grave S2-V?
10	Swifterbant-S2	femur	female?	adolescent or adult	4300-4000	
11	Swifterbant-S2	tuber calcanei?			4300-4000	
12	Swifterbant-S3	mandibula	female (2)	adolescent	4300-4000	
13	Swifterbant-S3	tibia			4300-4000	
14	Swifterbant-S4	skull fragment	male?	20-40 yrs	4300-4000	
15	Urk-E4-2	skull	female?	35-45 yrs	4300-3400	
16	Urk-E4-6	skull		23-40 yrs	4300-3400	

Table 5.2 Overview of skeletal elements outside graves.

In the north of the Swifterbant region is the Kamperhoek nature reserve. Shortly after the reclamation of the polder, sand extraction took place as part of the construction work for the modern village at Swifterbant. An adult skull has recently been recovered from this sand. Even though this find consists exclusively of a skull, it has not been interpreted as a skull burial. Since it was known that the sand had been extracted from a river dune where archaeological remains had previously been found (S71) and supported by the expounded hypothesis that cemeteries are often found on river dunes, the discovery of the skull is interpreted here as a possible first indication for a cemetery on the location.[456] A femur and two skull fragments have been recovered from the site of Emmeloord-J97. The publication doesn't provide enough information to determine if the bones were originally from a disturbed grave or graves, or whether they were unstratified finds from within the settlement.[457] Human skeletal material found outside the graves has also been documented on Schokland-P14. The assemblage consists of seven finds, three of which are assumed to come from a disturbed grave because of the proximity of the cemetery.[458] Given the presence of settlement material in the same area, however, an interpretation as settlement material is equally plausible.

Because of the lack of provenance data and reliable ^{14}C dates, these human remains are not discussed further here. On S3 the discovery of eight milk teeth will also not be discussed further, since the finds are not evidence enough to establish that children were ever at the site.[459]

What information is then available for the manipulation of human remains? Selected skeletal parts have been found in two contexts. Discussed briefly above, multiple burials have been found in which a few adult bones have been found in association with the remains of children. In addition, skull burials have been found at Swifterbant-S22 and Schokland-P14: in burial pits that were too small to have ever contained a complete skeleton, and from which no other bone material was recovered (table 5.1). In Schokland-P14-3 the evidence might even suggest that two skulls had been buried in the same pit.[460] The skull burials do however fit into the same pattern of age at death as observed for the regular burials.

Secondly, human skeletal material has also been found outside of graves (table 5.2). What stands out in this regard is that most of the remains are either skulls or skull fragments. Whilst these bones, because of their size and thickness are more likely to be better preserved as well as more readily identifiable, a large quantity of unburnt animal bone has been recovered from the sites of Swifterbant-S2, Swifterbant-S3 and Swifterbant-S4. The state of preservation of this bone was such that

---

456 The skull is ^{14}C dated (GrA-53171: 5555 ± 35 BP, c. 4460-4340 cal. BC). With thanks to D. Velthuizen, Nieuwland Erfgoed Centrum.
457 Ten Anscher 2012, 695.
458 Ten Anscher 2012, p 348-349.
459 Meiklejohn & Constandse-Westermann 1978, 73-74, 82.
460 Ten Anscher 2012, 318-319.

any other human skeletal parts should also have been preserved. All the bone material has been analysed which means that any other human bone would not have gone unrecognised. The significance of the skulls is therefore not a consequence of either poor preservation or determination, but a prehistoric reality.

Both types of observations show that human skeletons were manipulated. Parts of the skeleton, skulls in particular, but also long bones, circulated in the society of the living before they were at some point taken out of circulation, either by being added to a regular grave or by being interred as a separate skull burial ((in both cases as a *pars pro toto* burial). The age and gender of the individuals represented by these bones does not appear to deviate from that of the skeletons of individuals who were given a regular burial. It remains uncertain what criteria were applied to warrant an inhumation burial on the one hand, or the circulation (possibly over a long period?) from a skull or long bones on the other.

It is by no means certain to what sort of death ritual the loose bones can be connected. One possibility is that certain skeletal parts of a normal inhumation could have been disinterred once the bones had been defleshed. However, the absence of inhumations without a skull suggests an alternative explanation needs to be sought. Another possibility could be that the corpse was laid out in the open air until the flesh had been stripped (excarnation). Such a location may well have been excavated at Hekelingen III: six split oak posts had been placed in a rectangle measuring 1,5 x 0,8 m, it is important to state that this is the only evidence for the use of oak at this wetland site. Bones of arms, legs, shoulder, lower jaw and teeth of an upper jaw were found around and between these posts. The excavators have suggested that the poles represent the remains of a platform on which the corpse would have been laid out to undergo the process of excarnation.[461] The Flevoland dataset has one piece of archaeological evidence that may fit in with this: a tibia from Swifterbant-S3 shows evidence for gnaw marks.[462] These marks could have been made during an excarnation process or thereafter. The absence of any further supporting archaeological evidence on the site could indicate that such death rituals took place outside the settlement area.

Even if the selection criteria are not known, it is a fact that there were at least three different ways of dealing with the deceased. Next to the formal burial practice of inhumation and cremation, a number of bodies were not buried. Instead, after undergoing an unknown ritual, their bones were circulated for a time in the community of the living. In addition, it is likely that the bodies of young children were treated in a way that has let no archaeological trace.

5.4 Depositions

Special treatment was not just reserved for human bones. Other materials were also treated or used in a manner that cannot be explained purely in terms of functionality.[463] How can we differentiate depositions from other finds assemblages? For this study, was decided to use a definition that fits the available data. The precondition for an interpretation as deposition is that the provenance must be known, In the absence of any information it is difficult to determine with certainty whether the original provenance was a settlement site, a grave or, for instance, an object deliberately deposited as a votive offering in a river valley. This proposition can best be defended by using research carried out in the Noordoostpolder as a case study.[464]

The research differentiates between various types of sites, namely (temporary and permanent) settlements, fields, graves and ritual depositions. The latter category consists of finds selected on the basis of three criteria.[465] Firstly, complete objects that have been chosen with care. In other words, objects that are not yet at the end of their useful life in a functional sense. Secondly, the significance of wet contexts is often pointed out in relation to depositions, but dry locations could also have been used. This means that, in fact no location can be excluded on the basis of landscape characteristics. The third criterium is that of 'archaeological indicators': the context in which the objects were found. The interpretation of finds from a wet context as representing a deposition is seen as quite plausible, whilst the same finds recovered from a settlement context could equally be interpreted as the displaced contents of a grave. This third criterium is of critical importance: "The finds [the votive depositions] are not or hardly distinguishable from indicators for a temporary or permanent settlement and will in practice be seen as such, especially when they are found lying on the ground".[466] This means that the isolated, unstratified discovery of, for example, a flint axe, is impossible to interpret without information about the original provenance: would this have been a deposition, a (disturbed) grave or a settlement? Such isolated finds cannot be included with any certainty in any corpus of deposition sites. To look at this from the perspective of the dataset from Flevoland, all so-called depositions have been found in or near a settlement.

---

461 Verhart 2010, 170-171 and personal communication 2014.
462 Meiklejohn & Constandse-Westermann 1978, 74.
463 See also Raemaekers 2019.
464 Ten Anscher 2012, 503-536.
465 Ten Anscher 2012, 506.
466 Ten Anscher 2012, 506.

There are three further criteria used in the interpretation of finds as depositions. Firstly, a deposition can be recognised by the presence of (almost) complete objects. Objects that have clearly not become unusable, but by being included in a deposition have deliberately been taken out of use. The interpretation of such objects as depositions is commonplace.[467] The second criterium is the composition of an assemblage made up of different objects in a deposition. A clear understanding of provenance is essential. An assemblage of finds found in a feature are more readily interpreted as deposition rather than a concentration of finds from a settlement layer. The interpretation of a group of objects as a deposition rather than as refuse completely depends on the type of objects that make up the assemblage. A concentration of flint debitage would obviously not be seen as a deposition, whereas a concentration of flint tools most probably would (due to the first criterium). The third criterium is that of repetition. The spatial association of two or more objects found together once could be seen as a coincidence, but if these objects are frequently found together, then the spatial association may well be a reflection of a more conceptual association. As such, these types of finds also represent a deposition.

On the basis of these criteria, three groups of depositions can be distinguished in the Flevoland dataset (table 5.3). These contexts should be seen as examples, because little attention was paid to the identification of depositions in the older excavations. Even in the recent publications on Swifterbant[468] the lack of detailed field documentation means that there are few starting points to support the identification of depositional activities.[469] The three suggested pottery depositions from Swifterbant-S3 are based on the illustrated pottery fragments and the added description of the spatial distribution of the pottery sherds.[470] The three pots are to a significant extent complete, whereas the sherds have been recovered from one collection unit in the excavation trench. It is important to note that in the Swifterbant area, apart from postholes and graves, hardly any other features have been documented so that hardly any closed archaeological contexts have been found.

The first group of depositions consists of flint assemblages dating from the Late Mesolithic to Classic Swifterbant. These have been selected because of the inclusion of apparently still functional tools (criterium 1) and because they represent a finds composition that differs from the general picture (criterium 2). In four cases these are assemblages of material that was found in or near the settlement area (fig. 5.17). The fourth deposition was found in a pit. It is important to state that in the absence of any closed context such as a pit, we are actually dealing with four concentrations of flint. These flint concentrations are interesting because their composition differs from the general picture from finds assemblages. Striking by this group is that, despite the large-scale research that has taken place in the Swifterbant area, there is hardly any evidence for such observations in the later sites. The flint concentration from Swifterbant-S4, comprising seven very similar blades, was found in an excavation unit measuring 50 x 50 cm. There are several possible interpretations for this find. It could have been a votive deposition, the result of a specific activity on this location or else a spatial association that was completely coincidental.[471] The small number of similar assemblages recorded in the Swifterbant area could simply be the result of very little attention being given to this type of finds context in excavations (see above). It is also possible that this sort of deposition did not take place, or at least less frequently in the later periods.[472] In remains difficult to determine whether such assemblages should be interpreted as just stock put aside (a 'cache') or as depositions.

The second group are the pottery depositions (fig. 5.18). Pottery depositions can be differentiated on the basis of functional criterium (1) and the fact that this sort of context occurs repeatedly (criterium 3). Pottery depositions are found on all the larger excavations, with the exception of Hoge Vaart-A27. Relatively little, but extremely fragmented pottery was found on this site, sometimes in dense clusters. It is therefore not clear that such depositions were actually absent on Hoge Vaart-A27. Evidence for pottery depositions in this phase (Early Swifterbant) has been found on sites further afield, such as Hardinxveld-Giessendam De Bruin[473], Hardinxveld-Giessendam Polderweg[474] en Bronneger.[475] Pottery depositions are also frequently recognised in the Classical Swifterbant/pre-Drouwen period. From later phases, evidence for such observations is only available from Schokland-P14, and then only in the form of a wooden bowl. This is included in the category pottery depositions because it is a vessel. The latest pottery deposition was found on Emmeloord-J97. A so-called collared, or inverted

---

467 E.g. Koch 1998 (Neolithic pottery, Denmark), Wentink 2006 (Neolithic axes, The Netherlands) and Fontijn 2002 (bronze objects, the Netherlands).
468 De Roever 2004; Devriendt 2013.
469 Devriendt is of the opinion that the field documentation makes it impossible to distinguish depositions from other sort of contexts (personal communication 2014).
470 De Roever 2004, figs 9-26 and appendix.

471 Devriendt 2013, p 198-199.
472 Devriendt sees this as a real possibility (personal communication 2014).
473 Raemaekers 2001a: fig. 5.5.
474 Raemaekers 2001b: fig. 5.5.
475 Kroezenga *et al.* 1991: fig. 3.

Nr	Site	Deposition type	Context	Functionality (criterium 1)	Assemblage (criterium 2)	Repitition (criterium 3)	Date range (calBC)	Remark
1	Dronten-N23	Circa 41 flint artefacts including three unworked pieces, three cores, four tested pieces, six flakes and three blades	Within settlement	x	X		6600-6250	
2	Hoge Vaart-A27	21 tested nodules and pre-cores, associated with four oak trunks	In peat layer near settlement	x	X		4790-4540	
3	Hoge Vaart-A27	5 exhausted blade cores and 4 large refitting flakes, associated with one core and standing oak post	Near settlement	x	X		4790-4540	
4	Hoge Vaart-A27	100 flakes	In pit in peat leayer near settlement	x	X		4790-4540	
5	Swifterbant-S4	Seven long blades	Within settlement	x	X		4300-4000	
6	Schokland-P14	Complete pot	Within settlement	x		x	4300-4200	
7	Swifterbant-S3	Complete pot	In gully near settlement	x		x	4300-4000	De Roever 2004: fig. 9d
8	Swifterbant-S3	Complete pot	Within settlement	x		x	4300-4000	De Roever 2004: fig. 9b
9	Swifterbant-S3	Complete pot	Within settlement	x		x	4300-4000	De Roever 2004: fig. 9g
10	Schokland-P14	Complete pot	In gully near settlement	x		x	4300-4000	
11	Urk-E4	Complete pot	On peaty slope near settlement	x		x	4230-3980	
12	Zeewolde-OZ35/36	Complete pot	Within settlement	x		x	3630-3360	GrN 26612: 4660 ± 40 BP
13	Schokland-P14	Complete wooden bowl	Within settlement	x		x	3400-2800	
14	Emmeloord-J97	Complete pot	Around wooden pole, part of fish-weir	x		x	2300-1900	
15	Hoge Vaart-A27	Two auroch skulls associated with standing oak post	Bank of gully zone near settlement			x	4950-4460	
16	Hoge Vaart-A27	Auroch skull	On gully bottom near settlement			x	4950-4460	
17	Hoge Vaart-A27	Anterior parts of wild boar mandibles with lacking tusks	In gully near settlement			x	4950-4460	

Table 5.3: Overview of depositions.

pot beaker[476] was found around a wooden pole that was part of the structure of a fish weir.

The third group consists of three depositions of animal skulls in the channel near the site of Hoge Vaart-A27 (fig. 5.19). These have been selected because of their strong mutual similarity (criterium 3). Interesting is the association with an oak post and tree stumps that was not only observed by two flint depositions from Hoge Vaart-A27, but also by the deposition of two aurochs skulls at this site. The lack of comparable observations from other site must be seen at the very least as a result of the lack of attention given to depositions on older excavations. There may also be a link here with a more local tradition in which the oak played a role in a conceptual relationship with a spiritual world.

Depositions make it clear that material culture was not only used to perform the more mundane domestic activities. Material culture played an important role in shaping the conceptual relationships with ancestors, gods and/or spirit world. On the basis of available data, there appear to be few regional differences and developments to be seen in depositions. This can be because the dataset is too small, but given the fact that comparable

---

476 Bell Beaker Culture.

Figure 5.17a-b: Almere Hoge Vaart-A27: deposition of exhausted flint cores and several refitted flakes (number 3 in table 5.3) (source: Peeters 2007).

depositions are also common in other periods and areas, a more restrained interpretation is appropriate. It is quite possible that whilst the methods used to depict these conceptual relationships are similar, the specific content and significance of the ritual practices could have been very diverse.[477]

5.5 Materiality

The central theme of this chapter could be paraphrased to be the interaction of prehistoric man with the notions of a spirit world. Archaeological research into the cosmological relationships between man and conceptual identities can be based on aspects such as burial practice and depositions. In the archaeology of Flevoland, there also exist examples of 'non-functional' human actions that could be interpreted as examples of the conceptual organisation and shaping, or formalisation, of the material culture.

The first example of this is the treatment of the pottery on the sites of Swifterbant-S2, Swifterbant-S3 and Swifterbant-S4. An analysis of the pottery from these three sites made it clear that, at sherd level, there is a correlation between the type of tempering used, the wall thicknesses and the occurrence of decoration. A possible functional significance for this correlation could be investigated thanks to the presence of organic residues on pottery from Swifterbant-S3. This residue was analysed using the SEM method (*Scanning Electron Microscope*) and the DTMS method (*Direct Temperature-resolved Mass Spectrometry*). The study showed that two functionally distinct groups of pottery were present on Swifterbant-S3. The first group was usually tempered with organic material and used

---

477 Raemaekers 2019.

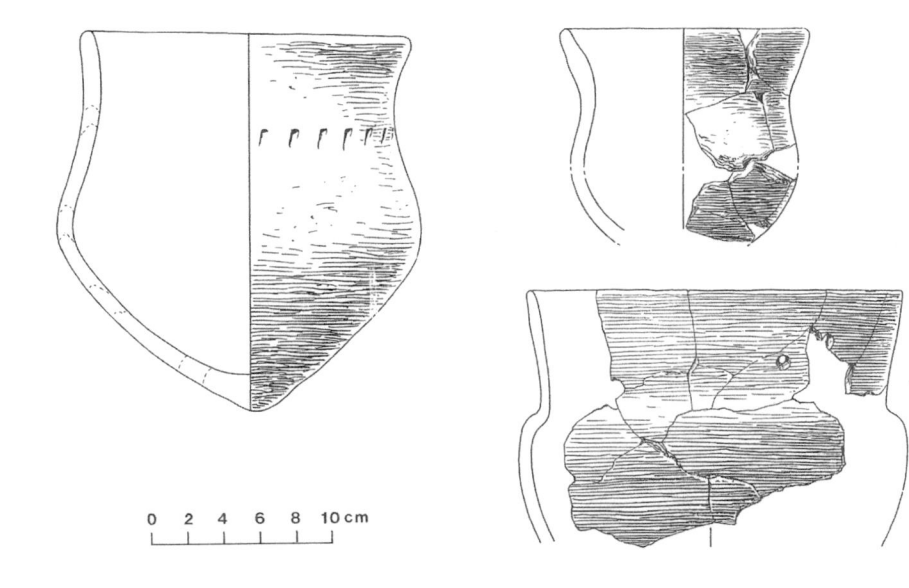

Figure 5.18: Swifterbant-S3: pottery depositions (number 7-9 table 5.3) (source: De Roever 2004).

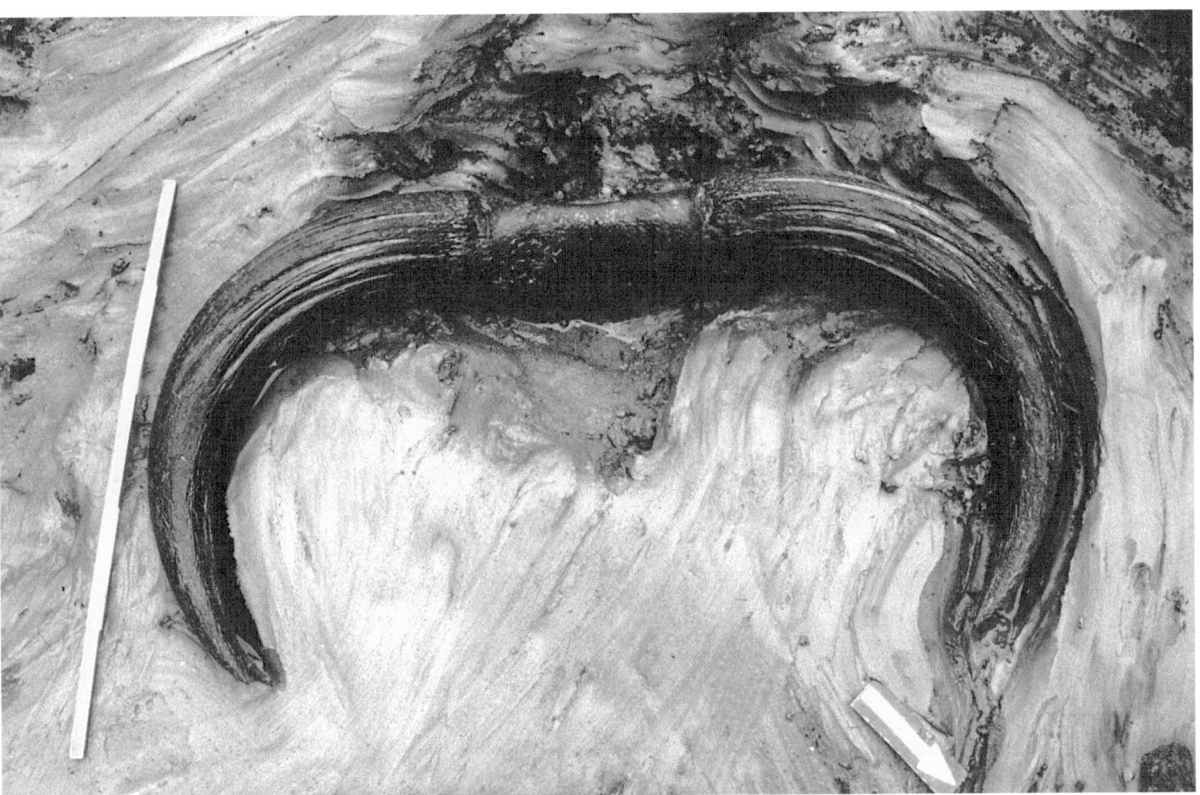

Figure 5.19: Almere Hoge Vaart-A27: one of the three aurochs skulls in situ (number 16 in table 5.3).

PEOPLE, RITUAL AND MEANING

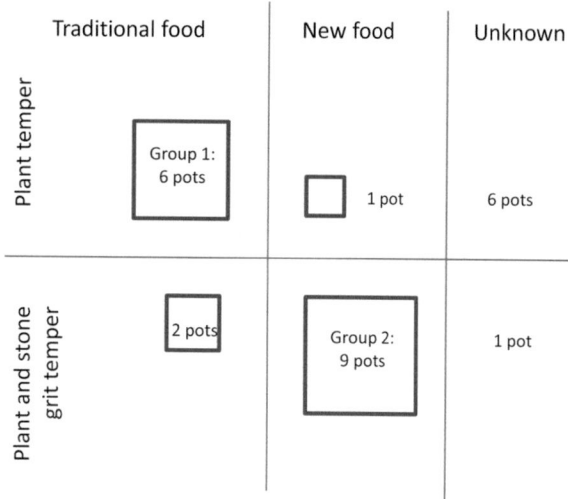

Figure 5.20: Functional groups of pottery as found at Swifterbant-S3 (source: Raemaekers, Kubiak-Martens & Oudemans 2013).

for meals without emmer wheat. The second group was usually tempered with organic material and grit and was used for meals with emmer wheat (fig. 5.20). The analysis shows that the inhabitants of Swifterbant-S3 made use of different pots or vessels for different meals.[478]

Since evidence for the oldest grain remains in this region date to the period in which Swiferbant-S3 was in use, the hypothesis was tested as to whether grain consumption was introduced during the build-up of the finds layer on the site. This would mean that pots with emmer deposits should only have been found in the upper part of the finds layer on the site and that thereby the introduction of cooking with grain can be dated on Swifterbant-S3. Unfortunately, the hypothesis could not be tested because no pots found in the lower stratigraphical layers of the site were included in the study (most of the later pottery sherds were found in the top of the finds layer). To enable further investigation of the hypothesis, pots from Swifterbant-S3 were ordered on a scale between two extreme stereotypes. On the one hand, thick-walled pots tempered with organic material and of poor quality, on the other hand thin-walled pots that were also tempered with grit and of good quality. All the illustrated pottery[479] could be ordered on the basis of these two extremes. The result of this ordering showed that both extreme stereotypes were actually present during the whole period of occupation of Swifterbant-S3, as well as pots that combined the characteristics of both stereotypes (fig. 5.21). Significant is that during the occupation of the site, the higher quality, thin-walled, grit-tempered pottery increased in importance at the cost of pots with intermediate characteristics. This development is interpreted as the conscious design of two contrasting technological groups. Thanks to the residue analysis, these two groups can be linked to function. This means that the preparation of meals using wheat developed hand in hand with the development of these two technological groups. It can be concluded that the potters from Swifterbant-S3 conceptualised the incorporation of grain consumption into the community: the preparation of meals with emmer was apparently something fundamentally different to meals without emmer.[480]

Quern-stones also apparently played an important role in prehistoric Flevoland, at least on the sites in Swifterbant.[481] More fragments of quern-stone have been found than of any other implement made of natural stone. On Swifterbant-S2 as many as 19 of the 25 quern-stones found were fragmented.[482] This is the first indication that this category of implement, or tool, had a special significance. This impression is strengthened by the discovery of fragments from one quern-stone made of gneiss discovered on the sites of Swifterbant-S2 en Swifterbant-S3. A small fragment from Swifterbant-S2 actually fits onto a larger fragment found at Swifterbant-S3. It is, however, possible that three other fragments from Swifterbant-S2 and six others from Swifterbant-S3 also belong to the same quern-stone.[483] Such discoveries indicate that quern-stones were deliberately broken, after which they were transferred from one location to the other. Clearly, from the perspective of functionality, unusable fragments of quern-stones still had an important role to play in the community. The importance of deliberately broken material culture has gained a lot of attention in the profession since the publication of *Fragmentation in Archaeology*.[484] It seems that two aspects of this process are important: the destruction of the objects and then the taking of parts of the broken objects. The fragments then function as a reminder of an activity in a different place, as 'pieces of place'.[485]

Flint has been found on various sites that has clearly been brought from a great distance (see chapter 4). Given their small number and there functional equivalence

---

478 Raemaekers, Kubiak-Martens & Oudemans 2013; for new lipid analyses see Demirci *et al* 2020.
479 De Roever 2004, figs 9-20 and appendix.
480 Raemaekers 2015.
481 Devriendt 2013. Quern-stones are found after the Neolithic. Ten Anscher presents a brief summary of the stone found in the excavation trench 89-17 (2012, p 419-420). The remaining natural stone is unstratified and cannot therefore be placed with certainty within any phase of the long occupation history on P14. Ten Anscher comments that a total of 51 lithic implements were found in the excavation trench, including ten quern-stones. Evidence for (deliberate) fragmentation is not discussed.
482 Devriendt 2013, 66.
483 Devriendt 2013, 94.
484 Chapman 2000.
485 Van Gijn 2010, 166-167.

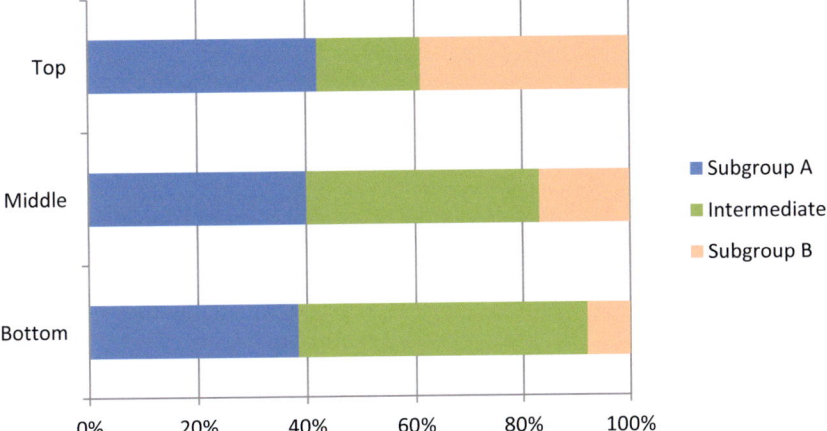

Figure 5.21: Schematic overview of the development in pottery characteristics at Swifterbant-S3 (after Raemaekers 2015).

with more readily available flint material, these finds can be seen as indications that prehistoric man in Flevoland attributed a significance to these exotica.

These three illustrations of the meaningful treatment of material culture are given as examples. They represent the archaeologically tangible remains of human behaviour. They make it clear that prehistoric people gave meaning to the material world with which they surrounded themselves. The three examples each give a different perspective.

The example of the pottery shows how the introduction of emmer wheat as a new food source led to the development of new characteristics in the pots used to prepare meals. In this way Swifterbant pottery developed into Pre-Drouwen pottery.[486] The destruction of quernstones and the taking of the fragments is an indication of the great conceptual significance the quern-stones had, in contrast to other categories of lithic artefacts.[487] Both of these finds groups underline the importance of the same new food source: grain. Its importance lay not so much in the great contribution that this food source made to the diet (chapter 4.6), but in the fundamental reordering of the dependency relationships between prehistoric man and the environment in which he lived. The exotic flint is not just evidence for the existence of exchange networks over long distances, but shows that prehistoric man recognised these types. As such these exotic objects were imbued with the same symbolic association as a 'piece of place' as the fragments of quern-stone.

5.6 Conclusions

The archaeological record in Flevoland has provided many points that together paint a picture of prehistoric cultural life. Since the available data is unevenly distributed through the different periods of prehistory, it is impossible to follow developments through prehistory as a whole. The most information we have comes from the period 5000-3400 cal. BC (Early Swifterbant up to and including Pre-Drouwen). Another limitation lies in the fact that the data comes, almost without exception, from settlement contexts. As a result, any meaningful behaviour in other parts of the landscape are not documented by research.

The aim of this chapter was to look for archaeological evidence that could be used to show that prehistoric man in Flevoland was more than just a walking meal machine, but rather in his behaviour and actions gave shape to the conceptual relationships that surely existed with ancestors and the imagined spirit world. Evidence for such relationships are proposed on the basis of the remains of the burial practice, the different attitude and treatment of human bone, depositions and the attitude towards and shaping of the material culture. Although it remains impossible to give substance to the thought processes behind these conceptual data, it is certainly clear that these relationships played an important role in these prehistoric communities.

---

486 Raemaekers 2015.
487 Devriendt 2013, 66.

Chapter 6

From land to water

Geomorphological, hydrological and ecological developments in Flevoland from the Late Glacial to the end of the Subboreal

L.I. Kooistra & J.H.M. Peeters

6.1 Introduction

The previous chapters have discussed the prehistoric landscape of Flevoland in various ways. Chapter 3 showed that the dynamics of the landscape had a major impact on the texture of the subsurface, the formation of the buried archaeology and the potential for identifying and investigating sites. Chapters 4 and 5 present an account of the prehistoric inhabitants, who utilised a wide range of naturally available resources, and relied on a diverse range of landscape components for their survival. The close relationship between human and environment evidenced by the archaeological remains brings us, in this chapter, to the question of what the landscape of Flevoland looked like and how it changed over time.

The changes in the landscape could hardly be greater than those that occurred in Flevoland between the Late Glacial and the end of the Subboreal (c. 12,000 to 1100 BC). The climate, hydrological processes and developments in vegetation drove geomorphological processes, and vice versa. As a result, Flevoland developed during this period from a fluvioglacial and coversand region with a dry Arctic climate, through a number of transitional stages, into a region of extensive mires (including bogs) with a temperate climate.[488] Thanks to a wealth of geomorphological, hydrological and palaeoecological data collected since the 1930s, we can gain an impression of the landscapes that prehistoric man would have witnessed between roughly 12,000 and 1100 BC.

This chapter concentrates on describing the contemporary "living landscape" comprising the flora and fauna as components of the environment in which prehistoric man lived. By combining existing data on the landscape with an emphasis on the available palaeobotanical data, a new interpretation will be presented here that focuses on the character of the vegetation, and how it changed. This chapter therefore goes a step further than the previous chapters, where the main objective was to present a synthesis of *archaeological* information contained in reports, books and academic papers. The

---

488 Marshes (wet areas with herbaceous plants) and swamps (wet areas with trees) occur on mineral soils; mires are wet areas which develop on peat (incompletely decomposed organic matter). There are three types of mires mentioned in this chapter, fens (mires with herbaceous plants), carrs (mires with trees) and bogs (ombrotrophic mires with mainly bog-moss). For definitions see for example Pons 1992, 7-12; Casparie & Streefkerk 1992, 84-86; Van der Linden & Kooistra 2019.

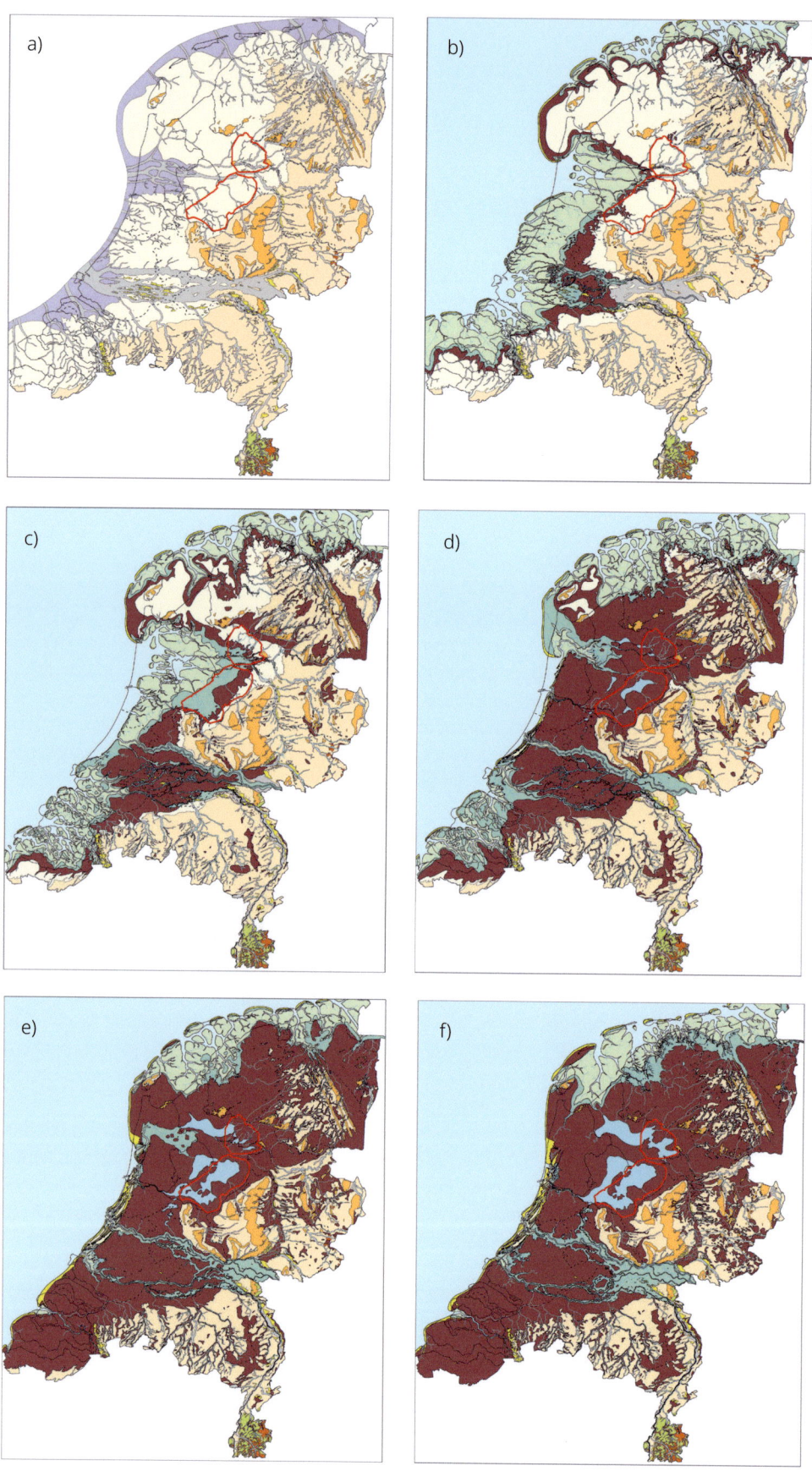

158 RESURFACING THE SUBMERGED PAST

Figure 6.1 (previous page): Palaeogeographical maps of the Netherlands in the Holocene, with Flevoland in the centre (outlined in red). a) 9000 cal. BC; b) 5500 cal. BC; c) 3850 cal. BC; d) 2750 cal. BC; e) 1500 cal. BC; f) 500 cal. BC (after Vos & de Vries 2018; Vos *et al.* 2020).

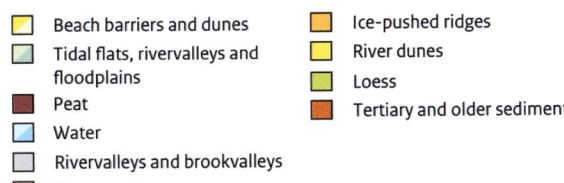

- Beach barriers and dunes
- Tidal flats, rivervalleys and floodplains
- Peat
- Water
- Rivervalleys and brookvalleys
- Coversand
- Ice-pushed ridges
- River dunes
- Loess
- Tertiary and older sediments

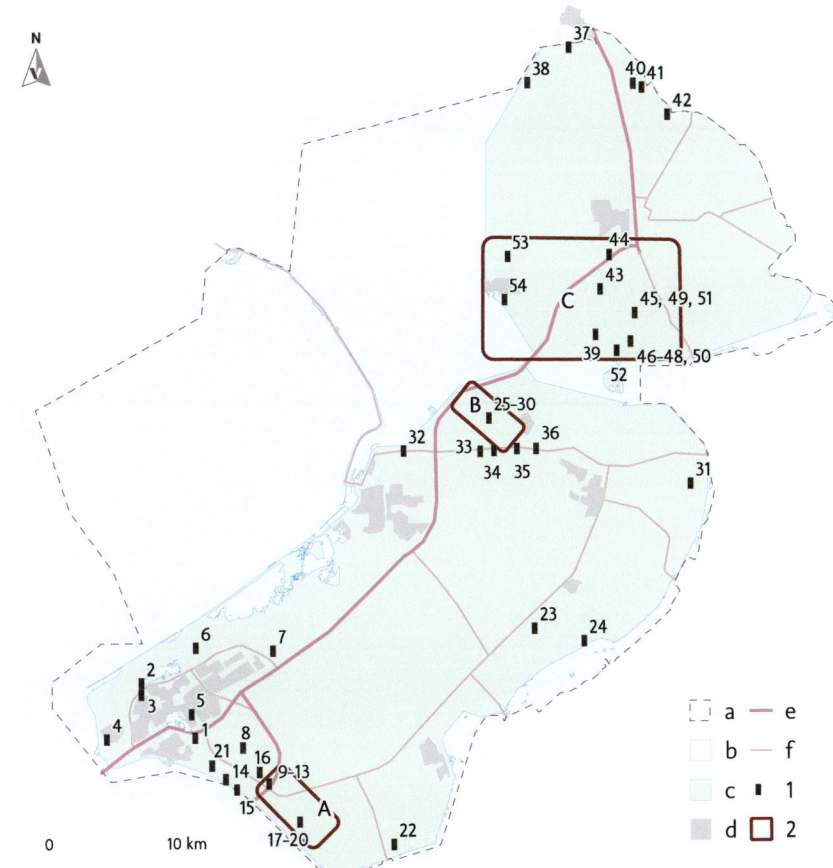

Figure 6.2: Topographical map of the province of Flevoland, with the investigations discussed in this chapter indicated. The numbers refer to the sites/locations in table 6.1. Legend: a) province border, b) water, c) rural areas, d) urban areas, e) highways, f) main roads, 1) sites/locations (table 6.1) and 2) microregions.

palaeobotanical data, which have largely been interpreted on the basis of very general models of vegetation succession, need, in contrast, to be reconsidered in preparation for the comprehensive synthesis of information on the prehistory of Flevoland that will be presented in the final chapter.

6.2 Study of the history of the Flevoland landscape in broad outline

This chapter relies on books, reports and academic papers on the geology, pedology and palaeobiology of Flevoland, the majority in Dutch, published between the 1930s and 2014 (table 6.1). The information refers to four spatial scales. In order to better understand the development of the landscape in Flevoland, the reconstructed global palaeogeographical development of the Netherlands (the supraregion) has been used as a basis. This development is summarised in a number of maps showing the distribution of the subsurface in each period, including the course of rivers and the bodies of water present (fig. 6.1).[489] In the 1990s and in the first decade of the 21st century, the development of the landscape in Flevoland as a whole (the

---

[489] Vos *et al.* 2009; Vos *et al.* 2020.

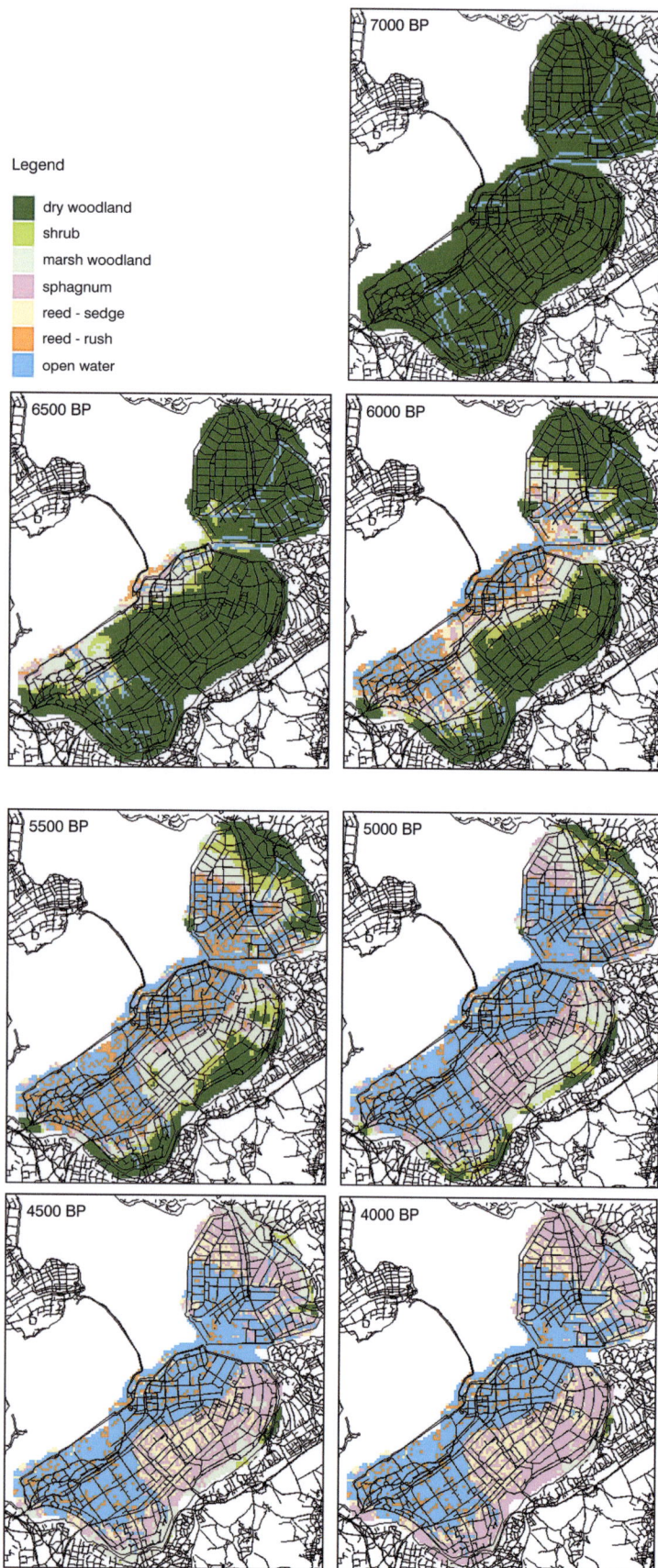

Figure 6.3: Landscape developments in Flevoland according to the 'Capillary-Growth-Erosion' (CGE) - model with intervals of 500 years (source: Peeters 2007).

Site number	Site name	Toponym/place name	Reference	Late Glacial	Preboreal	Boreal	Early Atlantic	Middle Atlantic	Late Atlantic	Atlantic	Subboreal	x-coordinate	y-coordinate
Zuidelijk Flevoland													
1	Almere kasteel	Kasteel, Fongerspad	Van Smeerdijk 2002	X	x	x	x	x	.	x	.	145.700	484.600
2	Almere-Noorderplassen-West	Plangebied 2X3, Noorderplassen-West	Van Smeerdijk 2006	.	.	.	.	x	.	.	.	141.413	488.886
3	Almere busbaan	Busbaan, Boring Almere 1	Bunnik & Verbruggen 2010	.	x	.	.	x	.	x	.	141.402	487.976
4	Almere Poort	Nederlandstraat / De Geest	De Moor et al. 2013	X	x	x	x	x	x	.	.	138.631	484.514
5	Almere-Veluwedreef	Veluwedreef	Makaske et al. 2002a	.	.	.	.	x	x	.	.	145.432	486.438
6	Almere-De Vaart	De Vaart (Gordingweg)	Van der Linden 2010	.	.	.	x	x	x	.	x	145.804	491.626
7	Almere-Kotterbos	Kotterbos	Van Heeringen et al. 2014; Kooistra 2014	.	.	.	.	.	x	.	x	151.900	491.355
8	Almere-Hout	Zwaanpad	Van Smeerdijk 2003	.	.	.	.	x	.	.	.	149.456	483.820
9	Almere-Hogevaart A27	Hogevaart-A27: opgraving	Hogestijn & Peeters (eds) 2001; Brinkkemper et al. 1999; Laarman 2001; Peeters et al. 2001; Peeters 2007; Van Rijn & Kooistra 2001; Visser et al. 2001	.	.	x	.	x	x	.	.	151.520	481.000
10	Almere Hoge Vaart A27	Hogevaart-A27	Spek et al. 2001a&b	.	.	.	x	x	.	.	.	151.520	481.000
11	Almere Hoge Vaart A27	Standaardkern	Gotjé 2001	.	.	.	.	x	x	.	x	151.520	481.000
12	Almere Hoge Vaart A27	Eem1-kern	Gotjé 2001	.	.	.	.	.	x	.	.	151.520	481.000
13	Almere Hoge Vaart A27	Eem2-kern	Gotjé 2001	X	x	.	.	.	x	.	.	151.520	481.000
14	Almere-Musweg	Musweg	Makaske et al. 2002a	.	.	.	.	.	x	.	x	148.079	481.359
15	Almere-Gooimeerdijk	Gooimeerdijk	Makaske et al. 2002a	.	.	.	.	.	.	.	.	148.946	480.549
16	Almere-Tureluurweg	Tureluurweg	Makaske et al. 2002a	.	.	.	.	.	x	.	.	150.753	481.906
17	Almere-Rassenbeektocht	Rassenbeektocht	Makaske et al. 2002b	X	.	.	.	.	.	.	x	153.500	478.800
18	Almere-Winkelweg	Winkelweg	Makaske et al. 2002b	.	.	.	.	.	x	.	.	155.867	477.500
19	Almere-Eemhof	Eemhof	Makaske et al. 2002b	.	.	.	.	.	.	x	x	155.200	475.267
20	Almere-Eemmeerdijk	Eemmeerdijk	Makaske et al. 2002b	.	.	.	.	.	.	.	x	154.500	475.333
17-20	Almere-bodembeschermingsgebied	Bodembeschermingsgebied	Gotjé 1997b; Makaske et al. 2002b	X	.	.	.	.	x	x	x	154.000	478.000
21	Almere MMM	Maatweg – Meesweg – Meentweg	Opbroek & Lohof (eds) 2012; Kooistra 2012a; Bos & Verbruggen 2012	.	x	x	x	x	.	.	x	147.020	482.428
22	Zuidelijk Flevoland-OZ-43	Scheepswrak OZ43	Van Smeerdijk 1989	.	.	.	.	.	x	.	x	161.533	476.200
A	Microregio Hoge Vaart		Peeters 2007	.	.	x	x	x	.	.	.		
Oostelijk Flevoland													
23	Biddinghuizen/Biddingringweg	Biddinghuizen/Biddingringweg	De Jong 1974	X	x	.	.	.	.	.	.	173.000	493.000
24	Flevoland I, II, III	Flevoland I, II, III	Havinga 1963	.	.	x	x	x	x	x	.	177.000	492.000
25	Dronten-N23	N23 / Hanzelijn	Hamburg (eds) 2012; Bouwman & Bos 2012; Kooistra 2012b; Van der Linden 2008; 2012	.	x	x	x	x	.	.	.	171.320	506.973
26	Swifterbant-S2	S2 (Kavel G41-G42)	Huisman et al. 2009; Prummel et al. 2009	.	.	.	.	.	x	.	.	168.133	510.867
27	Swifterbant S3	S3 (Kavel G43)	Casparie et al. 1977; Van Zeist & Palfenier-Vegter 1981 (1983); Cappers & Raemaekers 2008; Raemaekers et al. 2013; Schepers 2014a/b; Zeiler 1986	.	.	.	.	.	x	x	x	168.133	510.133

Table 6.1 (continued on the next page): List of literature references.

Site number	Site name	Toponym/place name	Reference	Late Glacial	Preboreal	Boreal	Early Atlantic	Middle Atlantic	Late Atlantic	Atlantic	Subboreal	x-coordinate	y-coordinate
28	Swifterbant S4	S4 (Kavel G43)	Bakels & Zeiler 2005; Brinkhuizen 1976; Clason & Brinkhuizen 1978; Hullegie 2009; De Jong 1966; Out 2009; Wolf & Cleveringa 2009; Zeiler 1997	.	.	.	.	.	x	.	.	168.140	510.140
29	Swifterbant S5	S5 (Kavel G43)	Van Rooij 2007; Van der Veen 2008	.	.	.	.	.	x	.	.	168.133	510.133
30	Swifterbant S25	S25 (Kavel H45)	Raemaekers et al. 2010	.	.	.	.	.	x	.	.	172.178	510.935
31	Dronten-tunnel Drontermeer	Tunnel Drontermeer	Prangsma (eds) 2009; Bos et al. 2009; Kooistra et al. 2009	.	.	.	x	.	.	.	.	185.723	504.223
32	Hanzelijn deelgeb. I&II	Hanzelijn deelgeb. I&II	De Moor et al. 2009	.	.	.	.	x	x	.	x	162.500	506.867
33	Hanzelijn deelgeb. VI	Hanzelijn deelgeb. VI	De Moor et al. 2009	.	.	.	.	.	x	.	.	168.667	506.800
34	Hanzelijn deelgeb. VII	Hanzelijn deelgeb. VII	De Moor et al. 2009	.	.	.	.	x	.	.	.	169.800	506.867
35	Hanzelijn deelgeb. VIII	Hanzelijn deelgeb. VIII	De Moor et al. 2009	.	.	.	.	x	x	.	x	171.667	507.000
36	Hanzelijn deelgeb. IX	Hanzelijn deelgeb. IX	De Moor et al. 2009	.	.	.	.	x	x	.	x	173.200	507.000
B	Microregio Swifterbant		See literature sites 25-36	.	.	x	x	x	x	.	.		
Noordoostpolder													
37	Kavel A3, A58	Kavel A3, A58	Wiggers 1955, 34-44	x	x	x	x	.	.	.	.	176.000	538.267
38	Kavel A19, A48	Kavel A19, A48	Wiggers 1955, 34-44	x	.	.	.	.	.	.	.	172.667	535.533
39	Kavel E155	Kavel E155	Wiggers 1955, 40-42	x	x	.	.	.	.	.	.	178.067	515.867
40	Kavel K24	Kavel K24	Wiggers 1955, 47-51	.	x	x	x	.	.	.	.	181.233	535.433
41	Kavel K25	Kavel K25	Wiggers 1955, 47-51	.	.	x	x	.	.	.	.	181.933	535.133
42	Noordoostpolder I, II, III	Noordoostpolder I, II, III	Havinga 1963	.	.	.	.	.	.	x	.	184.000	533.000
43	Emmeloord-J78	Nagelerweg, Kavel J78	Gehasse 1995; Zeiler 1997	.	.	.	.	.	x	.	x	178.467	519.400
44	Emmeloord/Nagele J97	J97	Bulten (eds) 2002; Van der Heijden 2000; Van Rijn 2002; Rompelman 2003	.	.	.	.	.	x	.	x	179.230	522.090
45	Schokland-I	Profiel I	Polak 1936	.	.	x	.	.	.	x	x	181.133	518.333
46	Schokland-II	Profiel II	Polak 1936	.	.	.	.	.	.	x	.	180.733	515.600
47	Schokland-III	Profiel III	Polak 1936	.	.	.	.	.	.	x	x	181.200	515.667
48	Schokland-VI	Profiel IV	Polak 1936	.	.	.	x	.	.	x	x	181.067	515.133
49	Schokland-SRW1	Sectie SRW1	Gotjé 1993	.	.	.	.	.	.	x	x	181.467	516.667
50	Schokland-ZP	Scectie ZP	Gotjé 1993	.	.	.	.	.	.	x	x	180.933	515.333
51	Schokland-P14	Kavel P14	Anscher & Gehasse 1993; Gehasse 1995; Cappers & Raemaekers 2008; Lauwerier et al. 2005; Luijten 1986 (Van Haaster 2010; Vernimmen 1999; 2004)	.	.	.	.	.	x	.	x	181.467	518.000
52	Schokland-Schokkerhaven-E170	Kavel E170	Cappers & Raemaekers 2008; Gehasse 1995; Weijdema et al. 2012; Luijten 1987; Van Haaster 2010; Vernimmen 1999; 2004	.	.	.	.	x	.	.	x	179.811	514.625
53	Urk-D56	Boorsectie D56	Gotjé 1993	.	.	.	.	.	x	.	x	171.000	522.000
54	Urk-E4	E4 (Domineesweg)	Peters & Peeters (eds) 2001; Cappers & Raemaekers 2008; Oversteegen 2001; Van Smeerdijk 2001; Vermimmen 2001	.	.	.	.	.	x	.	.	170.731	518.615
C	Microregio Noordoostpolder		Gehasse 1995; Gotjé 1993; Ten Anscher 2008	.	.	.	.	x	x	x	x		

Table 6.1 (continued): List of literature references.

Zone	Relation to (ground)water table
Dry woodland zone	1 m above
Shrub zone	1 – 0.5 m above
Marsh woodland zone	0.5 – 0.1 m above
Sphagnum-peat zone	0.1 – 0.05 m above
Reed-sedge zone	0.05 m above – 0.5 m below
Reed-rush zone	0.5 m – 1 m below
Open water zone	> 1 m below

Table 6.2: Boolean classification of dominant vegetation zones relative to the water table (source: Peeters 2007).

macroregion) attracted great interest. A relief map of the subsoil at the start of the Holocene was recreated on the basis of geomorphological knowledge and primary coring descriptions.[490] In response to the creation of the Flevoland polder in the previous century, many geomorphological studies were carried out at the mesoregion level, compiling information on the soil structure of the Noordoostpolder, Oostelijk Flevoland and Zuidelijk Flevoland.[491] Archaeological research was finally carried out, with a few exceptions, on a microregional scale, whereby each mesoregion contained an archaeological microregion: the Hoge Vaart-Eem region in Zuidelijk Flevoland, Swifterbant in Oostelijk Flevoland and Schokland-Urk in the Noordoostpolder (fig. 6.2).[492]

Models of groundwater level rise play a key role in palaeogeographical reconstruction. Around the turn of the century, several studies were devoted to developments in the groundwater level curve in various parts of Flevoland.[493] These studies combined with geomorphological data, were used to develop a computer model of vegetation development in Flevoland between 7000 and 4000 BP.[494] The model linked the presence of vegetation zones to the depth of the groundwater level (table 6.2) and also took account of groundwater capillary rise, the accumulation of peat and clay, and a number of geologically determined phases of erosion. The result was a series of maps showing the shift in vegetation zones, in stages of 100 years, between 7000 and 4000 BP (fig. 6.3). The computer simulation shows that an initially wooded, dry landscape was replaced over the course of 3000 years by a landscape with open water and vegetation zones defined by fens and bogs.

---

490 Peeters 2007.
491 Wiggers 1955; Ente *et al.* 1986; Menke *et al.* 1998 respectively.
492 Gotjé 1993; Ten Anscher 2012.
493 Gotjé 1997a, 1997b, 2001; Makaske *et al.* 2002a, 2003; Van de Plassche 1982; Van de Plassche *et al.* 2005; Roeleveld & Gotjé 1993.
494 Peeters 2007, 56-74.

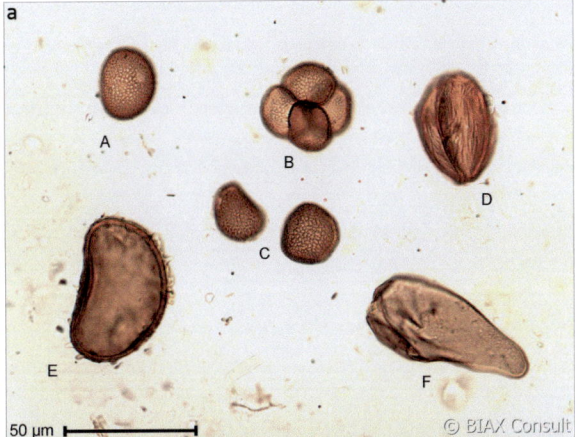

Figure 6.4 Examples of paleobotanical remains that were used to reconstruct the vegetation at the Late Atlantic and Subboreal at Almere-Kotterbos (location 7). a) Pollen of the branched bur-reed type (A), common bulrush (B), lesser bulrush (C), bogbean (D), male-fern / buckler-fern (E) and saw-sedge (F) (photo: BIAX *Consult*); b) Seeds of bogbean (A), saw-sedge (B) and grey club-rush (C) (photo: BIAX *Consult*).

Vegetation models have proved to be a highly useful way of revealing trends and gaining a better understanding of the connection between landscape processes and human activity.[495] Another approach involves data-driven reconstructions of these living landscapes, concentrating on palaeoecological, alongside geomorphological and hydrological data (fig. 6.4). This method is often used to address specific questions, such as: what did the dry coversand regions really look like in the first half of the Holocene? How varied were the marshes and mires in the Late Atlantic and the Subboreal? Was there really no longer any dry land available to humans at the end of the Subboreal?

---

495 The sensitivity of the model to various parameter settings was recently discussed by Peeters & Romeijn (2016).

Did the lakes that existed in the Subboreal remain in one location, or did they 'wander' around the huge fens and bogs that was Flevoland several thousand years ago?

Information on the palaeovegetation and the fauna associated with it has been drawn from dozens of, often small-scale, palaeoecological studies. Like excavations, these are windows, with certain temporal and spatial limitations (fig. 6.2 and table 6.1), but because there are so many of them, in combination they provide some overall information about the period, beginning in the Late Glacial. For the purposes of this chapter, the palaeoecological data has been translated in a traditional manner into vegetation types. This has the advantage of allowing very small studies with a limited dataset to be included in the overall picture. Because the entire palaeoecological dataset has been reinterpreted, this chapter does not include any summary of previous interpretations. Before turning to the three microregions, a rough outline of landscape development from the Late Glacial to the Subboreal will be presented. The information on these microregions covers much shorter chronological periods, because the palaeoecological information obtained at this scale relates to the archaeological research carried out. For the Hoge Vaart-Eem microregion (Zuidelijk Flevoland), therefore, we primarily have information about the landscape in the period 7000 to 4000 cal. BC,[496] from the Swifterbant microregion (Oostelijk Flevoland) information relates to the period 8300 to 3700 cal. BC[497] and from the Schokland-Urk microregion (Noordoostpolder) to the period 5000 to 800 cal. BC.[498]

6.3 Landscape dynamics in Flevoland

6.3.1 Late Glacial: c. 12,500 – 9800 cal. BC (Late Palaeolithic)

Geology, climate and hydrology
In the Late Glacial, Flevoland was part of a coversand region extending from the southern Netherlands via central and northern Germany into Poland and the southern part of the North Sea basin. The sea level was more than 80 metres below that of today and what is now the southern part of the North Sea was land through which rivers flowed. The Rhine, Meuse and Thames drained in a southerly direction into the Channel River that flowed into the sea to the southwest of what is now Great Britain (Strait of Dover) (fig. 6.5).[499]

There were no major rivers in and around Flevoland. There were, however, local and regional rivers and streams that drained to the west in the summer. In Zuidelijk Flevoland, water drained from the sandy soils of the Utrechtse Heuvelrug ice-pushed ridge, the western Veluwe, and the Gelder Valley, via the Eem and its tributaries. The literature mentions two river systems in Oostelijk Flevoland and the Noordoostpolder.[500] The Hunnepe drained the northeastern part of the Veluwe region and parts of the eastern Netherlands coversand region. The main channel was situated along the southern edge of the Noordoostpolder. In the west it curved to the southwest and flowed through the northwestern part of Oostelijk Flevoland. It is likely that the streams in Oostelijk Flevoland branched onto the main channel of the Hunnepe. The Overijssel Vecht, to the north of the Hunnepe, crossed into the Noordoostpolder to the south of the outcrop of glacial till known as De Voorst, flowing to the north and south respectively of the outcrops of glacial till at Schokland and at Urk. The Overijssel Vecht drained the Pleistocene areas of the eastern part of Gelderland, Overijssel and the southern part of the Drenthe Plateau. It is assumed that the Hunnepe and Overijssel Vecht were connected to each other to the west of Flevoland. The main direction of flow of the Eem in southern Flevoland was to the northwest, and it is not unlikely that this system joined the Hunnepe and the Overijssel Vecht in Noord-Holland in the Early Holocene. Although it would appear that elevation in Flevoland in the Late Glacial generally differed by no more than 20 metres, water will have flowed through a multitude of streams outside the regional river systems mentioned above.

Warm and cold periods alternated in the Late Glacial, in the sequence of the Bølling interstadial, the Older Dryas, the Allerød interstadial and the Younger Dryas. During the cold periods the subsurface was permanently frozen, sand began to drift and the river systems took on a braided character. In both interstadials, vegetation was more abundant, which prevented sand from drifting and caused rivers to meander.[501]

---

496 Hogestijn & Peeters 2001; Peeters 2007.
497 Cf. Dresscher & Raemaekers 2010; De Moor *et al.* 2009; Hamburg *et al.* 2012.
498 Cf. Bulten *et al.* 2002; Gotjé 1993; Gehasse 1995; Peters & Peeters 2001; Ten Anscher 2012.
499 Cf. Jelgersma 1979; Bourillet *et al.* 2003; Ménot *et al.* 2006.
500 Ente *et al.* 1986; Koopstra *et al.* 1993; Menke & Lenselink 1991; Menke *et al.* 1998; Lenselink & Menke 1995; Wiggers 1955; Vos *et al.* 2020.
501 Nowadays, the Late Glacial climate is defined as being made up of one warm and one cold period (Björck *et al.* 1998, 288). The warm period encompasses the Bølling interstadial, Older Dryas and Allerød interval, and is known as the Greenland Interstadial 1 (GI1: 14,700 to 12,650 years BP). The Older Dryas was a colder period within this interstadial. The cold period that followed, which is often referred to in the literature as the Younger Dryas, is also known as the Greenland Stadial 1 (GS1: 12,650 to 11,500 years BP).

Figure 6.5: The course of the major rivers at the end of the Late Glacial (adapted from Bourillet et al., 2003; Ménot et al., 2006).

Vegetation development in the Late Glacial

Palynological material from nine locations in Flevoland has been analysed (fig. 6.2 and table 6.1). Five of the locations are in Zuidelijk Flevoland, on the edge of the Eem valley, one is on the eastern edge of Oostelijk Flevoland, two are in the northern and one in the southern part of the Noordoostpolder.

The water table in the coversand landscape in the south of the Eem valley rose during the Bølling interstadial (fig. 6.2, table 6.1, no. 17). The soil-forming vegetation that grew here on the coversand transformed halfway through this period into marsh vegetation dominated by mosses and sedges.[502] Thanks to the water-saturated conditions, the moss and sedge did not decay, but accumulated as peat. Unfortunately, no palynological analysis was performed on the peat, so we have no information as to what the Late Glacial landscape would have looked like.

In the low-lying southwest of the polder (fig. 6.2, table 6.1, no. 4) soil formation occurred in the second half of the Allerød when open pine woodland with some birch and juniper developed.[503] Dryas (fig. 6.6), spikemoss (*Selaginella*) and adder's tongue grew in the undergrowth. Although the presence of cold-resistant plants such as dryas and spikemoss suggest the soil contained chalk, the presence of heather and crowberry indicate that there were also soils that were lower in nutrients and more acidic. Willow and species of the sedge family complete the picture of a landscape where variation in relief caused hydrological differences, resulting in a varied vegetation (fig. 6.7).

---

502   Makaske et al. 2002b, 35, 72, 74. Dated to between 12,342 and 12,178 cal. BC (12,240 ± 50 BP, GrA-17230).

503   De Moor et al. 2013, 17, 24, 27, 31-32. Soil in coring 7 dated to between 11,446 and 10,904 cal. BC (11,268 ± 118 BP, Ua-44518).

FROM LAND TO WATER

Figure 6.6: Dryas, Greenland (photo: J.A. de Raad).

Figure 6.7: Impression of Flevoland in the second half of the Allerød: a slightly hilly landscape with marshland, birch and pines, Leersumse Veld 2012 (photo: J.A. de Raad).

At the transition from the Allerød to the Younger Dryas, around 11,000 cal. BC, the average summer temperature fell and the climate started to show arctic characteristiscs. Information on the vegetation in this period has been obtained from a location (fig. 6.2, table 6.1, no. 1) to the south of the Eem river valley.[504] The landscape consisted of low sandy ridges with shallow lakes, small rivers and streams in the depressions between them (fig. 6.8). The landscape changed over this period from open parkland vegetation with trees and shrubs into tundra vegetation. All kinds of mosses, grasses and herbaceous plants grew on the dry soils, including mugwort and rockrose, interspersed among low shrubs such as dwarf birch, crowberry and juniper.[505] Trees were scattered around the landscape,

---

504 Van Smeerdijk 2002. Dated to between 11,053 and 10,772 cal. BC (10,980 ± 60 BP, GrA-17645).

505 In the temperate climate zone juniper develops into a small tree, but in the mountains, on the tundra and on the Scandinavian fells it grows as a small, low shrub, no taller than the low vegetation around it. It is likely that juniper also occurred as a small shrub in the Late Glacial.

Figure 6.8: Impression of Flevoland in the Younger Dryas: a gently undulating landscape with sandy ridges, lakes and rivers. Taymyr Peninsula (Russia), late June 2006 (photo: J.A. de Raad).

including birch and occasionally pine. The small lakes contained algae, species of the Charophytes and cold-resistant water plants. The presence of species of the Charophytes suggests the water was fairly calcium-rich in this period. The marshy low-lying depressions in the landscape were colonised by a wide range of species of the sedge family. However, bogbean, marsh marigold, and species of bedstraw, buttercup, dropwort, meadow rue and mint would have provided a splash of colour in the marshes and along lake shores and streams in the summer. As on the dry soils, shrubs and trees, mainly birch and willow, would have been scattered throughout this marsh landscape. It is not known whether the willow was the short Arctic type, or took the form of shrubs and trees.

At this location the boundary between the Pleistocene sand and the Preboreal peat above it is clear and well-defined. This, combined with the absence of soil formation, has led to an assumption that there was a hiatus in the stratigraphical development.[506] It is possible that the impact of frost or the flow of meltwater during the short summers in the Younger Dryas, eroded the sandy soil that lay at the surface, enabling only short vegetation with shallow roots to grow there.

In the relatively high northern part of the Noordoostpolder a peaty horizon from the Late Glacial was found in the subsurface.[507] This peat appears to have formed in the Allerød and the start of the Younger Dryas. Peat formation is generally an indication of stagnant water. That this relatively high area was so wet during this period was probably due to the presence of an impermeable glacial till layer just below the surface. The poor drainage would have attributed to the wet appearance of this area, certainly in the summer months. Mosses of the Hypnaceae family and all kinds of sedges grew in the fens, with localised occurrences of rushes, bogbean and marsh cinquefoil.

At a certain point, drainage conditions seem to have deteriorated and the fens transformed into shallow lakes in which water plants such as pondweed could grow. These shallow lakes were able to form due to a combination of factors. This development occurred at the transition between the Allerød and the Younger Dryas.[508] As the average summer temperature fell the vegetation changed, giving low-growing herbaceous plants (grasses and sedges) the upper hand, and reducing the abundances of trees and shrubs. There was probably a precipitation excess, the consequences of which could have been enhanced by the decline in vegetation. The combined effects of wind, water and melting ice would have led to the development of open areas with little vegetation, which allowed the sand to drift (Young Coversand). The coversands created barriers that prevented the flow of water. Nor could the water soak into the soil because of the buried glacial till and the permanently frozen soil, causing shallow lakes to form.

At the same time, the Overijssel Vecht was active in the southern part of the Noordoostpolder. This river was either braided or meandering, depending on the climate.[509] The river drained from the east to the west and northwest.

---

506   Van Smeerdijk 2002.
507   Wiggers 1955, 34. The Late Glacial peat extended as far as the modern villages of Creil and Bant, but has also been found elsewhere in the polder.
508   Wiggers 1955, 37 (no. 37 and 38 in table 6.1 and fig. 6.2).
509   Wiggers 1955, 40-42 (no. 39 in table 6.1 and fig. 6.2).

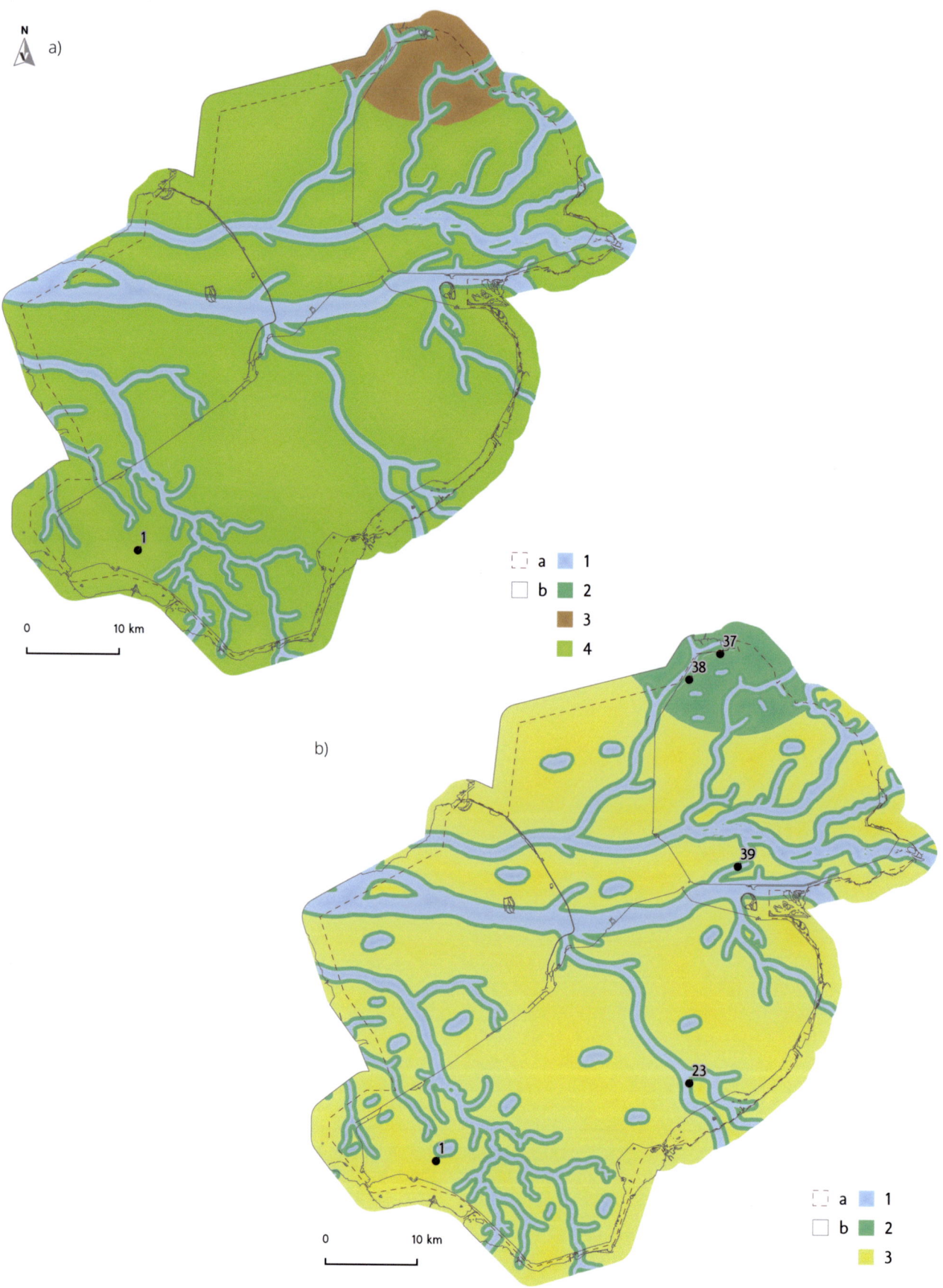

168   RESURFACING THE SUBMERGED PAST

Figure 6.9a-b (opposite page): Reconstruction of the vegetation of Flevoland in the Allerød (a) and in the Younger Dryas (b) based on palaeoecological information and after the palaeogeographical map of 9000 BC of Vos & De Vries (2018). Legend (a) Allerød: a: topography; b: outline of research area; dot: locations of palaeoecological information (table 6.1); 1: rivers and lakes; 2: marshes and swamps along the rivers with willow, species of the sedge family and herbaceous plants; 3: mires in the north with mosses and species of the sedge family (after Wiggers 1955); 4: open woodland / wooded grassland on dry soils with pine, birch, juniper, species of the heath family, species of the grass family, and other herbaceous species. Legend (b) Younger Dryas: a: topography; b: outline of research area; dot: locations of palaeoecological information (see table 6.1); 1: rivers and lakes; 2: marshes along rivers and lakes and in the north (after Wiggers 1955) with mosses, different small willow species, species of the heath family, species of the sedge family and some herbaceous species; 3: tundra on the drier soils with dwarf birch, (mountain) pine, (small) juniper, species of the heath family and herbaceous species.

Changes in climate also gave rise to different types of vegetation. Outside the Overijssel Vecht river basin, however, dry soils were present on some scale, covered with grasses, mugwort and species of the heath family. In the Allerød there was a much higher proportion of birch and pine in the vegetation than in the Younger Dryas. The subsurface in the southern part of the Noordoostpolder, also became unstable during the Younger Dryas due to the absence of robust vegetation. As a result, the existing marshy depressions disappeared beneath a layer of loam on top of which sandy river dunes began to form.

As in the northern part of the Noordoostpolder, the Younger Dryas also saw the stagnation of water flow in the higher-lying eastern part of eastern Flevoland.[510] Shallow, relatively nutrient-rich lakes formed in which algae, spiked water-milfoil, pondweed and water-crowfoot grew. The surface of the water would have been covered with the floating leaves and flowers of white and yellow water lilies in summer. Around the lakes there were dry soils showing evidence for the same type of vegetation found elsewhere in Flevoland: open vegetation with grasses, mugwort, rockrose, juniper and species of the heath family, including crowberry. Tree species like birch and pine would have been scattered about the landscape.

In conclusion, the Late Glacial, a period spanning some 2700 years, was characterised by the alternation of warm and cold periods. Most of the information we have for Flevoland comes from the Allerød and Younger Dryas (fig. 6.9), which both lasted roughly 1000 years. The landscape of Flevoland was remarkably wet in this period. Extensive marshland existed during the Allerød. Due to the deposition of Young Coversand and a frozen subsurface, this marshland developed into a series of interlinked lakes during the Younger Dryas. More or less the same cold-resistant plant species grew in the area in both periods, although the vegetation structure and proportional composition of the species changed. In the Allerød the dry soils were covered with open woodland consisting of birch and pine, with grasses and herbaceous plants in the undergrowth. In the Younger Dryas the birch and pine population declined sharply, and at a certain point only woody shrub-like vegetation would have grown there, including dwarf birch, mountain pine and the willows that grow nowadays in Arctic regions (Arctic willow, dwarf willow, net-leaved willow and polar willow). The landscape was no longer home to any of the tree species, and parts of the ground were devoid of vegetation, or covered with herbaceous vegetation consisting of grasses and sedges.

6.3.2 Preboreal: c. 9800 – 8200 cal. BC (Early and start of Middle Mesolithic)

Geology, climate and hydrology

After the Younger Dryas the average summer temperature rose to between 15 and 20°C, heralding the start of a warm period, the Holocene, in which several phases have been distinguished on the basis of the predominant climate conditions and the occurrence of certain plants. In the first phase, the Preboreal, large quantities of water were stored in land ice. The sea was far away, which meant a continental climate prevailed, with cold winters and hot, dry summers. As in the period discussed above, Flevoland was still part of an extensive coversand region and, although the temperatures were similar to those of today, many cold-intolerant plants (and animals) were still confined to southern Europe.

It could have been a combination of landscape and climatological factors that caused the marshes and lakes of the Younger Dryas to make way for a landscape more accessible to humans. The frozen subsurface thawed, which lead to improved water drainage. It is also likely that, as a result of the higher temperatures, the rate of evapotranspiration exceeded the precipitation in summer, as is generally the case today. As the temperature rose, the biomass of the vegetation increased. The Preboreal vegetation will therefore have required more water than the tundra vegetation of the Younger Dryas. Influenced in part by the more robust vegetation, the shallow, braided rivers transformed into meandering rivers that cut into the landscape. These rivers drained water from large

---

510 De Jong 1974, Biddinghuizen profile (no. 23 in table 6.1 and fig. 6.2).

areas of the Drenthe Plateau and the coversand regions of the eastern and central Netherlands, flowing via low-lying Flevoland to the west and northwest (fig. 6.1a).

Vegetation development in the Preboreal

Botanical remains dating to the Preboreal have been analysed from nine locations in Flevoland (fig. 6.2, table 6.1): five from southern Flevoland along the edge of the Eem valley, one from the eastern edge of Oostelijk Flevoland and three from the Noordoostpolder.[511]

Information from Zuidelijk Flevoland comes once again from south of the Eem valley, from an area with low coversand ridges alternating with valleys (fig. 6.2, table 6.1, no.1).[512] Two ^{14}C dates taken from the peat base in one of these valleys show that it developed in the Preboreal.[513] The peat may have accumulated within a lake environment.[514] Water plants like alternate-flowered water-milfoil and species of pondweed were gradually replaced in the Preboreal by fen vegetation featuring reeds, various sedge species and other herbaceous plants. The botanical remains show evidence for several species of marsh and water plants. Some of these would have preferred water with high levels of oxygen in relatively nutrient-poor conditions, but some species also indicate conditions with higher nutrient levels. It has been assumed, on the basis of the composition of plant remains, that nutrient-poor rainwater accumulated on top of relatively nutrient-rich groundwater which contained minerals from a deeper loam layer.[515] Birch was initially dominant in the dry regions of the landscape. The remains of seeds and catkins clearly show that this was downy or silver birch, the two native tree species within the birch genus. The birch grew with pines in open spaces featuring sea-buckthorn, juniper, crowberry, grasses and other herbaceous plants in the undergrowth. In the course of the Preboreal the proportion of birch declined in favour of pine, as was also the case elsewhere in the Netherlands.[516] Two woody taxa have not yet been mentioned: willow and poplar. The willow taxon probably include shrub species or species of small trees. The poplar may have been aspen, a tree species known to have grown in the Late Glacial in, or on the edge of marshes and swamps.[517]

Palynological material from a buried Pleistocene soil just to the east of this location was also analysed (fig. 6.2, table 6.1, no. 3).[518] A pine woodland grew on the coversand here.[519] This must have been a fairly dense woodland on dry coversand, as barely any remains of herbaceous plants or indicators of water have been found. However, heather, male-fern / buckler-fern and stag's-horn clubmoss probably grew in the undergrowth (fig. 6.10). Male-fern / buckler-fern include various fern species that are specific to very different types of locations. In peat, for example, remains of marsh fern, crested buckler-fern and/ or narrow-buckler fern have been found. It is likely that other fern species were found here, as the palynological material comes from a coversand soil. Given the evidence for relatively dry pine woodland, it could be male-fern or broad buckler-fern. These species are still commonly found in woodland on sandy soil. Stag's horn clubmoss in combination with heather suggests there were glades in the woodland. This clubmoss species prefers sandy soil with good drainage but that does not dry out.[520]

That pine was a dominant component in the vegetation can also be assumed by the discovery of pine charcoal in the coversand to the east of Hoge Vaart (fig. 6.2, table 6.1, no. 21). Charcoal dating from the second half of the Preboreal has been found in the coversand in the most southwesterly part of Zuidelijk Flevoland (fig. 6.2, table 6.1, no. 4).[521] The palynological material suggests that at least some of the charcoal is of a more recent date and comes from the younger peat cover. This may indicate the landscape in this area was covered in pines with heather and crowberry undergrowth, as was common in the second half of the Preboreal.

Palynological material found in peaty sandy soil at the Hoge Vaart-A27 site (fig. 6.2, table 6.1, no. 13) in the riparian zone of the Eem probably dates from the Preboreal.[522] In the Preboreal, the bank of the river was saturated with water at this spot, so organic material did not decay completely. The palynological analysis has revealed that willows, grasses, sedges and bulrushes grew on the riverbank and that birch and pines grew on drier soils, possibly with heath in the undergrowth.

---

511 Bunnik & Verbruggen 2011; Gotjé 2001; De Jong 1974; Kooistra 2012a; Van Smeerdijk 2002; Wiggers 1955, 47-51.
512 Van Smeerdijk 2002.
513 Van Smeerdijk 2002. The dates are 10,050 ± 50 BP (GrA-16838) and 10,010 ± 50 BP (GrA-16753).
514 In the geological terminology of the Netherlands peat accumulated during the Holocene is attributed to the Nieuwkoop Formation (*Formatie van Nieuwkoop*).
515 Van Smeerdijk 2002, 19.
516 E.g. Hoek 1997a, 1997b.
517 Van Smeerdijk 2002, based on Hoek 1997b, 92.

518 Bunnik & Verbruggen 2011.
519 In coversand soils palynological material migrates downwards from the surface (cf. Dimbleby 1985; Havinga 1963, 1974; Van Mourik 2001; 2003; Van Mourik *et al.* 2010, 2011, 2012; Van Smeerdijk 1989). Roughly speaking, the oldest material will lie deepest in the soil. The level at which it is found does not therefore necessarily represent the level at which the material was deposited. This contrasts with palynological material in peat and clay, which was deposited at the depth at which it is found.
520 Weeda *et al.* 1985, 15.
521 De Moor *et al.* 2013, 16, 25. Dated on basis of charcoal to between 9194 and 8296 cal. BC (9389 ± 172, Ua-44511).
522 Gotjé 2001, 23, 42. The Eem2 core.

Figure 6.10: Stag's-horn clubmoss, a common species of clubmoss on dry Pleistocene coversands (photo: J.A. de Raad).

The lake in the higher-lying eastern part of Oostelijk Flevoland disappeared in the course of the Younger Dryas and Preboreal as sedge vegetation developed (fig. 6.2, table 6.1, no. 23).[523] The peat that accumulated from this vegetation also contained pollen from vegetation in the vicinity of the fen. This suggests that pine woodlands grew here on the dry soils in the second half of the Preboreal. This woodland appears to have been dense initially, but the presence of hazel and birch pollen suggests that there must have been open patches, possibly at border zones between the dry coversands and the sedge fens.

After the Younger Dryas, the low tundra vegetation on the coversands in the northern part of the Noordoostpolder seem to have become colonised by birch and pine, and sometimes also willow.[524] The coversand is humic here, which could be an indication of the continued presence of stagnating groundwater. Reed fens with large quantities of sedge species were to be found in the lowest-lying parts of the coversand region (fig. 6.2, table 6.1, no. 40). Stagnating water led to the development of peat, which trapped pollen from the surrounding vegetation. As in many other parts of The Netherlands, pines dominated the woodland in the drier parts of the landscape at the end of the Preboreal.[525]

The age of the peaty and sandy deposits in the southern Noordoostpolder is not known for certain.[526] Peat found beneath a river dune (fig. 6.2, table 6.1, no. 39) has been dated to between 10,990 and 9444 cal. BC. This would have accumulated in the Younger Dryas or the first half of the Preboreal.[527] However, the pollen spectrum found would appear to be more consistent with vegetation from the final phase of the Allerød, when the amount of pine pollen declined and birch pollen increased.

In the initial period of the Holocene the subsurface of Flevoland was drier than it had been for a long time. In the course of the Preboreal, sand ceased drifting as rising temperatures caused an expansion in vegetation which held the subsurface more firmly in place. The fact that river dunes formed along the Hunnepe and Overijssel Vecht during this period suggests that sand was still drifting along the rivers, however (fig. 6.11).[528] The herbaceous, grass- and sedge-rich tundra vegetation on the coversand ridges and flanks made way for a wooded landscape with birch and pines. Pine woodland came to dominate during the Preboreal. Pines produce large amounts of pollen, this suggests that the woodland may not have been as dense as one might assume on the basis of the high percentages of pine pollen recovered. In a natural situation, pines grow in open woodlands, as they need light to survive. The presence of birch and hazel, which also require plenty of light, suggests that the woodland that did form was not dense.[529] Even in the most recent period of the Preboreal, in which the palynological signal for pine woodland is

---

523 De Jong 1974.
524 Wiggers 1955, 50.
525 Wiggers 1955, 47-48.
526 Wiggers 1955, 40-42 (plot E155). Interpretation of the diagrams published by Wiggers is difficult because the dates ascribed to the Late Glacial and Early Holocene periods have changed since the 1950s.

527 Wiggers 1955, 42. Dating: 10,500 ± 280 BP (GRO 375) calibrated using OxCal v4.2.3 (Bronk Ramsey 2013) and the IntCal13 atmospheric curve (Reimer et al. 2013).
528 Cf.. Wiggers 1955, 38-42.
529 Our image of pine woodlands is influenced by commercial forestry, in which trees of equal age are planted equidistant, and as close to each other as possible.

Figure 6.11: Reconstruction of the vegetation of Flevoland in the Preboreal based on palaeoecological information and after the palaeogeographical map of 9000 BC of Vos & De Vries (2018). Legend: a: topography; b: outline of research area; dot: locations of palaeoecological information (table 6.1); 1: rivers; 2: marshes and swamps along the rivers with willow, poplar, species of the sedge family and species of the grass family; 3: mires in river valleys and in the north with species of the sedge family; 4: open woodland on the dry soils with pine, birch, species of the grass family, species of the heath family, and in the second half of the Preboreal also hazel.

strongest, heather and stag's-horn clubmoss still occurred locally. These two species would have grown in sunny or semi-shaded spots.

Fens with sedges and reed formed in the lowest-lying valleys in the coversand landscape, such as along the Eem and in the northern Noordoostpolder. Here, and elsewhere along rivers and the edges of low-lying valleys, grew willows and probably also aspen.

6.3.3 Boreal: 8200 – 7000 cal. BC (Middle Mesolithic)

Geology, climate and hydrology
During the Boreal there was barely any change in the geological landscape of Flevoland. The ice caps of northern and central Europe continued to melt however, causing the lowest coversand regions of the North Sea basin to disappear under water. Flevoland was not yet affected by the rising sea level and, as in previous periods, was situated in an extensive coversand region. The climate therefore still had a fairly continental character.

Vegetation development in the Boreal
Botanical remains found at eleven sites and dating from the Boreal have been analysed: four from Zuidelijk Flevoland, two from Oostelijk Flevoland and five from the northern Noordoostpolder (fig. 6.2, table 6.1).

During this period, the mire in a low-lying valley to the south of the Eem (fig. 6.2, table 6.1, no. 1) transformed into land.[530] This process was instigated by reed and species of the sedge family. This vegetation may have developed so

---

530 Van Smeerdijk 2002.

rapidly that it caused the mire to dry out. The higher sandy soils around the valley were home to pine trees and other vegetation during this period. Given the pollen spectrum which includes birch, juniper and a small amount of heath, this could have been open or semi-open woodland, rather than expanses of dense pine woodland. The large quantity of pollen recovered indicates, however, that pine was the most predominant tree species on the dry soils. This may have reduced the amount of water flowing into the valleys, and could have been a factor in the drying out of the mire. Unlike the many deciduous trees in a temperate climate zone, pines grow all year round and therefore require water all year round.[531] An increase in the number of pines in the landscape would therefore have led to the ground becoming drier, if the supply of water remained constant. The large quantities of pine pollen, as well as the clear presence of pollen from cold-intolerant deciduous trees are typical of the Boreal. Cold-intolerant deciduous trees had reached the Netherlands by this time and had established themselves in the vegetation.[532] In the swampy hollows of the Eem valley willows were joined by alder in this period. Woodland on drier soils was more varied, with hazel, oak, elm and lime.

At the same time, the coversand in the far southwest of Zuidelijk Flevoland (fig. 6.2, table 6.1, no. 4) still supported open pine woodland in which the first deciduous trees established themselves.[533] Pine also appears to have dominated the vegetation in the riparian zone of the Eem in the higher-lying southern area of Zuidelijk Flevoland (figure 6.2, table 6.1, no. 9), as suggested by the discovery of a pine log and pine cones in a peaty layer (fig. 6.12).[534] Just to the east, at the end of the Boreal, peat formed from the decayed remains of sedges as well as other fen vegetation was laid down in a coversand hollow (fig. 6.2, table 6.1, no. 21).[535] The fact that species of the sedge family are the ones most commonly identified in the fen vegetation suggest relatively nutrient-poor, possibly acidic soil conditions when peat formation began. Pine, birch, oak, hazel and elm all grew in the vicinity of the mire. Again, there are indications of heath in the undergrowth.

Three layers containing palynological material found in the east of Zuidelijk Flevoland and dating to the Boreal were analysed in the mid-20th century (fig. 6.2, table 6.1,

Figure 6.12: Seeds and a cone of Scots pine (*Pinus sylvestris*) found in a channel of the Eem at Hoge Vaart-A27 (location 8) (photo: Cultural Heritage Agency of the Netherlands).

no. 24).[536] Pines and an undergrowth of ferns and stag's horn clubmoss appear to characterise the vegetation here in this period. The presence of hazel, birch and oak again suggests an open landscape with trees, open woodland or woodland with open patches. In the north of Oostelijk Flevoland information on the vegetation comes from the analysis of charcoal from 14 pit hearths dug by the Mesolithic inhabitants (fig. 6.2, table 6.1, no. 25).[537] The charcoal recovered from the hearths was indisputably the product of human activity but, if the wood had been deliberately selected for use, it could give a distorted picture if used as an indicator of the woody vegetation growing in the Boreal. It is, however, very likely that the wood was gathered in the vicinity of the Mesolithic camp. As could have been expected on the basis of palynological analysis of samples from elsewhere in Flevoland, pine was the species most strongly represented. As well as charcoal, carbonised fragments of pine cone scales have regularly been found. The oldest pit hearths contained only pine. The hearths dating from 7600 cal. BC onwards contained other species: birch, oak, alder, elm and willow. Carbonised remains of hazelnuts have also been found. Charcoal of species of the Malaceae subfamily cannot be determined with a high degree of accuracy, but may be from crab apple, mountain ash, hawthorn or wild pear. Again, a picture emerges of a landscape featuring lots of pine, but with enough room for other species of tree that needed lots of light. In summary, therefore, an open landscape with plenty of trees, but not dense pine woodland.

The build-up of peat that started in the Preboreal continued in the northern Noordoostpolder (fig. 6.2, table 6.1, no. 40-41).[538] The wet low-lying depressions in the landscape became overgrown with vegetation that included reed, sedge, saw-sedge, bogbean, marsh cinquefoil and ferns. Vegetation such as this indicates the

---

531 Each type of vegetation has its own water requirements. Herbaceous vegetation generally transpires less water and therefore needs less than woodland. The water requirement or level of transpiration in a woodland depends on the type of trees growing there.
532 Janssen 1974, 55-57; Van Geel *et al.* 1981, 411; Hoek 1997a, 21.
533 De Moor *et al.* 2013. No purely Boreal palynological samples were analysed, as all appear to have been contaminated with material from more recent periods (see 23, 25, 28-29, 32).
534 Spek *et al.* 2001a, 2001b.
535 Bos & Verbruggen 2012.

536 Havinga 1963, 75-77.
537 Kooistra 2012c.
538 Wiggers 1955, 47-49.

Figure 6.13: Impression of Flevoland in the Boreal with open coniferous woodland on the highest spots in the landscape and fen meadows and marshes at the contact zone with water. Omolon river area (Russia), July 1993 (photo: J.A. de Raad).

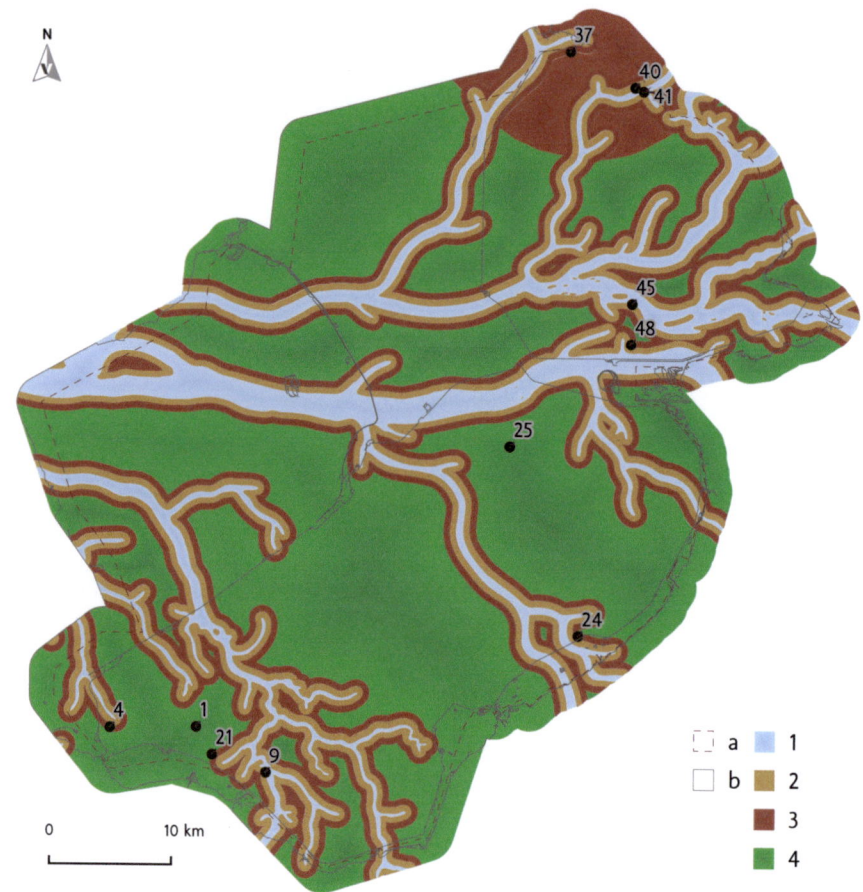

Figure 6.14: Reconstruction of the vegetation of Flevoland in the Boreal based on palaeoecological information and after the palaeogeographical map of 9000 BC of Vos & De Vries 2018. Legend: a: topography; b: outline of research area; dot: locations of palaeoecological information (table 6.1); 1: rivers; 2: marshes and swamps along the rivers with willow, poplar, alder, species of the sedge family and species of the grass family; 3: mires in the river valleys and in the north with species of the sedge family; 4: open woodland on the dry soils with pine, birch, hazel, oak, elm, (lime), species of the grass family, species of the heath family.

presence of a moderately nutrient-rich soil. The presence of saw-sedge indicates the availability of calcium. The dry coversands supported pine, birch, hazel and oak, as elsewhere. Finally, a palynological sample from a humic coversand soil in the far north of the Noordoostpolder (fig. 6.2, table 6.1, no. 37) indicates that the location itself was fairly wet in the Boreal, as evidenced by the presence of species of the sedge family, alder and willow. The dryer soils in the vicinity were however dominated by pine, birch, oak and hazel.

As in the Preboreal, the Boreal landscape would have been quite accessible to the inhabitants. It seems there were fewer mires in the coversand hollows. Many of these mires may have dried up during this period, causing the peat to partially degenerate and decay. It is not clear whether this drying out can be explained solely by climatological factors (temperatures higher than in the Late Glacial and a more continental climate than at present). It is, however, clear that there was an abundance of pine trees in the dryer parts of the landscape, and these would have drawn more water from the surrounding environment than other sorts of woody vegetation (fig. 6.13).

It has been suggested in the past that the dry coversand areas of Flevoland were covered with dense pine woodlands, but this was not in fact the case, although pine was certainly the most dominant tree species. The landscape of the dry coversands must have been much more open, as suggested by the presence of other trees and shrubs that required lots of light, and the widespread evidence for low vegetation featuring ferns, stag's head clubmoss and heath (fig. 6.14).

As was the case elsewhere in the Netherlands, an increasing number of cold-intolerant deciduous trees and shrubs began to establish themselves among the vegetation in the course of the Boreal.

6.3.4 Early Atlantic: 7000 – 6000 cal. BC (Middle and early Late Mesolithic)

Geology, climate and hydrology

The Early Atlantic was the last period in which Flevoland formed part of the relatively high-lying coversand regions through which regional rivers drained water from the Utrechtse Heuvelrug ice-pushed ridge, the Veluwe, eastern Gelderland, Overijssel and the southwestern part of the Drenthe Plateau. Whilst Flevoland was at the centre of an extensive coversand region during the Boreal, in the Early Atlantic it came to lie on the western edge of the region. The infilling of the North Sea basin created an Atlantic climate, with less cold winters and less warm and dry summers. The general consensus regarding the entire Atlantic is that annual precipitation levels were higher than in previous periods, and that the average annual temperature reached its highest level for the entire Holocene (disregarding current temperature increases).

Although many parts of Flevoland still had the typical appearance of a dry coversand area, the hydrology and the composition of the surface water changed. Initially, only rivers transported clay and relatively nutrient-rich water in from the wider region. The vegetation in the dry parts of coversand ridges had to survive mainly on rainwater containing few nutrients. The rising sea level prevented the free drainage of fresh water. This effect in turn caused the water level in Flevoland to rise. Groundwater that had been in contact with old river deposits was relatively rich in nutrients, and the rise in the water level meant that, at a certain point, the groundwater became accessible to the vegetation. The higher moisture levels in the soil and the availability of water containing these higher nutrient levels caused the composition of the vegetation to change. Freshwater marshes and mires developed in places where the groundwater came to the surface, such as in river valleys, the western part of Flevoland and in low-lying parts of the coversand landscape.

Vegetation development in the Early Atlantic

Botanical material possibly dating from the Early Atlantic was analysed from twelve locations: four locations near the Eem in Zuidelijk Flevoland, five in Oostelijk Flevoland and three in the northern Noordoostpolder (fig. 6.2, table 6.1).[539] Plant remains from coversand soils were analysed from most of these locations. A few samples came from a channel fill, with analysed samples from peat deposits coming only from the south of Zuidelijk Flevoland and the northern Noordoostpolder. A large number of observations regarding the composition of the Atlantic peat exist for the Noordoostpolder. These date from the first half of the twentieth century, just after the polder had been reclaimed.[540]

One of the peat locations analysed in Zuideliijk Flevoland was situated in the valley mentioned previously, to the south of the Eem valley (fig. 6.2, table 6.1, no. 1). The picture that emerges from the botanical remains is one of vegetation growing in nutrient-poor soil. The drier areas in the vicinity were covered with wooded parkland-like vegetation, featuring hazel and oak on a nutrient-poor soil. The peat appears to be developed from bog-mosses and other moss species. In combination with the remains of hare's-tail cottongrass and heather, this suggests an environment resembling bog. Birch seeds and alder pollen show, however, that this was not an extensive area. Birch, specifically downy birch, and heather are known to grow

---

539 Bos & Verbruggen 2012; Havinga 1963; Kooistra et al. 2009; Kooistra 2012a, 2012b; De Moor et al. 2009; Van der Linden 2010; Van Smeerdijk 2002; Wiggers 1955.
540 Wiggers 1955, 51-53.

Figure 6.15: Impression of an open landscape with heathland, where pine, birch and oak grew. Edense hei (te Netherlands), August 2015 (photo: BIAX *Consult*/L.I. Kooistra).

in relatively dry mesotrophic fens, mainly on slightly ripening peat. Alder, on the other hand, is a tree found in eutrophic fens and carrs. It can grow happily with the foot of its trunk in water. Since only alder pollen (no seeds, no wood) has been found, it can be assumed that the wetter, more nutrient-rich landscape lay further away. The seeds of birch and heather, in contrast, suggest there was relatively dry, nutrient-poor peat in the immediate vicinity, though we cannot rule out the possibility that the birch pollen is from a close relative of downy birch – silver birch – which prefers dry, nutrient-poor positions on coversand. Given that there was open woodland, including lots of hazel, growing on coversand, it is likely that the heather was part of the undergrowth.

It is not clear how the vegetation in southwest Zuidelijk Flevoland developed in this period, since the palynological material is mixed (fig. 6.2, table 6.1, no. 4).[541] Large quantities of heather pollen have been found in the sandy peat deposited on the southern banks of the Eem (fig. 6.2, table 6.1, no. 6).[542] The peat (containing no sand) found above this dates to between 5800 and 5662 cal. BC.[543] We can deduce from this that the sandy peat therefore accumulated slightly earlier, possibly during the transition from the Early to Middle Atlantic. The presence of heather pollen together with pollen from birch, hazel, oak, pine and elm makes it likely that in this period the dry coversands were covered with heathland on which grew hazel, birch and, to a lesser extent, oak, pine and elm. This evokes similarities with today's parkland-type heathlands on nutrient-poor, acidic sandy soils (fig. 6.15). In the period in question the sampling point itself was situated in a water-saturated landscape, as evidenced by the sandy peat and the waterlogged plant remains it contained. The main evidence for vegetation are the reed remains and a high proportion of alder pollen. Carr may have developed at this spot, largely comprising alder, with reed as undergrowth. Dense alder carr does not normally admit much pollen from plants growing on dry soils. However, 76 heather pollen grains and 65 hazel pollen grains were counted for every 100 alder pollen grains, so clearly the alder did not form a complete physical barrier to pollen from the surrounding environment.[544] This means that though there was a lot of alder pollen, there were not so many alder trees that they formed dense woodland. The presence of sand in the peat in fact suggests an open vegetation including alder, rather than dense woodland.

Further inland, pine charcoal was found in the coversand (fig. 6.2, table 6.1, no. 21).[545] It was ^{14}C-dated, and a calibrated age of 6648 to 6482 cal. BC established.[546] A pollen sample from the coversand soil at this location gave high pollen percentages for heather, hazel, alder and willow.[547] Pine and oak were found in small proportions. The low percentage of pine makes it likely that the pollen reflects a vegetation younger than the above date from the middle of the Early Atlantic . However, the possibility that the pine

---

541 De Moor *et al.* 2013.
542 Van der Linden 2010.
543 6847±32 BP (KIA-43244).

544 An upland tree pollen sum was used (pine was not included as upland tree pollen). The percentages of other species, such as alder and heather, were calculated on the basis of this pollen sum.
545 Kooistra 2012a.
546 Opbroek & Lohof 2012, 55.
547 Bos & Verbruggen 2012, 39, 42. The reference is to the results from the pollen sample from the coversand ridge, pit 4 (AO-515).

charcoal is the result of human selection cannot be ruled out, despite the fact that little evidence of human activity has been found at this location. Whether the palynological material formed in the Early Atlantic, at the transition from the Early to the Middle Atlantic, or in the Middle Atlantic, it is clear that there was no dense deciduous woodland there at that time. Again, there was dry heathland with hazel, oak, birch and perhaps a few pines. Alder and willow grew in the lower-lying parts of the landscape.

The botanical material from a channel deposit in the north of Oostelijk Flevoland is difficult to interpret (fig. 6.2, table 6.1, no. 34).[548] The material originates from washed out deposits, which means it could be a mix of material from different periods and different geographical origins. A podzol was analysed just to the east of this sample location (fig. 6.2, table 6.1, no. 35).[549] The dry soils appear to have supported open oak woodland with hazel. Birch, pine, elm or lime may have grown here and there, although the latter is probably not all that likely. Lime needs relatively nutrient-rich, calcareous soil, and the undergrowth of this open woodland included heather and possibly also crowberry – indicators of dry, nutrient-poor, acidic sandy soils. Furthermore, there is clear evidence of bracken. Nowadays, this fern species indicates old woodland soils low in calcium and nutrients, and the same would have been the case in the past. At the sample location itself wet conditions predominated, as evidenced by the presence of alder pollen, and also of pollen from several aquatic and marsh plants such as pondweed, bogbean and possibly celery. There is no ^{14}C date that can give a definitive age for the botanical material, but the researchers assume the Middle Atlantic, given the pollen spectrum.[550] A dating in the Early Atlantic period cannot however be ruled out. A Mesolithic site was excavated close to this location (Dronten-N23, fig. 6.2, table 6.1, no. 25). Charcoal from three pit hearths dating to the Early Atlantic shows a broad spectrum of tree species, with barely any pine.[551] The variety of species in the charcoal suggests that the inhabitants of the site did not select specific wood species, but simply gathered what was available nearby. This reinforces the idea that the pollen spectrum at Location 35 dates from the Early Atlantic.

While pine had disappeared from the landscape in the relatively low-lying coversand area around Locations 25 and 35 by halfway through the Early Atlantic, it appears that it was still prominent on the higher coversand in the east of Oostelijk Flevoland (fig. 6.2, table 6.1, no. 31).[552]

Several dozen Mesolithic pit hearths have been excavated in this area. On the basis of the analysis and ^{14}C dating of charcoal from a number of the pits, a date in the first half of the Early Atlantic can be established. The pits contained almost exclusively pine charcoal and carbonised pine cone scales. Only the most recent pit hearth (6360-6070 cal. BC; 7350 ± 40 BP) contained a few pieces of oak charcoal.[553] We should however bear in mind that the picture may be distorted as a result of selective wood use.

The palynological material from another location in the east of Oostelijk Flevoland (fig. 6.2, table 6.1, no. 24) was not ^{14}C dated. It is therefore not clear whether any palynological information from the Early Atlantic has been preserved.[554] If pine continued to grow on the dry coversands of Flevoland, it could be that some of the palynological data traditionally attributed to the Boreal due to a dominance of pine actually date from the Early Atlantic. Interestingly, heath barely played any role here in the transition from a landscape dominated by pine to a landscape of largely deciduous woodland. Perhaps the soil contained more nutrients here, or the hydrology was different.

The peat in the northern Noordoostpolder became less nutrient-rich in the course of the Boreal, or perhaps the Early Atlantic – there is no ^{14}C date available (fig. 6.2, table 6.1, nos. 40 & 41).[555] The peat initially consisted of reed, remains of species of the sedge family, marsh cinquefoil and bogbean. At the start of the Atlantic, probably in the Early Atlantic, the reed disappeared from the fen and a transitional mire with mesotrophic sedge peat developed. The pollen shows that pine was gradually making way for hazel, oak and birch in the vicinity. Interestingly, there was an increase in heath pollen at the same time, making it likely that heath provided ground cover on the dry coversands in the area. A mesotrophic peat also developed at a nearby location (fig. 6.2, table 6.1, no. 37) on the humic sand from the Boreal, because of an increase in birch pollen right from the start of the peat formation process.

During the Atlantic – it is not clear exactly when – peat started to develop in several of the lower lying areas in the coversands of the Noordoostpolder.[556] In almost all cases this was oligotrophic peat, that is peat which formed in a very wet, acidic, nutrient-poor environment. In the western Noordoostpolder peat that developed from a sedge mire has been found on top of the coversand. Given the elevation of this coversand, this peat probably did not accumulate until the start of the Late Atlantic, from 5000 cal. BC onwards.

---

548 De Moor *et al.* 2009, 67-68.
549 De Moor *et al.* 2009, 72-74.
550 De Moor *et al.* 2009, 74.
551 Kooistra 2012c.
552 Kooistra *et al.* 2009.

553 Poz-29482.
554 Havinga 1963, 75-77.
555 Wiggers 1955, 48-49.
556 Wiggers 1955, 51.

Figure 6.16: Impression of open oak woodland, Mijnweg (the Netherlands), July 2010 (photo: J.A. de Raad).

No Atlantic woodland in Flevoland in the Early Atlantic

In considering the relatively scarce data from the Early Atlantic, it is clear that the vegetation in Flevoland developed in a slightly different way than was previously thought. The generally accepted picture is that the Boreal pine woodland was replaced in the Atlantic by a mixed deciduous woodland that included elm and lime.[557] This would indeed have occurred in the nutrient-rich loess areas and broad valleys of the major rivers like the Rhine and the Meuse, although the label 'pine woodland' does require some qualification (see § 6.3.3). The surface of Flevoland for the first 4000 years of the Holocene was made up of Younger Coversands. These contained little loam, so water and nutrients would quickly leach out of a dry coversand soil, becoming inaccessible to vegetation. By the start of the Atlantic (7000 cal. BC), pine had dominated the vegetation on the dry coversands for over a thousand years. At the start of the Early Atlantic the coversands were fairly dry and acidic and had lost a large proportion of their nutrients. This had implications for the vegetation. In the course of the Boreal more and more cold-intolerant plant species, including many deciduous trees and shrubs, reached northwest Europe, after surviving in refugia in southern Europe during the previous ice age. Temperature is not however the only factor determining where plants establish themselves. Soil conditions, the hydrology, the availability of nutrients and of course the vegetation already present are also key factors. It therefore comes as no surprise that mixed oak and lime woodland did not develop in a landscape such as this. Lime trees need moist to dry soil rich in calcium. Such conditions did not exist in Flevoland in the Early Atlantic, although the vegetation did however change in the course of the period. The open pine woodlands slowly made way for a landscape in which open woodland alternated with vegetation featuring heather, hazel, oak, birch with only the occasional pine (fig. 6.15). Depending on local soil conditions, there would have been either more heather or more oak, or alternatively pine would still dominate. This variation was probably driven by differentiation in hydrological processes in the coversand subsurface. While pine continued to dominate in the higher and dryer areas in the east of Oostelijk Flevoland until well into the Early Atlantic, the vegetation in the low-lying central and western areas transformed into open woodland featuring oak and hazel (fig. 6.16). One explanation for these differences probably lies in the rising sea level, which hampered the drainage of excess water via the regional rivers in the central and western part of Flevoland. This probably also led to changes in groundwater flows and the hydrology of the coversand subsurface. Minerals that previously lay inaccessible, deep in the subsurface, may have been transported to the surface and therefore became available for the vegetation. The differentiation in the vegetation increased further in the second half of the Early Atlantic (fig. 6.17). Groundwater probably played a key role in this, although it had not yet reached the surface.

---

557 Cf. Janssen 1974; Van Geel *et al.* 1981.

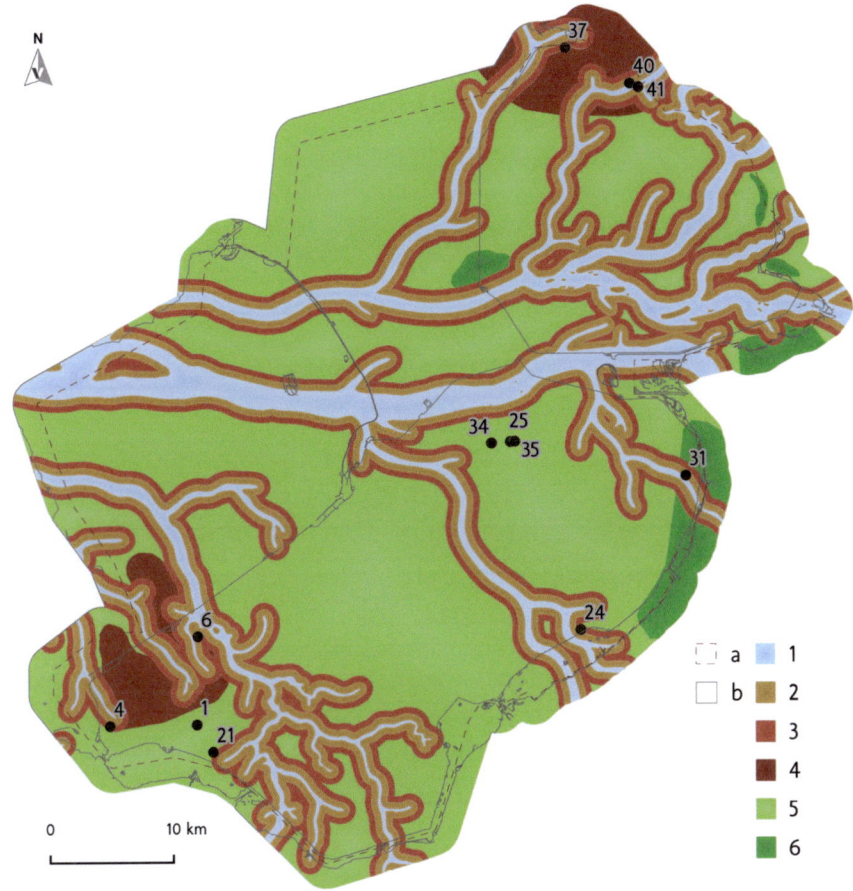

Figure 6.17: Reconstruction of the vegetation of Flevoland in the Early Atlantic based on palaeoecological information and after the palaeogeographical map of 9000 BC of Vos & De Vries (2018). Legend: a: topography; b: outline of research area; dot: locations of palaeoecological information (table 6.1); 1: rivers; 2: marshes and swamps along the rivers with reed and alder; 3: fens in the river valleys with reed and alder carr; 4: bog like vegetation in the river valley of the Eem and in the north; 5: open woodland on the dry soils with deciduous species (oak, birch, elm and possibly lime), species of the grass family and species of the heath family; 6: open woodland on some of the dry soils with pine, birch, hazel, oak, elm (and lime), grasses, species of the heath family.

6.3.5 Middle and Late Atlantic : 6000 – 3700 cal. BC (Late Mesolithic and Early Neolithic)

Geology, climate and hydrology

We have no detailed information about the climate in the Middle and Late Atlantic. Generally speaking, it would have been an Atlantic climate, with relatively mild winters and cool summers, due to the proximity of the sea. With prevailing winds from the west, there would have been quite a lot of rainfall.

During this period, the hitherto relatively uniform coversand landscape of Flevoland began to show more differentiation. The drivers behind this development were the rising water level and land subsidence, two processes connected with the last ice age. The rising sea level was the result of the further melting of land ice in northern Europe and other regions. This caused the subsurface to rise in these areas, and led to land subsidence elsewhere, in Flevoland for instance, in a process known as glacio-isostasy (see also chapter 3). Although, in an absolute sense, the sea level rose less rapidly in this period because there was less remaining land ice, as a consequence of land subsidence it actually rose by several decimetres a century. Around 3700 cal. BC the water level in Flevoland had risen to approximately five metres below NAP.[558] Large areas of Flevoland's Pleistocene coversands were covered with water, clay and/or peat (see fig. 6.1b/c, fig. 6.2).

The river valleys and low-lying parts of the coversand area in the west were the first to be exposed to the effects of the rising water. Towards the end of the Atlantic (c. 3700 cal. BC), only the river dunes of Oostelijk Flevoland

---

558 Makaske *et al.* 2002b, 2003; Van de Plassche *et al.* 2005; Roeleveld & Gotjé 1993.

Figure 6.18: Start of terrestrialization at a flood basin of the Kasari River (Estonia) just before it ends up in the Matsalu Bay of the Baltic Sea (photo: BIAX Consult/L.I. Kooistra).

and the Noordoostpolder, the outcrops of glacial till in the Noordoostpolder and the higher coversand areas in the east of Zuidelijk and Oostelijk Flevoland and the northeastern Noordoostpolder were still visible on the surface, as 'relics' of the Pleistocene landscape.

The first beach barriers developed in the western Netherlands at the end of the Late Atlantic (between 4200 and 3700 cal. BC). Over the subsequent millennia, they would grow into effective barriers against the sea, although during the Middle and Late Atlantic Flevoland still maintained an open connection to the sea via Noord-Holland (fig. 6.1b/c). The North Sea is a shallow sea, especially in the flooded areas of western Netherlands. Given that the sea level in the Atlantic was lower than it is today, the North Sea would also have been shallower. Furthermore, it is assumed that the sea bed was less steep. Like today, the difference between ebb and flood along the coast of Noord-Holland would have been smaller than along the rest of the western and northern coast of the Netherlands.[559] These factors give reason to assume that though the sea level was rising, the influence of the tides was not strong in Flevoland, although this would have had some influence on the hydrology of the region, resulting in a much more constant marine water pressure than would have been the case had the amplitude between ebb and flood been greater.

Water pressure was also exerted by fresh water from the landward side. The coversand areas of the central Netherlands drained water to the sea via regional rivers: the Eem, Hunnepe and Overijssel Vecht. Exactly how much fresh water flowed into the sea each year is not clear. In addition to the physical properties of the subsurface, the amount of fresh water would also have depended on factors like the annual rainfall and – very importantly – the types of vegetation in the region.[560] Given the source area for each of these regional rivers in Flevoland (the Drenthe plateau, Overijssel, Gelderland and the Utrechtse Heuvelrug ice-pushed ridge), it is, however, likely that substantial amounts of fresh water were discharged.[561] It is therefore reasonable to assume that large volumes of marine and freshwater collided, perhaps even forming layers, whereby the heavier marine water would have sunk beneath the fresh water.

The fact remains that many researchers have struggled to explain the presence of fine-grained sand and clay deposits of marine origin, which were largely deposited under water, while there are also indicators of fresh water present.[562] Not only the clay itself, but also diatoms and foraminifera in the clay provide indisputable evidence of its marine origin.[563] However, what was originally

---

559 Vos 2015.

560 The hydrology of a landscape can be modelled on the basis of knowledge of the palaeovegetation and the reconstructed groundwater curve for Flevoland, using simulation models. However, no such studies have been performed.

561 Much less, however, than in the Zuid-Holland coversand area, where two major European rivers – the Rhine and Meuse – enter the sea. It is not therefore surprising that the marine influence here did not extend further to the east, despite the greater tidal amplitude.

562 This is known, in the current geological typology of the Netherlands, as the 'Wormer Member' (*laagpakket van Wormer*). For arguments supporting deposition of clay under water, see Ente *et al.* 1986, 52, 54; Menke *et al.* 1998, 38.

563 Ente *et al.* 1986; Menke *et al.* 1998.

Figure 6.19: Examples of accumulation of washed up material by a. wind and weak currents (Texel, the Netherlands), and b. by wind, ice and strong currents (Pripyat River, Belarus), (photo: BIAX *Consult*/L.I. Kooistra).

marine clay also contains shells of freshwater mussels (Unionidae), which in the past led geologists to refer to the clay as 'Unioclay'. Plant remains found in the clay are also typical of freshwater marshes. Plant remains from brackish or salt water, and plant remains that represent saline marshes have been found only occasionally, if at all, dating to this period in Flevoland. This is in contrast with the coastal region of Zuid-Holland, for example, where a period of saline marshes has clearly been identified.[564] It was commonly assumed by geologists that in Flevoland, as in the western coastal area, the clay had been deposited in an intertidal zone, which fairly rapidly evolved into a freshwater tidal zone.[565] However, the botanical data discussed in the following section show that there was no tidal influence whatsoever.

The inundation of the Pleistocene landscape followed a more or less fixed pattern in the Middle and Late Atlantic. Initially, localised water levels existed in the coversand region, a result of differences in relief and the diversity in stratification of the subsurface. As the sea level rose, the localised water levels became part of a groundwater regime linked to sea-level rise. Where the fresh water subsequently reached the level of the Pleistocene surface as drainage water stagnated, peat generally began to accumulate. The ongoing rise in the sea level and stagnation in the drainage of fresh water lead to inundation of the peat, and the covering of the peat with a layer of marine clay deposited under water.[566] When, towards the end of the Atlantic (between 4200 and 3700 cal. BC), the process of clay sedimentation in Flevoland slowed down, peat could again begin to form and accumulate. This occurred in part on top of the clay deposits, on top of the low-lying coversands further inland and on the higher coversands where no clay had been deposited. The latter were not affected by the rising water until the end of the Late Atlantic. On these higher grounds, mires formed directly on the coversand. In some places, however, shallow lakes with a clay bed formed, which then became overgrown with vegetation from their shores. Reed and marsh fern played a particularly important role in this process. Such plant species grow into the water from the bank forming floating mats that provided a place for other mire vegetation to establish itself. Over time a shallow area of water can become covered by a floating layer, or mat, of peat formed in this way (fig. 6.18). Mire vegetation then grows on top of the peat, whilst water from the original lake is still underneath. A process of continuous accumulation results in an increasingly thicker and heavier layer of peat, the base of which will sink, due to its weight, and eventually touch the original lake bed.[567]

As far as is known, the first clay deposits in Flevoland occurred in the Eem estuary between 5300 and 5000 cal. BC.[568] The first clay was deposited in channels of the

---

564 E.g. Kubiak-Martens 2006.
565 Ente *et al.* 1986, 49; Menke *et al.* 1998, 38; Vos *et al.* 2020 (see also fig. 6.1b/c).
566 The Wormer Member (*laagpakket van Wormer*).

567 Dating the base of peat formed in this way will give a start date for accumulation, but the level at which it is found is not the level at which it is formed. Such bodies of peat are not therefore suitable for reconstructing the rise in the water level or for determining the moment at which the landscape was inundated. Only peat that is known to have accumulated on coversand and that was not subject to sinking or subsidence – where it can be shown that peat plants were rooted in the coversand, for example – is suitable for this type of analysis.
568 Derived from Van der Linden 2010. Menke *et al.* (1998, 40) published a date for sedge-reed peat with humic clay that falls between 4538 and 4337 cal. BC (5585±60 BP; GrN 6716).

Hunnepe and Overijssel Vecht shortly afterwards.[569] In the western part of Flevoland, the clay transported via the river valleys spread over the low-lying coversand. Over 1000 years later, between 4300 and 4000 cal. BC, the process of clay sedimentation and the inland expanding water surface temporarily ceased. Some of the clay dried out during this period, and marshy areas developed with channels, levees and flood basins. The fine-grained texture of the clay on the levees was the same as that in the flood basins behind them, suggesting that the clay was deposited in calm waters. The clay on the levees was firm, however, while the layer deposited under water in the flood basins was of a softer consistency.[570] At the boundary between water and land bands of detritus formed consisting of organic remains (generally of plant origin) that had been washed up by water.[571] Such accumulations of washed up material can be caused by the wind and weak currents (fig. 6.19a-b). These bands of detritus are generally richer in nutrients than the surrounding environment, enabling a lush vegetation to develop on it. Over the course of time, as more and more organic material accumulates, these bands of detritus can even develop into slight elevations in the landscape.

The marshes probably developed on the deposit of clay in western Flevoland as sea-level rise slowed. In the coastal region of the western Netherlands this resulted in the formation of more solid complexes of sandy beach barriers. The deposit of clay and the nutrient-rich water allowed a rapidly growing marsh vegetation to develop in the low-lying part of western Flevoland. The dead plant remains formed eutrophic peat, the accumulation of which kept pace over several centuries with the rising water level.

Vegetation development in the Middle and Late Atlantic

For the previous periods vegetation development has been discussed per location. The developments at the different locations are very comparable, so to avoid repetition, this paragraph will suffice with a single overview of the types of vegetation that occurred in the Middle and Late Atlantic. At locations where the Pleistocene coversand lay more than five metres below NAP, the changing conditions meant that the climax vegetation that developed there was forced back. The types of vegetation described below not only succeeded each other, but also occurred simultaneously, depending on the depth of the Pleistocene surface. Four maps show the distribution of the reconstructed vegetation types in Flevoland over time (fig. 6.20).

There is evidence for heathland on the dry, relatively nutrient-poor coversands. Depending on the availability of nutrients in the subsurface and the hydrology, the heathlands would have been overgrown with trees. Birch and oak were scattered across the nutrient-poor, dry areas. Pine also occurred locally at the start of the Middle Atlantic, though it disappeared from the vegetation during the course of the Atlantic. The widely encountered remains of hazel probably came initially from bushes that grew in places with nutrient-rich (river) deposits in the shallow subsurface. At the start of the Middle Atlantic these locations were more likely to be in river and stream valleys, low-lying areas and on the flanks of coversand ridges rather than on the tops of the higher coversand ridges. In the more nutrient-rich areas with lots of hazel, the density of deciduous trees would also have been higher, Oak and birch being the most common. The proportion of elm and lime increased in the course of the Middle and Late Atlantic. Of all these deciduous trees, lime is the only one that creates and prefers dense, shady woodland. The increase in lime during the Atlantic suggests the soil was richer in nutrients and less acidic, and that the hydrology of the subsurface improved. A rising water level no doubt made the climate in Flevoland more attractive for lime, which requires a water level that is not too deep, but equally not too close to the surface. As long as the groundwater remained within this particular bandwidth it would have led to improved hydrological conditions in the coversand. The groundwater probably transported minerals from the deeper substrate. An increase in the number of deciduous trees thus creating open, or even dense, woodland, would, however, also have improved the soil conditions. A different, thicker layer of leaf litter would have developed in deciduous woodland, partly because in the temperate climate zone of Europe deciduous trees shed their leaves in winter. The litter layer would have attracted a greater variety of mesofauna and microfauna. The decomposing leaves would have contributed to an increase in the humus content of the soil which meant that rainwater could be retained in the soil for longer and minerals would have leached out less quickly.[572]

Depending on the hydrology, the level of nutrients, the acidity and the amount of incompletely decomposed

---

569 Date 6200 to 6250 ^{14}C-years BP (Ente *et al.* 1986, 123-126; Gotjé 1993, 109; Wiggers 1955). Improved calibration programmes have put back the start of clay formation to over 500 years earlier than these authors assumed.

570 Ente *et al.* 1986, 126; Huisman *et al.* 2009; Dresscher & Raemaekers 2010.

571 Detritus is generally regarded as an accumulation of coarse organic material, including dead plants and animals that lived in the water (see Ente *et al.* 1986, 55 and others). It is deposited in a watery environment. The nature of the material in the bands complies with this definition: coarse organic material deposited at the water's edge.

572 Kooistra & Pulleman 2010.

organic matter, the marshes in the river and stream valleys of the Pleistocene area took the form of either swamps or carrs, reed, sedge marshes or fens, which would eventually develop into bogs.

The composition of the deciduous woodland on the coversand changed depending on how close the water table came to the surface. Oak was probably able to survive the longest. This is indicated by the many, fairly intact oak trunks and root systems found at location 10 (Hoge Vaart-A27) in Zuidelijk Flevoland (fig. 6.2, table 6.1). Such remains are only preserved if they end up in anaerobic conditions fairly soon after they die, *i.e.* under water or embedded in peat. Around 4900 cal. BC the water table at location 10 was so high that the soil of the coversand ridge began to develop hydromorphic characteristics. Palynological data, dendrochronological measurements and two ^{14}C-dates for oak roots suggest that the last oaks disappeared from the landscape soon after 4600 cal. BC.[573] The same development has been observed at location 25 (Swifterbant-N23) in Oostelijk Flevoland (fig. 6.2, table 6.1). Here, several dozen oak trunks and root systems have been found on the Pleistocene coversand. One of the trunks has been dendrochronologically dated to after 4799 cal. BC.[574] Interestingly, the wood at locations 10 and 25, which were found at more or less the same depth – 6.00 and 6.70 metres below NAP respectively – gave similar dates.

Tree ring patterns indicate the conditions in which trees grew. If broad rings form, the conditions are good; narrow rings suggest poor growing conditions. The tree ring patterns of the oaks at Dronten-N23 (location 25) show an alternation between very narrow and slightly wider rings (fig. 6.21). No increase in the number of narrow rings has been observed during the lifetime of the trees, which might mean that the water level was rising. The periods with narrow rings occurred when the trees were still young. There seems to be a pattern of several poor years followed by a period of slightly better growing conditions. Since there was no bark or sapwood, it is not known what the growing conditions were like in the final years of the trees' growth. Almost a hundred growth rings have been counted on two tree trunks. They were small trees, with estimated trunk diameters of 20 and 50 centimetres. Despite the slightly broader growth rings, these were therefore trees that experienced at least a hundred years of poor growing conditions, undoubtedly due in part to the high water level.

In many places the deciduous woodland transformed into carr with alder and birch as well as reed and sedge. Where peat accumulation did not keep pace with the rise in the water level, alder carr transformed into sedge reed fens where bulrush, bur-reed, marsh fern, bogbean and yellow iris grew. In places where the water was more than a metre deep, the fen vegetation declined and eventually disappeared entirely. The land disappeared under open water with no vegetation. Whether this water was fresh, saline or layered is not known.

The first phase of clay sedimentation appears to have occurred in open water in many places, given the virtual absence of plant remains. It was not until sea-level rise slowed at the end of the Late Atlantic that marsh flora re-established themselves. The ongoing sedimentation of clay had made the water shallow, and channels with levees and flood basins had started to develop. This means that the water was largely fresh by this time, leading to a succession of different types of vegetation. However, the continuing rise in the water level halted this process in a number of places, allowing the same type of vegetation to grow there for a much longer time. In other places, where other factors prevented plant growth from keeping pace with the rising water (for example, places with less nutrient-rich water), the succession either reversed, or open water took the place of vegetation. In some places, the marsh vegetation developed to a climax, which in these succession series meant a development from reed marshes to reed fens, to carrs, sedge fens and, eventually, raised bogs. Over the long term, factors such as rising water and the nutrient content of the soil, lead to the

---

573 Peeters *et al.* 2001, 37; Spek *et al.* 2001a, 86 & 137. Dates of oak roots: one between 4909 and 4730 cal. BC (5940±46 BP, UtC-5062) and one between 4785 and 4693 cal. BC (5856±46 BP, UtC-5061).
574 Kooistra 2012b, 373-374 (RING-report 2011068: RING Dendrocode: DRT00050).

Figure 6.20a-d (following spread): Reconstruction of the vegetation of Flevoland in the Middle and Late Atlantic in 5500 cal. BC (a), 5000 cal. BC (b), 4500 cal. BC (c) and 4000 cal. BC (d) based on palaeoecological information and after the palaeogeographical map of 3850 BC of Vos & De Vries (2018). Legend 6.20a-c: a: topography; b: outline of research area; dot: locations of palaeoecological information (table 6.1); 1: rivers and open water; 2: reed fens, alder carrs, birch carrs; 3: bog like vegetation; 4: deciduous woodland with oak and alder; 5: deciduous woodland on moisture soils in the coversand area with oak, lime and elm; 6: open woodland on the dry soils in the coversand area with deciduous species (oak, birch, elm and possibly lime), species of the grass family and species of the heath family. Legend 6.20d: a: topography; b: outline of research area; dot: locations of palaeoecological information (table 6.1); 1: rivers and open water; 2: reed marshes and swamps on clay deposits; 3: reed fens, alder carrs, birch carrs; 4: bog like vegetation; 5: deciduous woodland with oak and alder; 6: deciduous woodland on moisture soils in the coversand area with oak, lime and elm; 7: open woodland on the dry soils in the coversand area with deciduous species (oak, birch, elm and possibly lime), species of the grass family and species of the heath family.

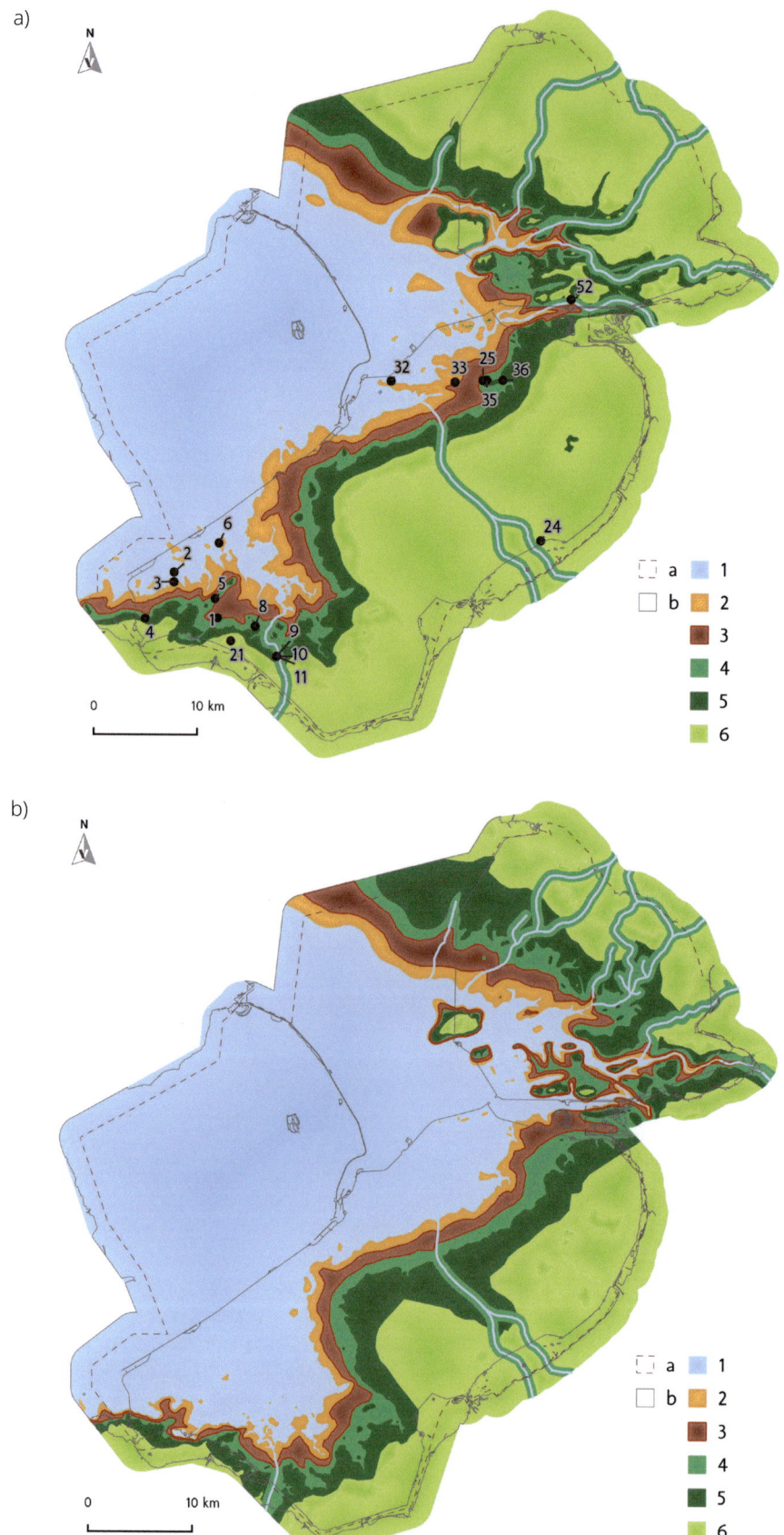

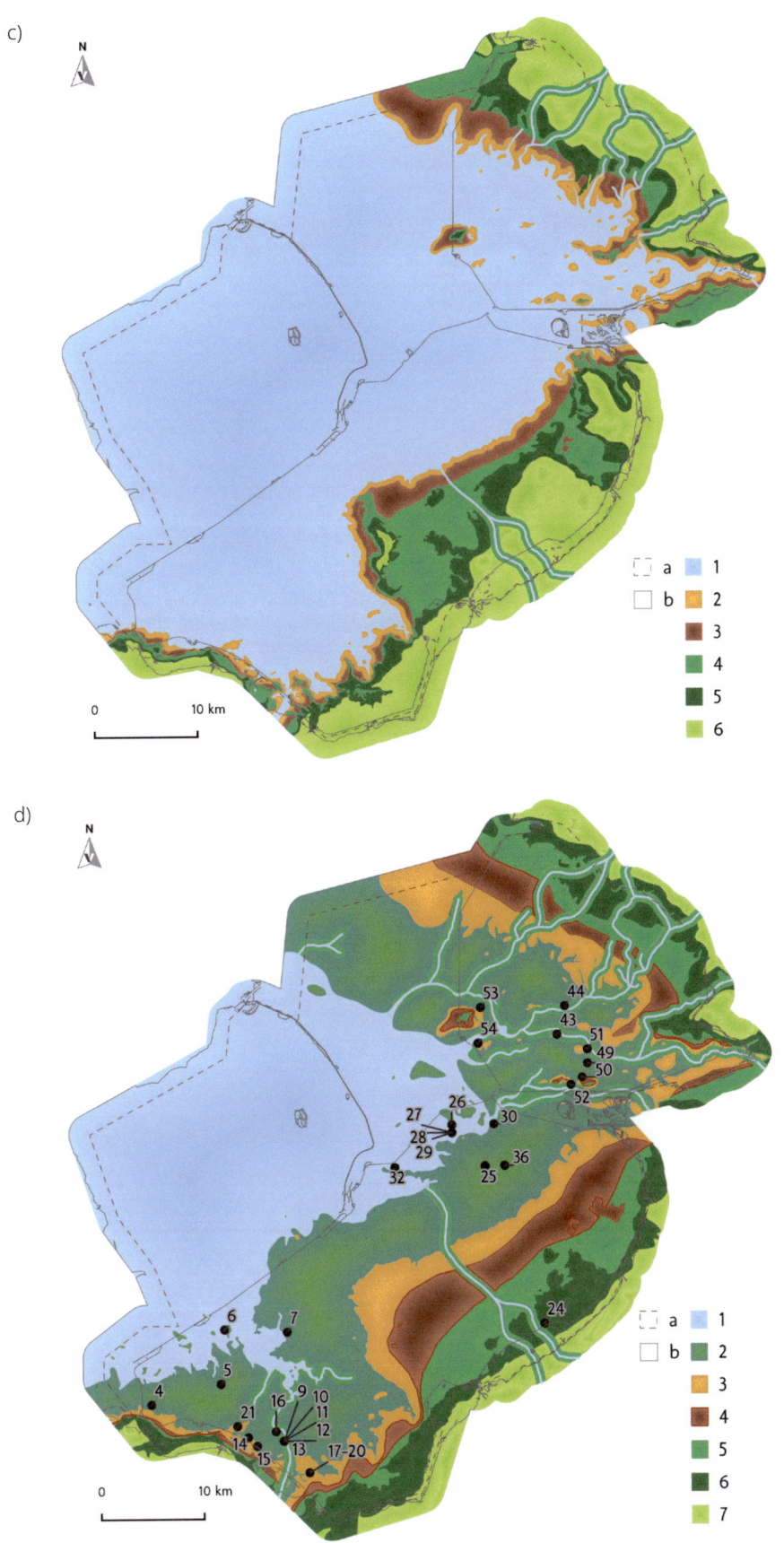

FROM LAND TO WATER

Figure 6.21: a) Oak trunk in situ at Dronten-N23 (location 25), b) the growth ring pattern of one of the oaks from Dronten-N23 (location 25), (photo: BIAX *Consult*/L.I. Kooistra).

Figure 6.22: a) Impression of vegetation with reed and saw-sedge (Nieuwkoopse Plassen, the Netherlands), b) flowering bogbean (near the Nieuwkoopse Plassen, the Netherlands) (photo: J.A. de Raad).

development of a mosaic of different types of vegetation and open water.

Reed roots have been found in the clay deposits at a number of locations, indicating that reed was an important component of the Flevoland marshes after the flooding of the Pleistocene landscape in the Middle and Late Atlantic. This is remarkable, given the specific requirements that need to be met for reed to become established. The seeds can namely only germinate in water-saturated locations on land or on levees. The seedlings can neither withstand flooding, nor survive drought.[575] The water level must therefore be constant during germination and initial establishment – the growing season, in other words. Once reed has established itself, it can withstand a certain amount of fluctuation in the water level and a degree of salinity in the water. The reed colonisation of the clay area of Flevoland during the period in question suggests that there can have been little or no tidal movement occurring at that time.

The analyses present a picture of relatively nutrient-rich marshes in which reed, saw-sedge (fig. 6.22a), lesser bulrush, common bulrush, grey club-rush, common club-rush and yellow iris dominated. Further from the open, nutrient-rich water, and beyond the reach of clay sedimentation, these marshes developed into wet fens. The spectrum of fen vegetation remained to some extent the same, but also including other species associated with slightly acidic environments, such as marsh fern and bogbean (fig. 6.22b). Where terrestrialization continued, the vegetation transformed to alder carr, and then to birch carr and raised bog. Such successions

---

575   Weeda *et al.* 1994, 192.

in vegetation development mainly occurred far from open waters.

Trees and shrubs grew on the driest parts of the levees and in the clay-rich flood basins at the end of the Late Atlantic. Alder and willow grew in the wettest places. This again suggests there were no tidal movements since, whilst alder is adapted to wet environments, it cannot withstand changeable water levels during the growing season. Oak, hazel, ash and elder grew in higher spots.[576] As peat continued to accumulate, space became available for birch.

6.3.6 Subboreal: c. 3700 – 1100 cal. BC (Middle Neolithic – Late Bronze Age)

Geology, climate and hydrology

The sea level and the groundwater level in Flevoland rose from approximately five metres below NAP to between one and two metres below NAP during the Subboreal. Halfway through the Subboreal, around 2300 cal. BC, the water level had risen to slightly above three metres below NAP.[577] Virtually the entire Pleistocene landscape was covered with either peat, clay or water (fig. 6.1c-e). Detritus gyttja was deposited in water over two metres deep.

Dry coversands occurred on the surface only in the Noordoostpolder, in the west near Urk and in the east near Voorst, and in the far east of Zuidelijk Flevoland. There were levees along the Overijssel Vecht that were dry enough to allow habitation, such as evidenced by human habitation remains at Emmeloord J97 (fig. 6.2, table 6.1, no. 44). There may also have been dry levees along the Hunnepe, though no evidence of this has yet been found. Parts of the Eem valley became overgrown with peat, and lakes formed in the former Eem estuary.

Geological developments in the Subboreal were dominated by the closing off of the coastline in the western Netherlands and the formation of peat. At the end of the Late Atlantic, as soon as sea-level rise began to slow, wave action perpendicular to the western Netherlands coastline deposited sand, which resulted in the formation of sandy tidal flats.[578] The sand that lay above the water began to drift, creating systems of beach barriers and dunes that in combination eventually formed a single barrier to the sea. Behind them, peat formation started on a large scale. Initially, there were many openings in the beach barriers, particularly at the mouths of the major rivers, the Rhine, Meuse and Scheldt.

Although the Noord-Holland coast closed off entirely during the course of the Subboreal, in northern Noord-Holland a connection between the sea and the hinterland extending as far as Flevoland. Known as the Bergen Inlet, this connection continued to exist for some considerable time (fig. 6.1e). At the start of the Subboreal, the formation of the Bergen Inlet and the associated channel system changed the course of the river Overijssel Vecht in the Noordoostpolder. The Overijssel Vecht had originally flowed to the south of the outcrop of glacial till at Tollebeek and Urk. A channel of the Bergen Inlet penetrated into the Noordoostpolder to the north of these outcrops and joined the Overijssel Vecht to the north of the outcrop of glacial till at Schokland, after which the Overijssel Vecht no longer followed its southern course. Somewhere to the west of Flevoland a branch of the river Hunnepe presumably joined the Overijssel Vecht or the Bergen Inlet channel system, though less is known about this. In southern Flevoland peat rivers and lakes drained into the sea via the Oer-IJ estuary. Halfway through the Subboreal (around 2000 cal. BC) saline and brackish water penetrated Flevoland via channels and rivers. Sea water penetrated Zuidelijk Flevoland via the Oer-IJ estuary. The Noordoostpolder and Oostelijk Flevoland became subject to marine influences that spread via the Bergen Inlet, the Overijssel Vecht and the Hunnepe.[579]

Paradoxically, as the sea inlets in Noord-Holland narrowed, the start of large-scale peat accumulation in the Subboreal probably also allowed saline and brackish water to penetrate into Flevoland. Peat accumulation not only reduced the storage capacity for sea water,[580] but also led to stagnation in the drainage of fresh water. Research has shown that raised bogs in eastern Poland store large quantities of water. Forty million cubic metres of raised bog can hold 34 million cubic metres of water. Only 5% of this water circulates; the other 95% remains in the bog as long as it does not oxidise or erode.[581] If an area of peat is stable, water discharge will remain more or less constant. In the Subboreal, however, peat developed on a massive scale in large areas of the Netherlands (fig. 6.1c-e). This included the formation of raised bog, which expanded not only in area, but also in thickness. Peat also accumulated on a large scale in the source regions of the Eem (to the south of Zuidelijk Flevoland), the Hunnepe (to the east of Flevoland) and the Overijssel Vecht (to the east and northeast of Flevoland). While the raised bog was expanding in this way, each year it would have retained

---

576 Incl. De Jong 1966; Kooistra 2014; Raemaekers *et al.* 2010.
577 Based on Makaske *et al.* 2002a; Van de Plassche *et al.* 2005. As regards the water table in 2300 cal. BC, see Van Smeerdijk 1989, 478. In southern Flevoland, at location 22 (see figure 6.2; table 6.1) peat that developed at 2.72 m below NAP has been dated to between 2561 and 2141 cal. BC (3870±70 BP, GrN-13035).
578 Though most of the land ice had melted, subsidence was still occurring as a result of glacio-isostasy (see chapter 3).

579 Incl. Menke & Lenselink 1991.
580 For more information on the concept of 'accommodation space', see Coe 2003, 58-61.
581 Tobolski 2000, 28-30.

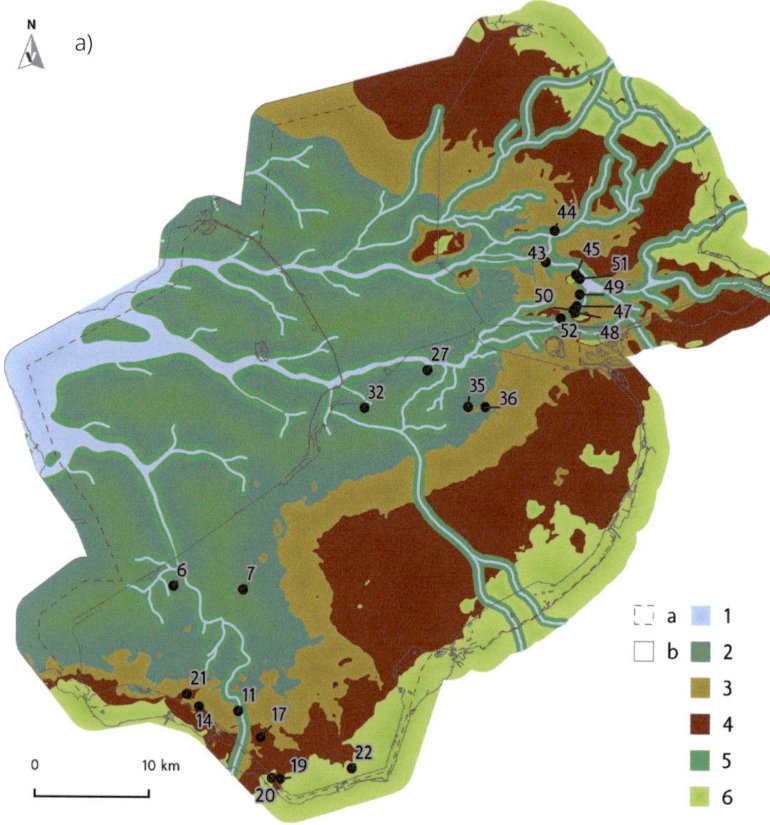

Figure 6.23a-c: Reconstruction of the vegetation of Flevoland in the Subboreal in 3500 cal. BC (a), 2500 cal. BC (b) and 1500 cal. BC (c) based on palaeoecological information and after the palaeogeographical maps of resp. 3850 cal. BC, 2750 cal. BC and 1500 cal. BC of Vos & De Vries (2018).

Legend 6.23a: topography; b: outline of research area; dot: locations of palaeoecological information (table 6.1); 1: rivers and open water; 2: reed marshes and swamps on clay deposits; 3 reed fens, alder carrs, birch carrs; 4: bog like vegetation; 5: deciduous woodland with oak and alder; 6: open woodland on the dry soils in the coversand area with deciduous species (oak, birch, elm and possibly lime), species of the grass family and species of the heath family. Legend 6.23b-c: a: topography ; b: outline of research area; dot: locations of palaeoecological information (table 6.1); 1: rivers and open water; 2: reed fens, alder carrs, birch carrs; 3: bog like vegetation; 4: deciduous woodland with oak and alder; 5: open woodland on the dry soils in the coversand area with deciduous species (oak, birch, elm and possibly lime), species of the grass family and species of the heath family.

more water than it drained. It is not impossible, therefore, that this might have caused the counter-pressure of the fresh water in the regional river systems to decrease to such an extent that marine water was able to penetrate further inland via tidal channels and rivers.

In the second half of the Subboreal more saline to brackish water reached Flevoland than in the Atlantic, as suggested by the presence of foraminifera, dinoflagellates and molluscs from brackish waters in clay layers deposited in this period. Bivalve molluscs have been found, particularly *Cerastoderma glaucum*. This mollusc used to be known as *Cardium edule*. The clay in which it was found has also been referred to as *Cardium* clay in the past.[582] The presence of doublets of this mollusc species means that these invertebrates lived at the location. Large shells suggest a mainly marine environment, small shells suggest a more brackish one. Doublets of *Cerastoderma glaucum* have been ^{14}C dated in order to determine the age of the clay.[583] Doublets found in Zuidelijk Flevoland proved to be 4220±90 and 3975±35 ^{14}C-years BP old.[584] In Oostelijk Flevoland they dated to 3995±40 BP,[585] and in the

---

582 In the literature this clay deposit is regarded as part of the Wormer Member (*Laagpakket van Wormer*).

583 *Cerastoderma glaucum* is a bivalve species. Finding doublets in a clay deposit means that the creatures lived at the location. If only single shells or broken shells of the species are found in a sediment, the origin of the invertebrate is unclear. It may have lived in the clay at the location found, or, alternatively, it might have been transported in with the clay. In such a situation, the shell might come from an older sediment.

584 Menke *et al.* 1998, 40. The doublets in the clay deposit at Tureluurweg dated to 4220±90 BP (GrN-11498); those from Almere plot Az 122/123 dated to 4010±100 BP (GrN-18358); those on Wulpweg dated to 3975±35 BP (GrN-19513).

585 Ente *et al.* 1986, 57, in the Swifterbant region plot G42c (GrN-7082).

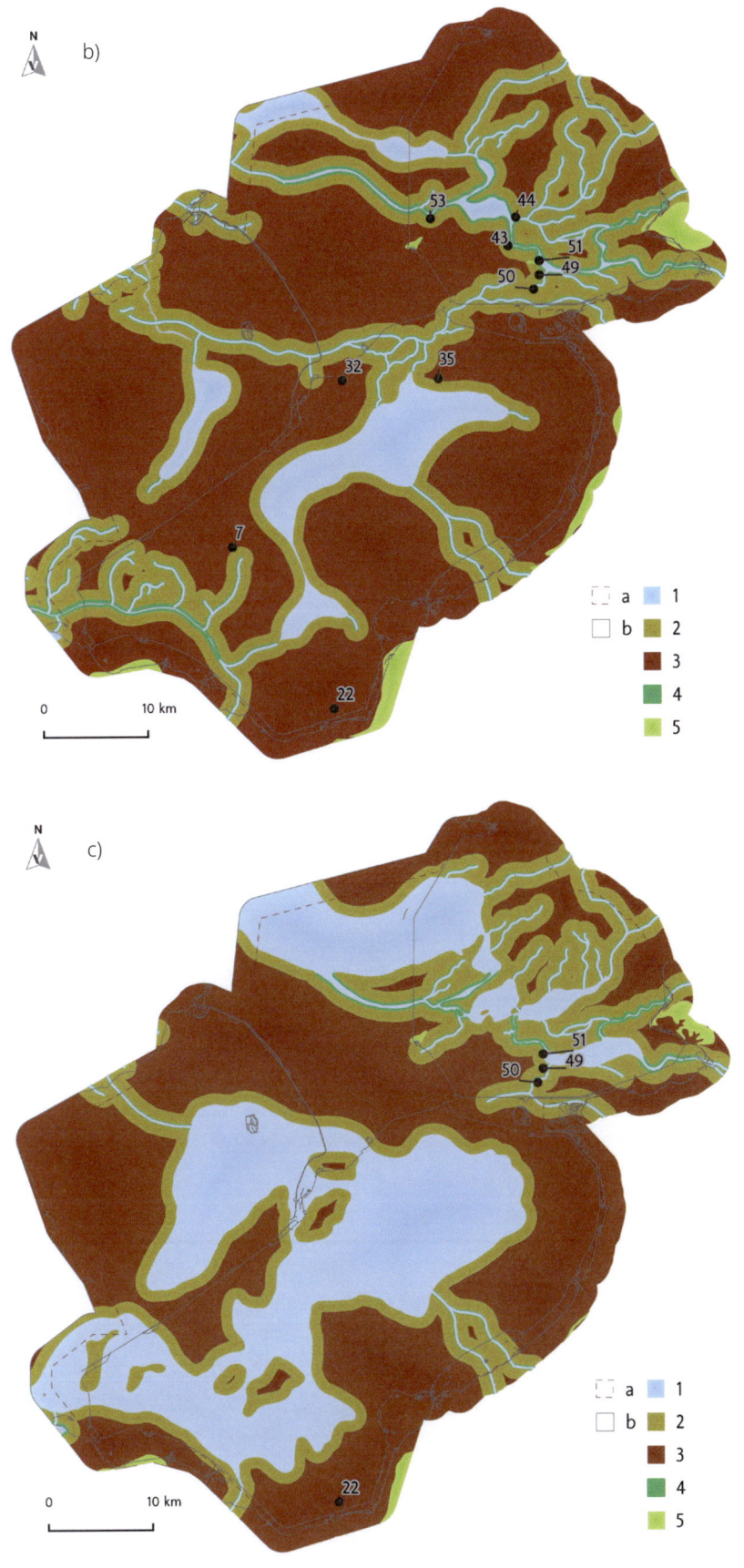

FROM LAND TO WATER

Figure 6.24a-f: The vegetation development from open water to peatbogs in the Natura2000 area Nieuwkoopse Plassen van Natuurmonumenten (the Netherlands), a) open water, b) development of floating vegetation on shores and development of detritus, c) floating reed fens and development of a detritus layer, d) alder carr, e) transition to raised bogs, f) raised bogs (drawing by Wim Dasselaar).

Noordoostpolder to 3920±60 BP.[586] However, given that aquatic organisms derive carbon not from the atmosphere but from water, we must consider the reservoir effect, which can mean that the age measurement turns out older than the actual age. The reservoir effect of carbon in marine environments for Dutch coastal waters is approximately 400 years, which must be deducted from the dates which are calibrated with atmospheric carbon.[587] The carbon in brackish water, however, comes partly from a marine environment and partly from river systems. The reservoir effect can differ markedly from one river system to another, and depends on the layers through which the river flows on its way to the sea. The difference can be as great as 1000 years.[588] From the ^{14}C datings of *Cerastoderma glaucum* can therefore only be deduced with any certainty that the molluscs lived in the Flevoland clay sometime after 3000 cal. BC.

---

586 Koopstra 1981, at Tollebeek II (GrN-10623).
587 Reimer & Reimer 2001.

588 Cf. Culleton 2006, Philippsen 2012; 2013.

Figure 6.25: Impression of a raised bog. Fochteloërveen (the Netherlands) (photo: BIAX *Consult*/L.I. Kooistra).

Dated peat from Zuidelijk Flevoland and the Noordoostpolder accumulated at the same time as the clay sedimentation took place. On Schokland, a date comes from a small layer of peat embedded in the '*Cardium* clay', which formed between 1876 and 1416 cal. BC.[589] In the east of Zuidelijk Flevoland, peat formation dates between 2550 and 1750 cal. BC coincide with the estimated age of the *Cardium* clay.[590] Together, these data suggest that clay deposition occurred between 2500 and 1500 cal. BC. The influx of brackish or saline water in this period caused localised erosion, as evidenced by the presence of detrital plant remains and spoiled coversand in the Noordoostpolder.[591] Given the limited impact of these saline and brackish water incursions on the vegetation in Flevoland, it would appear that these were probably isolated events, rather than a protracted period of prevailing saline and brackish conditions in the area.

Between 1500 and 1350 cal. BC the Bergen Inlet silted up and the marine influence in the Noordoostpolder and Oostelijk Flevoland finally came to an end.[592] In Zuidelijk Flevoland the connection with the Oer-IJ estuary continued to exist, but the sea water no longer penetrated here. The connection served mainly to drain fresh water from Flevoland and the Pleistocene area partially overgrown by peat that lay beyond (Utrechtse Heuvelrug ice-pushed ridge, eastern Gelderland, Overijssel and the Drenthe plateau) (fig. 6.1f).

Vegetation development in the Subboreal

The landscape of Flevoland in the Subboreal can be described, on the basis of the palaeoecological analyses (fig. 6.2, table 6.1), as comprising extensive mires with many interspersed lakes, through which rivers flowed (fig. 6.23b-c). There were only a few areas in the east of Zuidelijk and Oostelijk Flevoland where (initially) more or less dry Pleistocene coversands lay at the surface. There was probably mixed deciduous woodland growing on these subsurface at first, although information is scarce. In the course of the Subboreal, however, this woodland made way for mire vegetation. In the Noordoostpolder, mixed deciduous woodland was found not only in the north and east, but also on the outcrops of glacial till at Voorst, Schokland, Tollebeek and Urk, and possibly also on a river dune to the north of the Hunnepe.[593] Softwood and hardwood riparian woodland would also have grown on the levees of the Overijssel Vecht, certainly into the Middle Bronze Age (1800-1100 cal. BC), as evidenced, among other things, by the wood used to make fish traps found in a small tributary of the Overijssel Vecht (see chapter 4).[594] Botanical information for the last 1000 years of the Subboreal is scarce, but as long as rivers flowed through the Flevoland mires, there would have been higher silt levees supporting riparian woodland. Depending on the height of the levees relative to the water, these would have been softwood woodlands comprising willow, alder and poplar, or hardwood woodlands with oak, ash and elm.

There is no doubt that there was only a small area of dry subsurface in what is now Flevoland, and that this area reduced in size even further during the Subboreal.

---

589 Wiggers 1955, 64-65; 3315±90 BP (GrO-377).
590 Van Smeerdijk 1989, 488-490; 3870±70 BP (GrN-13035); 3780±60 BP (GrN-13034); 3570±60 BP (GrN-13033).
591 Wiggers 1955, 60-62.
592 Cf. Berendsen 2011, 293.

593 Gotjé 1993; Weijdema *et al.* 2011; Ten Anscher 2012.
594 Van der Heijden & Hamburg 2002; Van Rijn 2002.

The landscape was largely a patchwork of lakes and mires. No land formation occurred during the transition from the Late Neolithic to the Early Bronze Age (probably between circa 2400 and 1800 cal. BC), at a time when the lakes contained saline or brackish water. Logically also no land formation occurred in the large freshwater lakes, where wave action eroded the shoreline and the peat.[595] Although the edges of large lakes can be subject to erosion, new land can also form on the shoreline. On the leeward side of a lake this occurs when reed beds grows into the water; on the windward side organic material can accumulate to form bands of detritus. Although it would seem that the deep lakes in Flevoland continued to exist, possibly 'wandering' through the mires, it is likely that shallow freshwater lakes grew over after a time. This process would certainly have occurred in part along the shoreline, as floating reed vegetation grew into the water (fig. 6.18).

Over the last hundred years of the Subboreal, the silting up of the Bergen Inlet and the fact that brackish and saline water disappeared from the channels and rivers of Flevoland, meant that seawater no longer entered the region via the Oer-IJ estuary. A mosaic of different types of mires developed in Flevoland and the surrounding region. The type of mire depended to a large degree on the hydrology, *i.e.* the quantity and origin of water, as well as its nutrient content and acidity. The types of mires that occurred when and where is not clear, as by no means all the vegetation zones in the layers of peat examined have been ^{14}C dated, and in many places the peat has been lost due to (natural or artificial) drainage. However, it is known that a variety of mire types existed, as indicated by the various stages of land formation on the peat (fig. 6.24). There is, for example, evidence for various types of reed fens (with or without saw-sedge, species of the sedge family, marsh fern and bogbean), alder carr, birch carr, transition mires with heaths and species of the sedge family and raised bogs (fig. 6.25).

6.4 A new view of the landscape

6.4.1 Pine woodlands and heathlands in the Atlantic (7000 – 3700 cal. BC)

It was long assumed that vegetation developed according to a fixed pattern in the Holocene. During the Preboreal (between 9800 and 8200 cal. BC), woodland dominated by birch initially developed on the Pleistocene subsurface of the Netherlands. In the second half of the Preboreal the woodland was dominated by pine. Open pine woodlands with undergrowth consisting of fern species, stag's-horn clubmoss and heath defined the look of the landscape in the Boreal (between 8200 and 7000 cal. BC). However, more and more cold-intolerant deciduous trees from Southern and Central Europe became established in the pine woods at the end of the Boreal. In the Atlantic deciduous trees displaced the pines and a mixed oak deciduous woodland with elm and lime developed. Swamps dominated by alder grew in the river and brook valleys. Humans increasingly influenced the vegetation from the Subboreal (3700 to 1250 cal. BC) onwards, and were a factor in the disappearance of elm and lime from the deciduous woodland. During the course of this period the area of woodland decreased, to be replaced by meadows, pastures and heathland. Beech first established itself in Dutch woodlands in this period.[596]

This interpretation of the palynological data, in particular, was first called into question towards the end of the twentieth century.[597] It was argued that the proportion of hazel and oak in the pollen diagrams was too high for dense woodland. Hazel is a shrub which these days grows on the edges of woodland and in glades. Oak requires a lot of light and is not therefore likely to be found growing in dense woodland with lime trees. The explanation for these results was said to be that large herbivores such as aurochs, elk, red deer and horses prevented the development of woodland after the last ice age, and that in the Atlantic the Netherlands had a varied landscape consisting of open woodlands with pastures and meadows in between.

An examination of the pollen diagrams for Flevoland from the first half of the Holocene (from 9800 to circa 5000 cal. BC) shows that this alternative interpretation is only partially correct. It looks very much like there was no dense pine woodland in Flevoland in the Preboreal and Boreal. The clear evidence of heath vegetation, stag's-horn clubmoss and ferns suggests that the pine woods of Flevoland were much more open. The diagrams containing data from the Early Atlantic (7000 to 6000 cal. BC) mainly show large amounts of pollen from hazel, oak, birch and locally also pine. This pollen, combined with palynological material from the heath family and grass family, suggests that in this period the landscape consisted of open oak woodland with heath in the undergrowth or heathland with scattered trees such as oak, hazel, birch and pine. Since the water level lay relatively far beneath the surface in Flevoland at that time, it is likely that the soil conditions determined these types of vegetation. A lack of water and leaching of minerals probably produced the kind of nutrient-poor coversands that are still found in the Veluwe and Noord-Brabant today.

The open landscape was not created by large herbivores, but by a loss of nutrients from the soil. However, it is not impossible that large grazing animals inhabited this landscape, though they were not the primary reason for the vegetation structure. This can be

---

595   Raemaekers & Hogestijn 2008.

596   Cf. Janssen 1974; Van Gijssel & Van der Valk 2005, 61-62.
597   Vera 1997.

seen from the vegetation development that followed this phase of heath and open deciduous woodland. The rising water level improved the hydrology of the coversand, and the deciduous woodland expanded. This caused the formation of humus, which improved the soil quality. Eventually, dense oak woodland with lime developed, as described in the classic version of vegetation development in this region. If large herbivores had been the cause of the vegetation structure at the start of the Atlantic, the later development of dense woodland could not have occurred.

There is differential relief in the Pleistocene coversand of Flevoland. Roughly speaking, the coversand in the west is lower than that in the east. The transformation from open woodland with birch, pine, oak and hazel, with heath, to dense oak woodland with lime first began in the low-lying west. Over the centuries this vegwetation zone shifted eastward. Palynological analysis combined with ^{14}C dating has shown that successive vegetation types could occur simultaneously on a spatial scale. This implies that the general picture of vegetation succession on a regional or sub-regional scale does not have any chronological basis.

6.4.2 No salt marshes or tides in the Late Atlantic (5000 – 3700 cal. BC)

By around 5000 cal. BC the water level had risen to approximately eight metres below NAP. Seawater moving via Noord-Holland towards Flevoland met the fresh water draining from a large proportion of the Netherlands via the rivers Eem, Hunnepe and Overijssel Vecht. It is not known which source dominated. It is however clear that the clay deposited on the inundated coversand region had marine origins. It did not however contain large numbers of marine organisms, but nor did it, at first, contain plant remains from freshwater marshes. It is generally assumed that the clay was initially deposited under water.[598] To this day, no plant remains from halophyte species have been found in any botanical survey. This could be because no botanical surveys have been performed in order to identify locations that were along the edges of the rising water. However, several studies have been performed on the coversand landscape around the Eem in southern Flevoland, which was gradually becoming inundated. None has produced information that might suggest the presence of saline marshes. For now, therefore, it seems likely that the saline marsh zone did not reach as far as Flevoland. Under the influence of the rising water, predominantly nutrient-rich reed marshes with saw-sedge developed along the edges of freshwater marshes. One explanation for the marine clay deposits in freshwater marshland might lie in the fact that the counter-pressure from the rivers in Flevoland was so great that the heavier saline and brackish water disappeared beneath the fresh water of the rivers in Flevoland.

Since Flevoland was connected by water to the sea, it has always been thought that there was a difference between ebb and flood along the coast of Flevoland. This has now been called into question on the basis of the botanical data. Willow swamps and reed marshes are typical of freshwater tidal areas. Though they did occur in the coastal zone, which was shifting inland, the low shores at Swifterbant were covered with riparian woodland, and alder was the most important tree species in the swamp. Alder has only limited capacity to withstand fluctuating water levels. In the winter it tolerates higher water levels because the tree is then dormant, but it cannot survive fluctuations in the summer, let alone differences in ebb and flood.

All in all, the botanical data have led us to assume that the coastal zone of Flevoland, which was shifting inland in the Late Atlantic, consisted of freshwater marshes, and that evidence for tidal movement was absent.

6.4.3 Lakes and large-scale peat accumulation in the Subboreal (circa 3700 – 1100 cal. BC)

The characteristic landscape features of the Subboreal in Flevoland are lakes and extensive mires. Layers of peat several metres thick accumulated during this period, not only in Flevoland, but in large parts of the Netherlands. However, fairly little is known about how the prehistoric peat landscapes of the Netherlands functioned. This is surprising, given the fact that peat accumulated in the Subboreal on a much greater scale than the sedimentation of clay during both the Atlantic and Subboreal.[599] In a sense, the lack of focus on peat is understandable. The clay deposits are still present below the surface, whereas most of the peat that formed in the Subboreal had by now disappeared, due to natural or man-made erosion, or because it has been cut and burnt as fuel.

Nevertheless, the huge expansion of the peat in the Subboreal, in terms both of area and of thickness (thus in volume) brought about major changes in the Dutch landscape and the way it was exploited. It therefore deserves a great deal of attention. Peat can only develop in a wet environment. At the same time, peat draws in water and retains it for a certain length of time, dramatically altering the hydrological processes in the landscape. In a sense, an expanding area of peat acts like a reservoir, particularly if raised bogs form. Though a lot of water enters the area, only a fraction of it drains away (see 6.3.6).

At the start of the Subboreal sea-level rise levelled off (and along with it the rise in the groundwater level in

---

598 Wiggers 1955; Ente *et al.* 1986; Menke *et al.* 1998.

599 See Van der Linden & Kooistra (2019) for the history of peat accumulation in the Holocene.

Flevoland), and there were large lakes in Flevoland. The largest and deepest of these lakes continued to exist. Most of the shallow lakes filled with vegetation. It is likely that most of them grew over from the shores inwards, as real aquatic plants generally produce little organic matter. Marsh plants, particularly reed, do however produce a lot of organic matter every year, which is converted to peat in water-saturated conditions. Reed colonises a body of water from the shore, however, with long offshoots that grow into the water. Only a few marsh plants (*e.g.* common club-rush and lesser bulrush) produce seeds that can germinate in shallow water, thus contributing to this process. Given this fact, it is likely that shallow lakes grew over from the shores inwards in the Subboreal. After a time, this would create a floating mat of peat. Progressive accumulation causes the peat layer to grow thicker, until it eventually reaches the lake bed.

Plants need nutrients from the environment to grow. Peat-forming plants draw nutrients from groundwater and surface water. If there are no minerals in the water, the environment will supply fewer nutrients. Minerals are absorbed by peat-forming plants and when they die they do not decay, so do not therefore release the minerals back into the environment. Instead, they remain in the peat. Development of peat without any supply of minerals via hydrological processes (surface water in the event of flooding, groundwater flows) thus depletes the nutrient content and causes the environment to become more acidic. The vegetation adapts to this. A nutrient-rich environment has different plants than an acidic, nutrient-poor environment. In short, therefore, we can say that the longer peat accumulation continues, the fewer free nutrients the peat soil will contain, and the more the vegetation will change.

Peat accumulation in the Netherlands, and particularly in Flevoland, had major implications for the landscape. If a peat landscape develops due to a rise in the groundwater level, a dry landscape will transform into a mire. Landscape development of this type occurred in the Atlantic. It did continue into the Subboreal, although in large parts of Flevoland the opposite occurred, with bodies of water (lakes) transforming into mires. As a result of these two developments, extensive mires developed consisting of a large variety of vegetation types, from which peat accumulated. In the expanding peat landscape the drainage of water stagnated. The origin of the water entering the area determines the nutrient content of the peat. Variation in the nutrient content of the water produced a variety of mire types in the peat landscape.

6.5 Three windows of observation

6.5.1 Zuidelijk Flevoland: Hoge Vaart-Eem microregion between 7000 and 4000 cal. BC

The site excavated at Hoge Vaart-A27 (location 9) consisted of a coversand ridge oriented northwest-southeast, the highest point of which lies some 5.60 metres below NAP. To the east, the ridge bordered the river Eem (fig. 6.26). The coversand ridge was inundated between 4400 and 4000 cal. BC. Traces of human activity there date to between 6900 and 4200 cal. BC.[600] The landscape underwent dramatic changes during the approximately 2500 years within which identifiable traces of human activity have been discovered on the coversand ridge and on the adjacent levee and in the channel, transforming from a dry inland coversand area to a freshwater marsh connected via a lagoon to the North Sea. The vegetation and fauna also changed radically during this period, mainly as a result of sea-level rise and the associated stagnation in the drainage of fresh water. The migration of cold-intolerant plants from Central and Southern Europe would also have affected the look of the landscape.

Exploitation phases 1 and 2: 7000 to 6000 cal. BC

At the transition from the Boreal to the Early Atlantic, the coversand ridge was high and dry, both literally and relatively speaking. The sea was some distance away and the water level lay far below the surface of the coversand ridge. During this period, the climate changed from continental, with hot, dry summers and cold, dry winters, to Atlantic, with more rainfall and less extreme summer and winter temperatures.

At the start of the Early Atlantic, pines still grew on a large scale (fig. 6.26a), as evidenced, among other things, by pine trunks found along the edge of the Eem valley. Pine cones found with one of the trunks could be dated to between 7451 and 6642 cal. BC.[601] In what sort of setting the pines grew is not clear, as no other botanical material from this period have been analysed. Given the developments elsewhere in Zuidelijk Flevoland, however, it is likely that there was open woodland with lots of pine, plus oak and birch, with an undergrowth consisting of heathland species. Hazel probably grew on soils containing more moisture and nutrients in the river valley and its flanks. In this dry landscape where lots of pines grew, hunter-gatherers dug pit hearths, one of which has been dated to between 6817 and 6477 cal. BC.[602]

---

600   See chapter 4.
601   Peeters & Hogestijn 2001, 163, 8060 ±140 BP (GrN-25487).
602   7800±60 BP (UtC-5709).

Figure 6.26a-d: Impression of the vegetation during inhabitation of the Hoge Vaart-Eem microregion for a) the period around 7000 BC: open pine woodland (Nationaal Park de Hoge Veluwe, the Netherlands, photo: BIAX *Consult*/ L.I. Kooistra); b) the period between 5500 and 5000 BC: deciduous woodland dominated by lime and oak (Savelsbos, the Netherlands, photo: Ecologisch Adviesbureau Maas); c) the period between 5000 and 4500 BC: nutrient-rich marshland (Oostvaardersplassen, the Netherlands., photo: BIAX *Consult*/L.I. Kooistra); d) the period around 4200 BC: stagnant water with water soldier (Nieuwkoopse Plassen, the Netherlands)(photo: J.A. de Raad).

Exploitation phase 2: 5500 to 5000 / 4900 cal. BC

The hunter-gatherers who made their fires in pit hearths on the coversand ridge over a thousand years later lived in an entirely different landscape. In the Eem valley there was water surrounded by swamps and carrs containing alder. On the coversand ridge there was deciduous woodland dominated by lime and oak (fig. 6.26b). Although lime commonly signifies dense woodland, the vegetation of the Hoge Vaart-Eem microregion was more open, as evidenced among other things by the pollen of species of the heath family found in the soil. The heath species may well have been a remnant of older vegetation from a period with a lower water level, when the coversand was more acidic and contained fewer nutrients as a result of leaching.[603] However, it seems that there was still an open type of woodland with oak and birch, and an undergrowth of heath, on the higher, drier and generally less nutrient-rich coversand soils to the south and east of location 9 (Hoge Vaart-A27).

Exploitation phase 3: 5000 / 4900 to 4500 cal. BC

At the transition from the Middle to the Late Atlantic the banks of the Eem were beginning to erode. People frequently visited the Hoge Vaart-A27 (location 9) site during this period. The landscape setting and wealth of botanical and zoological remains suggest that they lived on a spit of land in nutrient-rich marshland (fig. 6.26c). In the northwest the marsh made way for an area of open water (largely fresh water). In the south and east there was high, dry coversand. This transitional zone between land and water may have been attractive to humans for several reasons. The water provided good transport opportunities and the abundance and variety of plants and animals supplied a range of food and other resources that people needed to survive.[604] At the end of this period the coversand ridge was subject to regular flooding, which would have hampered continued habitation at this location.

Prior to being inundated, the coversand ridge was either covered with open deciduous woodland, or was an open area in mixed oak woodland fringed by a variety of shrubs that required large amounts of light. Over the course of 500 years, lime was the first tree species to disappear from this woodland landscape. Around 4500 cal. BC the area also became too wet for oak. As the environment grew wetter, the woodland gradually came to be dominated by alder, willow and birch. Club-rush species, reed and sedge species grew at the wettest spots in the marsh.

An open connection developed between the sea and the water-rich Eem valley. The remains of at least two seals, anadromous fish (eel and houting), sea fish that tolerate brackish water (flatfish and mullet) and one actual sea fish species (sea bass) are evidence for this connection.[605] Interestingly, only sporadic plant remains of saline marshes have been found, these mostly belong to the salt-tolerant freshwater species beaked tasselweed. Although there may be many reasons for the low proportion of marine fauna in the food spectrum, the virtual absence of brackish and salt marsh plants is a clear indication that marine influence was minimal in this microregion at this time. Furthermore, the evidence for freshwater marshland is overwhelming. It can be concluded from this that substantial amounts of fresh water must have drained into this area from the Gelder Valley and the Utrechtse Heuvelrug ice-pushed ridge.

This landscape at the boundary between land and water was home to a great variety of mammals. Some of them – red deer, wild boar, elk and squirrel – support the idea this was an environment rich in water and woodland. Though aurochs are now extinct, research has found evidence to show that these animals preferred river areas with nutrient-rich reed beds and meadows.[606] Marshes were also home to beavers and otters. Both beaver bones and gnaw marks on wood have been found.

The presence of horse bones suggests that, even at the start of the Late Atlantic, there were still open landscapes where woodland alternated with low-growing vegetation. The animals may have lived on the dry coversands to the south and east of the Hoge Vaart-A27 site, where there might still have been open oak and birch woodland with a heath undergrowth. The river Eem would have provided them with water. It is not impossible that horses co-existed with aurochs on the reed beds and meadows along the Eem, perhaps in the summer, when there was little to eat on the dry coversands and vegetation would have been abundant in the marsh.

Exploitation phase 4: 4300 to 4200 cal. BC

Around 4300 cal. BC sedimentation in the Eem channel ceased and the channel grew over with water soldier.[607] It seems likely that the drainage of water stagnated when a barrier formed in the Eem, possibly as a result of beaver activity.[608] The water soldier suggests the water was fresh to slightly brackish, and was either completely stagnant, or flowed only slightly. This species also indicates that the water was half a metre to two metres deep (fig. 6.26d).[609]

---

603   See section 6.3.4.
604   See chapter 4.
605   Laarman 2001.
606   Van Vuure 2005.
607   Gotjé 2001, 41 (Eem core-1).
608   Peeters 2007, 55.
609   Weeda *et al.* 1991, 232.

Figure 6.27: Hoge Vaart-A27 (location 9), post of a fish trap made of alder, probably from a beaver dam, as beaver gnaw marks can be seen on the tip (photo: Cultural Heritage Agency of the Netherlands/T. Penders).

Although the coversand ridge was no longer habitable, fish weirs with fish traps were still being placed in the open water. The fish weirs were made mainly of alder from the surrounding area. The wood was hardly worked, and small logs gnawed through by beavers were also used (fig. 6.27).[610] After a time the open water became grown over again, bringing an end to human activity at this location.

6.5.2 Oostelijk Flevoland: Swifterbant microregion between 8300 and 3700 BC

The Swifterbant microregion covers the area between Dronten to the east and the IJsselmeer coast (fig. 6.2). In the Late Glacial and Early Holocene there is evidence for a valley, through which the Hunnepe flowed from east-northeast to west-southwest (fig. 6.28). River dunes formed on either side of the river in this period, in a coversand landscape that lay six to eleven metres below NAP. Only a few of the tops of the river dunes extended higher, to four metres below NAP.

From 5400 cal. BC the coversand landscape, which had hitherto been dry, grew wetter as the sea level rose and the drainage of fresh water stagnated.[611] Peat initially developed, but from around 5000 cal. BC much of the area became covered by a large expanse of water. Although the river system was incorporated into this water body, its position could still be identified by the river dunes that flanked its original course. Marine clay was deposited under the water. The composition of the clay suggests that sedimentation occurred in calm conditions.[612]

Between 4300 and 4000 cal. BC some of the land became dryer. Channels cut into the clay and a landscape formed consisting of a main channel with side channels, levees and flood basins. The highest river dunes protruded above the clay and the lower-lying flood plain.[613] Despite the marine origin of the clay, very few brackish tolerant plants or remains of vegetation typical of saline marshes have been found.[614] The marine diatoms would have come from outside the area and were deposited with the clay. Many plant remains from freshwater marshes have been found, however, and alder grew in the flood basins. This means that the landscape at this period consisted of freshwater rather than saline marshes.[615]

Around 4000 cal. BC the channels and levees disappeared beneath a layer of detritus and gyttja, an indication that the area was again gradually being inundated by water on a large scale. Sea-level rise slowed down, however, and the water was colonised from the shoreline by marsh plants, signalling the start of peat accumulation and land formation in this microregion. Around 3700 cal. BC the river dunes were the last features to disappear beneath clay, gyttja detritus or peat.[616] In the eastern and higher parts of the microregion peat began to develop from around 4800 cal. BC, depending on the elevation relative to current NAP. At a certain stage, this peat would have been covered with water.[617]

Exploitation phase 1: 8300 to 7000 cal. BC

The oldest traces of human activity have been found on a parabolic dune in the east of the microregion, at the Dronten-N23 site (fig. 6.2, table 6.1, no. 25). The river dunes along the Hunnepe (fig. 6.2, table 6.1, nos. 26-30) were also visited by humans during this phase, though there is no known contemporaneous palaeoecological data.[618] Hunter-gatherers visited the dune at the Dronten-N23 sites with

---

610 Van Rijn & Kooistra 2001, 13.
611 De Roever 2004.
612 Ente *et al.* 1986.

613 Incl. Dresscher & Raemaekers 2010; Schepers & Wolteringe 2020.
614 Casparie *et al.* 1977; De Jong 1966; Van Zeist & Palfenier-Vegter 1983.
615 Schepers 2014a/b.
616 Incl. Dresscher & Raemaekers 2010.
617 Known as the Almere layer.
618 Devriendt 2013.

Figure 6.28a-d: Impression of the vegetation during inhabitation of the Swifterbant microregion for a) the period between 8300 and 7000 BC: an open grassland area with heather and pine trees and a pine woodland in the distance (Nationaal Park De Hoge Veluwe, the Netherlands, photo: BIAX *Consult*/L.I. Kooistra); b) the period between 7000 and 6000 BC: heathland with isolated oak and birch trees on coversand (Vasserheide, the Netherlands, photo: BIAX *Consult*/H. van Haaster); c) the period between 6000 and 5000 BC: nutrient-poor fen woodland with mainly birch (Nationaal Park De Groote Peel, the Netherlands, photo: BIAX *Consult*/L.I. Kooistra); d) the period around 4300 BC: reed marsh as one of the nutrient-rich wetland types with clay underneath (Danube delta, Rumania, photo: BIAX *Consult*/L.I. Kooistra).

some regularity, leaving evidence for flint working and pit hearths at the site.[619] The hearths contained mainly pine charcoal. At the end of this period deciduous species also appear in the charcoal spectrum, including oak, willow, birch and elm.[620] A concentration of carbonised hazelnuts from the beginning of the Boreal found in the depression of the dune suggests that hazel also grew there at that time.[621]

It is assumed on the basis of charcoal from Dronten-N23 (location 25), as well as data from elsewhere in Flevoland, that Middle Mesolithic humans in this microregion lived in a dry, relatively nutrient-poor coversand area on which open pine woodlands grew, with heath in the undergrowth (fig. 6.28a). During the course of the Boreal, more and more deciduous trees and shrubs began to grow there. However, heath continued to account for a substantial proportion of the vegetation, given its pronounced presence in palynological samples from the Early Atlantic (7000 to 6000 cal. BC).[622] Although no zoological material was found – due to taphonomic processes (see chapter 4) – it is not impossible that the open pine woodlands and heathlands were home to horses and red deer, as remains of these animals dating from later periods have been found in Flevoland.[623]

Exploitation phase 2: 7000 to 6000 cal. BC

Traces of Middle and Late Mesolithic hunter-gatherers from this period have been found on the parabolic dune (location 25) in the east of the microregion. The feature density on the site is low compared with the previous and subsequent exploitation phase. Like their predecessors, the humans of this period dug pit hearths, but activities associated with flint working seem to have disappeared. The charcoal spectrum from the pit hearths differ from the preceding phase, containing virtually no pine. The charcoal assemblage in these pits consisted of oak, willow and alder. From 6800 cal. BC the river dunes along the Hunnepe seemed to have become increasingly attractive for human exploitation. These dunes also show evidence for activities associated with fires in pit hearths.

Between 7000 and 6000 cal. BC the coversand area of this microregion appears to have featured dry, relatively high, nutrient-poor spots alternating with moist to wet, lower-lying, more nutrient-rich locations. This can be deduced from the palynological assemblages from the Dronten-N23 site (location 25), in combination with charcoal data.[624] Heathland with isolated oak and birch trees dominated the higher areas of coversand (fig. 6.28b). In the lower-lying parts, as well as in the river and brook valleys, the hydrology was probably more favourable for denser woodland featuring a greater variety of deciduous trees, including hazel, elm and lime, as well as oak. It is not impossible that heath formed part of the undergrowth here. Alder and willow dominated the woody vegetation in wet locations.

As in the previous phase, hunter-gatherers in this microregion lived in a dry but nevertheless hydrologically varied coversand landscape. Water was more widely available and the vegetation was diverse, which naturally gave the inhabitants a greater variety of plant and animal foods as well as resources.

Exploitation phase 3: 6000 to 5000 cal. BC

The parabolic dune at the Swifterbant N23 site (location 25) was frequently visited during this period. Pit hearths were being dug again, and the composition of the charcoal that they contained remained unchanged relative to exploitation phase 2.[625] Towards the end of this phase the environment probably became too wet for these activities. The river dunes in the west of the microregion (fig. 6.2, table 6.1, nos. 26-30) did remain in use, however. The landscape became more differentiated, continuing a trend that had started in the previous phase. Low-lying parts of the coversand landscape became so wet that fairly nutrient-poor sedge-reed marshes and marsh woodland with alder and birch developed (fig. 6.28c). These types of vegetation provided a basis for the formation of peat. The zone with deciduous woodland shifted to higher-lying areas of the coversand landscape as a result of the rise of groundwater. The woodland became more varied, and deciduous trees preferring a moist environment and slightly clayey subsurface, such as ash, grew on the coversand. The area of open heathland with oak and birch probably reduced in size, persisting only in the highest parts of the landscape.

In this phase, there was still a great variety of landscape types, although towards the end the coversand area had turned into marsh and only the highest coversand ridges and river dunes were dry and accessible all year round.

Exploitation phase 4: 5000 to 4300 cal. BC

Alder carr in the east of the microregion spread over the parabolic dune at the Dronten-N23 site (location 25). The last oaks, which had held out on the highest parts of the parabolic dune, disappeared shortly after 4800 cal. BC. As a result of the high water level, plant remains underwent little if any decay, and the dune was covered with peat. More open waters developed in the west of

---

619 Hamburg et al. 2012.
620 Kooistra 2012a.
621 Hamburg et al. 2012; Cunningham 2012.
622 Bouman & Bos 2012; Van der Linden 2012.
623 Incl. Laarman 2001; Zeiler 1997; Gehasse 1995.
624 Bouman & Bos 2012; Kooistra 2012c; Van der Linden 2012.

625 Kooistra 2012c.

the microregion. Only the higher river dunes were still visible as small islands in this wetland area and were only occasionally visited by humans.

The landscape altered radically during this phase. With the exception of the east of the microregion, a very nutrient-rich wetland developed, which was probably home to a large variety of fish, birds and mammals (fig. 6.28d). However, little is known about this phase, and these assumptions are based on finds from the following phase.

Exploitation phase 5: 4300 to 3700 cal. BC

When a marsh landscape with channels and flood basins developed in the west of the microregion around 4300 cal. BC, the inhabitants of the region began to exploit the levees. As we saw in chapter 4, the economic base of hunting, fishing and gathering expanded to include small-scale crop cultivation and stockbreeding. The mires in the east of the microregion near the Swifterbant N23 site (location 25) no longer appear to have been used for settlements, though the picture could be distorted here due to less intensive research being carried out at, and less favourable preservation conditions.

The high proportion of alder pollen and wood from both the levee and the main channel (fig. 6.2, table 6.1, no. 26), the levees of the side channels (figure 6.2, table 6.1, no. 27 & no. 30) and the flood basins, in the nutrient-rich marsh in the west of the microregion is striking.[626] A range of wood types and the remains of other sorts of woody vegetation have been found in various places. This indicates the presence of a variety of trees, with willow, oak, ash, hazel, elm and poplar alongside alder. This spectrum suggests the various types of softwood and hardwood riparian woodland typical of river areas.[627] The composition of the woodland varies, depending on the location, the water level and the flow rate of the river. Willow is best able to withstand water currents and flooding, and is found in wet spots along the river.[628] In drier places, slightly higher up the bank, today's riparian woodland features black poplar as well as willow. If the ground level at a particular location rises through sedimentation, the composition of riparian woodland dominated by softwood species can shift more towards more hardwood varieties (*Alno-Padion*). Hardwood riparian woodlands include woodland along rivers growing on young, nutrient-rich mineral soils that occasionally flood. Nutrients are supplied on an incidental basis by flooding or via groundwater. Litter on the surface is rapidly converted to soil.[629]

The channels and some parts of the flood basins were dominated by nutrient-rich reed marshes and open water.[630] Given the abundance of beavers in the mammal spectrum, the gnawing marks on wood and the lack of dynamics in the water system, it is likely that beavers built dams in this marsh in order to create a safe habitat. Shallow parts of the flood basins may have developed into alder carrs over time.

During this final habitable period in the Swifterbant region, people lived in an exceptionally rich and varied freshwater marshland with an open connection to the sea in the west and access to the dry coversands in the east. This is evidenced not only by the botanical remains. Zoological remains also show a broad spectrum of fish, birds and mammals.[631] Remains of seal, fish that can live in brackish water and anadromous fish reveal the open link to the sea. However, the majority are from animals that are most at home in freshwater marshes, such as ducks, geese, otters, beavers and elk. The presence of horse and auroch bones is unusual. Aurochs are assumed to have been able to adapt to river meadows.[632] It is possible that horses came to this area occasionally from the dry coversand. This would presumably have been in the summer months, when the coversands dried out and water was in short supply. Equally, as areas of the marsh became drier, they provided nutrient-rich wet meadows and reed beds.

Over time, the Swifterbant region developed into a mire, and even the higher river dunes became too wet for habitation. The final visible features related to human activity date from around 3700 cal. BC, though this does not necessarily mean that people disappeared from the region entirely.

6.5.3 Noordoostpolder: Schokland-Urk microregion between 5000 and 1250 cal. BC

The Schokland-Urk microregion comprises mainly of the Overijssel Vecht river valley (fig. 6.2). The western boundary is the present-day IJsselmeer. The southern boundary is the Hunnepe. To the north and east are coversand landscapes that currently lie less than seven metres below NAP. The Early Holocene courses of the Overijssel Vecht and Hunnepe were flanked by river dunes formed in the Late Glacial and possibly also the Preboreal. In Flevoland the Overijssel Vecht flowed past four outcrops of glacial till that formed at the end of the Saalian and were covered by coversands during the Weichselian. The outcrops of glacial till – Voorst, Schokland, Tollebeek and Urk, from east to west – would have been very prominent hills in a low-lying coversand landscape in the first half of the Holocene. The river valley and adjacent coversand landscape

---

626 Casparie *et al.* 1977; De Jong 1966; Prummel *et al.* 2009; Raemaekers *et al.* 2010.
627 Van Beurden 2007; Margl & Zukrigl 1981. Wolf *et al.* 2001; Schepers 2014.
628 Van der Werf 1991, 242.
629 Stortelder *et al.* 1999, 302.
630 Schepers 2014; Van Zeist & Palfenier-Vegter 1983.
631 Prummel *et al.* 2009; Zeiler 1997.
632 Van Vuure 2005.

Figure 6.29a-d: Impression of the vegetation during inhabitation of the Schokland-Urk microregion for a) the period around 4000 BC: flooded landscape (Pripyat area, Belarus, photo: BIAX *Consult*/L.I. Kooistra); b) the period between 3700 and 3400 BC: hardwood riparian woodland in spring with an exceptionally high water level (levee of the Pripyat river, Belarus, photo: BIAX *Consult*/L.I. Kooistra); c) the period between 3400 and 1900 BC: sedge mire with common cottongrass (Kamerikse Nessen, the Netherlands, photo: BIAX *Consult*/L.I. Kooistra). d) the period between 1900 and 1250 BC: alder carr (Müritz National Park, Germany, photo: BIAX *Consult*/L.I. Kooistra).

were affected by the rising water level from 5000 cal. BC onwards. The area became a landscape featuring mires and lakes, through which the Overijssel Vecht meandered, and in which outcrops of glacial till, river dunes and higher coversand ridges in the south and northeast became dry 'islands' suitable for habitation. There are many traces of human activity in this microregion dating from the Late Palaeolithic (incl. Schokland-P14, fig. 6.2, table 6.1, no. 51) and the Mesolithic (incl. Urk-E4, fig. 6.2, table 6.1, no. 55) into the Bronze Age (incl. Schokland-P14, figure 6.2, table 6.1, no. 51).[633] Since there are barely any finds relating to the earliest human activity, the emphasis in this chapter is on traces of habitation that can be linked to the landscape. Although ten habitation and landscape phases have previously been identified on the basis of the archaeological and palaeoecological remains,[634] some phases have been combined here to reduce complexity.

Exploitation phase 1: 4900 to 3700 cal. BC: Swifterbant and the first half of the Pre-Drouwen

From 5000 cal. BC onwards the water level in the broad, shallow valley of the Overijssel Vecht rose, and over the following centuries the surrounding coversand flooded (fig. 6.29a). Initially, this rise in the water level caused peat to form on the coversand, but peat accumulation, that created overwhelmingly nutrient-poor sedge peat, did not keep pace with the rising water. When flooding occurred some of the peat washed away and clay was deposited under water. At first, there were very few marine elements in the clay. These did not appear until later.[635]

By around 4400 cal. BC, so much clay had been deposited that active channels were carved into the clay deposits in the west and a differentiation began to appear between the levees and clay-filled lower-lying depressions. From the levees and the dry Pleistocene coversands the depressions in the west became overgrown with reed marshes. Here, saw-sedge began to grow due to the increased amount of calcium-bearing clay. Further inland (towards the east) the reed marsh zones along the Overijssel Vecht were narrower, and there was a lateral transition to swamps containing alder, with some interim form of vegetation dominated by sedge species. The presence of swamps with alder makes it clear that, as in the Swifterbant microregion, there were no tidal influences in this part of the Noordoostpolder. The clayey levees of the channels became overgrown with willow. More varied deciduous woodland grew in places where the levees were broader and higher. In the south, just to the north of the Hunnepe, a high, east-west orientated coversand ridge with river dunes formed, providing a land connection to the coversand landscape further east until circa 4100 cal. BC. Until that time, the coversand ridge and the outcrops of glacial till of Schokland, Tollebeek and Urk appear to have been covered with deciduous woodland which initially also contained lime. However, a surprisingly large amount of grass pollen has been found in the sandy soil of the river dune at Schokkerhaven (location 52), suggesting that the woodland there was open rather than dense.[636]

During this period, humans mainly exploited the river dunes, outcrops of glacial till and the dry coversands in the northeast and south of the microregion. From approximately 4400 cal. BC they appear to have created more permanent places of habitation at a number of locations, including Schokland-P14 and Schokkerhaven-E170/171.[637] There was a broad spectrum economy, with crop cultivation and livestock breeding alongside hunting, fishing and the gathering of edible plants.[638] At Schokland (P14) people settled on the east side of the outcrop of glacial till on the levees of the Overijssel Vecht. The animal bone material and plant remains gathered during the excavations provide a glimpse of the landscape during the period for which there is evidence for human activity.[639] Beaver bones occur regularly, suggesting swamps or carrs. Fragments of water chestnut suggest there were shallow flood basins. Red deer bones and possibly wild boar bones indicate that the landscape also had drier wooded areas, although red deer are also known to spend a lot of time in wet areas. All in all, the faunal remains and remains of plant foods support the results from the palaeobotanical landscape study, which was largely performed outside archaeological sites.

Between 4400 and 3700 cal. BC the influence of the sea declined, although the water level continued to rise, to around five metres below NAP. The high coversand ridges of the microregion therefore lost their land connection to the dry areas of the north and east and, like the river dunes and outcrops of glacial till, they became 'islands' in the mire. The character of the mires gradually changed. The transition mires with sedges expanded, and bogbean grew in abundance there, a sign that the peat was becoming less rich in nutrients. In the west the reed fens that were richer in nutrients, and contained saw-sedge, continued to exist.

Exploitation phase 2: 3700 to 3400 cal. BC: Pre-Drouwen and Funnel Beaker culture

The start of this period was the development of the Bergen Inlet. A channel of this penetrated into the

---

633 Ten Anscher 2012.
634 Gotjé 1993; Gehasse 1995; Ten Anscher 2012, 507-536.
635 Wiggers 1955, 53-58.

636 Weijdema *et al.* 2011, 36.
637 Ten Anscher 2012, 510.
638 See chapter 4.
639 Gehasse 1995, 37-68.

Noordoostpolder via northern Noord-Holland and joined the Overijssel Vecht to the north of Schokland (fig. 6.29b). The strong dynamics in the channel caused peat erosion, and created a large lake by the Tollebeek outcrop of glacial till.

As a result of the new water regime and the supply of nutrient-rich water, reed marshes developed along the lake shores. The clay deposits around the tidal creek caused the reed marshes to become overgrown with saw-sedge. The old course of the river Overijssel Vecht to the west of the connection with the tidal creek grew over. The reduced dynamics and the declining supply of nutrient-rich water allowed sedge mires to develop, from which peat accumulated.

Though the course of the Overijssel Vecht altered radically and the connection with the Hunnepe, which must have been to the west of the Noordoostpolder, was severed, by and large the same landscape features continued to exist. The landscape remained attractive to humans for hunting and pasturing livestock, for fishing and for the collection of edible and other useful plants. The dry grounds on river dunes, outcrops of glacial till and the remains of dry coversand ridges were also used to cultivate crops.

Exploitation phase 3: 3400 to 1900 cal. BC: Late Neolithic

After the tidal channel broke through the coastline at Bergen, a period of stability in the landscape set in. The only change was in the water level, which rose, albeit less rapidly, from around five metres below NAP to just over two metres below NAP. The rising water reduced the area of habitable dry land. Around this time, for example, habitation at location 55 (Urk-E4) ended because the river dune grew too wet.[640] As the supply of nutrient-rich water via the sea reduced, the level of nutrients in the mire landscape of the region also declined. This process was probably exacerbated by the fact that less river water drained out of the region via upstream stretches of the Overijssel Vecht (eastern Gelderland, Overijssel and the southern part of the Drenthe Plateau) and its tributaries.[641] The mires of the Noordoostpolder reflected the changes in the hydrology. The mesotrophic sedge mires developed into oligotrophic raised bog. The fens with reed became alder carrs. The alder carrs then developed into birch carrs and, eventually, raised bog.

In this period people lived along the Overijssel Vecht, the Hunnepe, along tribruaries and on the shores of the large lake (fig. 6.29c). Although the dry area was shrinking, people continued to live in this mires and use what the landscape had to offer. They cultivated crops on the increasingly scarce dry land, and used the surrounding sedge mires as pasture for their livestock. Wild animals, including red deer, aurochs and horses may also have grazed there. The wealth of game, birds, fish and wild plants would certainly have been a reason to continue living in the area, rather than relying exclusively on crop cultivation and livestock breeding. Nevertheless, this abundance of wild fare would have declined over time, as more and more raised bog developed and the landscape became more monotonous and nutrient levels fell.

Exploitation phase 4: 1900 to 1250 cal. BC: Early to Middle Bronze Age

From circa 2400 cal. BC flooding from the sea became a more frequent occurrence. Around 1900 cal. BC this also began to impact on the Noordoostpolder. Nutrient-rich water infiltrated the area from the west via the Overijssel Vecht. The adjacent peatlands eroded and marine clay was deposited in large parts of the microregion. The influx of nutrient-rich water containing clay undoubtedly led to a greater abundance of flora and fauna. Although it is likely that brackish water penetrated as far as the Noordoostpolder, it did not lead to the development of saline marsh vegetation. Fields of common club-rush and reed beds developed in shallow water with clay deposits. Common club-rush is one of the few marsh plants whose seeds can germinate under water. This species can therefore grow in the middle of a shallow lake. Reed, on the other hand, has to colonise the water from the shore. Eventually, they form floating mats which, at a certain point, are robust enough to serve as a subsurface for other species of marshes and mires and, eventually, for trees. After a time the thickening layer of peat becomes attached to the lake bed.

The Bergen Inlet closed between 1500 and 1350 cal. BC. The influence of the sea had not been felt in the microregion for several centuries. Open water became land and water nutrient levels fell. Fens again developed into raised bogs, though they did not cover the entire area. Large lakes continued to exist. Although the drainage of water to the sea stagnated, some river water will have entered from the east. It is not therefore likely that the raised bogs extended to the levees of the Overijssel Vecht and the route through the large lakes.[642] The levees of the Overijssel Vecht and the lake shores are more likely to have been covered with softwood riparian woodland, and locally with hardwood riparian woodland, and edged with alder and birch carrs and transitional mires with sedge species and reed. Evidence of this was found in the investigation at location 44 (Emmeloord-J97), where alder, birch, willow, poplar, ash, oak and elm were all found in

---

640 Peters & Peeters 2001, 20.
641 See section 6.3.6.

642 As suggested by Ten Anscher's (2012, 527) reconstruction for this period.

fish weirs and traps from the period 1900 to 1600 cal. BC (fig. 6.28d).[643] Bundles of reed had also been positioned against one of the fish weirs.

Towards the end of the Bronze Age there was very little Pleistocene landscape left at the surface. Only the outcrops of glacial till at Schokland and Urk and possibly river dunes along the Hunnepe in the south were suitable for habitation. New habitable locations may have developed on the floating peat along the edges of the lakes. Until they became attached to the lake bed, these mats moved up and down with the water level. Once they grew thick enough, and therefore more stable, they became accessible, and possibly habitable. We can therefore by no means exclude the possibility that people still inhabited the area, though there is little evidence of this due to later peat oxidation.

6.6 Conclusions

This chapter has presented an overview of landscape development in Flevoland between circa 12,500 and 1100 cal. BC, based on numerous investigations performed between the 1930s and 2014. Key factors in landscape investigations are developments in climate, texture of the subsurface and relief, hydrology and vegetation.

The Late Glacial (12,500 – 9800 cal. BC) marked the final phase of the last ice age. Warm periods alternated with cold periods. Flevoland had a coversand landscape that was still developing at this time. Three rivers flowed through the area, bringing in water from the wider region: the Eem in the south and the Hunnepe and Overijssel Vecht in the north. The subsurface was permanently frozen during the cold periods. The vegetation consisted of cold-tolerant species. In stadial periods Flevoland was covered with tundra vegetation, and in the interstadial periods it had open birch and pine woodlands.

Around 9800 cal. BC a protracted warm period began, known as the interglacial Holocene. In the Preboreal (9800 – 8200 cal. BC) coversand formation came to an end as the subsurface became covered with dense vegetation. River dunes continued to form only along the rivers. The vegetation in the Preboreal and Boreal (up to circa 7000 cal. BC) consisted mainly of cold-tolerant trees and shrubs. Contrary to the commonly held view, the landscape did not feature dense woodland, but rather open birch woodland, and later pine woods. In the coversand area local groundwater levels determined conditions, and peat developed in some places, depending on the extent to which water could permeate the subsurface.

From the Atlantic (7000 – 3700 cal. BC) onwards, the North Sea basin filled with water. This changed the climate from a land to a sea climate. The rising sea water caused stagnation in the drainage of water, which in turn caused the groundwater level to rise in Flevoland. This allowed peat to accumulate over large areas of coversand. Initially, the mires had fairly mesotrophic vegetation types (transitional mires) with features typical of raised bogs. Pines grew on the dry coversands until well into the Atlantic. Lime woodland occurred for only a short period. Initially the coversand were too dry and lacking in nutrients to sustain lime trees, and by the end of the Atlantic large parts of the coversand area were inundated with water and clay, and the conditions were too wet for this species.

Between 5500 and 4500 cal. BC the western part of Flevoland was inundated by water and clay of marine origin which was deposited under water. Although we have few botanical data from this period, we have the impression that Flevoland did not have saline marshes with halophyte plant species, and that there were no tidal movements there. Freshwater marshes developed in the low-lying parts of the Flevoland landscape. From 4500 cal. BC clay accumulated to such an extent that it protruded above the surface of the water, and the Eem, Hunnepe and Overijssel Vecht cut new channels through it. A river area with clay deposits became overgrown with eutrophic reed marshes, willows and alder carrs. Mixed deciduous woodland with oak grew on the levees of the rivers.

From 5500 cal. BC marshes, mires and lakes dominated the landscape of Flevoland. A mosaic of marsh and vegetation types ranging from nutrient-rich to nutrient-poor developed, from extremely wet clubrush and reed marshes to slightly drier alder carrs, depending on the level of nutrients in the environment. In the course of the Subboreal (from 3700 cal. BC) the Eem was transformed into an area of peat and lakes. The Overijssel Vecht still flowed from east to west through the landscape, but its course changed due to incursions from the Bergen Inlet in Noord-Holland. Less is known about the Hunnepe.

From 3700 cal. BC there was little dry land in Flevoland. It is likely that only the outcrops of glacial till at Voorst, Schokland and Urk were accessible by the end of the Subboreal (circa 1100 cal. BC). The levees of the Overijssel Vecht, Hunnepe and other peat rivers and brooks may still have been dry. Although no remains have been preserved, floating peat with riparian woodland may also have been accessible at that time.

---

643 Van der Heijden & Hamburg 2002, 34-40.

Chapter 7

Transformations in a forager and farmer landscape
A cultural biography of prehistoric Flevoland

J.H.M. Peeters, L.I. Kooistra & D.C.M Raemaekers

7.1 Introduction

The previous chapters have provided some insight into various facets of prehistoric occupation and characteristics of the former landscape that we now call Flevoland. The data from the selected sites (see Appendix I) shows that Flevoland holds a wealth of information about activities carried out by the Mesolithic and Neolithic hunter-gatherers and early farmers (Chapters 4 and 5) in a landscape that had changed dramatically since the end of the last ice age (Chapter 6). At the same time, we have to realise that the information we have is fragmentary and unbalanced in all kinds of ways and is a direct result of the focus of research in the area (Chapter 3) as well as the formation of the landscape and the possibilities there are to uncover the remains of prehistoric occupation of the area. In this last chapter, all these insights shall be assessed in relation to each other and shall be placed and interpreted in a wider geographical framework. The themes of the previous four chapters will be guiding in this.

We have seen that the landscape completely changed in the course of the Mesolithic and Neolithic periods. This change had far-reaching implications for the opportunities to make use of the landscape. The question to be answered is what picture we see when we combine all the archaeological and palaeolandscape data from Flevoland. On the one hand, we want to look at the exploitation of the landscape from an economic perspective: what relationship existed between the characteristics of the structurally-changing landscape and the use of food and non-food sources? On the other hand, we want to look at the landscape from a social-cosmological perspective: what relationship existed between the roles of places – and the connection between these places – in the dynamic landscape and how did that give structure to the historical significance of the landscape for the prehistoric inhabitants? To conclude, a discussion will be had about the social-cultural relationships, as interpreted from the data.

Although the emphasis of the discussion will be on Flevoland, this chapter will also explicitly look beyond the provincial borders. During prehistory, the present area that is Flevoland formed part of an extensive landscape which was exploited by man in various ways and in which social relationships were maintained in a cultural context that was subject to change over time. How does the story of Flevoland fit into the bigger picture or, viewed from another perspective, how does Flevoland archaeology influence the image of the past beyond its borders?

7.2 The landscape as a source of subsistence

It seems increasingly clear that during the warmer climate conditions of the Late Glacial, especially in the Allerød interstadial, the subsurface was saturated which enabled the development of peat on a large scale. In the Noordoostpolder in particular, this development coincides with the occurrence of glacial till on or near the old surface at that time which made water drainage difficult, but there were also wet conditions elsewhere. The landscape would have been made up of a succession of lakes, marshland and mires, whilst in the drier parts of the landscape there were open birch-pine woodlands with an undergrowth of grasses and herbs. The (very) limited number of archaeological sites dating to the Late Palaeolithic hardly makes it possible to develop a picture of the late glacial landscape as a subsistence source for hunter-gatherers in this period. For this reason, the Late Palaeolithic has been left more or less outside the scope of this book.[644] However, whilst the palynological data and evidence for peat formation indicates that the landscape must have been dominantly wet during the Allerød interstadial, there is no reason to assume that there was no or limited human activity. The region may have seen an abundance of fish, waterfowl and mammals that felt at home in these wetlands. The open birch and pine woodland in the higher parts of the landscape were not only attractive as a habitat and source for animals (deer, furred animals), but also because of the plant resources they offered. Birch bark and wood would have probably been very important for the building of constructions, and the making of canoes, containers and other utensils and implements, whilst it also served as fuel and possibly for the extraction of resin and pitch.

During the interstadial, hunter-gatherers from the Federmesser Gruppen who were active in large areas of The Netherlands, including Flevoland, appear to have exploited a wide range of animal sources. Not only the remains of mammals, but also remains of fish and waterfowl have been found on Federmesser sites in The Netherlands and neighbouring countries.[645] If the lakes in the Flevoland region were interconnected by navigable watercourses, this would have made the marshy areas and mires easily accessible, which would have been beneficial for the exploitation of these animal and plant resources. This was certainly the case via the river systems of the Overijsselse Vecht, Hunnepe and Eem. It is, however,

Figure 7.1 (opposite page): Geographical location of place names/sites in the Netherlands mentioned in this chapter. The background map represents the top of Pleistocene surface and drainage systems at the start of the Holocene. A: brook/river valley; B: Pleistocene sandy area deeper than -16 m Dutch O.D.; C: Pleistocene sandy area between -16 m and 0 m Dutch O.D.; D: Pleistocene sandy area above 0 m Dutch O.D.; E: river dune; F: ice-pushed area; G: Loess area; H: Tertiary and older deposits. Place names. 1: Slootdorp; 2: Leeuwarden-Hempens; 3: Hoogkerk-Groningen; 4: Nieuwe Pekela; 5: Anloo; 6: Bronneger; 7: Urk; 8: Schokland; 9: Schokkerhaven; 10: Swifterbant; 11: Dronten-N23/Hanzelijn Area VIII; 12: Kampen; 13: Hattemerbroek; 14: Zwolle; 15: Dalfsen; 16: Mariënberg; 17: Raalte; 18: Almere-Hoge Vaart; 19: Soest; 20: Uddel/Uddelermeer; 21: Deventer/Epse-Olthof; 22: Zutphen; 23: Neede; 24: Rotterdam; 25: Brandwijk; 26: Hardinxveld-Giessendam Polderweg/De Bruin. Drainage systems. I: Overijsselse Vecht; II: Hunnepe; III: Eem; IV: Tjonger/Linde; V: Drentse Aa; VI: Hunze; VII: Regge; VIII: Berkel; IX: Rhine-Meuse.

completely unclear what the poorly drained areas of land between the larger water courses exactly looked like and how these would have been accessible. The ease of accessibility may well have been affected by seasonal differences: there may well have been better accessibility to the marshy areas in winter than in other seasons.

The few Federmesser sites we know from Flevoland are situated in higher parts of the landscape, in the vicinity of rivers or major streams: Schokland-P14 is situated on the Overijsselse Vecht and Kuinderbos on the Tjonger/Linde (fig. 7.1).[646] Further upstream in Friesland and Overijssel, the number of known sites increases rapidly, due to the fact that late glacial sediments and the substrate within them lie close to the surface. Traces of Federmesser hunter-gatherers have also been found upstream along the river Hunnepe. Near Hattemerbroek, in the coversand under peat, large scale excavations in the route corridor for the Hanzelijn railway uncovered various scatters of flint artefacts belonging to the Federmesser Gruppen. An assemblage of material – it is not entirely clear whether it should be counted to the Federmesser Gruppen or the Hamburgian – may represent a 'workshop' where blades were produced from one or two nodules of flint.[647] The Federmesser material from Schokland-P14 – here also, the cultural assignment is not very clear – comes from a similar

---

644 The scarcity of Late Palaeolithic sites in Flevoland is most probably the result of the fact that contemporary land surfaces from the Bølling and Allerød interstadials lay much deeper in the coversand and are almost never investigated as part of a borehole survey or excavation (see chapter 3).

645 Lauwerier & Deeben 2011.

646 The finds from Schokland-P14 are not published. An association with the Federmesser Gruppen is not certain because of the absence of diagnostic artefacts, such as projectile points; it is equally possible that the material belongs to the Hamburgian. The finds from Kuinderbos have been incorporated in Hogestijn (1986).

647 Verbaas et al. 2011.

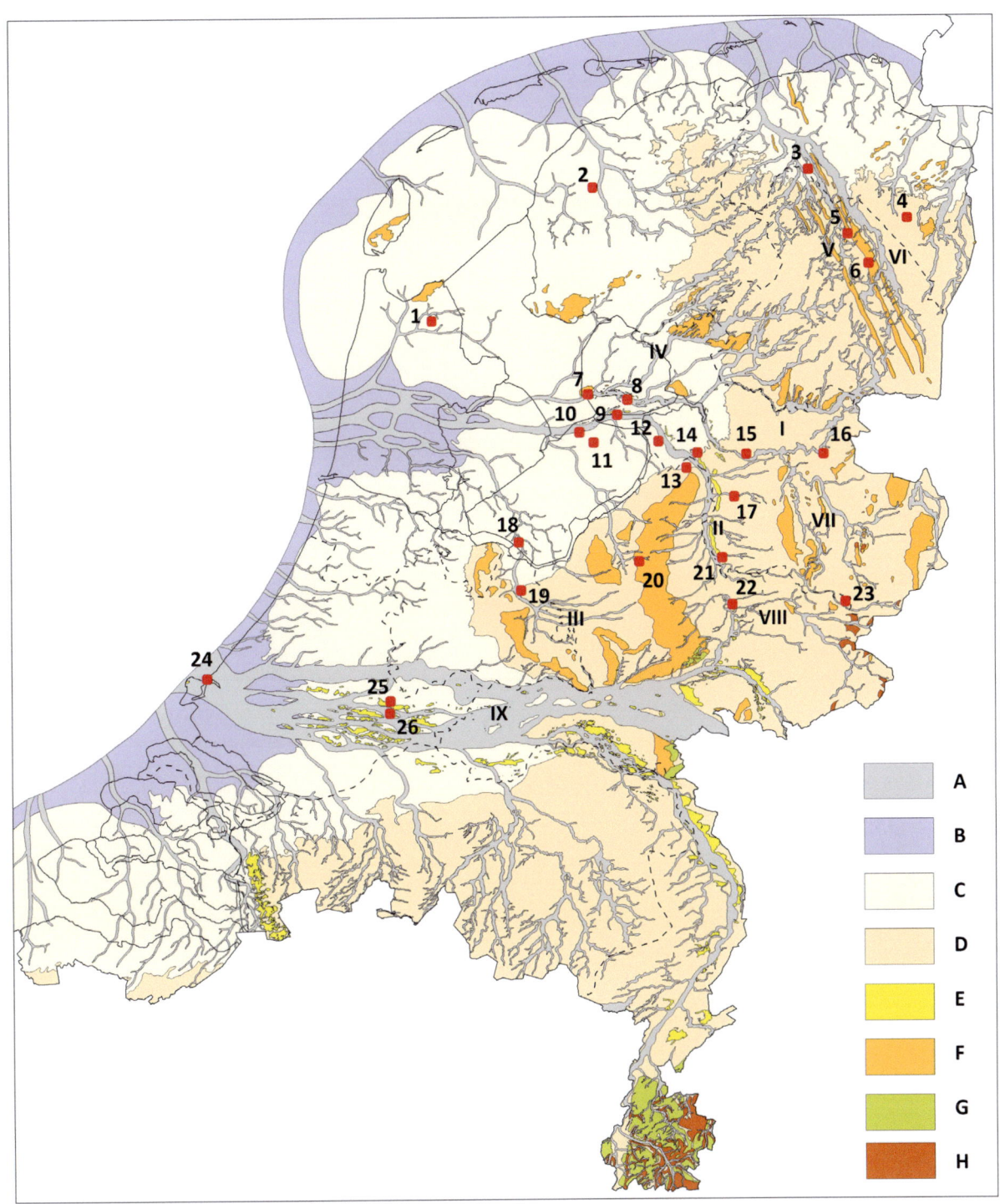

context, but on this site all the flint fragments appear to come from one single flint nodule that could have been collected out of the locally occurring glacial till. There are no local deposits of flint by Hattemerbroek however, so the flint nodules must have been transported over an unknown distance before (any further) flintknapping took place and the waste products (debitage) was left behind. It is likely that Flevoland was part of this larger Federmesser landscape, but we are still in the dark about the character and exploitation of the region.

The ensuing cold conditions of the Late Dryas meant that the Federmesser Gruppen disappeared from the landscape. The region entered a short, unstable phase where there was hardly any vegetation with trees. River

TRANSFORMATIONS IN A FORAGER AND FARMER LANDSCAPE 207

dunes formed in and along the valleys of the Overijsselse Vecht and Hunnepe in places were vegetation was present, whilst outside of these river systems coversand was deposited on a large scale. Due to the rapid rise in temperature that heralded the beginning of the Holocene, the landscape is typified once more by the expansion of vegetation. Open pine woodlands appeared that offered habitat for light-loving heath shrubs and club moss as well as some birch and hazel. In view of recent insights that show that river dunes were still being formed in the Rhine – Meuse river valleys into the Early Holocene,[648] the same process cannot be discounted for the valleys of the Overijsselse Vecht and Hunnepe. At the same time, in other low-lying parts of the landscape localised marshes and mires were created in which peat started to develop.

The open character of the Preboreal vegetation, with an important place for pine, appears to have been maintained in Flevoland into the Boreal. Conditions were relatively dry, so that peat could only start to develop in the very poorly-drained parts of the landscape where glacial till layers were part of the subsurface (Noordoostpolder). Even in the Early Atlantic, the landscape of Flevoland was dominated by dry conditions and a nutrient-poor subsoil. As a result, a mixed oak woodland, containing elm and lime trees, barely had the chance to get established. As has been discussed in Chapter 6, such a detailed picture does not correspond well with the more generalised models that exist for vegetation succession. We therefore need to start thinking of a much more differentiated landscape.

These differentiations in landscape dynamics may well be reflected in the activity pattern of hunter-gatherers, who had meanwhile (re)appropriated their place in the environment. At the transition from the Late Dryas to the Preboreal, hunter-gatherers of the Ahrensburg culture are known to have been active in the southeastern parts of The Netherlands and sporadically in the northern Netherlands.[649] In Flevoland, no traces of human activity have been found dating to this time.[650] Activity is only recorded from the Preboreal. It is interesting to note that remains that can be dated with certainty to the Early Mesolithic are found in zones where the coversand deposits were on the surface. Traces of Early Mesolithic activity appear to be absent on river dunes. This might suggest that the river dunes were still relatively young, and date from the Preboreal. If hunter-gatherers were ever active on such unstable dunes, then this activity has not left any recognisable archaeological record. Such evidence only starts to exist from the end of the Preboreal and the first half of the Boreal, when we start to see different sites for Middle Mesolithic activity. One of these sites is associated with surface hearths, flintworking and the consumption of (at least) hazelnuts and another site with pit hearths (see 4.9.1). This contextual differentiation continues into the first half of the Early Atlantic.

In relation to this, it is important to view the dataset from Flevoland in a broader geographical context. Outside the catchment areas of the Hunnepe and the Overijsselse Vecht we see comparable patterns to that in Flevoland. In the coversand areas of the northern Netherlands, numerous sites have been discovered with clusters of pit hearths.[651] Closer to Flevoland, we have evidence for Mesolithic activity on river dunes upstream along the Overijsselse Vecht and Hunnepe near Zwolle and on coversands near Kampen and Soest.[652] In a chronological sense these sites show a similar pattern to what has been recorded in Flevoland. Here too, for instance at Hattemerbroek and Mariënberg, we see sites that have been occupied repeatedly over a period of more than 1500 years. Further upstream along the Hunnepe and in the adjacent basin of the Berkel, between Deventer and Zutphen, it is interesting to note that on the basis of available ^{14}C dates, the remains of Mesolithic activity are, on average, older than those sites recorded downstream. In the flint assemblage, Early and Middle Mesolithic traditions are both clearly represented. The flint assemblages show only incidental typological and technological evidence for Late Mesolithic activity,[653] whilst the use of the pit hearths continues into the Late Mesolithic.[654] There is also evidence for a different sort of exploitation of the environment. In Zutphen, for instance, where Late Mesolithic wood remains with traces of cutting marks have been found in a river valley and indications for harvesting plant roots have been found in a Late Mesolithic fen.[655] What we seem to be witnessing in the upstream region of the Hunnepe, is a (dis)continuous pattern of Mesolithic activities in which sites relating to the production and use of flint tools only play a limited role in the Late Mesolithic.

The continuity that seems to exist relating to the use of pit hearths, begs the question as to what the possible link could be with the environment. As has been discussed in Chapter 4, there are indications for the use of pit hearths for the production of tar, for which pine wood and birch

---

648 Cohen & Hijma 2008; Hijma *et al.* 2009.
649 Crombé, Van Strydonck & Deeben 2014.
650 But in relation to this, we should also be aware that this recorded absence could be the result of the intensity of research and chance discovery.
651 For example sites in Nieuwe Pekela, Stadskanaal (Groenendijk 1997, 1999, 2004; Niekus 2006), Leeuwarden-Hempens (Noens 2011) Kampen (Geerts *et al.* 2019).
652 Soest (Woltinge *et al.* 2019), Kampen (Geerts *et al.* 2019)
653 Made with indirect percussion using a punch, characteristic elements include regular blades and trapezoidal or transverse arrowheads.
654 Hermsen, Van der Wal & Peeters 2015.
655 Groenewoudt *et al.* 2001; Peeters 2007; Bouwmeester, Fermin & Groothedde 2008.

bark are the most suitable resource. Let us assume that the local availability of birch and pine was a prerequisite for tar production on any specific location in the landscape. For Flevoland, interpretation of the botanical evidence suggests dry and nutrient-poor conditions prevailed into the first half of the Atlantic and that the pine remained a component of the woodland vegetation for a long time. A wetter landscape only started to develop in the second half of the Atlantic under the influence of a structural rise in the sea level, in which nutrient-rich groundwater caused changes in the vegetation, ensuring temporarily favourable conditions for deciduous woodlands with oak, lime and elm. The pine disappeared out of these woodlands. Birch, alder and willow were found in the boggy, marshland plains. The rising water and difference in relief in the coversand surface – on average the surface is lower in the west – caused the vegetation zones to move eastwards in the course of the Atlantic. It could be precisely the effect of these vegetation zones shifting through the landscape, that meant that the Flevoland region remained more favourable for tar production for a relatively longer period; on the basis of ^{14}C dates from pit hearths this production would have lasted until the end of the Middle Atlantic. If this were indeed the case, then the trend seen in the archaeological record that the use of pit hearths continued longer in the area downstream along the Hunnepe and in the catchment area of the Overijsselse Vecht into Flevoland, could indicate geographically differentiated changes in the landscape.

The 'disappearance' of the Mesolithic occupation in the upstream region of the Hunnepe has also been associated with the increase in density of the vegetation in the later part of the Preboreal, the related reduced areas of open water and the decrease in density of game for hunting.[656] It has, however, been established that specific forms of landscape exploitation continued for longer. Mesolithic people had not gone from the area. Another question is how dense the vegetation cover actually was. The lack of palynological data means that no answer can be given at this moment. We have been able to establish that there was no continuous, extensive forest in Flevoland. The landscape had an open structure in which different type of vegetation alternated under the influence of localised soil conditions. Such a situation would have offered a range of possibilities, but account must be taken of an even greater diversity that would have probably existed on a more localised spatial scale. In part, the differences may have been caused by human activity. For instance, interventions in the vegetation (for example, deliberate firing) could have led to the creation of open spaces, such as is proposed for Hanzelijn Area VIII, in the direct vicinity of Dronten-N23.[657] The burning of vegetation bordering rivers may also have led to small-scale differences. In Almere, evidence for this has been found, although a date for the activity could not be estimated precisely.[658] Animals such as the beaver could also have played a role in the creation and maintenance of localised differences.[659]

The variation in the archaeological record is maybe better understood if we start by looking at the diversity in the data instead of starting out with overarching and somewhat monolithic landscape models. In the open and differentiated landscape of Flevoland, it would not have been strange, for instance, to find animals such as horses. The assumption that horses disappeared from the northwest European landscape early in the Holocene, only reappearing as domesticated animals at the end of the Neolithic is at odds with the evidence from bone remains found in the Swifterbant contexts of Hoge Vaart-A27, Swifterbant-S3 and Schokkerhaven-E170.[660] A more or less similar context is that of horse remains found on the Danish Ertebølle site of Ringkloster.[661] An open and differentiated landscape with an alternating pattern of dry and wet milieus was perhaps a biotope *par excellence* for animals that did not need to have lived in a large herd, such as aurochs and elk. The variation in fauna can also have a link to the plant world. Plant resources may have played a much greater role in the Mesolithic than is often assumed. The firing of (levee) vegetation, a Mesolithic practice that has been confirmed outside of Flevoland in Groningen, Zutphen and Rotterdam, may have been a strategy to create small-scale conditions that attracted wildlife and offered space for the regeneration of various plant species.[662] It may, however, also have been carried out to make it easier to reach open water, thus improving the accessibility of a specific location.

The importance of small-scale differentiation in the dynamics of the landscape and the possibilities that this would have offered to inhabitants of that landscape, is evident from the post-Mesolithic occupation of Flevoland. Chapter 4 makes the case that inhabitants designated as Swifterbant culture essentially had a Mesolithic existence, but they did not live in the same landscape as their Mesolithic predecessors. The structural paludification of the area created different hydrological conditions and, in comparison with the previous period, provided a nutrient-rich environment. Not only did nutrient-rich marshland arise, but also the availability of plant and animal

---

656 Waterbolk 1985; Bos *et al.* 2005.
657 De Moor *et al.* 2009
658 Woltinge 2006.
659 Coles 2006.
660 Zeiler 1997; Gehasse 1995.
661 Andersen 1974.
662 Mellars 1976; Zvelebil 1994; Simmons 1996; Dark 2004; Innes, Blackford & Simmons 2010; Bos *et al.* 2005; Woldring *et al.* 2012; Moree & Sier 2015.

resources became more varied. Tidal and other marine influence remained limited in the area. No evidence for coastal vegetation has been found (Chapter 6) nor does the archaeozoological evidence reveal a clear marine component in the diet (Chapter 4). A similar picture emerges from the Mesolithic sites along the coast of the western Netherlands (Rotterdam, Yangtzehaven), where it is clear that freshwater fish were an important component of the diet.[663] Stable isotope analysis on Mesolithic human remains found in the North Sea gives the same picture.[664] The stable isotopes from human remains from the Swifterbant context point to the importance of (freshwater?) fish as a source of subsistence.[665] That fish played a role is also evidenced by the discovery of fish weirs and fish traps found in freshwater channels and made out of locally available wood. Certainly the use of fish weirs combined with fish traps fits well in such a setting. It is likely that the influence of tidal activity was of short duration in the Flevoland region. Channels in which fish weirs and traps have been found, seem to have silted up within 100-200 years. That there was a certain connection with the coast is demonstrated by the occurrence of flint, and possibly also by amber that would have been collected on the beaches by people from the Swifterbant culture. In addition, bone remains from seal have been found, although actual salt-water fish are virtually absent.

The relatively rapid changes that occurred in this landscape were particularly evident in the river valleys of the Overijsselse Vecht and Hunnepe, where channel beds shifted and the Overijsselse Vecht at a certain point completely changed its course. Small field systems were laid out on the low, clayey levee deposits along the Hunnepe near Swifterbant, These would have easily flooded. There is also evidence for small-scale farming on the higher dunes along the Overijsselse Vecht. The cultivation of emmer wheat and naked barley was a novelty and meant a marked broadening of the food spectrum, to which livestock had previously been added.[666] Keeping livestock in a dynamic landscape dominated by wetter areas is perhaps not surprising. Cattle and pigs can be kept well in such an environment. The cultivation of non-indigenous crops, on the other hand, is a different story. Whether this was also the case in the river valley of the Eem is not known. The clay levee deposits that are present in this area have hardly been the subject of research. However, the fish weirs found at Hoge Vaart-A27 are 'contemporary' with the fields in the Swifterbant area, which strongly suggests that the southern part of Flevoland was certainly exploited. Small-scale farming – horticulture in a Swifterbant context, is also supported by palynological research on the Drents Plateau, in a region where wet conditions also dominated in the same period.[667]

Recently, research based on an assessment of the soil characteristics, has suggested that the limited fertility of the soil in the southern part of Flevoland was unfavourable for cultivation, in comparison to the river valleys of the Overijsselse Vecht and Hunnepe, and which would explain the absence of indications for cultivation in the sandy river valley of the Eem and at Hoge Vaart-A27 in particular.[668] This suggestion implies that the soil characteristics in the Classical Swifterbant phase, formed a consideration when deciding where to lay out field systems. The first evidence for cultivation in the Dutch wetlands dates to this phase. It implies that there must already have been some understanding about the sort of ground and soil characteristics that would have been suitable or unsuitable for cultivating emmer wheat and naked barley. If we, however, recognise that in the river valley of the Eem, Hoge Vaart-A27 is the only site that has been excavated to any scale and that the sand ridge on which the site lies disappeared under peat at about the time that the first evidence for cultivation is found along the Overijsselse Vecht and the Hunnepe, as well as outside Flevoland, then the question has to be asked whether the suggested importance of soil characteristics in sandy substrate is entirely relevant.

More important is the fact that small-scale cultivation on a regular basis was possible in a predominantly wet landscape and that at a certain 'moment' the people of the Swifterbant culture made the choice to extend their range of foods by this means. The cultivation of two sorts of grain with different properties on the prepared fields was possibly also seen as a form of risk management: if it didn't work with one sort, then perhaps it would with

---

663 Moree & Sier 2015.
664 Van der Plicht et al. 2016.
665 Smits & Van der Plicht 2009.
666 It should be noted, however, that the 'exact' timing of emergent animal husbandry in the Netherlands is still uncertain. The Groningen Institute of Archaeology has just recently started a project (EDAN) to investigate processes of early domestication of pigs and cattle on the basis of aDNA, isotopes and chronological modelling.
667 Bakker 2003.
668 Van den Biggelaar et al. 2014. There are several methodological and interpretative problems with this study. The relevant substrate characteristics for this discussion are based on very different lithological units that have an extremely complex formation history. Spek, Bisdom & Van Smeerdijk (2001a, 2001b) have confirmed. on the basis of detailed analysis of the profiles at Hoge Vaart-A27, that in the top of the coversand, hydromorphic horizons with a clear AEBC-profile have developed from a brown woodland soil that was leached by the strongly fluctuating water level due to structural groundwater rise. A similar picture has been revealed in geoarchaeological research at other locations in Flevoland (Opbroek & Lohof 2011), as well as outside the region (Exaltus 2007). The hydromorphic AEBC profiles have formed in a relatively short period of time and are not the result of gradual podzol formation as assumed by Van den Biggelaar et al.

the other.[669] The importance may not even have been in the quantitative share of these crops in the diet; the cultivation of grain did not lead to the rapid abandonment of the traditional way of life.[670] As has been discussed in Chapter 4, the preparation and consumption of this food source seems to have been surrounded by other activities and manipulations than were the case for the 'traditional' component: different pots were used for grain and traditional foodstuffs, and quern-stones were deliberately broken with the fragments divided over various locations. In other words, a socio-ideological dimension appears to be attached to these cultivated crops which contrasts with the treatment of the traditional, 'natural' crops.[671]

We have also been able to show that cultivation on a greater scale took place in the Neolithic, notably at Schokland-P14, using a hoe (the earliest form of plough) to till the fields. This expansion could indicate that the cultivated crops became increasingly important in the diet.[672] It is, however, not clear how this increase in scale came about. As far as we can see, cultivation in a Swifterbant context was small scale but by the Late Neolithic the scale of production had increased and was based on other technology. What took place in the periods in-between, in the pre-Drouwen and younger Funnel Beaker phases is not known. Outside of Flevoland there is also little understanding of this development. The plough would have been introduced, but when this happened cannot be dated more accurately at present. The – as yet – earliest ploughmarks in The Netherlands have been found in Groningen and have been cut by a pit containing Drouwen Funnel Beaker pottery.[673] A survey of ploughmarks under Funnel Beaker burial monuments in Denmark has made clear that the earliest ploughmarks date to the pre-Drouwen phase (northern Europe: Early Neolithic),[674] a phasing similar to that for the northern German Flintbek.[675]

We need of course to realise that Flevoland was not isolated from other areas. It is, however, very difficult to determine to what extent 'remote' resource areas actually made up a part of the landscape in which one was active. In the extensive wet landscape of which Flevoland became increasingly a part, the possibilities for the exploitation of animals and plants that were bound to the dryer areas decreased sharply. Deer and wild boar, for example, would have been found less and less over time in the territory of present-day Flevoland. In order to hunt these animals, one was therefore forced to scout the higher lying areas that surrounded Flevoland. Smaller, high-lying landscape areas within Flevoland which protruded out of the peat for longer, such as Schokland and Urk, may have formed suitable biotopes for these animals, but given the human presence in these areas, it is difficult to see how such animal populations could have maintained themselves. At the same time, fields were attractive foraging sites for game, especially when surrounded by scrubland and woods. The changes in the landscape that were caused by structurally wetter conditions, could have been compensated in this way by a symbiotic relationship between human intervention in the landscape and, on a more local scale, the occurrence of animal populations that would otherwise have been under pressure.

7.3 Cultural structuration of the environment

The significance of the Mesolithic and Neolithic landscape was not only determined by the availability and exploitation of all sorts of subsistence resources. Social, historical and cosmological connections were at least of equal importance. Flevoland is a fascinating area in this regard because the prehistoric landscape here underwent radical transformation. This transformation took place under the influence of different factors (climate, hydrology) and on different spatial and temporal scales. From our current research perspective, we naturally have various options for mapping different scales of landscape dynamics and relating this to human behaviour. But for the hunter-gatherer and early farmer who were active in this region, the relevance of landscape dynamics lay primarily on the scale of the individual lifespan, where the temporal perception of the environment was determined by personal experience on the one hand and collective memory on the other. Individuals moving through a landscape would have made that environment their own, whilst events at places in the landscape provided the opportunity to create time and place-related stories and markers that communicate with a world view shared by group members.[676] The anchoring of places in the landscape through repeated use and the passing on of narratives from generation to generation creates a historical connection, by which the relevant timescale for changes in the landscape can span several generations. Anthropological research has shown that a period of approximately 200 years is realistic for

---

669 Cappers & Raemaekers 2008.
670 Raemaekers 1999; Amkreutz 2013.
671 Van Gijn 2010, 2013; Raemaekers, Kubiak-Martens & Oudemans 2013. Recent analysis of lipids in Swifterbant pottery also indicates a shift in the composition of animal foodstuffs prepared in these pots, notably the addition of porcine foodstuff next to fresh-water fish (Demirci et al. 2020).
672 We should note, however, that there is increasing evidence for the existence of fields (i.e. crop cultivation) on and close to levees. If this is indeed the case, crop cultivation seems to have been a structural aspect of the Swifterbant broad-spectrum food economy.
673 Overeem 2005.
674 Thrane 1989.
675 Mischka 2011.

676 Ingold 1993.

the oral communication of information that has direct meaning for those involved.[677]

From the perspective outlined above, mobility and historical connectedness form an important basis for the creation of a significant structure in the landscape that consists of a network of places and zones connected by routes. Of course, as archaeologists, we only have small pockets of data, that as we have seen in the previous chapters, often provides fragmented information about activities carried out locally. This makes it difficult to gain any insight into the cultural structuration of the landscape and any changes therein. Yet there are some patterns that may shed light on this: continuity and discontinuity in the use of specific locations in the landscape and the relation between place and material culture.

In the discussion about the exploitation of resources in the Mesolithic – Neolithic landscape, archaeological evidence supports a long period of continuity on many sites. For a site such as Dronten-N23, a chronology of site use that covers two millennia has been established on the basis of a long series of ^{14}C dates. On the basis of ^{14}C dates and archaeological indicators, sites such as Hoge Vaart-A27, Swifterbant S21-S25, Schokland-P14 and Urk-E4 also had long histories of use. It seems that sites, as long as they were not swallowed up by the encroaching marshland and bogs, remained in use despite the changes in the palaeolandscape conditions of the place. It is of course true that we cannot see the frequency with which sites were visited through the generations. Sufficient chronological resolution is simply lacking for this. In Chapter 4 it is pointed out that there may be different contexts of activity that are associated with the surface hearths and the production and use of lithic tools coupled with a narrow or wide spectrum of activities, or associated with the use of pit hearths. In the case of Dronten-N23, activities associated with pit hearths carry on longer than flint-related activities. At Hoge Vaart-A27 there is an abrupt transition from activities associated with pit hearths to activities associated with flint and surface hearths. Timespans measuring more than 1000 years of 'continuous' use on the basis of numerous ^{14}C dates are available for sites outside Flevoland, for example for Hattemerbroek, Mariënberg, Kampen, Soest, Zwolle, Nieuwe Pekela, Leeuwarden-Hempens and Epse-Olthof.

The picture forming is that of various sites in the constantly-changing landscape that continued in one way or another to play a role as a meaningful place for many generations. From an economic perspective, it is often assumed that these places lay in attractive landscape zones or at strategic positions. The question is whether such an assumption can be verified without the availability of more detailed information over the palaeolandscape conditions and changes therein on different spatial and temporal scales. In most cases, such information is simply not available.

Where we do have information, such as for the Swifterbant and Hoge Vaart-A27 areas, we see that the palaeolandscape can change fundamentally, but people nevertheless returned to the sites. Changes in the environment may, of course, have offered new economic exploitation opportunities. But why did people keep returning to those places, while there were undoubtedly many other locations that offered the same or comparable possibilities?

In this context, it is important to realise that Mesolithic-Neolithic hunter-gatherers and early farmers did not survey the landscape from a helicopter in order to be able to make the 'optimal' choices. Individuals moved through the landscape and observed the environment from a certain cultural perspective, in which the perception of how people were related to (elements of) the environment was of great significance. Information would be collected that played a role in making choices to do with the exploitation of resources or the construction of camps and settlements.[678] Such information would not always have been generally shared within the group, for example as a result of gender-related division of tasks.[679] Many tasks that were carried out in the landscape would have been connected to widely varying travelling distances and for which information about the environment would have been retained to a variable level of completeness or reliability, depending on the purpose. Routes used for movement would probably have stayed in use for a long time and would have formed part of the structure of the cultural landscape. In wet areas streams, rivers and lakes would have played an important role in allowing people to travel through the landscape.[680]

Movements of individuals and groups of individuals would not have been arbitrary and would have led to the creation of structures in the landscape that lasted for generations. Undoubtedly shifts occurred in this structure, for example as a result of changes in the landscape itself, or simply because new routes could be taken and new information over the environment could be added. In that sense, landscape structures would not have been static, but would have consisted of dynamic networks of places and zones connected by routes. The locations used for a long time – persistent places – that we can recognise in the archaeological record, may have formed historical anchoring points ('time nodes') if viewed from this

---

677 Ingold 1993.

678 Whallon 2011.
679 E.g. Funk 2011.
680 Lovis & Donahue 2011.

perspective.[681] People continued to visit these places in the landscape, not necessarily because they offered the best locations for the exploitation of specific resources, or were even strategically placed – for example an observation post – but maybe because these places were bound to a world of experience that had been gathered and shared across group members for generations.

This connection of places in the landscape with a world of experience shared over generations, is even more evident in the role of locations in a ritual context. Chapters 4 and 5 discuss the treatment of the dead and related objects. What has been emphasised is that, at least in the Late Mesolithic and Neolithic settlements, evidence has been found of ritualistic activities. Because no excavations have taken place outside the established settlements, we cannot say at this point whether what we have already found in settlements actually represents the tip of the iceberg, or to what extent the phenomena encountered are representative for the 'ritual world' of which Flevoland formed a part. Nevertheless, the research carried out in Flevoland does offer opportunities to view the significance of places in the landscape in other ways and not only from an economic perspective.

The fact that the dead were buried within a settlement seems to indicate that there was a connection between the site and the individual. However, the chronological relationship between the settlement activity and the burial of the dead in a settlement area is difficult to establish. In the case of Dronten-N23 and the river dune location of Swifterbant-S21-S25, it seems very likely that the graves were situated on locations where settlement activities had not taken place for several hundred years, at least insofar as these are archaeologically visible. However, the question is whether this means that there was no relationship whatsoever between the Mesolithic settlement site and the place where a death ritual was performed. The 'secular' activities carried out in these areas in the Mesolithic, as far as can be determined, seem to have been linked to different behavioural contexts. In that regard, the use of sites in a later phase for burying the deceased can be conceived as another aspect of the differentiated use of location in a landscape consisting of meaningful places and zones. The relatively short use of levee locations in the river valley of the Hunnepe by Swifterbant might also reflect the same behaviour. Here too, we see different behavioural contexts – small-scale cultivation; settlement activities; burial of the dead – on locations that appear to have fallen out of use relatively shortly after the emergence of the low levees.[682]

That the use of places and zones in the landscape can be very complex but also structured at the same time, is evident from ethnographic research among hunter-gatherers. For example, groups of Khanty and Evenki living in Siberia appear to have a dual perception of space. The contrast between life and death plays an important role in relation to a world of experiences that consist of a 'wild' world, inhabited by animal spirits and a 'domestic' world inhabited by human spirits.[683] By inhabiting places or zones for a long time, it is believed that the spirits linked to animals can be gradually driven away by the spirits of the deceased. In this perception, the landscape consists of a space that is completely inhabited by spirits of different character. The physical marking of places, zones and routes in the landscape play an important role and is related to specific events.[684] The place where a person died would, for example, be marked using possessions from the person that have been rendered unusable, whilst other possessions would remain untouched. Only after the deceased has been buried would the location of the camp be abandoned and not visited for several years in order to allow the spirit time to go to the 'other world' inhabited by living spirits. The tent belonging to the deceased is not removed, but stays on its pitch to mark what is now seen as a dangerous place. The departing group chooses a new route to avoid returning too quickly to the old camp. Burial sites that are marked by objects hanging in the trees are also seen as dangerous places that should be avoided.[685] Other aspects of life are also surrounded by cosmological rules and taboos: the location of the former tent can only be entered via the spot where the original entrance stood; when moving around the fire, one shouldn't step over it, nor walk completely around it.

Such examples cannot, of course, be used as models to explain the fragmentary patterns of prehistoric cultural landscapes. The idea that ethnography can be used directly as an explanatory framework for the past has been long abandoned.[686] What such examples can do, is to show us that all kinds of aspects of the daily use of space on a local and regional scale are inextricably linked to cosmologically-anchored rituals. It may very well be the case that the 'mixing' of the behavioural contexts that we think we have identified on many sites in Flevoland is actually a reflection of such a complex interplay. The various evidence we have for the treatment of (skeletal

---

681   Peeters 2007.
682   In how far the locations had any or no role in the Late Neolithic landscape is a difficult question to answer. The large-scale erosion of the younger landscape has made it impossible to understand,

for instance, Late Neolihic occupation. That Bell Beaker pottery has been found on Swifterbant-S2 as a residue of an eroded landscape is perhaps writing on the wall (Van der Heide 1965; Raemaekers & Hogestijn 2008).
683   Jordan 2003; Lavrillier 2011.
684   Haakson & Jordan 2011; Lavrillier 2011.
685   Jordan 2003; Haakson & Jordan 2011; Lavrillier 2011.
686   Lane 2014; Bettinger, Garvey & Tushingham 2015.

parts of) the dead and the treatment of objects could be seen as part of this.

Flevoland is not alone in this. The deposition of objects in other parts of The Netherlands can help us to determine to what extent the Flevoland finds were typical for that region, or were part of a geographically and temporally wider tradition. The examples outside of Flevoland also give us a better picture of the possible spatial location of depositions, especially outside the settlements.

Flint depositions have been documented on other Mesolithic sites in The Netherlands. As a rule, these are assemblages of a specific nature, whereby either mainly tools, or cores or chunks predominate. A *cache* is therefore equally seen as a valid interpretation. A Mesolithic deposition of cores in a pit in Hoogkerk (Province of Groningen) is one of the exceptions.[687] In this instance, an interpretation as votive deposition is plausible. The same is true of a deposition consisting of cores and flakes excavated in Uddel (Province of Gelderland).[688]

A large number of axes of lithics including flint have been found in The Netherlands, especially in areas of the country where the prehistoric level has not been covered by later sedimentation. The long tradition of depositing complete axes in lower-lying parts of the landscape is particularly striking. This tradition seems to span the whole of the Neolithic.[689] The almost total absence of such finds in the Flevoland dataset is very likely to be a consequence of the sedimentary deposits covering the Neolithic landscape: the probability of chance finds out this period is very small.

Depositions of pottery are also not exclusive to Flevoland. These are found from the beginning of the Swifterbant culture. A very similar context has been discovered on the site of Hardinxveld-Giessendam De Bruin.[690] This settlement was also later covered by sediment. An almost complete Swifterbant pot was found here in a small pit. The excavation also uncovered a number of Blicquy sherds in association with Swifterbant pottery. An almost complete Blicquy pot was found in the levee zone directly next to the settlement.[691] The find from Bronneger makes clear that pottery depositions from the Swifterbant period also took place in river and brook valleys: here a large part of a pot was found together with two red deer antlers (and a fragment from a third) during dredging activities.[692]

A second period in which we find a relatively large number of pottery depositions is in the Funnel Beaker and Vlaardingen-Stein cultures, dating approximately to the second half of the fourth millennium and the beginning of the third millennium cal BC. Pottery depositions from the Funnel Beaker Culture are known from five sites in the Drenthe fenlands.[693] These depositions consist of either single or several pots. The contexts in which these pottery depositions were found is very different from that of Flevoland. In the middle of The Netherlands the discoveries are of a single pot buried upside down within a settlement site at Neede, but also four pots that seem to have been buried in a seemingly empty space. Of interest is that three of the pots were missing the base.[694]

Pottery depositions have also been discovered dating to the Late Neolithic, again in the Drenthe fenlands. These sometimes consist of several sherds; sometimes it is unclear whether the finds represent (almost) a complete pot. The most intriguing find derives from a peat bog in Kooiker. A Single Grave culture sherd was discovered together with three almost complete pots from the Bell Beaker culture and a complete pot dating to the Early Bronze Age. According to the documentation, all the finds came from the same location and were found under 3 m of peat.[695] This means that the deposition took place in the Early Bronze Age and included pottery from earlier periods that still might have been circulating in the community.

Twelve wooden cartwheels have been discovered in the Drenthe fenlands. The seven available ^{14}C dates suggest a date for these depositions between *c.* 2900 and 2200 cal. BC.[696] A large number of cattle horns, originating from 99 different contexts, were also found in the Drenthe fenlands.[697] The numbers makes it clear that such deposition of cattle horns was not uncommon. At least as important to note is that there are no comparative datasets for the deposition of other animals. (with the exception of red deer antlers, see below).[698] The fenland finds fit well with the discovery of aurochs skulls at Hoge Vaart-A27. Both contexts endorse the ritual significance of bovids. Analysing the size of the horns from the Drenthe fenlands makes it clear that depositions were made of both aurochs and domesticated cattle.[699] The 14 ^{14}C-dated cattle horns show that the depositions all took place over a long period between the Neolithic and the Middle Ages.

The pottery deposition from Bronneger also included three red deer antlers. This discovery formed the

---

687 Kortekaas 1998.
688 Groenewoudt *et al.* 2006.
689 Raemaekers *et al.* 2011; Ter Wal 1995/1996; Wentink 2006, 2020.
690 Raemaekers 2001b: fig.5.10.
691 Raemaekers 2001b: fig. 5.5f.
692 Kroezenga *et al.* 1991; Ufkes 1993.

693 Bakker & Van der Sanden 1995.
694 Louwe Kooijmans 2010: 204.
695 Van der Sanden 1997: 136-138.
696 Van der Waals 1964; Lanting & Van der Plicht 1999/2000: 95-96.
697 Prummel & Van der Sanden 1997.
698 De Jong (2012) presents an overview of animal bones found in different stream valleys in the southern Netherlands. In many cases it is unclear whether these are actually archaeological finds, considering that cutting marks are only sporadically recorded.
699 Prummel & Van der Sanden 1997: 109-110.

starting point for an inventory that now comprises 21 red deer antler discoveries in the Drenthe fenland.[700] An association with clear anthropogenic material could not be further established. There are, however, persuasive arguments that these finds should be considered as depositions. For example, most come from animals with a highly-developed antler. This indicates non-natural selection. The fact of deposition also means that the antler have been preserved: with a natural death, the antlers would likely remain on the ground and would therefore not be preserved.[701] The eight available ^{14}C dates show a large temporal distribution, from the Mesolithic to the end of prehistory.[702]

In short, the treatment of objects that also had a function within the context of daily usage could have, just as skeletal remains, played a role in activities with cosmological and ritual connotations. What these activities actually were, can no longer be reconstructed, but the scanty information from Flevoland and the rest of The Netherlands indicates that depositions formed a structural part of the experiences of prehistoric communities.

7.4 Socio-cultural relationships

Prehistoric Flevoland was inhabited by people who maintained contacts, either on an individual basis or in groups. Information was exchanged, technological innovations found their way through networks of interaction and the landscape took on a significance because of the historical connection of people in their environment. But what did this look like? How should we place prehistoric Flevoland in the wider socio-cultural landscape that did not consist of the nuances of pots, flint tools and other utensils, but of interaction between people?

The discussion about the character, or the cultural affinity, of what we archaeologically-speaking have labelled as Swifterbant Culture is exemplary in this.

As described in Chapter 2, the discovery of pottery on the site of Swifterbant in the 1960s led to a search for 'cultural relationships'. Similarities and differences in the pottery from the Swifterbant sites and those of the Ertebølle culture are expected to provide arguments so that we can either place everything under one denominator, or so that we can emphasise the individuality of these sites. This last option is nowadays preferred by most researchers, so that the Swifterbant culture takes on the mantel of a *wetlands*

phenomenon with a distribution area that stretches from the Scheldt to the Elbe.[703] Strictly speaking, we know very little about the origin and transformation of this 'culture' which, it is worth emphasizing again here, is based solely on an archaeological definition of pottery characteristics.

An important problem lies in the fact that pottery is the most important diagnostic element in the attribution of a site to the Swifterbant culture. This pottery is generally fired at low temperatures, making it fragile and sensitive to erosion. On sites where such pottery has lain at the surface for some time, be this in the past or present, the sherds would have completely disappeared: only the quartz or granite tempering used in the clay would have survived, but would not be recognised as such. Other material remains, such as flint tools and debitage are less sensitive to erosion, but do not appear diagnostic on the basis of available data: the flint technology is essentially Late Mesolithic in character. Other categories of artefacts that we know from Swifterbant contexts, such as tools made of bone or antler, do not differ from Mesolithic specimens. If we look at the food remains, then we see an economy reflected in them that is primarily based on hunting, fishing and foraged plants; cultivated crops and some domesticated livestock do not appear until the Classical Swifterbant phase – as additions to the 'traditional' resources.

The consequence of all of this is that preservation conditions must be favourable in order to attribute a site to the Swifterbant culture. In the higher parts of the landscape, where sites were not covered with sediment shortly after their abandonment, and then preferably by peat or clay, these conditions are just not present. Outside the expansive marshlands of the coastal and river areas, the activities of those Swifterbant culture people living in, for example, the higher-lying parts of the northern and eastern Netherlands are hardly visible, mainly because they have been identified as Mesolithic. Isolated finds, such as the sherds of Swifterbant pottery out the river valley by Bronneger (Drenthe), where the localised conditions were favourable for preservation, show that the higher-lying areas did form part of the regularly-used landscape. More evidence for the use of the higher parts of the northern Netherlands comes from palynological data that indicate cultivation activities.[704] A couple of sites in the eastern Netherlands (Zutphen Ooijerhoek; Raalte-Jonge Raan) with dates that fall in the early phase of Swifterbant can also be added to this, but the absence of pottery prevents a more reliable association.[705]

On the basis of the data currently available to us it cannot be concluded with any certainty that the

---

700 Ufkes 1997.
701 Ufkes 1997: 169.
702 Depositions in wet contexts do not exclusively date to the period under discussion, The exhaustive analysis of Drenthe bog finds carried out by Van der Sanden and others make it clear that such depositions carried on until the medieval period (E.g. Van der Sanden & Taayke 1995).

703 Ten Anscher 2012; Amkreutz 2013; Crombé et al. 2015.
704 Bakker 2003.
705 Groenewoudt et al. 2001.

Swifterbant culture was principally connected to the extensive wetlands in the coast and river areas.[706] Our perception of the palaeolandscape is influenced by the actual contrasts in the Dutch landscape, between low-lying areas (read: wet) and high-lying areas (read: dry). In historical times, the large-scale reclamation and successive levelling of peat areas in The Netherlands is well-documented. It is clear that in the course of the Holocene, sizeable parts of what we now call the ' high hinterland' would have become increasingly saturated and overgrown with peat. This process involved a complex interplay between changes in local and regional hydrological systems, which led to peat formation under different conditions and in different periods. Research in the Drentse Aa valley, for example, has provided a good picture of what influence the Holocene sea level rise had on the accumulation of peat in different parts of the valley: peat formation occurred initially in the stream channel under the influence of local drainage conditions and later on the valley flanks under the influence of sea level rise.[707] Interestingly enough, peat formation seems to have started first on valley flanks in the upstream section rather than in the middle and downstream sections.

It is very likely that large areas of the high hinterland was also wetter than initially assumed. If that is indeed the case, then it is probable that waterways played an important role in the forming of socio-cultural networks. People did, after all, have to move around and travelling over water might have been easier than over land. Following on from this assumption then it is conceivable that communication networks also followed these water routes to a considerable extent. For Flevoland, the Overijsselse Vecht, the Hunnepe and the Eem were the most important navigable routes through which the hinterland and the coast were accessible. The number of navigable tributaries were, however, very limited; there may not have even been one. Water courses such as the Tjonger and Regge that connected to the Overijsselse Vecht could only take a limited volume of water as local drainage systems. The geographical reach of these rivers was also quite limited. The Eem didn't reach further than the area of the Gelderse Vallei, an area bounded by ice-pushed ridges, and the Hunnepe reached the border area of the eastern Netherlands and North Rhine-Westphalia. The Overijsselse Vecht formed the longest of the rivers, with the source in North Rhine-Westphalia, only a short distance (*c.* 10 km) from the much bigger Ems river, that flowed in a northerly direction.

If the Overijsselse Vecht was an important route used by the Flevoland members of the Swifterbant culture – a realistic expectation if we take into account the evidence for earlier and later settlement along its banks[708] – then we can expect to see that socio-cultural contact with related groups would have been in an easterly line. From this perspective it is interesting to note that the Early Neolithic adzes with a drilled-through hole found in The Netherlands are mostly found between Flevoland and the upstream areas around Hanover,[709] not far from where farming communities of the Linear Bandkeramik lived in this period. In addition, coastal routes would also have been of great significance. These may be apparent because of the origin of part of the flint found, and possibly also the amber. Coastal routes from Flevoland may well have had a greater significance than the southerly and northeasterly orientated lines in the socio-cultural network. At the same time the Rhine, Meuse and Scheldt formed important axes in the southwesterly areas of the Swifterbant distribution area, as well as the Ems, Weser and the Elbe in the northeasterly areas.

During the millennium spanned by the Early and Classical Swifterbant phases (*c.* 5000-4000 cal. BC), there would have been some necessary shifts in the socio-cultural constellation in the bordering areas. The LBK and its successors, the

Grossgartach, Planig-Friedberg, Rössen, Bischheim and Michelsberg cultures formed the predominant units in the south-southeasterly range,[710] whilst the Ertebølle culture lay in an northeasterly range. Looking at the evidence, it seems that socio-cultural contacts with groups from this Danube tradition influenced developments within the Swifterbant culture, which in turn influenced developments in the Ertebølle culture.[711]

New elements in the Swifterbant food economy – small-scale cultivation of crops and the keeping of some livestock – point to contact with food-producing groups. On account of the date these aspects appeared, between 4200-4100 cal. BC, this would most probably have been with the Michelsberg culture that were settled to the south and southwest of the Swifterbant culture distribution area. This is considerably later than the agricultural activities of the Linear Bandkeramik and its direct successors in and on the edge of the loess area. That does not mean that there was no contact between the food-producing groups and the groups of hunters-fishers-gatherers, including those from the Early Swifterbant culture. Several 'exotic' elements as far as the distribution area of the Swifterbant were concerned, such as perforated adzes and wedges or '*Rössener Breitkeilen*', flint (including LBK projectile points) from South Limburg or bordering Belgium, and

---

706   Contra Amkreutz 2012.
707   Makaske *et al.* 2015, 75.
708   Van Beek 2009.
709   Raemaekers *et al.* 2011: fig. 12.
710   These groups are often counted as part of the Danube tradition (*Donauländische Tradition*).
711   Ten Anscher 2015.

Wommersom quartzite from Belgium show that contact did exist.[712] Such 'exotics' for the Early Swifterbant – the earlier mentioned Blicquy sherds may also fall under this category – are mainly known from the central Dutch river area (Hardinxveld-Giessendam De Bruin and Brandwijk).[713] Recent aDNA research also showed that the 'native' populations had contact with Linear Bandkeramik groups.[714] There is less insight into how Flevoland had a place within the contact lines. We know few exotic items that can be dated with certainty to a Swifterbant context, but there are 'imitations' of *Breitkeilen* and a single small fragment of a proper *Breitkeil*.[715]

Again, the pottery plays an explicit role in the identification of influences. The Swifterbant pottery certainly has morphological and technological similarities with pottery from the Danube tradition and the Ertebølle culture, but there are also differences.[716] Pots with pointed bases, typical for Swifterbant and Ertebølle cultures are absent in the whole Danube tradition pottery range. At the same time, there are explicit technological and morphological differences between Swifterbant and Ertebølle pottery: lamps are absent in the Swifterbant and the pottery is constructed and finished in a different way. The morphology of Swifterbant pottery, with the exception of the pointed bases, connects better with the Danube tradition pottery. This is also true for the decorative patterns used on pottery, although the Swifterbant pottery does appear to be slightly more conservative, retaining its own specific character.

It therefore seems likely that the Swifterbant culture built on a Late Mesolithic hunter-gatherer tradition in which new elements were incorporated, such as the pottery and food production. The question is whether or not these additions actually resulted in a fundamentally different socio-cultural perspective. Although Mesolithic hunter-gatherer traditions in Northwest Europe had no agricultural activities incorporated in their system, we do have increasing evidence to suggest that from an early date there was intervention in the environment, for example the deliberate firing of vegetation, possibly in an attempt to increase the local mass of plant and animal resources.[717] Such practices, that form an important link in the development of complex ecosystems,[718] have many documented ethnographic examples.[719] The cosmologically-anchored relationship between people and the elements of their environment was, next to economic motivations, perhaps of fundamental significance. If we assume that Mesolithic hunter-gatherers lived in the context of an 'animistic ontology' in which the world is inhabited by human and non-human spirits,[720] then the influencing of the environment could be related to activities aimed at the regeneration of the spirit world. By creating conditions under which plant and animal resources could renew themselves, then perhaps the survival of the world was guaranteed.

The inclusion of new elements in such a system would not have been at odds with existing traditions. Evidence for this is the continuity in deposition practices of red deer antlers and bovid horns in the Neolithic: the essence seems to be in the symbolic role of antlers and horns.[721] Ethnographically it is know that antlers and horns are used – *e.g.* by shamans among the Siberian Ket – to achieve 'soul flight', *i.e.* to spiritually transform into other beings.[722] The long timelines of continuity that can be recognised during the Mesolithic and Neolithic in the use of the landscape, in combination with more or less explicit changes in material culture, can therefore in this respect indicate long-lasting ideological links that were not easily undermined or did not come under pressure by shifts in the socio-cultural constellation. The destruction and successive distribution of fragments of quern-stones for example, fits in with the image of the deposition of divided pieces of flint. The ritual deposition of broken pottery could also fit with this line of thinking. But at the same time, the use of different pottery vessels for the preparation of 'traditional' and 'cultivated' food could be indications for the changes in the understanding of its socio-cultural significance.[723]

In this context, it is interesting to see what seems to have happened after the Classical Swifterbant phase. As already discussed in Chapter 2, the Funnel Beaker Culture (TRB) has long been assumed to be the successor of the Swifterbant Culture. The area of origin is thought to have been in the distribution area of the southern Ertebølle culture (northern Germany), where the characteristic TRB pottery apparently developed under the influence of the Michelsberg culture.[724] From there, the TRB expanded in a westerly direction and became the TRB West Group. An analysis of the pottery from Flevoland sites has, however, led to the realisation that prior to the 'classic' TRB pottery (Brindley horizons 1-7), other pottery was produced in the Swifterbant tradition, but with elements inspired by Michelsberg culture pottery. This was then continued

---

712 Verhart 2000, 2012; Raemaekers *et al.* 2011; Amkreutz 2013.
713 Raemaekers 2001a/b; Raemaekers 1999.
714 See E.g. Nikitin *et al.* 2019.
715 Devriendt 2013.
716 Raemaekers 1999; Ten Anscher 2015.
717 Mellars 1976; Zvelebil 1994; Simmons 1996; Bos & Urz 2003; Bos *et al.* 2005, 2012; De Moor *et al.* 2009; Woldring *et al.* 2012.
718 Delcourt & Delcourt 2004.
719 Mellars 1976; Scherjon *et al.* 2015.
720 Descola 2010.
721 Raemaekers 2003, 2019.
722 See E.g. Vajda (2010) and Little *et al.* (2016).
723 Raemaekers, Kubiak-Martens & Oudemans 2013.
724 E.g. Schwabedissen 1979.

into the TRB. This so-called Pre-Drouwen phase can now be seen as the basis for the further development of a TRB West Group.[725] The TRB is therefore seen as a local development of Swifterbant, incorporating material and economic elements from the south.

The appearance of the same elements in the TRB North Group and the continuity of specific traditions in both the TRB West- and North Group, such as the ritual deposition of pottery and axes for example, seems to signify a shared socio-cultural connection. Both the people of the Swifterbant and the Ertebølle cultures maintained contact with people from the Michelsberg culture. Contact between groups from the Swifterbant culture and the Ertebølle culture cannot be ruled out – such contact did not have to be expressed in the pottery. It seems that the Swifterbant culture was receptive to innovations, although the continuation of hunter-gatherer traditions possibly meant that in the end the Ertebølle world also eventually accepted innovations that were introduced via a network of existing connections.

Depending on the flexibility with which innovations could be assimilated through the socio-cultural constructions, different trajectories for change could have been followed. Neither did one new 'monolithic' situation arise, but rather a new constellation of more or less related traditions within a wide socio-cultural network. The broad range of ritual and symbolic expressions in the TRB world – diversity in the burial ritual, the deposition of objects, the decoration of pottery -perhaps gives expression to a process of profound changes.

The previous chapters show that we know very little about the TRB occupation of Flevoland. The evidence is very fragmented and is limited to a few sites in the Noordoostpolder, including Schokland-P14 and Schokkerhaven-E170 as the most important. That the region, that in this period was dominated by extensive marshlands, did form part of the TRB cultural landscape is evident. It is possible that the Overijsselse Vecht and Hunnepe acted as major arteries connecting all the different parts of the TRB landscape that extended to the east, north, west and south. Nearby Hattemerbroek on the Overijsselse Vecht, traces of settlement activity and a palisade enclosure can be attributed to the TRB.[726] The function of the area enclosed by the palisade is not known, since only a small part of it could be excavated. Indications for a TRB palisade have been found at Schokkerhaven-E170 in Flevoland, but here too we have very little information. This is also the case for the more intensively researched palisades from Anloo (Drenthe) and the Uddelermeer (Gelderland), for which the interpretations vary from a livestock corral to a defence work.[727] The latter interpretation has far-reaching socio-cultural implications as it assumes conflict between (local?) groups.

We do not know how TRB society was organised. The megalithic collective tombs that we know primarily in the northeastern Netherlands are iconic remains of this period, but it is simply not known how the ritual related to these sites was connected to the social structure within TRB society. A cemetery and adjoining settlement excavated at Dalfsen on the Overijsselse Vecht offers the first opportunity to carry out detailed research into aspects of the death ritual.[728] The location shows that a part of the death ritual was directly linked to settlements, which immediately raises the question as to what the role was of the megalithic collective tombs. The Flevoland TRB sites situated along the Overijsselse Vecht and Hunnepe, undoubtedly had connection with the upstream TRB world, but also with the downstream area where the Overijsselse Vecht via the fenland and salt marsh landscape of Noord-Holland connected to the North Sea. Here traces of TRB sites have also been found at Slootdorp-Bouwlust.[729] Does the picture show that small, relatively autonomous communities were active here, who expressed their cultural connections via their material culture and rituals? Or, is the picture that of a larger, overarching socio-cultural structure in which collectivity had a more compelling meaning?

The connection between the peat-rich TRB and post-TRB landscape of Flevoland and the Noord-Holland coastal area is interesting from a social-cultural perspective. There is evidence for TRB occupation in the coastal zone, but also evidence for the contemporary Vlaardingen Group, mostly in the western coastal zone and in central and southern Netherlands. The Vlaardingen Group would continue to exist as a separate tradition in these areas, whilst in the northern coastal area the Corded Ware tradition left its mark as the successor to the TRB. Recent research into pottery characteristics has shown, however, that there are no substantial differences in the chronological development of Vlaardingen and Corded Ware traditions in either the southern or northern parts of The Netherlands respectively.[730] The thick-walled vessels in particular seem to build on local traditions, whilst the thin-walled, decorated vessels can be linked to supra-regional social-cultural structures.

---

725 Ten Anscher 2012, 2015.
726 Knippenberg & Hamburg 2012.
727 Waterbolk 1960; Harsema 1982.
728 The cemetery contained at least 120 inhumation graves, many with grave goods; skeletal remains were not preserved, but in several graves a silhouette of the skeleton remains in the soil. The excavated area of the settlement contained features relating to at least one house structure. The post-excavation analysis is still being carried out (for advance information, see Van der Velde, Bouman & Raemaekers 2019; 2021).
729 Hogestijn & Drenth 2001; Beckerman 2015.
730 Beckerman 2015, 189-193.

The few assemblages found in Flevoland fit this picture. As a result of the large-scale erosion of the Late Neolithic landscape, we have found hardly any evidence for the occupation of Flevoland by members of the Corded Ware tradition and ensuing(?) Bell Beaker tradition. It speaks for itself that sites such as Schokland-P14, where the most evidence has been found, was not an isolated incident. There would have been links to the western coastal area of Noord-Holland – a large number of Corded Ware settlement sites are known here [731] – and, for example, the more easterly-lying landscape zones in the river valleys of the Overijsselse Vecht and the Hunnepe. In this context, the site of Hattemerbroek re-enters the discussion as a location where a number of Corded Ware and Bell Beaker graves have been found.[732] However, we have to note that it is still not at all clear where the settlement had been. It is very possible that the Late Neolithic occupation of the area, as was the case in Flevoland, had a similar character to that of Noord-Holland, even though the differences in the landscape might have led to accentual differences in the sites.[733]

The fragmentary picture we have of Late Neolithic occupation continues into the Bronze Age. As has been stated in Chapter 4, the information for this period is just as scarce and comes exclusively from sites in the Noordoostpolder. Nevertheless, the fragmentary evidence does support a picture of structural occupation in the Noordoostpolder at least, especially in the Early Bronze Age, With regard to the material culture, it is also clear that this occupation was embedded in a larger network where, for example, typological developments in the pottery were followed.[734] Links would undoubtedly have existed with the areas in the hinterland, along the Overijsselse Vecht and Hunnepe. At Hattemerbroek, a large scale Bronze Age site has been excavated, consisting of a settlement and an extensive 'agricultural' area within which was found a network of fencing.[735] A number of sites with features dating to the Early Bronze Age have been discovered further upstream by Zwolle, Deventer and Zutphen,[736] but also in other parts of the eastern and northern Netherlands, as well as the wet coastal area of the western Netherlands.[737] Flevoland represented only a small part of an extensive cultural landscape that, in comparison to the preceding period, seems to have been characterised by a more explicit spatial layout. Whilst agricultural activity formed an important dimension of this cultural landscape, it is also clear that this landscape was not limited to the dryer ground. The wetland areas such as Flevoland were also part of this landscape, but how the settlement activities in these wetter regions related to the surrounding areas is not yet clear.

7.5 Conclusions

In the previous sections we have attempted to arrive at a synthetic interpretation of the prehistoric occupation of Flevoland from different thematic perspectives.

It is obvious that the 'cultural biography' of this area cannot (yet) be written except in general terms. The unbalanced availability of data for different prehistoric periods means that there is a lot more to be said about certain subjects than about others. This is naturally not a problem unique to the archaeology of this province; it is a general problem that affects all archaeology research. This book has hopefully been able to show that there is a lot that can be said about the prehistoric occupation of Flevoland, but that the data from this area cannot be seen in isolation to data collected beyond its boundaries.

The perception of the prehistoric occupation of The Netherlands – including Flevoland – and the processes of cultural change, seem to be largely determined by the current geographical situation combined with extremely generalised models of landscape change being influenced by climate change. The landscape to the east of the southern North Sea had become increasingly wet and marshy since the end of the last ice age and as such, quickly dismissed as a marginal area, on the edge of a world in which the 'real' cultural developments took place.[738] The prehistoric inhabitants of this marshland apparently had to make do with whatever nature had to offer. However, we have seen that the history of the Late Glacial and Holocene landscape development in Flevoland is difficult to capture with the general models of vegetation succession that are generally used. The data does lend itself to another interpretation, giving insight into a complex interplay of climatology, hydrology, soil science, geo-chemical and biological processes. All this had an impact on the potential exploitation possibilities available to prehistoric inhabitants of the region.

This does not, however, mean that the environment determined what people did or did not do. Choices would have been made in which cultural traditions played a fundamental role. The long chronology that has been recognised on some sites, shows that, over many

---

731 Three settlements belonging to the Corded ware culture have been published in detail. These are Keinsmerbrug (Smit *et al.* 2012), Mienakker (Kleijne *et al.* 2013) and Zeewijk (Theunissen *et al.* 2014).
732 Drenth & Meurkens 2011.
733 Ten Anscher 2012, 461.
734 Ten Anscher 2012, 485.
735 Lohof, Hamburg & Flamman 2011; Lohof, Hamburg & Quadflieg 2012.
736 Clevis & Verlinde 1991; Groenewoudt, Deeben & Van der Velde 2000; Bouwmeester, Fermin & Groothedde 2008.
737 See also Fokkens, Steffens & van As 2016.

738 The title of L. Louwe Kooijmans' article published in 1976 "Local developments in a borderland" is significant in this regard.

generations, this played a role in the exploitation of the landscape that in the Mesolithic, Neolithic and Bronze Age itself underwent change. Places were the deceased were buried, skeletal parts of group members (family members?) that were left behind, or the ritual deposition of votive objects, all gave meaning to the landscape. It could have been that the emphasis on the historical relationship with the landscape fulfilled at least as big a need as the necessity for food resources – indeed, humans also need to eat – that was based on a broad spectrum economy. The incorporation of domesticated livestock and cereals in the food resources in the course of the Classical Swifterbant period – a result of interaction with farming communities to the south and east of Flevoland – demonstrates a flexible approach to food sources on the one hand, possibly associated with changes in the socio-cultural role of food on the other.

Evidence from Flevoland indicates that this region occupied an explicit position in all kinds of cultural developments in northwestern Europe. The Swifterbant period formed an important time window in this respect. Changes in the Swifterbant pottery can be linked to interaction with groups with a different social-cultural background. This ultimately resulted in the development of a new social-cultural constellation in the Middle Neolithic. The later succession and dissemination of archaeological cultures in the Late Neolithic and the Bronze Age shows that the social-cultural dynamic continued on different scales. The occupation in the wetlands of Flevoland appears to have built on local traditions, but was an integral part of supra-regional structures.

The insights presented in this book are based on data collected over a decennia-long tradition of combined archaeological and earth science research in Flevoland. As has been emphasised several times, the nature and quality of this data varies enormously, but in combination all the data delivers an important contribution to international discussions, which are often based on only a handful of 'top sites'. For example, the discussion on the exploitation and perception of landscapes by Mesolithic hunter-gatherers in Northwest Europe has traditionally been based on sites from the England (Star Carr, Howick, Bouldnor Cliff, Goldcliff, Mount Sandel), France (Téviec, Hoëdic), Germany (Düvensee, Hohen Viecheln, Friesack, Bedburg) and South Scandinavia (Tybrind Vig, Skateholm). In discussions about the introduction of animal husbandry and arable farming as the fundamentals for 'Neolithisation' sites of the LBK, La Hougette, Blicquy and the Ertebølle culture invariably play a central role. The Swifterbant culture hardly gets a mention in many overviews.[739] This is actually not so strange as it seems. We started the book by saying that the Dutch data – not only that from Flevoland – has only rarely been made available in another language other than Dutch.

It may also be related to the fact that sites with an excellent state of preservation are rarely found, let alone excavated and extensively published. The value of the archaeological record from Flevoland for the international discussion does not lie only in the 'sites'. These are just pin pricks, miniscule jigsaw parts of the 'landscape puzzle' exploited by man, a landscape that in itself is not a 'fossilised surface' buried in the subsoil. Prehistoric landscapes, just like human behaviour, need to be reconstructed. In fact, data must be interpreted that leads to the construction of a picture or a narrative, or indeed several pictures or narratives. Taking a landscape perspective is, in our opinion, a better way to approach an interpretation of the varied archaeological information and give it meaning in a narrative about the past. With 'landscape perspective' we do not mean that 'landscape' guided what prehistoric man did or did not do, but rather delivers a perspective that takes into account as much as possible a potentially infinite range of behaviours in a more or less differentiated landscape. This book shows what insights can be gained from such an approach, but also that there are all sorts of problems involved in this. The majority of the data comes from sites inhabited over a long time span. We do not possess a 'representative' set of data, but it is unlikely that we ever shall.

We can, however, set ourselves the goal of focussing on building up a more differentiated database. We have seen that cultural heritage management, despite various attempts to change it, is still focussed on 'sites' in the traditional sense of the word. These sites can, of course, provide valuable information about the lives of the prehistoric inhabitants of Flevoland. But this does not speak for itself. Broadening the horizon of research seems to us to be a prerequisite to gain new insights. That equally doesn't mean that even more research must be done. It means that well-considered choices should be made with regard to what and how we undertake research in the future.

---

739 See E.g. Cunliffe 2011.

Appendix I. Site Atlas Windows of observation

The quality, nature and context of excavated prehistoric sites in Flevoland: site atlas

T. Hamburg & B.I. Smit

I.1 Introduction

As has been discussed in Chapter 2, a number of 'windows' of observation have been selected in order to better describe the prehistory of Flevoland from the Mesolithic to the Middle Bronze Age. The selection comprises twelve sites that have been researched in detail over the last 30-40 years (fig. I.1). These sites, together with a number of others discussed in Chapter 6, form the basis for our understanding of the human occupation history of Flevoland in the Early and Middle Holocene.

Investigations on all these selected sites have uncovered substantial amounts of archaeological remains. Multi-disciplinary research has also produced a great deal of information on the contemporary landscape in which human occupation took place. The sites included in this Atlas therefore provide the framework for the narrative of the prehistory told in the previous chapters of this book. All the sites described below have a strong phasing in the archaeological fieldwork carried out. For the majority, this phasing consists of consecutive borehole surveys in which selected parts of the site are subject to increasingly intensive surveys. With the exception of Hanzelijn Area VII, excavations took place as the final fieldwork phase on all the sites.

Flevoland has many more known prehistoric sites in addition to those described in this Atlas. Unfortunately, as a result of the nature and extent of the archaeological fieldwork carried out, research on these sites has generally delivered too little site specific data. That said, where significant results are available, these have been used to substantiate the information presented in this book. Fieldwork on the majority of these other sites has been limited to borehole surveys. As already stated in Chapter 3, this has implications for the analysis of the significance of the archaeological remains recovered. The lack of contextual reference means that it is not always clear how such remains should be interpreted in terms of past cultural behaviour.

As described in Chapter 3 the twelve selected 'windows' are divided over the three polders that make up Flevoland: Zuidelijk Flevoland, Oostelijk Flevoland and the Noordoostpolder. Four sites have been selected from each polder. These sites are related to occupation activity in parts of the landscape that in prehistory had their own distinctive characteristics. Zuidelijk Flevoland was dominated by the river system of the Eem, Oostelijk Flevoland by that of the Hunnepe and the Noordoostpolder by that of the Overijssel Vecht. In addition, the sites represent occupation activity on the Pleistocene

Figure I.1: Map of Flevoland showing the location of the twelve selected sites.

subsurface and within landscape units that evolved during the Holocene.

The main purpose of the Atlas is to provide the reader with information on the current state of knowledge and to highlight the most important results from each of the sites listed. The primary publications are listed for each site. Using these publications as reference, the sites are described according to geographical location and research methods, the geological and soil stratigraphy, dating, finds, features, the cultural context, the palaeoecological and geographical context, as well as aspects of taphonomy. This collected data forms the basis and background to Chapters 4, 5 and 6. In these chapters the archaeological remains and the research results are discussed and interpreted in greater detail and in relation to each other.

As has already been discussed in Chapter 3, the current landscape conditions in Flevoland play an important role in the discovery and assessment of prehistoric habitation remains and are a determining factor when planning excavations or indeed any other archaeological interventions. The sites have been researched using a wide array of different archaeological methods ranging from coring, test pits, trial trenches and excavations.

I.2 Zuidelijk Flevoland

I.2.1 Almere – Hoge Vaart/A27

References
Publications: Exaltus 1993; Hogestijn & Peeters 2001; Peeters 2007.
Data:[740] urn:nbn:nl:ui:13-0pr-of8

Geographical location and fieldwork
The site lies in the current route of the A27 motorway, directly to the north of the Hoge Vaart canal and was discovered during an exploratory borehole survey carried out in 1993 prior to the construction of the A27. Using a hand gouge auger, the top of a Pleistocene coversand ridge with a transitional zone covered by peat horizons was identified in the cores. Charcoal was found in both zones As a result of this discovery, the prospection was intensified by carrying out a further borehole mapping survey, but this time on the basis of a closer-spaced grid and using an auger with a larger diameter. Borehole core samples from this phase contained not only charcoal but also worked flint and burnt bone fragments. On the basis of the depth of the site (approx. 5.8 m -NAP/below average sea level) and data available relating to the process of rising groundwater levels in the area, it is possible to date the remains as older than c. 4100 cal. BC.

Based on the evidence for archaeological remains combined with geomorphological characteristics, a study area of 8600 m2 was enclosed with interlinked, watertight sheet-piling and the groundwater pumped out. A three metre-thick covering layer of sedimentary deposits was then mechanically removed to reveal the level of the site beneath. In order to have a better understanding of the distribution of archaeological material prior to actual excavation, a further borehole survey was carried out using a 20 cm diameter hand auger (*megaboringen*) within a 2 x 2 m grid. Results helped distinguish three main zones within the delineated area, prior to excavation (fig. I.2):

(1) the main concentration zone in the southern part of the north-west – south-east orientated coversand ridge, containing extensive and high-density concentrations of archaeological remains comprising occupation debris and anthropogenic features, (2) a 'peripheral zone', defined by the gradually westward and northward sloping flanks of the coversand ridge with a lower-density concentration of archaeological remains, (3) a 'gully zone' defined by the low-lying area infilled with peat, detritus and clay to the east of the coversand ridge, with evidence for a number of gullying episodes and remains associated with human activity.

Zone 1, the main concentration zone, was excavated in 50 x 50 cm grid cells, in stratigraphic units with a maximum thickness of 4 cm. The archaeological layer was systematically sieved over a 2 mm mesh. In Zone 2, a random sample of 2 x 2 m grid cells was excavated in the same way as Zone 1. The lower-density concentration of archaeological remains recorded in the core samples was excavated in 10 x 10 m grid cells.[741] The entire peripheral zone was finally machine-stripped in order to check for the presence of low-resolution and deeply buried remains. A 30 m long and 4-6 m wide trench was excavated across the flank of the coversand ridge and the gully in order to check for archaeological remains. Here again, archaeological remains were excavated by hand within grid cells of 50 x 50 cm, in 4 cm-thick stratigraphic units. Practical considerations meant that the rest of Zone 3 was excavated by machine.

Specialist research was also carried out in the following areas: geochemistry. lithology, soil, arthropods, diatoms, palynology, archaeobotany, archaeozoology, pottery, lithics, flint, and GIS spatial analysis of the distribution of finds.

---

740 This data code refers to the digital archive of the original project data on the website of DANS (Data Archiving and Networked Services).

741 Peeters 2007.

Geological and soil context

The site lies on a north-west – south-east orientated Late Pleistocene coversand ridge. Within the study area, the ridge gradually slopes westward and northward. To the east the ridge flank slopes steeply into a low-lying gully zone (fig. I.3). The coversand ridge forms part of a vast coversand landscape cut through by a number of rivers and streams that provided drainage for the Gelderse Vallei and the eastern Veluwe. Of these, the precursor of the river Eem was the most important.

Stratification within the coversand in the study area is made up of a succession of older and well-sorted younger coversand deposits intercalated with loamy soil horizons formed during the Bølling en Allerød interstadials. Evidence for peat formation in the Preboreal was found in the low-lying area to the east of the ridge, where, as a result of dryer conditions, the upper layers of the peat horizon had been exposed to oxidation and bioturbation (a process that probably took place in the Boreal and/or Early Atlantic). In the same period, probably as a consequence of human activity, localised erosion occurred on the flank of the ridge, resulting in a localised redeposition of sand into the low-lying area. A brown forest soil which could support a (dense) dry woodland and scrub zone developed on the top of the coversand.

During the Middle Atlantic, increasingly wetter environmental conditions due to rising water levels led to the formation of peat in the low-lying area and the deposition of detritus in the stagnant water. Towards the end of the Middle Atlantic, tidal activity cut a gully through these previously formed peat and detritus horizons in the low-lying marshland. In the same period, between 5000 and 4900 cal. BC, hydrological conditions triggered erosion of the more elevated coversand land surfaces. These areas were by now surrounded by expanding peat bogs. Under the influence of major fluctuations in the groundwater level, the brown forest soil transformed into a hydromorphic soil with an A-E-B-C profile. The continuing rise in the groundwater level led to more and more oak trees dying, suggesting increasing starvation due to flooding. By around 4500 cal. BC, even the most elevated land surfaces of the coversand ridge were completely overgrown by peat-forming vegetation.

After a period of relative environmental stability, which is evidenced by clay and detritus deposition in the low-lying area at the beginning of the Late Atlantic, tidal processes caused renewed gullying and the opening up of the low-lying marshland area to the east of the coversand ridge. Under the prevailing brackish conditions the new gully gradually infilled with clay and detritus deposits. Inundation of the landscape led to the now submerged coversand ridge being covered by a layer of clay. Decreased tidal activity meant that renewed detritus deposition occurred in the gully triggering peat growth. Subsequent dryer surface conditions led to the colonising of the area by a birch-dominated marsh woodland.

Dating/phasing

On the basis of ^{14}C-dates, traces of human activity in the study area can be dated between c. 6900 and 4200 cal. BC. Four different phases of activity can be distinguished, two (phase 1 and 2) dating to the Mesolithic and two (phases 3 and 4) in the Early Neolithic (fig. I.4):

Phase 1 can be dated in the Middle Mesolithic, but cannot be defined in any detail due to the limited number of dates or datable finds available.

Phase 2 dates to the Late Mesolithic and ends with the erosion of the coversand ridge between 5000 and 4900 cal. BC. This phase is defined by a number of ^{14}C-dates from pit hearths. The youngest dated pit hearth attributed to this phase was situated very close to the oldest dated surface hearth attributed to phase 3. The large-scale erosion of the Mesolithic land surface at the end of phase 2, the end of the Middle Atlantic, resulted in the 'decapitation' of the top layers of the pit hearths.

Phase 3 follows shortly after the end of the second phase and coincides with the Neolithic (Early Swifterbant). This phase of human activity begins after the earlier erosion of the land surface (see above) and ends with the saturation of the coversand ridge and its complete overgrowth by peat-forming vegetation around 4500 cal. BC.

Phase 4 involves a short period of human activity in the youngest dating gully and dates to the Neolithic (Classical Swifterbant), between 4300 and 4200 cal. BC.

Finds, features and cultural context

Phase 1 is represented by one Late Mesolithic deep pit hearth and possibly some flint material, including a number of microlithic triangular points. It is possible that the erosion identified on the eastern flank of the coversand ridge, which resulted in the redeposition of sand on the slope and low-lying area, can be associated with this phase and is the result of human activity which involved disturbance of the existing vegetation cover.

Phase 2 appears, on the basis of ^{14}C-dates, to be archaeologically characterised by deep hearth pits containing almost exclusively charcoal. However, because of the uncertainty with regard to the starting date of this phase, it cannot be ruled out that a number of the undated deep hearth pits should either be attributed to phase 1, or even that the combined archaeological remains actually represent one single phase of human activity. There are, however, strong stratigraphic and archaeological arguments to justify dating all the deep hearth pits (approx. 100) excavated on the coversand ridge to the Mesolithic. Such phenomena are typical for the Mesolithic in the northern Netherlands, where concentrations of tens or hundreds of deep hearth pits have been found. Dates younger than 5000 cal. BC have rarely been confirmed. The effects of erosion means that it is also not possible at present to exclusively attribute flint artefacts to phase 2. The same problems of attribution exist in relation to a number of quartzite flakes that were found buried relatively deeper in the coversand.

Phase 3 (Early Swifterbant) differs in many respects to the preceding phase or phases. Deep hearth pits are absent. Instead, a total of approx.120 surface hearths have been excavated on the coversand ridge. These surface hearths are associated with large concentrations of flint and other lithics, burnt bone (from mammals, birds, fish and including some human remains), charred hazelnut shells and other seeds, as well as pottery. The surface hearths were found on the higher parts and on the slightly lower, eastern flank of the ridge. Post holes and stake holes were also found in this zone, in some cases with the remains of wood still preserved *in situ*. Furthermore, a possible water hole was found as well as a pit with basketry imprints in association with natural, unfired clay.

Three flint hoards, interpreted as ritual depositions, and a wooden 'platform' consisting of four oak trunks, were found within a peat layer in the western and northern peripheral low-density zones of the coversand ridge. Three heavy oak posts, driven into the sand, and unburnt bone including two aurochs skulls were found on the eastern flank. More unburnt bone, including a third aurochs skull and large sherds of pottery, were found in a humic clay layer at the bottom of the Middle-Late Atlantic gully. The unburnt bone remains include a few dozen tools, mainly T-shaped antler mattocks, and associated production waste. The pottery is attributed to the early phase of the Swifterbant Culture. There is no evidence of cultivated crops or domesticated animals, with the exception of dog.

Phase 4 (Classical Swifterbant) is represented by stakes and wickerwork fragments from three fish weirs and associated fish traps. These were found in the top of the clay deposits in the youngest gully, in the transition between the clay and the overlying detritus. A fragment of a wooden paddle and a large pottery sherd were also found in the clay. Although hardly diagnostic, the pottery sherd is considered to be Swifterbant pottery.

Palaeoecological and palaeogeographical context

The long history of human activity within the study area runs parallel with dramatic changes in the landscape. This implies that the four different phases of human activity that have been chronologically distinguished on the basis of stratigraphic evidence should also be placed in different palaeoecological and palaeogeographical settings (fig. I.6).

Phase 1 coincides with a period in which the study area was located at a distance far from the coastline. The dry coversand ridge was dominated by a deciduous forest possibly interspersed with open spaces that may or may not have been the result of human activity. The eastern low-lying area was still fairly dry, possibly with a low-dynamic watercourse flowing through the lowest lying levels which were situated outside the study area.

Phase 2 occurred in a period in which the study area was subject to increasingly wet conditions as a result of the relative rise in sea level. Deciduous woodland still grew on the coversand ridge and the water table was still low enough to be able to light fires in the deep hearth pits. A marshy wetland environment had, however, already formed in the eastern low-lying area, with open stagnant water here and there.

Phase 3 is preceded by a relatively abrupt increase in the dynamic processes affecting the eastern low-lying area, resulting in the formation of a new gully under the influence of tidal activity. From this point on, the low-lying area became permanently water-carrying. The increasingly wetter conditions led to an expansion of marshy wetland habitats in other parts of the landscape surrounding the coversand ridge. The rising groundwater level meant that wetland vegetation zones dominated, even on the coversand ridge. Only moisture-tolerant species, such as bog oak communities, could survive in these conditions for any length of time.

Phase 4 coincides with a period in which wetland conditions dominated on a large scale. The coversand ridge was no longer visible in the marshland landscape, but may well have been recognisable in the vegetation. The active tidal gully was the most important dynamic factor in the area, but with the changing course of the gully the dynamics decreased and some land formation took occurred.

Taphonomy

In relation to the landscape evolutionary processes, the archaeological remains have not all been exposed to the same taphonomic processes. The remains from phases 1 and 2 in particular have been strongly influenced by pedological and physical processes. The predominantly dry conditions in the Mesolithic were detrimental for the preservation of any organic remains that may have been left behind on the coversand ridge. Only burnt/charred organic remains have been preserved. Whether or not other organic remains were originally present can no longer be confirmed. In any case, the erosion phase at the end of the Mesolithic resulted in the destruction and removal of the original land surface on the ridge, as well as the top layers of the deep hearth pits.

The wetter environmental conditions that existed during the periods of human activity attributed to phases 3 and 4 meant that organic remains from these phases have been better preserved. The rising groundwater level also ensured partial preservation of wooden stakes and posts *in situ* on the ridge. Surface hearths and other structures were also well preserved. Bonel remains recovered from on the ridge itself are almost without exception heavily burnt/calcined, whereas the remains recovered from the eastern flank and gully bottom are almost without exception unburnt. It is not exactly clear whether this contrast between the two assemblages is a result of differential preservation, whereby the unburnt bone material on the ridge has long since disintegrated as a consequence of chemical and biological processes, or whether human behaviour played a role. It is striking that no gradations in burning, from light to heavily burnt, can be recognised amongst the remains whilst evidence for this might be expected, even under differential preservation conditions. It is very possible that the picture, at least in part, reflects evidence for the deliberate burning of all sorts of refuse categories. The assemblage of unburnt bones recovered from the ridge is made up exclusively of fish bones. These were found directly below the deposits of peaty layers and probably represent a natural accumulation.

Tidal activity that led to the formation of new gully channels at different moments during the transition from the Mesolithic to the Neolithic and in the Early Neolithic also led to the clearing away of sediment layers that had been deposited in the low-lying area throughout the different phases of human activity. The stratigraphy along the coversand ridge therefore varies considerably and the different phases are unevenly represented. Only the younger units of phase 4 are present everywhere. These relate to the tidal gully channel and the successive terrestrialisation. This means that any refuse layers from the preceding phases of human activity have not been preserved everywhere.[742]

---

742 Peeters 2007, 89.

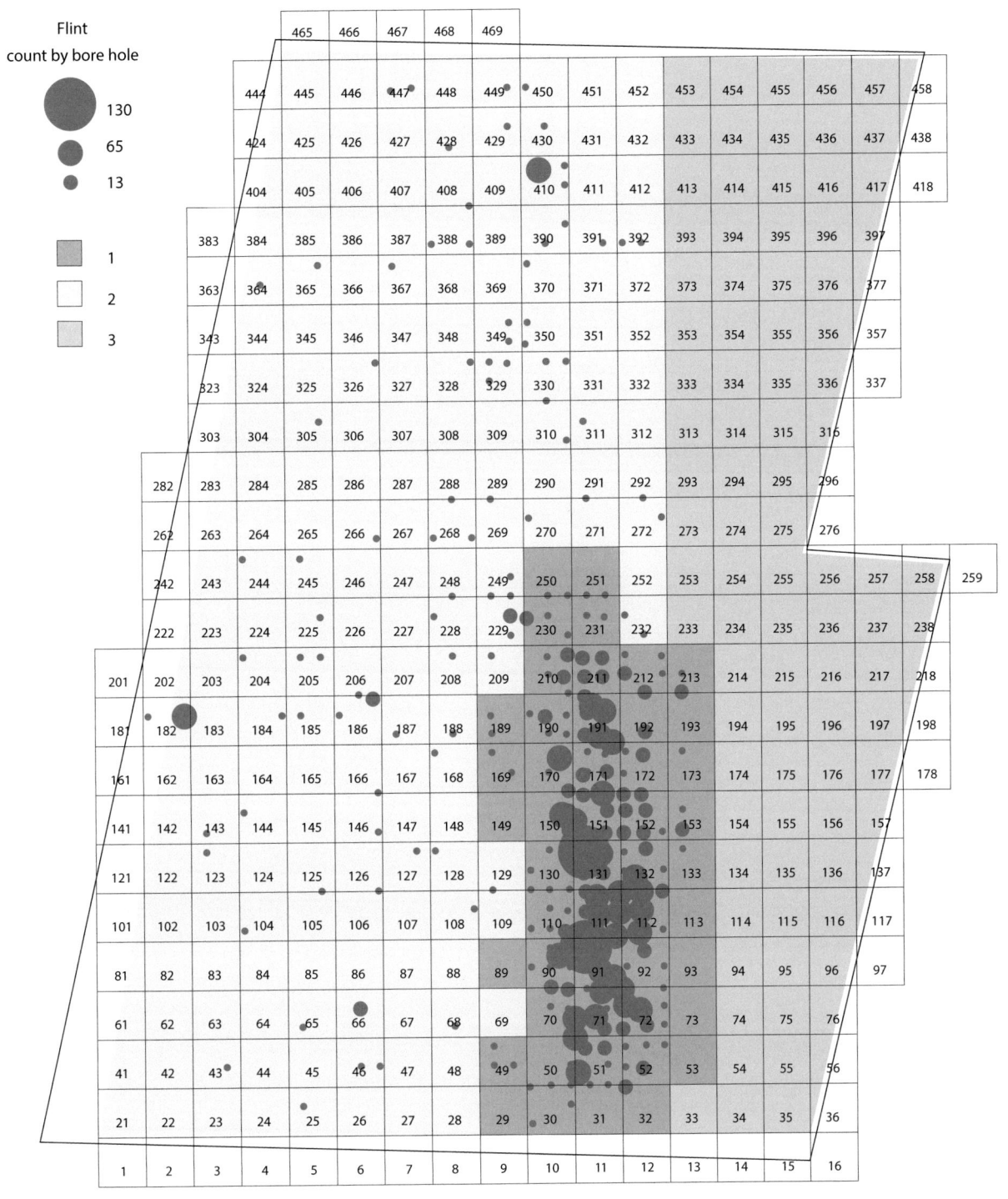

Figure. I.2: Zoning of research area of Almere Hoge Vaart/A27, 1: 'main concentration', 2: 'pheripheral zone', 3: 'gully zone'.

APPENDIX SITE ATLAS WINDOWS OF OBSERVATION

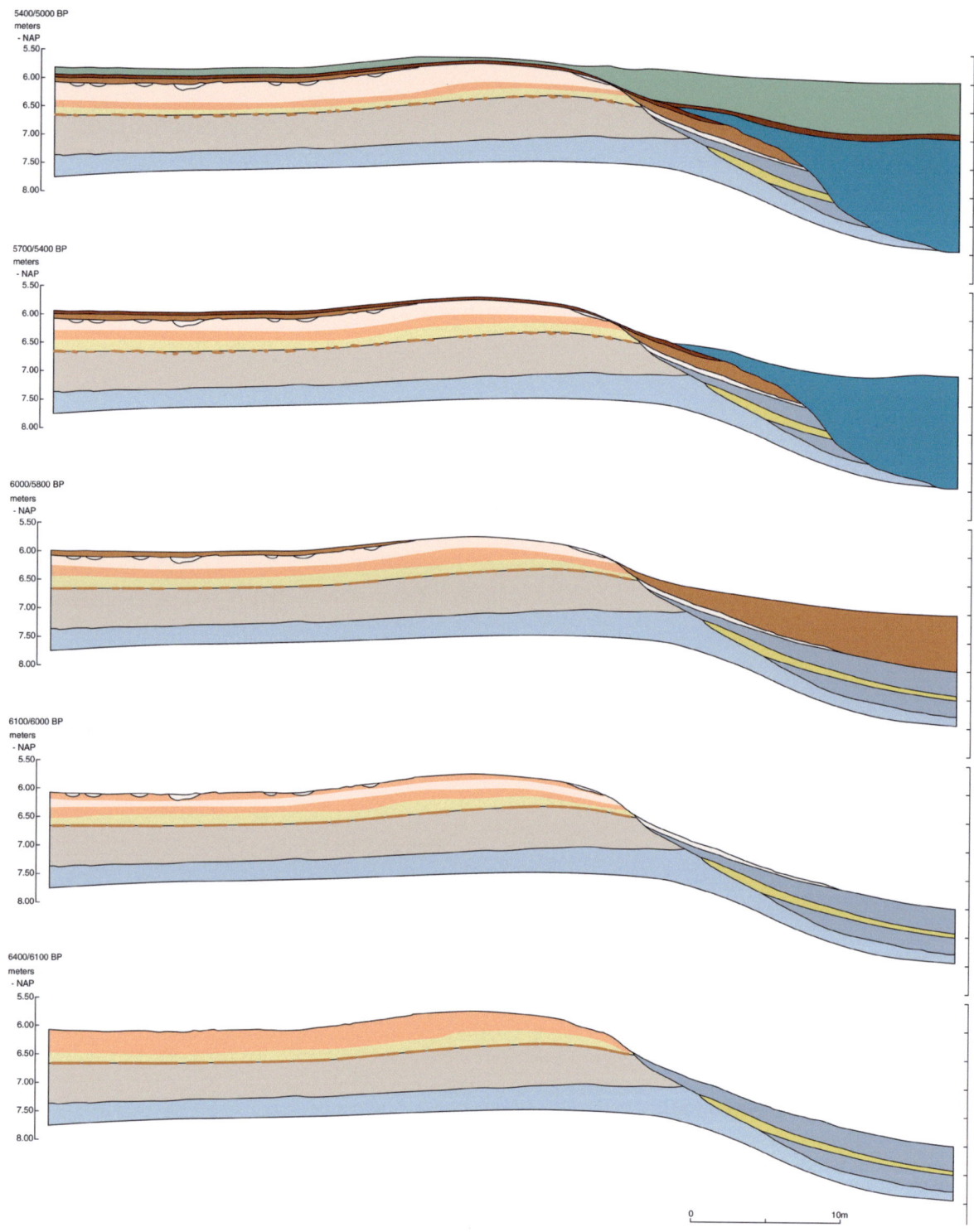

Figure. I.3a: Geological/pedological profile of the Almere Hoge-Vaart A27 (orientation East-West).

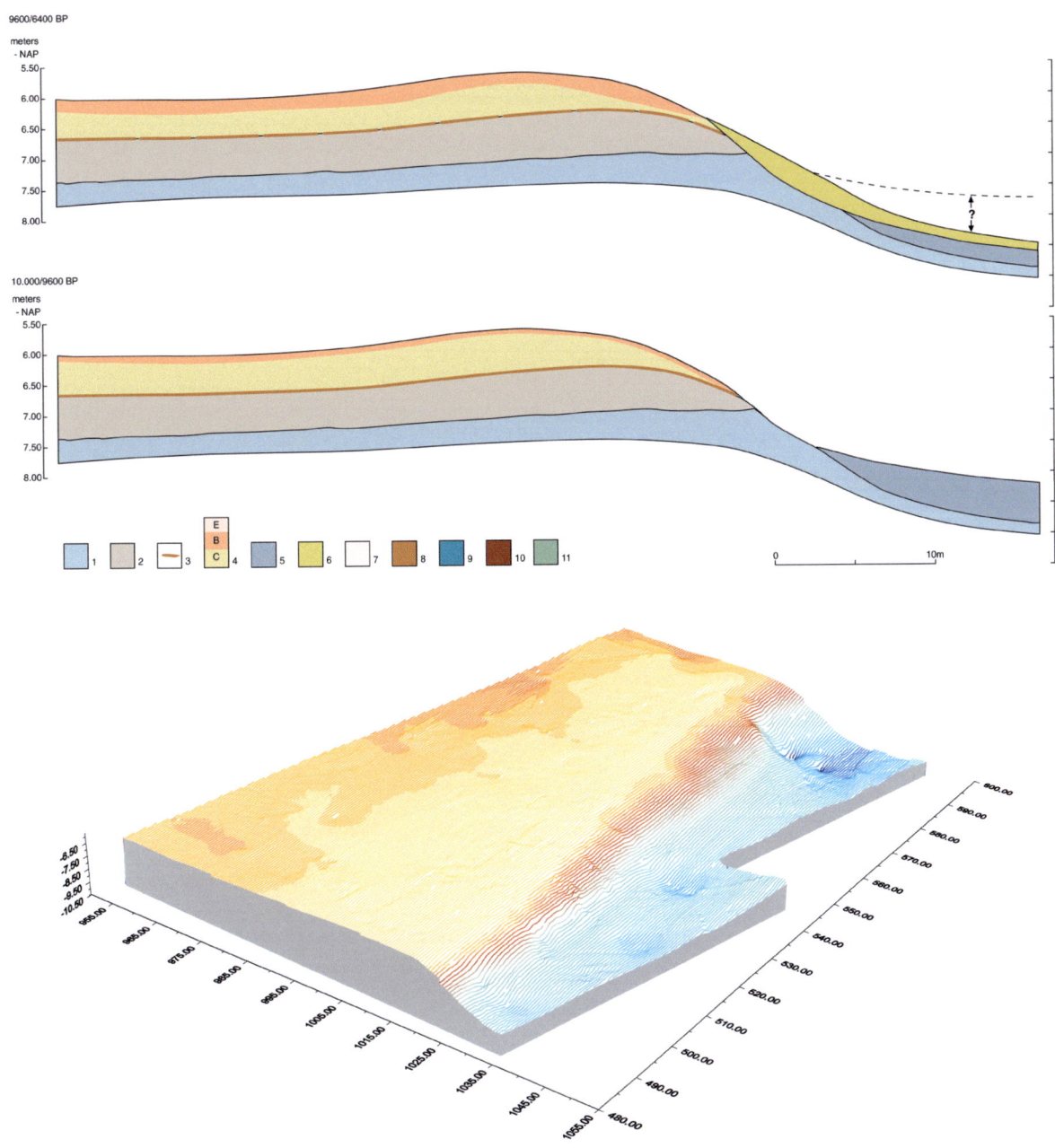

Figure. I.3b: Geological/pedological profile of the Almere Hoge-Vaart A27 (orientation East-West).

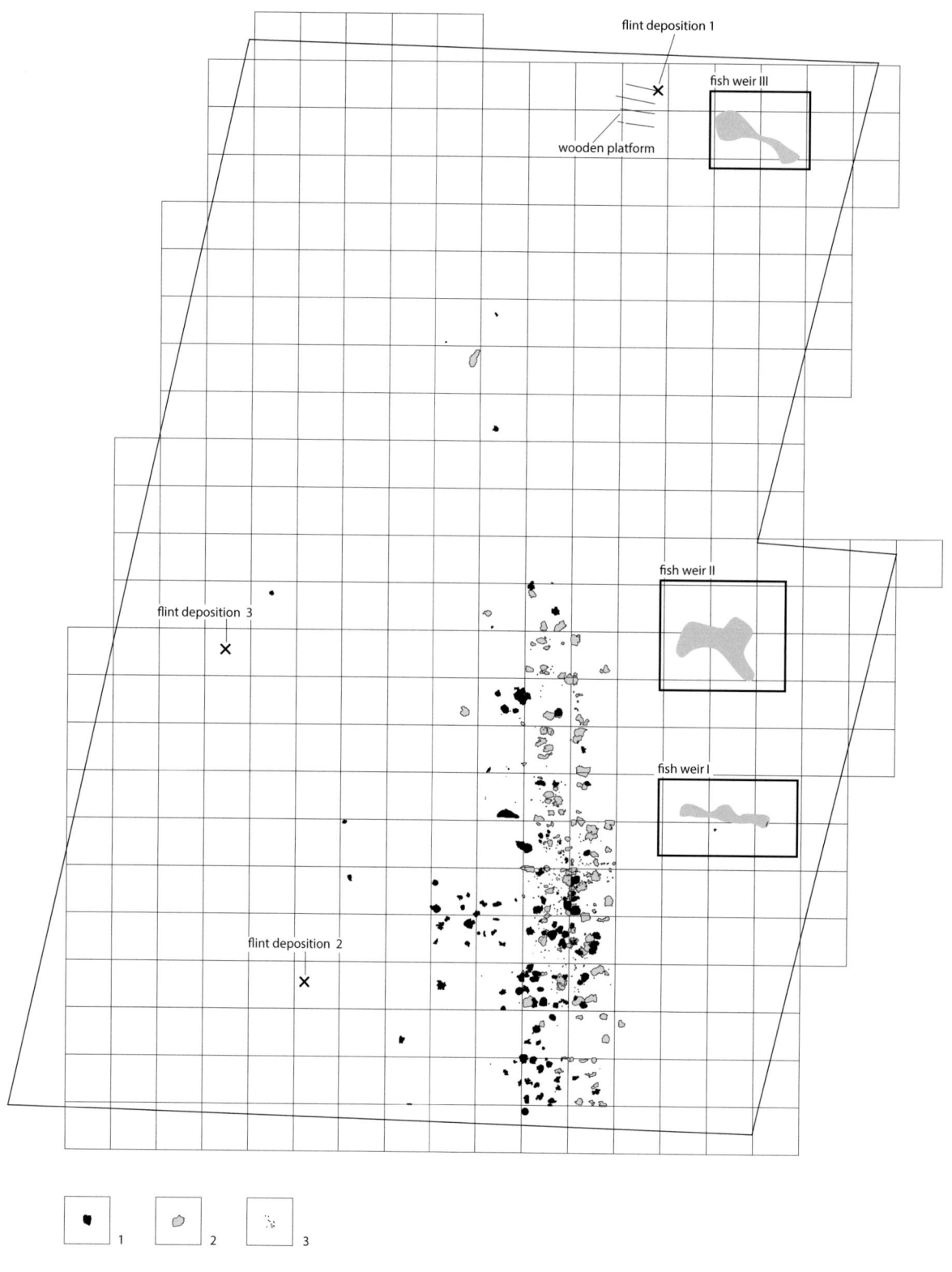

Figure. I.4 Map of all features: 1. Deep pits; 2. Shallow pits; 3. Postholes.

01-13-hv .fh8

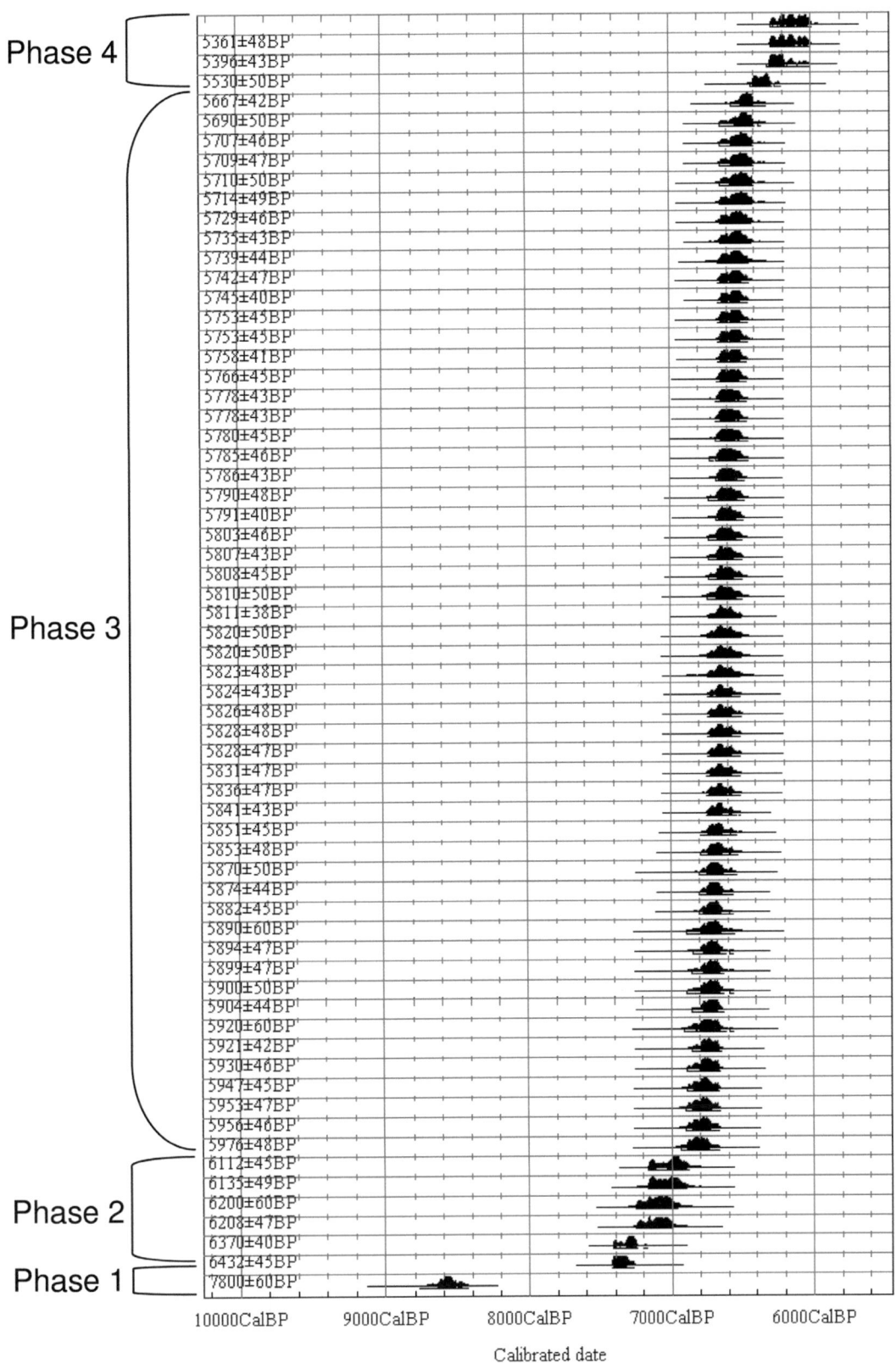

Figure I.5 AMS dates for Almere Hoge Vaart/A27 (source: Peeters 2007).

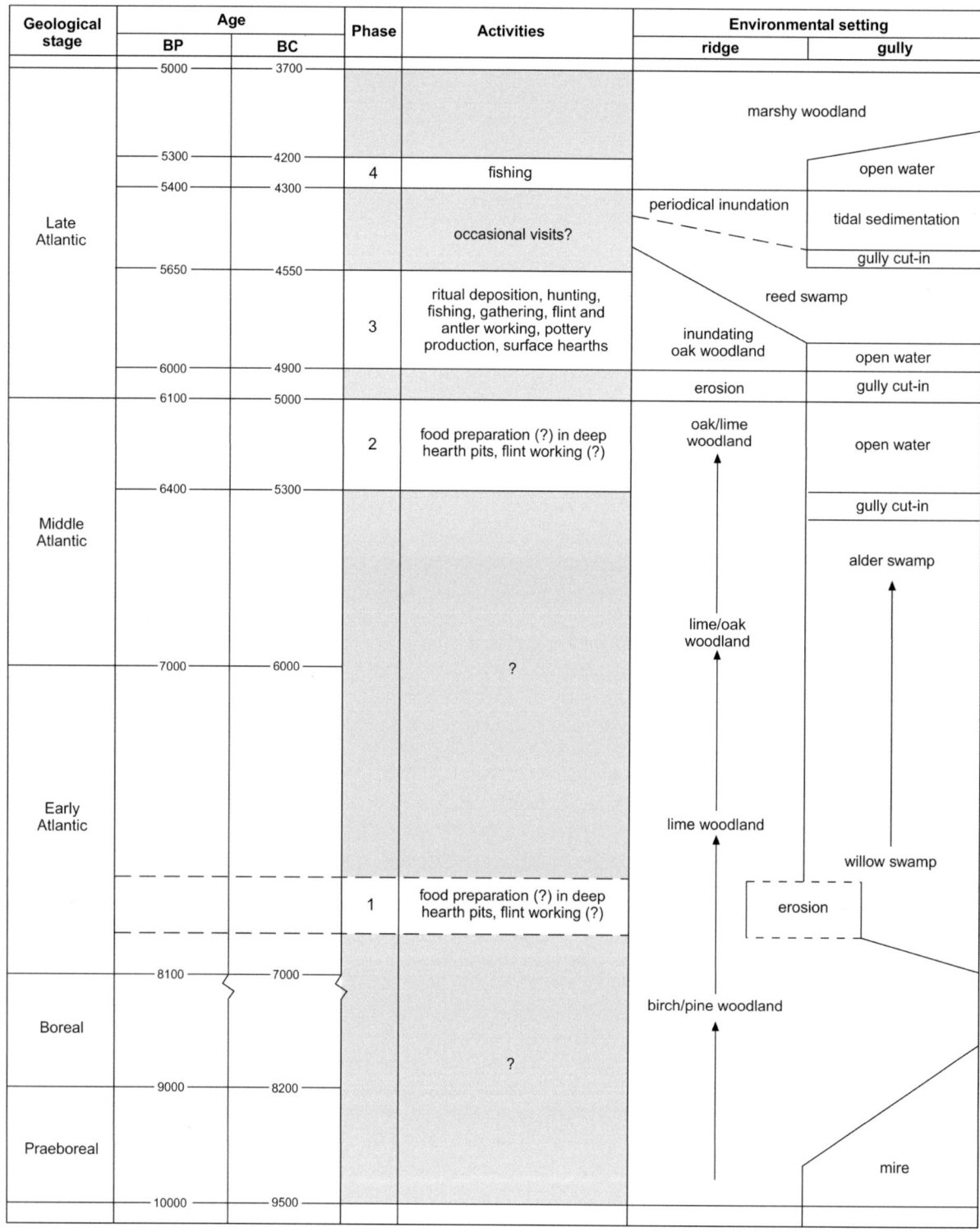

Figure I.6 Period diagram by phase for Almere Hoge Vaart/A27.

I.2.2 Almere – Europakwartier Site 7

References
Publications: Raemaekers *et al.* 2003; Hogestijn & Visscher 2011.
Data: urn:nbn:nl:ui:13-86i-cq4 and urn:nbn:nl:ui:13-fc7g-j6

Geographical location and research
Almere – Europakwartier Site 7[743] is one of several sites from which archaeological remains were discovered during borehole surveys carried out in the Europakwartier quarter within the district of Almere Poort, in the southwest of the municipality of Almere. Research into these sites formed an integral part of a much more extensive archaeological field evaluation that was carried out prior to construction work in an area of approx. 46 hectares planned for housing development.[744] An exploratory borehole survey was carried out across the entire development area between 1999 and 2001 in order to gain a better understanding of the geomorphology of the buried Pleistocene landscape. Data was collected from a total of 193 borehole cores within a 50 x 40 m grid using a 7 cm hand auger (*Edelman*) and a 3 cm hand gouge auger (*guts*). As well as landscape data, which identified the presence of two coversand ridges buried under layers of sediment and deposits, archaeological remains found in the borehole cores were also analysed.

Different phases of borehole mapping surveys, planned in accordance with construction work in the area, were carried out in order to determine the nature and extent of the potential archaeological sites. As in the previous exploratory phase, these surveys were also carried out using hand augers, but now within a closer-spaced 40 x 25 m grid in the southern part of the area and within a 20 x 25 m grid in the north. A total of 179 borehole cores were taken. Core samples from the top of the Pleistocene coversand were sieved over a 1 x 1 mm mesh. Additional evaluation in the northern part of the area was carried out consisting of mechanical boreholes within a 20 x 25 m grid using a mechanical screw auger (Avegaar) with a 14.5 cm diameter. The samples from the top of the coversand were again sieved over a 1x1 mm mesh. Particular attention was paid to the location where red ochre had previously been found in the coversand during the exploratory evaluation phase[745].

Numerous findspots were identified during the different phases of field evaluation due to the presence of archaeological remains within the borehole cores. Further intensive borehole investigations were carried out on three clusters of findspots on the northern coversand ridge as part of a valuation phase to determine the exact size and boundaries of the identified sites. Initially, 51 boreholes were sampled, some using a mechanical gouge auger (guts) with a diameter of 6 cm and others with a hand gouge auger with the same diameter. Core samples from the top of the coversand were then sieved over a 1x1 mm mesh. An additional 9 boreholes were made in Findspot Cluster 1 using an auger (Edelman) with a diameter of 12 cm. Core samples from the top of the coversand were again sieved using the 1x1 mm mesh.

The different field evaluation phases were subsequently extensively assessed. As a consequence, a number of identified sites were subject to another follow-up valuation phase. This included the location where red ochre had been found in previous core samples: Site 7. A total of 68 new boreholes were made in a grid of 2 x 1 m around the original ochre findspot using a hand gouge auger with a diameter of 3 cm. Once again, the core samples from the top of the coversand were all sieved. Further evidence for ochre was found in core samples from a number of these boreholes. Another 31 boreholes were made using the hand gouge auger in an area of 150 x 80 m around the findspot. The resulting core samples were again sieved over a 1 x 1 mm mesh. In a partially overlapping area of 50 x 75 m another 50 samples were taken using a mechanical screw auger (Avegaar) with a diameter of 14.5 cm. The samples were also systematically sieved over a 1 x 1 mm mesh. Although the location and surroundings of Site 7 were subject to intensive investigation during the different evaluation phases, utilising a number of different augering techniques and methods, no final excavation test pits or trial trenching was done. Instead, the integration of the location into the development plans for the area meant that the findspot could be physically protected and remain undisturbed *in situ*.

Geological and pedological context
The geomorphological evidence from the borehole surveys indicated the presence of two buried coversand ridges in the Europakwartier development area. The ridges were both orientated east-west and were separated from each other by a 1.5-2 m deep depression. The highest point of both ridges reached approx. 6 m -NAP/below average sea level. Site 7 was located on the northern slope of the most southerly coversand ridge, at a height of 7.15 m -NAP/below average sea level. Evidence from core samples indicates the presence of an intact podzol soil. Other areas of the coversand ridges indicated evidence for A/C soil profiles.

The site had been covered by a diachronous bed of Basal Peat growth a few decimetres thick. This Basal Peat

---

743 The RAAP site numbers have been renumbered in the report by Hogestijn *et al.* RAAP site no. 28 corresponds with Hogestijn *et al.* site 7.
744 The different phases of research were performed simultaneously in some parts of the area. Hogestijn & Visscher (2011, 14-24).
745 XRD analysis has shown this to be hematite, also known as red ochre.

Figure I.7: Section of a burial pit with red ochre on the bottom discovered at Mariënberg (source: Verlinde & Newell 2013).

in turn was covered by a 70 cm thick layer of tidal deposits (Wormer Member), on top of which were respectively Almere, Zuiderzee and IJsselmeer tidal and marine deposits (Naaldwijk Formation).

Dating/phasing
On the basis of the elevation of the site and the relative sea level curve, it can be established that the archaeological remains must predate c. 4800 BC.[746] There is no further data available to allow a more accurate dating of site.

Finds, features and cultural context
Finds recovered from the borehole samples included worked and unworked flint, charcoal and charred hazelnut shells. Red ochre was found in four boreholes, in core samples from the top of the coversand at depths varying between 3 and 42 cm below the top of the coversand. The data suggests that the ochre flecks found on Site 7[747] may be an indication for the presence of a burial (fig. I.7), like the the finds discovered in Mariënberg.[748] Charcoal was found in numerous borehole samples, sometimes in relatively large quantities. The spatial distribution pattern of the charcoal did not seem to have any relationship to the distribution pattern of ochre. Across the site, evidence for a patchy, mottled horizon (dirty layer), with a thickness varying between 15 – 45 cm, was found in a number of boreholes. It could not be determined whether this patchy, mottled layer was of anthropogenic or natural origin.

Palaeoecological and palaeogeographical context
The research design for the field evaluation was not focused on collecting data in order to place the site in a palaeoecological or palaeogeographical context.

Taphonomy
An intact podzol soil was identified on top of the coversand and under a bed of Basal Peat in two of the four boreholes containing red ochre. There appears to be no evidence of erosion of the top of the coversand deposits. A so-called A/C profile was identified in the other two boreholes which suggests some erosion took place, although the extent of this is uncertain. The origin of the "dirty layer" identified on the site are not clear. This may have been formed as a result of anthropogenic activity (such as ground disturbance or trampling), or as a result of natural phenomena such as surface erosion, for example, due to sediment runoff caused by flowing water.

---

746 Dates based on the relative sea level curve from Van de Plassche et al. (2005). .
747 Hogestijn & Visscher 2011.
748 See: Verlinde & Newell 2006, 2013; Louwe Kooijmans 2013 .

I.2.3 Almere – Zwaanpad

References
Publications: Raemaekers 2000; Cohen Stuart *et al.* 2006; Roovers *et al.* 2011. Niekus *et al.* 2012.
Data: urn:nbn:nl:ui:13-l0a-96s

Geographical location and research
The site of Almere-Zwaanpad lies in the southeast of the municipality of Almere in the Almere Hout district. As with other sites discovered in the municipality, Almere-Zwaanpad has a long research history comprising several consecutive phases of fieldwork. The first phase of archaeological prospection on the location was carried out in 2000 in order to test a predictive model for the archaeological potential within the development area of Almere Hout.[749] A single row of 83 boreholes spaced 25 m apart was made using a hand gouge auger (guts) with a 3 cm diameter. One of the outcomes of the survey was the realisation that the buried Pleistocene coversand landscape actually exhibited far more relief than had previously been assumed. Additional phases of borehole evaluation were then carried out in order to gain a more accurate picture of the coversand relief and the stratigraphic sequence in the soil profile. These phases utilised a mechanical sonic drill with an Aqualock piston type core sampler within various grid systems: a 100 x 80 m grid in the low-lying areas of the buried landscape and a 40 x 50 m and 20 x 25 m triangular grid for the more elevated parts of the landscape.

This more detailed evidence for differentiation in the buried Pleistocene relief provided the basis for the selection of a sample of 45% of the study area for further field evaluation. This was carried out using a larger diameter (14.5 cm) mechanical screw auger (Avegaar), within a closer-spaced grid varying between 40x50 m and 12.5x10 m. In addition, the mechanical screw auger was used to sample a few separate boreholes and some extra rows of boreholes across the study area. In each evaluation phase core samples from the top of the coversand were systematically sieved over a 1x1 mm mesh in order to recover any archaeological remains.

In 2003 a small excavation was carried out in cooperation with local amateur archaeologists on the location where archaeological evidence had been found in core samples from a depth of approx. 2 m below ground level. (fig. I.8). The excavation was divided over two small trenches and carried out in grid cells measuring 1 m² in 5 cm-thick stratigraphic layers, covering a total area of 30 m². The grid cells were partly excavated using trowels and partly using spades. Core samples from the coversand were systematically sieved using a 1x1 mm mesh, whilst the samples from the overlying detritus deposits were sieved over a 2x2 mm mesh. After the excavation, analysis was carried out on collected micromorphological and pollen samples.

Following on from the excavation, areas of the buried coversand ridge were subject to two further mechanical borehole surveys. Firstly, the 14.5 cm-diameter mechanical screw auger (Avegaar) was used within a closer-spaced grid of 5x6 m. Secondly, the sonic drill Aqualock was used for a double row of boreholes, with two boreholes per location, each with a diameter of 7.7 cm. The twin boreholes ensured a double quantity of sediment could be obtained from the core samples for further research.[750] It can be concluded that, although this site has a long history of research of consecutive phases, the most information regarding archaeological remains came from the excavation.

Geological and pedological context
Several coversand outcrops have been identified in the substrate under the Almere Hout development area. The site of Almere-Zwaanpad is located on the eastern flank of a more or less east-west orientated coversand ridge. The height of the top of the coversand varies between 5.75 and 10 m -NAP/below average sealevel.

Podzol had formed on the higher parts of the coversand landscape. This was then covered by deposits of peat or fine detritus which, in turn was covered by Almere and Zuiderzee marine deposits.

Dating/phasing
Several ¹⁴C-dates are available for this site (figure I.9). The dates cover a wide date range: 7734-4491 cal BC. It is not clear whether a part of the dated material can be attributed to the period of human activity on the site. Two dates from charcoal taken from a hearth pit place some human activity on the site in the Middle Mesolithic, between 7062-6661 cal. BC.[751]

Finds, features and cultural context
A large quantity of archaeological remains were recovered from both the various borehole surveys and the excavation. Those from the excavation were studied in more detail. The finds assemblage included more than 5000 fragments of worked and unworked flint, several fragments of sandstone, bone (fish remains), charred hazelnut shells, pips and seeds and a large amount of charcoal. An important observation is that all the flint artefacts, including so-called micro-triangles, are smaller

---

[749] Raemaekers 2000.

[750] Cohen Stuart *et al.* 2006.
[751] Niekus *et al.* 2012: dating from hearth pit GrN-28888: 8000+/-50 BP, UtC-12794: 7930+/-50 BP (weighed average 7965+/-35 BP).

than 4 cm in length. These finds were concentrated around the hearth pit.

Palaeoecological and palaeogeographical context

Evidence suggests the presence of a light, open deciduous forest around the site at the time of Mesolithic activity. The exclusive occurrence of fish remains indicates that open water would have been present in the vicinity. Expanding wetter conditions led to the formation of carr in the lower-lying areas of the landscape. The sand ridge finally became overgrown with peat around 5743 +/- 44 cal. BC.

Taphonomy

Micromorphological research indicates that the uppermost 14 cm of the coversand ridge surface had been exposed to trampling and root-turbation, resulting in charcoal particles ending up in the deeper sand layers. Erosion hardly seems to have taken place and most artefacts were probably still *in situ*. The fish remains comprise both burnt and unburnt material. If we accept that the ^{14}C-dates from the hearth pit charcoal samples are a reliable indicator for the period of human activity on the site, then it is unlikely that the unburnt fish remains are related to any Mesolithic activity. It would have taken at least another 1000-1500 years before the local environmental conditions were such that unburnt fish remains could be preserved. It is therefore far more probable to assume that the unburnt fish remains were the result of natural processes and that they originate from a later phase of sedimentary deposition.

Figure I.9 (opposite page, top): AMS ^{14}C dates for Almere/Zwaanpad.

Figure I.10 (opposite page, bottom): Selection of artefacts from the Zwaanpad site: 1-3. cores; 4-5. Retouched flakes; 6-20. (micro-) triangles; 21-22. Lancet points; 23. atypical point; 24-26. Triangular-backed bladelets; 27-28. Point preforms. Key: closed circle = percussion bulb present; open circle = percussion bulb removed / no longer present. An asterisk indicates the artefact is burnt (drawings by L. Johansen).

Figure I.8: Impression of the excavation of the Zwaanpad site.

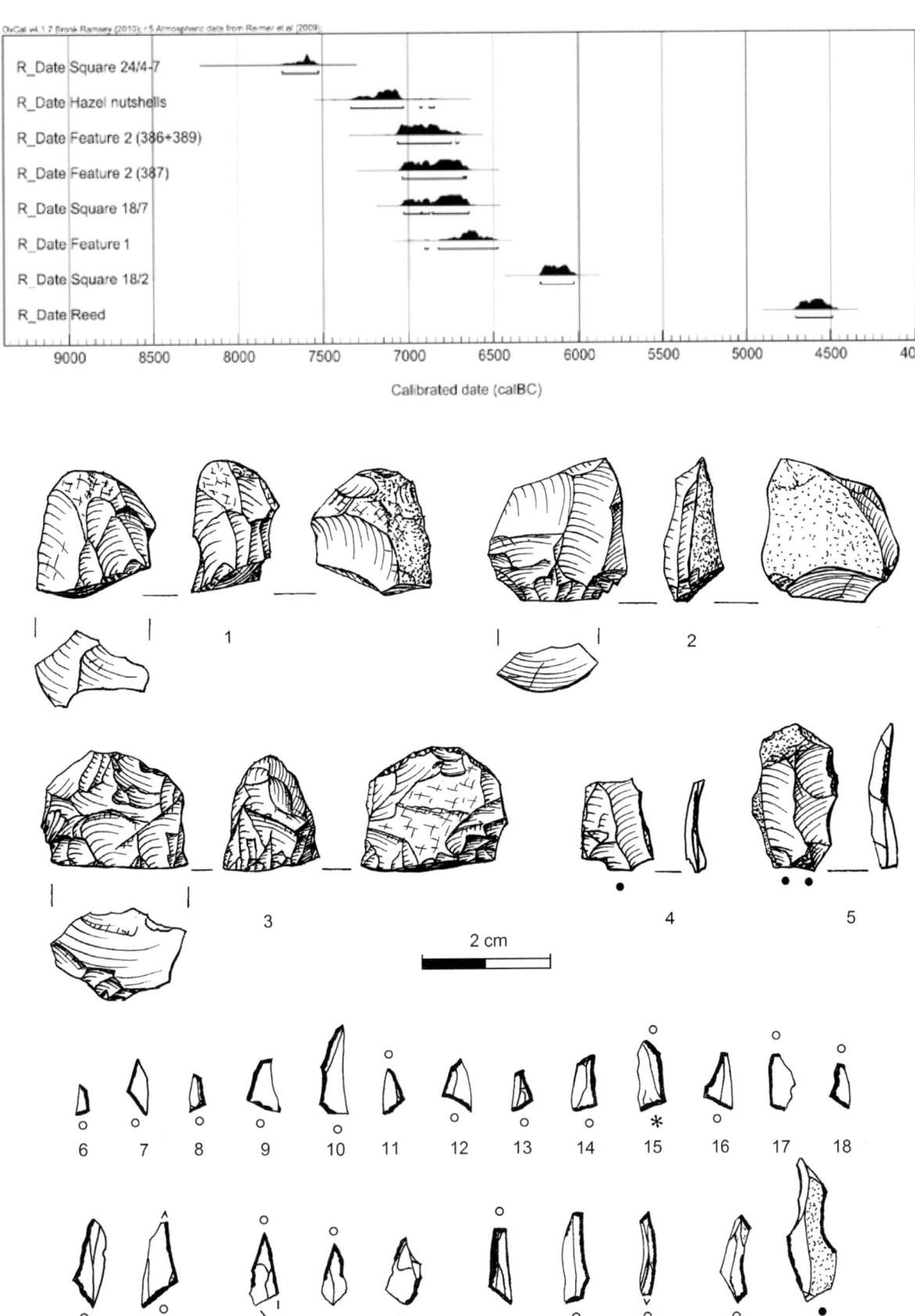

APPENDIX SITE ATLAS WINDOWS OF OBSERVATION

I.2.4 Zeewolde – Oz35/Oz36

References
Publications: Vlierman 1985.
Data: not available

Geographical location and research
The site Zeewolde-Oz35/Oz36 is located in the Hulkenstein Bos woodland in the south of the municipality of Zeewolde. The site was accidently discovered in 1981 during archaeological fieldwork by the State Service for the IJsselmeer Polders (*Rijksdienst voor de IJsselmeerpolders*: RIJP) to investigate the buried remains of a late medieval cog ship. The profile of a gully was exposed in the section of a trial trench across the site. The course of the gully was recorded by RIJP the following year using combined data collected from the palaeogeographical mapping of the sides of ditches and a borehole survey. Amateur archaeologists conducted a fieldwalking survey and excavated several test pits. In 1983, during the excavation of the cog, the course of the gully was mapped further over a length of approx. 500 m. The trial-trenching and excavations for the medieval cog also uncovered prehistoric finds. This assemblage included pottery, flint and a flint axe.

Geological and pedological context
The area around the site shows evidence for variable soil formation processes. This is mostly due to the highly undulating top of the Pleistocene surface, with height variations for the top of the coversand ranging between 0.25 and 2 m below the present surface. The presence of an A-horizon has been recorded over almost the whole coversand surface. In the lower-lying areas this soil had been covered by a peat layer. The peat was, in turn, covered by a 15 – 35 cm thick layer of Zuiderzee sedimentary marine deposits and in some parts by earlier Almere deposits (both Naaldwijk Formation).

The site was located on and in a southwest-northeast orientated gully that formed part of a larger gully system with numerous branches. On the basis of the finds assemblage from the gully fill, it would appear that the gully was active in the Mesolithic and Neolithic and was then partly reactivated in the medieval period.

Dating/phasing
Pottery sherds found *in situ* in the southeastern gully bottom all joined together to make an almost complete pot. The typological characteristics of the whole finds assemblage, combined with a ^{14}C-date for food residue on the pottery means that the pot could be dated to the Middle Neolithic.[752]

Finds, features and cultural context
On typological and technical grounds, the almost complete pot recovered from the excavated gully bottom can possibly be attributed to the (pre-Drouwen) late phase of the Swifterbant Culture (fig. I.11). Several wooden stakes were also recovered from the bottom of the gully. Unfortunately the relationship between the stakes and the pottery cannot be established. Therefore It is not clear whether these stakes can be dated to the same period as the pottery. Several prehistoric pottery sherds were collected from between the wooden stakes. Other prehistoric sherds, as well as a stone axe (a Fels-Ovalbijl, fig. I.12), charcoal and several worked and unworked flints were recovered during the fieldwalking survey carried out in the surrounding area. One of the sherds was decorated with fingernail impressions and another was pierced. Another sherd resembles Funnelbeaker pottery. The flint assemblage included several flakes and a scraper. A number of features were discovered in the coversands along the banks of one of the gullies.[753] These possibly date to the Mesolithic.

Taphonomy
The A-horizon in the coversand appears to have remained intact in some areas. From this it can be deduced that the Pleistocene surface had been either entirely or largely unaffected by modern agricultural practices. The stratigraphic sequence of the gully infill indicated that although the reactivation of the gully in the medieval period probably washed away part of the prehistoric remains, prehistoric deposits still remained *in situ* in the bottom of the gully.

---

[752] Raemaekers 2005, GrN-26612 4660±40 (3624-3602, 3525-3360 cal BC. The delta 13 C value indicates that in this case account must be taken of the reservoir effect.
[753] These features should possibly be identified as hearth pits, but this cannot be done with certainty on the basis of the description in the excavation report alone. The features are also not dated.

I.3 Oostelijk Flevoland

I.3.1 Dronten N23/N307 – Site 5

References
Publications: Mietes & Schrijvers 2005; Tol 2007; Van Lil 2008; Hamburg et al. 2012.
Data: urn:nbn:nl:ui:13-ntxf-4n

Geographical location and research
Dronten N23/N307 – Site 5 (=Dronten N23) lies to the south of Swifterbant and was discovered during archaeological fieldwork prior to the construction of the N307 provincial road. Borehole evaluation surveys using a mechanical sonic drill with AquaLock piston sampler with a 7 cm diameter were carried out within a 25 x 25 m grid at three locations along the planned route of the road. The top 30 cm of the Pleistocene coversand in the cores was systematically sampled and sieved over a 1x1 mm mesh. The fieldwork identified a number of different findspots within the three locations. One of these, Site 5, was threatened with destruction by the planned construction work.

The discovery of charcoal, burnt hazelnut shells and flint in numerous core samples was reason to carry out further evaluation on the location. Because the level with archaeological remains lies between approx. 2.5 – 4.5 m below the present surface meant that the excavation of test pits or trial trenches was not possible, therefore another borehole survey was carried out. The boreholes were set out in a very close-spaced grid based on equilateral triangles with sides measuring 5 m. The survey combined two different techniques. A mechanical sonic drill AquaLock with a 7 cm diameter was used to map the buried coversand relief. A mechanical screw auger (Avegaar) with a diameter of 14.5 cm was used to sample the top of the coversand in order to recover any archaeological remains.

Samples from all 563 boreholes were taken from 15 cm above to 30 cm below the top of the Pleistocene coversand. All these samples were sieved using a 1x1 mm mesh and the residue inspected for archaeological remains. Palynological research was carried out to shed light on vegetation development in the surrounding environment during prehistory. A number of ^{14}C-dates were carried out, not only for the purposes of vegetation reconstruction, but also to date the site. The results from the previous phases of evaluation highlighted the archaeological potential of the site. This potential, combined with an assessment of the destructive impact of the planned construction work, led to the decision to carry out further archaeological mitigation work on the site by means of an excavation.

Because of not only financial, planning, but also complex technical restraints, only a part of the site

Figure I.11: Almost complete pot (photo: D. Velthuizen).

Figure I.12: Fels-Ovalbijl (photo: D. Velthuizen).

could be excavated. The excavation area of 4750 m² was first surrounded by an interlocking sheet pile wall construction. This was driven into the ground sufficiently deep to intercept the groundwater flow. The sheet pile construction was then sealed below ground level by injecting a bentonite waterproofing membrane. (fig. I.13)

Prior to the excavation, five boreholes were made in the study area using a mechanical gouge auger (Begemann). The undisturbed core samples provided detailed information on nature and characteristics of the subsurface and the depth and thickness of archaeological levels. The covering deposits of clay and peat were then machine-stripped to a few centimetres above the archaeological level in the top of the Pleistocene coversand.

The excavation was carried out using a selective sampling strategy. In the first phase of fieldwork, a systematic grid of 1x1 m test pits was excavated in units of 50x50x5 cm. The fill from these units was wet-sieved over a 2x2 mm mesh. The number and weight of different categories of finds from each of the test pits were used to make interpolation distribution maps of archaeological remains for the whole excavation area. These maps were used to select nine zones with concentrations of different degree of archaeological finds, varying in size from 3x3 m to 14x12 m, to be excavated in the second phase of fieldwork. The nine zones were excavated using the same methodology applied in phase 1 (50x50x5 cm units, wet sieving).

The third phase of fieldwork concentrated on the features. The excavation area was divided into trenches 5 m wide and approx. 80 m long. The first 30-40 cm of the top of the coversand was machine-stripped down to the level where the features were visible (top of the C-horizon). The features were all recorded and the fill completely or partially wet-sieved over a 2x2 mm mesh.

During the post-excavation phase, a large amount of specialist analysis was carried out on the collected data, including: analysis of sections and stratigraphic sequences, material analysis, pollen analysis, use-wear and microwear analysis, charcoal analysis, ^{14}C AMS analysis, micromorphological and macrobotanical analysis, etc.

Geological and pedological context

The site is situated on the highest point of a parabolic coversand dune, at a depth of approx. 6.75 m -NAP/below average sealevel. The dune has a U-shaped depression in its centre approx. 1 m deep. The dune was formed during the Late Glacial and is made completely of Young Coversand (Boxtel Formation). A podzol soil developed in the top of the coversand. By around 4800 cal. BC, as a result of the rising groundwater the dune was completely covered with peat (Nieuwkoop Formation).

Dating/phasing

The site dates to the Middle and Late Mesolithic, from c. 8400 to 5300 cal. BC. Over 109 samples were dated using the ^{14}C-method and OSL. These dates can be used to phase the activities on the dune.

The oldest date – 8438-8251 cal. BC – comes from a concentration of charred hazelnut shells found in the central depression in the dune. The oldest pit hearth dates to c. 7900 BC, making it approx. 350 years younger than the concentration of hazelnut shells. This is followed by a period of approx. 1700 years during which the location was visited repeatedly. Around 6800 BC there appears to have been a short hiatus in the occupation, lasting some 50-100 years (it is not possible to conclude on the basis of the dating how long the site remained abandoned). A second hiatus, lasting approx. 200 years, took place around 6200 cal. BC. Interestingly, after this period another type of pit heath occurs, and no more charred hazelnut shells were deposited. After a period of 700 years during which the dune was in use again, there was another break in activity around 5500 cal. BC. The youngest date from a pit hearth is around 5300 cal. BC, marking the end of the use of the site for digging pit hearths.

The final indication of the use of the parabolic dune is an inhumation grave which can be dated on the basis of a combination of dendrochronological information, the relative age based on crosscutting (by the grave) of a pit hearth and an OSL dating, to around 5000 cal. BC.

Finds, features and cultural context

The archaeological remains at Dronten N23 consist of flint, lithics, charred hazelnut shells, pit hearths, pits and one inhumation grave. Over 100,000 flints were found. The flint assemblage was deposited over a very long period of repeated use, and is thus a palimpsest of approx. 3000 years. Nevertheless, the typological features of one of the flint concentrations can clearly be linked to the 'Rhine-Meuse-Scheldt complex', and in geographical terms therefore represents the most northerly occurrence of this complex.

The large quantity of stone processing waste found is fairly unusual for this period. As far as is known, no stone processing waste has been found at most other sites from the same period.[754] Use wear analysis of stone artefacts has shown, for the first time at a Mesolithic site in the Netherlands that natural stone was used to grind nuts (hazelnuts), as well as for grinding wood, bone or antler.

The majority of features were pit hearths; a total of 772 were found. It was possible to distinguish three different types of pit hearth, on the basis of their shape and size.

---

754 It could of course be that the processing waste has not been recognised as such at other sites; the focus tends to be on flint.

Carbon dating has shown that one of the types (type HAKB) did not occur until after 6000 cal. BC. (fig. I.14 and fig I.15)

One notable point is that the pit hearths and the flint are not associated in either a spatial or a temporal sense. They appear to relate to two separate activities that took place at different points in time. The distribution of charred hazelnuts shells can however be linked to the occurrence of flint.

As well as pit hearth, a further 20 pits were found which, given the absence of charcoal in their backfill, have not been interpreted as pit hearths. It was not possible to identify the function of these pits.

An inhumation of a woman was found in a deep pit at the top of the dune. (fig. I.15. The grave pit cross-cuts several pit hearths, and is therefore one of the youngest burials on the dune.

Palaeoecological and palaeogeographical context

Despite the covering with younger sediments, the state of preservation of palaeoecological remains was not good. Against all expectations, pollen was found in the fill of several pit hearths. Between c. 8100 and 7500 BC the landscape was dominated by pine. From c. 7700 BC, deciduous trees like oak and elm occurred, as well as common hazel, birch and pomaceous fruit trees. The presence of this last group suggests that there were also open patches in the woodland. There were wet areas in the vicinity where alder and willow grew.

Around 7000 BC there was a change in the vegetation. Although conditions in the wetter parts of the landscape remained more or less unchanged, the pines on the dune made way for deciduous woodland, mainly comprising oak. This situation pertained for a long time.

Pollen from one of the pit hearths shows that around 6000 BC there was an open patch near the hearth where heather, ferns and various herbaceous plants grew. Alder, willow and marsh plants grew in the lower-lying parts. There was also open water nearby. When the dune was inundated, oak stood on the higher parts, as evidenced by several preserved trunks distributed over the excavated area. One of the oaks has been dendrochronologically dated (to 4799 BC).

Taphonomy

After it was used, the top of the dune lay above the mean water level for several hundred years. As a result, virtually no unburnt organic material has survived. The grave probably lay for another 200 years or so in dry circumstances, as a result of which the skeleton was almost entirely disintegrated. The bone remains could not therefore be directly dated. As a result of the inundation of the dune the top of the sand was exposed to a slight degree of erosion in a few places. The podzol soil was found to be largely intact, though it had been impacted by water flow and plant growth. Some movement of sand thus occurred at the top of the soil layer, and the A-horizon is not fully intact in all places. Where it is still intact, there is evidence for treading. Some charcoal particles have penetrated deep into the soil, and some of them fragmented *in situ*. The site was also partially dug over in prehistory, particularly the parts where many pit hearths were found. This disturbed older features and displaced finds.

Figure I.13: Overview of the excavation of Dronten N23.

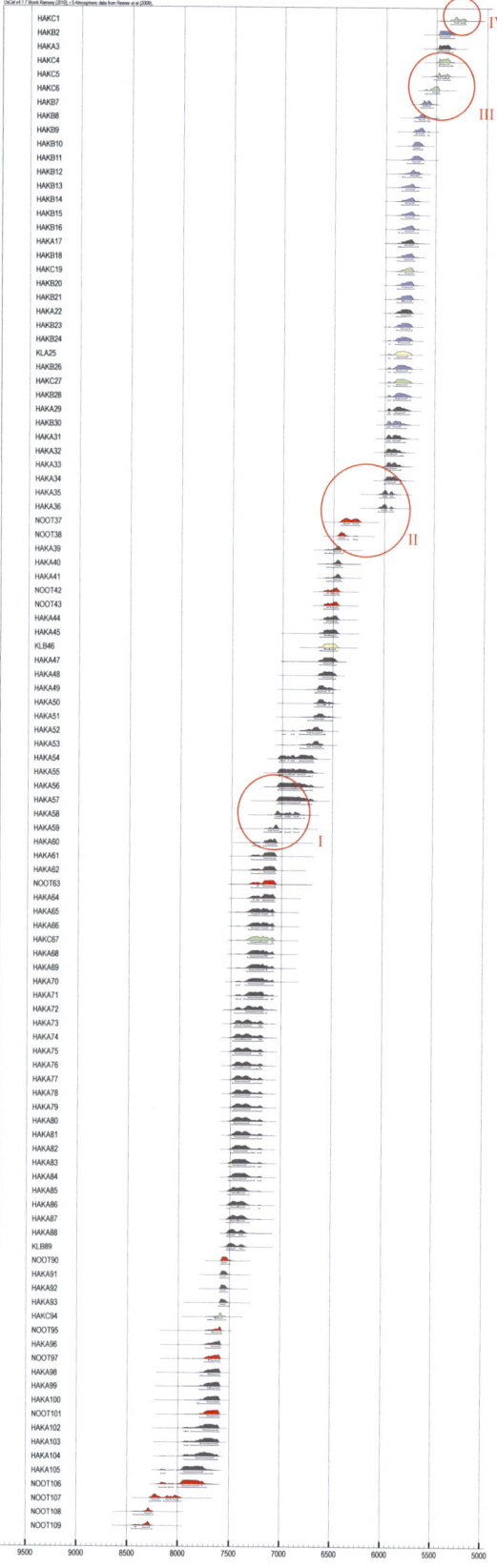
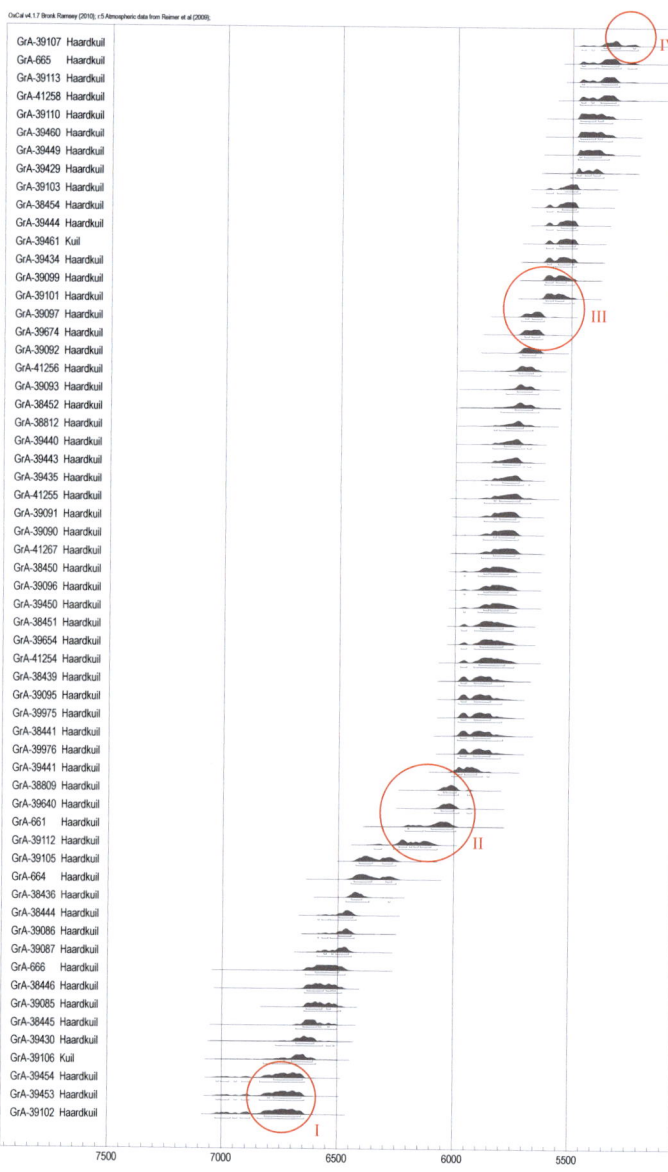

Figure I.14: AMS ^{14}C dates of the pit hearths.

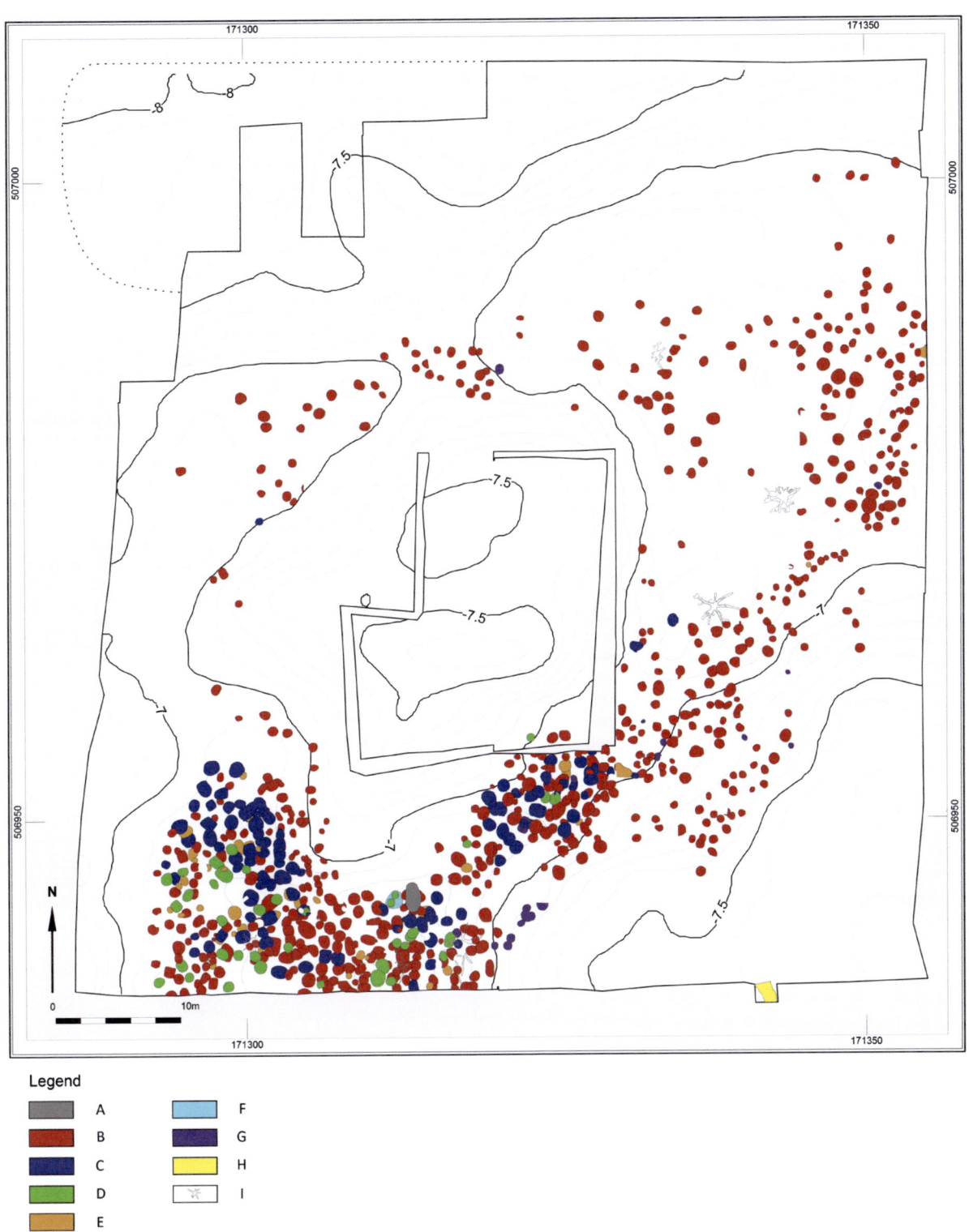

Figure I.15: Overview of features. Legend: A: burial; b: pit hearth type A; c: pit hearth type B; D: pit hearth type C; E: pit hearth indet.; F: waste pit?; G: other pits; H: trampling zone, path?; I: tree stumps in situ (from Hamburg et al. 2012).

I.3.2 Hanzelijn – Area VIII

References
Publications: Müller & Leijnse 2003, 24 (Coversand zone VII); Leijnse 2006; De Moor *et al.* 2009; Van Lil 2008.
Data: Dans-easy

Geographical location and research
Hanzelijn Area VIII is one of the areas along the route of the current Hanzelijn (a rail link between Lelystad and Zwolle) where an investigation revealed archaeological remains.

Hanzelijn Area VIII lies to the southwest of the village of Swifterbant and covers approx. 45 hectares. It was investigated in three phases from 2002 to 2009 by means of mechanical borehole surveys, plus palaeoecological analysis (fig. I.16). At first, the entire route of the rail link was investigated to identify archaeologically relevant zones and establish the intactness and characteristics of the coversand and the old tidal deposits (Wormer Member).

This research was performed using a sonic drill with Aqualock piston type (diameter 5 cm).[755] The boreholes were made every 50 m along a transect, in two offset transects 25 m from the centre of the railway line. The top of the Pleistocene coversand and the decalcified top of the Wormer Member deposits were sampled and then sieved over a 1 x 1 mm mesh.

On the basis of the results of the borehole survey, 16 areas, including area VIII, were further investigated in a denser borehole survey (phase 2),[756] with the goal of identifying and delineating any archaeological sites present. The same mechanical boring technique was used. The distance between the transects and the boreholes was however reduced to 25 m.

As in phase 1, the top of any archaeologically interesting layers were sampled and sieved over a 1 x 1 mm mesh. The results of this phase eventually led to the selection of nine areas for additional research, including Hanzelijn Area VIII.

In Area VIII the phase 3 research consisted of an evaluation based on boreholes and specialist analyses.[757] The boreholes were made using a high-quality mechanical gouge auger (Begemann). The cores were sampled for botanical (pollen and macro-remains), dating and soil micromorphological analysis.

Geological and pedological context
The subsurface of area VIII contains a Pleistocene coversand ridge orientated east-west and several smaller coversand hillocks. This layer of Pleistocene coversand overlies a layer of sandy, gravelly fluvial depositions from the Kreftenheye Formation that was reached by only one of the boreholes made using the mechanical gouge auger (Begemann). The top of the coversand lies between 2.5 and 5 m -Mv (6.5 – 9 m -NAP/below average sealevel); it slopes downwards towards the northeast.

A podzol soil is present over large parts of the area. The soil structure is intact in the higher-lying parts in particular. In the lower parts A/C and C profiles were observed that might suggest erosion and wet conditions.

The coversand ridge, the hillocks and the depressions are covered by a layer of reed or forest peat that formed during the Middle Atlantic (Basal Peat). The Basal Peat is covered by a layer of highly to moderately clayey detritus peat (the Flevomeer Layer). The whole thing is covered by Lelystad Complex deposits (the Almere and Zuiderzee Layer).

Dating/phasing
The borehole survey in Area VIII did not provide any clear date for the modest quantity of remains found, though it did give rough indications of when the area might have been occupied. It was assumed on the basis of the stratigraphical information from the cores that the area must have been suitable for occupation until the Early or Middle Neolithic. This date was calculated on the basis of the sea-level curve, based on the depth of the top of the coversand.[758] It was determined on the basis of pollen and macrobotanical research that the inundation of the area began during the Late Mesolithic. This date is confirmed by a ^{14}C-dating of the bottom of the peat, which gave a date of c. 5300-5000 cal. BC (6190 ± 55 BP). The area, and the immediate surroundings, were presumably inhabitable until (and into) the Late Mesolithic.

Finds, features and cultural context
During the borehole surveys worked flint was found in three boreholes, including one charred piece.[759] Unburnt fish bones and a possible fragment of quartz-tempered pottery were found in 31 boreholes. More than half the boreholes (461 of the 741) contained charcoal. In 114 cases, they contained large quantities. The charcoal and flint occur mainly in the more elevated parts of the landscape, in places where the soil profiles are still intact.

No flint was found in the five mechanical gouge auger (Begemann) boreholes made during the field evaluation. In these boreholes fragments of charcoal were found in

---

755 A small proportion of the boreholes could not be made mechanically because of local conditions. A 7cm hand auger (Edelman) and a 3 cm hand gouge auger (guts) were used to make these boreholes.
756 Leijnse 2006.
757 De Moor *et al.* 2009.
758 Gotjé 1993 (zie Leijnse 2006, 27).
759 The report does not state what the type of artefact.

the A-horizon. The charcoal particles are regarded as an indication of human activity in the immediate vicinity. The faecal fungi found (*Sordoria* and *Cercophora*) indicate the presence of large herbivores in the area. This is confirmed by the presence of tread resistant plants, and micromorphological analysis has shown that the top of the coversand does indeed display a large degree of treading in certain places. Furthermore, there appears to be a spatial correlation between the presence of faecal fungi and signs of treading. The treading may have been caused by animals and/or humans. The flint found suggests a human presence.

Palaeoecological and palaeogeographical context

During the Middle Atlantic there was mixed deciduous oak woodland that included elm, birch, lime and hazel on high grounds in the immediate vicinity. There were ferns and ivy in the undergrowth. Pine woodland could be found on the dry coversand ridges, which were open, particularly in area VIII. Heather also grew there, with hazel along the flanks. Further to the south there were alder groves, while saw sedge marshes dominated in the northwest. In the north of the area, where the Pleistocene coversand is at its lowest point, there was other salt marsh vegetation. The inundation of the area began in the Middle Atlantic, and continued in the Late Atlantic.

Taphonomy

Area VIII was barely subject to any erosion, except in the lowest parts. Coversand was deposited in a few places, but this had no impact on the top of the coversand. In several boreholes coversand deposited by water is present in a layer several centimetres thick between the A-horizon and the Basal Peat. No Holland Peat was found in area VIII. It may have been eroded when the area was inundated.

Micromorphological analysis has shown that the top of the coversand is trodden in several places. The presence of a quantity of charcoal that decreases as elevation declines and fragmented charcoal are also regarded as indications of treading.

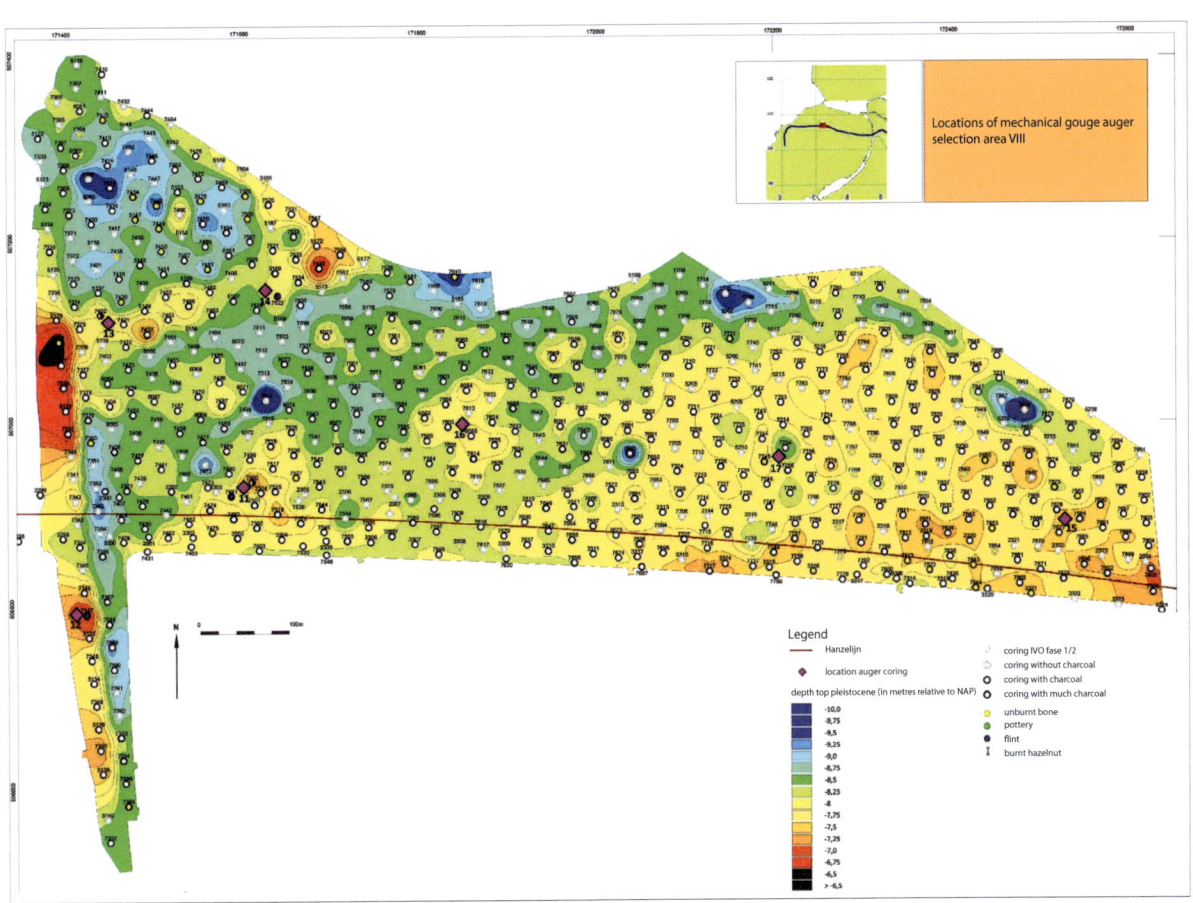

Figure I.16: Overview of boreholes in Hanzelijn – Area VIII.

I.3.3 Hanzelijn – Drontermeer Tunnel (area XVI)

References
Publications: Leijnse 2006; Prangsma & Gerrets 2009.
Data: urn:nbn:nl:ui:13-c7v-1z9

Geographical location and research
The Hanzelijn-Drontermeer Tunnel site is in the municipality of Dronten, immediately to the west of the railway tunnel under Drontermeer lake (fig. I.17). The site was discovered during an archaeological borehole evaluation performed for the construction of the Hanzelijn rail link.[760] It warranted further investigation because of large quantities of charcoal and the presence of an intact soil on the flanks of the coversand hillocks. An archaeological watching brief was to have been performed during the digging work for the construction of the rail link, but this did not occur.[761] By way of compensation, and to explore the significance of the large quantity of charcoal, several trial trenches were dug to the north of the route.[762] A number of excavation pits were dug on the basis of the results of the trial trench survey. The total area investigated was 1884 m^2.

Geological and pedological context
The site lies on a Pleistocene coversand ridge that is situated approx. 1 m below current ground level. The borehole survey suggested that the surface of the coversand (podzol) was intact in the lower-lying parts. This was not found to be the case anywhere, however. Overlying the coversand was a layer of silty sand transitioning up the slope to a detritus gyttja. The transition between the two layers is remarkably abrupt (erosive), and in many cases was marked by a layer of bleached sand grains, which was several centimetres thick in some places. This layer was covered by a layer of decalcified, moderately fine sand deposited by flowing water. Above this there was a silty clay containing shell remains which cut into the coversand layer, particularly in the highest parts. The researchers assume that the landscape was completely covered with peat, though this could be confirmed in only one profile. In the other parts of the area investigated, the peat was completely eroded by wave action. This process formed the gyttja and detritus deposit. In the final phase before the reclamation of the Flevopolder this layer and the highest coversand hillocks were again eroded by the formation of the Zuiderzee and IJsselmeer deposits (Naaldwijk Formation).

Dating/ phasing
On the basis of seven ^{14}C-dates for carbonised material from six pit hearths[763] the archaeological finds date to the Middle Mesolithic and the start of the Late Mesolithic (6650 and 6070 cal. BC).

Finds, features and cultural context
The excavation did not yield any finds other than two small pieces of unworked flint. The investigation did however find features, including 38 pit hearths, 14 pits and seven postholes.

The pit hearths are in three spatially separated clusters. They vary in depth from 4 to 40 cm and have a diameter between 44 and 110 cm. The pits differ from the pit hearths in the sense that no charcoal (or charcoal dust) is present in the fill. No function can however be identified for these pits on the basis of their shape and size. The postholes are distributed randomly over the excavation pits and do not form any recognisable structure.

Palaeoecological and palaeogeographical context
The preservation conditions for macroremains were found to be so poor that it was not possible to reconstruct the environment. The analysis of charcoal from the pit hearths showed that most of the wood was pine. A carbonised pine cone scale was also found in one of the pit hearths. This is striking, as extensive deciduous woodland was more typical of the Middle and Late Mesolithic.

Taphonomy
After it was used, the site was subject to erosion on a large scale, so any archaeological layer and the top of the pit hearths have disappeared. After the reclamation of the Flevopolder the uncarbonised organic remains oxidised to such an extent, under the influence of soil formation processes and drainage, that they were no longer suitable for analysis.

---

760 See Hanzelijn Deelgebied VIII for a description of the prospection and field evaluation.
761 Prangsma & Gerrets 2009, 9; Quadflieg 2007.
762 Prangsma & Gerrets 2009.
763 Two samples were dated from one pit hearth.

Figure I.17: Top: Aerial photo of the Drontermeer Tunnel. The site is on the opposite side of the water (photo: ProRail/Henk de Jong). Bottom: soil features related to the recent drainage of the Oostelijk Flevoland (photo: Hans Peeters).

I.3.4 Swifterbant Cluster

References
Publications: the research on this group of findspots performed in the period 1964-1969 has been published in 14 preliminary reports in Helinium (Swifterbant Contributions; 1976-1985) and four final publications in Palaeohistoria (Final Reports on Swifterbant; 1978-1981). Other aspects have been published in doctoral theses on the pottery (De Roever 2004), lithics and flint (Devriendt 2013), archaeobotany (Schepers 2014a/b) and archaeozoology (Zeiler 1997). The research by the Groningen Institute of Archaeology (GIA) (2004-2010) focused on S2 (2004; Prummel et al. 2009), S4 (2005-2007 Raemakers & De Roever 2020) and S25 (2008-2010; Raemaekers et al. 2014). Data: in archive of Groningen Institute of Archaeology, University of Groningen

Geographical location and research
The term 'Swifterbant Cluster' is used for a group of 18 findspots situated in an agricultural area to the northwest of the village of Swifterbant (fig. I.18).[764] They were discovered in 1961 in newly dug ditches, and lie approx. 1 m below the current surface. The findspots are located on the banks of a covered creek system and on river dunes. Both locations were further investigated by means of a manual borehole survey, which determined the size and the depth of the findspot (and thus also the relief in the subsurface).

The findspots on the banks can be seen in the borehole cores as a dark layer created by a combination of the ripening of the clay and the deposition of organic material, probably reeds. This material then partially decomposed, so the original structure has not generally been preserved.

The river dune findspots can also be seen in the core as a dark refuse layer, in this case caused by settlement waste (charcoal, organic remains etc.).

The findspots were excavated in three phases, each with a different strategy.[765] In the first phase (1964-1969) the State Service for the IJsselmeerpolders (RIJP) excavated small parts of the bank and dune. The excavation was performed by hand in grid cells of unknown size. No sieving was carried out.

The second phase (1972-1979) was performed by the Biological and Archaeological Institute of Groningen University (BAI), working in collaboration with a partner from the United States, the Museum of Anthropology of the University of Michigan. During this period the bank findspots S2 (fig. I.19), S3 and S51 were almost completely excavated, and partial excavations were performed on bank findspots S4 and S6 (fig. I.20). Dune findspots S11-S13 and S21-S24 (fig. I.21) were also excavated in this phase. A partial excavation of dune findspot S61 also took place. All the excavations on the bank in phase 2 were carried out in grid cells of 1 x 1 m with a stratigraphic thickness of 10-15 cm. Larger finds were plotted in 3D and the soil was then sieved (mesh size 3 mm). It is not clear how the excavations on the river dunes were performed.

In the third phase (2004-2010) there were partial excavations at bank findspots S2 and S4 and in depression S25, just beside the dune on which S21-S24 are located. The work was carried out in grid cells of 50 x 50 cm with a stratigraphic thickness of 5 cm. All the soil from S2 was sieved (mesh size 2 mm). This process turned out to be very time-consuming in the field, and also in terms of the examination of the residues. At other locations, therefore, 10% of the grid cells were sieved and the others were examined by trowel. A botanical sample was taken from every m^2 (four grid cells) at S2; at S4 and S25 a sample was taken from every sieved grid cell. These samples were sieved over a smaller mesh size to establish the presence of any botanical remains.

Specialist analysis was performed on the pottery, archaeobotany (macroremains and pollen), archaeozoology, diatomes, micromorphology, lithics and flint.

Geological and pedological context
The Holocene base of the Swifterbant area comprises the valley of the river Hunnepe, which flows roughly east-west here (fig. I.18). Late Pleistocene sandy ridges developed to the north and south of the valley. No research was performed on the coversand stratigraphy, so the precise age of these ridges is unclear.

Relative sea-level rise in the Holocene made conditions in the area wetter, and peat formed in the lower-lying parts of the Swifterbant area. From around 5000 cal. BC, the influence of sea-level rise was so great that clay was deposited. A dense network of creek systems formed in the clay landscape. The water level was generally low enough for soil to form along their banks, which were occupied by humans in this phase. The formation of soil is a clear indication that no sedimentation occurred for lengthy periods of time. Nevertheless, there was regular flooding, as evidenced by the presence of several layers of clay between the layers containing archaeological remains. This period of human occupation on the banks of the creeks dates to between 4300 and 4000 cal. BC.

During the long period when conditions became steadily wetter, the landscape around the dunes also changed. As the water table rose the dunes were covered with peat and clay. Around 3700 cal. BC the entire Swifterbant area was covered by a thick layer of peat, and human occupation there came to an end.

---

764 Different findspot numbers were used on the river dunes for parts of the dune. In this figure, the locations excavated on each river dune have been counted as one (Devriendt 2013, tabel 2.1).

765 See Devriendt 2013 for a detailed overview.

Dating/phasing

Human activities in the Swifterbant area can be divided into three phases. The first is the period before the creek system developed. In this phase activities were limited to the river dunes. Most date information for this phase, in the form of five ^{14}C dates, comes from S21-S25. Charcoal was found in 19 boreholes in the clay around this river dune. Its depth can be linked to the regional sea-level curve to approximate its age.[766] If both date indicators are taken together, it becomes clear that activities took place on the dune in the period 6800-4300 cal. BC. There is no evidence of any hiatus in these activities during this period.

The second phase is the period 4300-4000 cal. BC, when activities were taking place at all the bank findspots, as well as on the river dunes. It is difficult to determine how long the period of occupation lasted due to the lack of precise dating opportunities. The three centuries mentioned form a plateau in the calibration curve, and should be regarded as the maximum length of time that the bank findspots were used.[767]

The third phase is the period after activities at the bank findspots ceased, until the area was finally inundated. This phase dates to 4000-3700 cal. BC. Human activity was confined to the river dunes in this phase.

Finds, features and cultural context

The archaeology of the first phase of activities is confined to the river dunes, where findspots S11-S13 and S21-S24 were investigated most thoroughly.[768] S11-S13 have been published only in preliminary reports, so the information below is based largely on S21-S24. It is important to bear in mind that the remains from this phase are poorly preserved. These locations were not covered by sediment until several centuries later. The archaeological remains consists of a large number of pit hearths, found mainly on the higher parts of the dune (fig. I.20). Large quantities of flint and stone were also collected from these higher parts. This conclusion does not however result from the research history, as the higher parts of both dunes were investigated. It was in fact the lower part of S25 that was excavated. There, two pit hearths were also found, as well as flint, lithics and organic material (which has been preserved in the lower part of the dune).

Activities occurred in phase 2 at all findspots in the Swifterbant area. Interestingly, not all activities were performed at all findspots, or at least not to the same extent. Human graves were for example found at S2 (9x), S4 (1x) and S21-S24 (13x), but not at the largest location excavated, S3. There, however, many wooden posts and postholes were documented, and at S4 too, in a lower density, whereas at S2 only one row of postholes was found. The absence of postholes at S21-S24 was undoubtedly the result of poor preservation conditions. A structure (building) can be discerned in the spatial patterns of postholes, lithics and flint at S3 (see H4). There is also a lot of variation in the presence of surface hearths. At S3 several hearths were made on a clay plateau, which was reused several times. Furthermore, at S3 and S4, hearths were recognised from the presence of several lumps of burnt clay. No hearths were found at S2 and S21-S24. Evidence for fields was found at S2, S3 and S4. Interestingly, the periods of farming and occupation appeared to alternate. At both S2 and S4 fields were identified in thin-sections below, in and above the occupation layers.[769] Evidence of woodworking was found at S25, where in one layer approx. 80 pieces of processing waste were collected, including a discarded paddle.[770]

Evidence of human activity in phase 3 was found only at S25. This excavation documented a dump zone that was used in the period 4500-3800 cal. BC. This general picture does not do justice to the history of the findspot, however. For example, flint is the largest category of finds throughout the entire period of use. Lithics and flint were also found in similar quantities in the various trenches. The other categories of material (wood, bone and pottery) display clear temporal and spatial clustering. This large degree of internal variation suggests that there is a series of discrete dumps of material at S25 in a zone along the dune. During the period of use, these dumps shifted through the zone (horizontally) and were separated (vertically) due to sedimentation.

Palaeoecological and palaeogeographical context

Given the limited preservation of archaeological remains from occupation phase 1 on the dunes, and the lack of targeted investigations of landscape development away from the archaeological findspots, the landscape context and human activity in this phase can be discussed only in very general terms. The main landscape formation process was undoubtedly the increasingly wet conditions. The first indications of human activity date from around 6800 cal. BC. In this period the river valley will still have been a long way from the dunes. This means that there will have been deciduous forest over large areas of the landscape, dominated by trees that could tolerate relatively dry conditions. As conditions became wetter, the brook valley vegetation will have come to cover a growing proportion of the area.

We are much better informed about phase 2 thanks to an extensive archaeobotanical analysis of all excavated

---

766 Van der Plassche *et al.* 2005.
767 Raemaekers 2014.
768 Finds from this phase were also retrieved from S61, S71 and S81-S84 (Devriendt 2013: 34-37).
769 Huisman *et al.* 2008; Huisman & Raemaekers 2014.
770 Raemaekers *et al.* 2014.

locations and the study by Schepers *et al.* (2013) linking this data to current vegetation information. Schepers *et al.* distinguish several plant communities: reed, grass, pioneer and woodland communities. These communities cover the entire spectrum from creeks carrying water to undergrowth on the banks. The landscape will have been very open, as trees probably only grew on the banks and dunes. It is also important to note that there is a little evidence of marine influence. This means that saltwater could penetrate as far as the Swifterbant area during storm surges.

Our knowledge of phase three is based entirely on S25. Here, close to the dune, we see the same plant communities. The Swifterbant area was slowly overgrown with peat during this phase.

Taphonomy

The landscape formation processes had a significant impact on the quality of our image of the site. There was in fact no wetland archaeology in phase 1. The findspots remained above the water table for so long that no organic material was preserved. Furthermore, it is unclear whether any carbonised botanical material was collected. None has been published, at any rate.

The preservation conditions at the findspots on the banks were clearly better in phase 2, as several categories of organic material were preserved. The micromorphological analysis of the structure of the find layer at S4 is important. It shows that original reed layers have largely decomposed, and the archaeological layer consists largely of charred plant remains, small finds, lumps of earth and coprolites. The limited extent of preservation can also be seen in the macrobotanical remains. The grains found at S2 (2004) and S4 (2005-2007) were highly weathered, and the botanical investigation using finer meshes yielded virtually no results. The human skeletal remains were poorly preserved on the river dunes had to be impregnated to warrant further deterioration during the excavation so that they could be retrieved. No unburnt organic remains were preserved on the dunes in this phase, either.

The investigation of S25 was the only one that focused on phase 3. The excavation was confined to the clay and peat layers around the dune. It is worth noting that little bone material was found. This should not be interpreted as a result of the preservation conditions, but as an indication of the use of this location. Apparently, few activities were performed there that led to the discarding of bones.

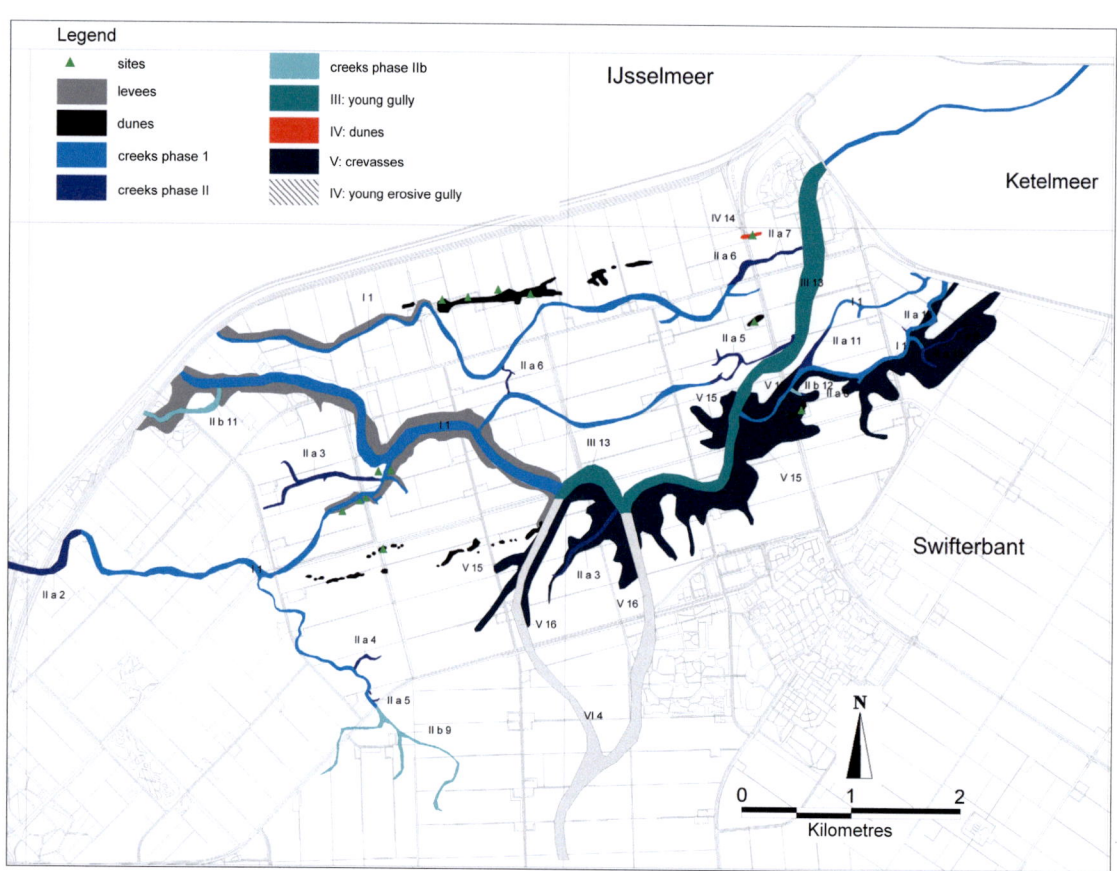

Figure I.18: Overview of Swifterbant area, green triangles are sites (source: Dresscher en Raemaekers 2010)

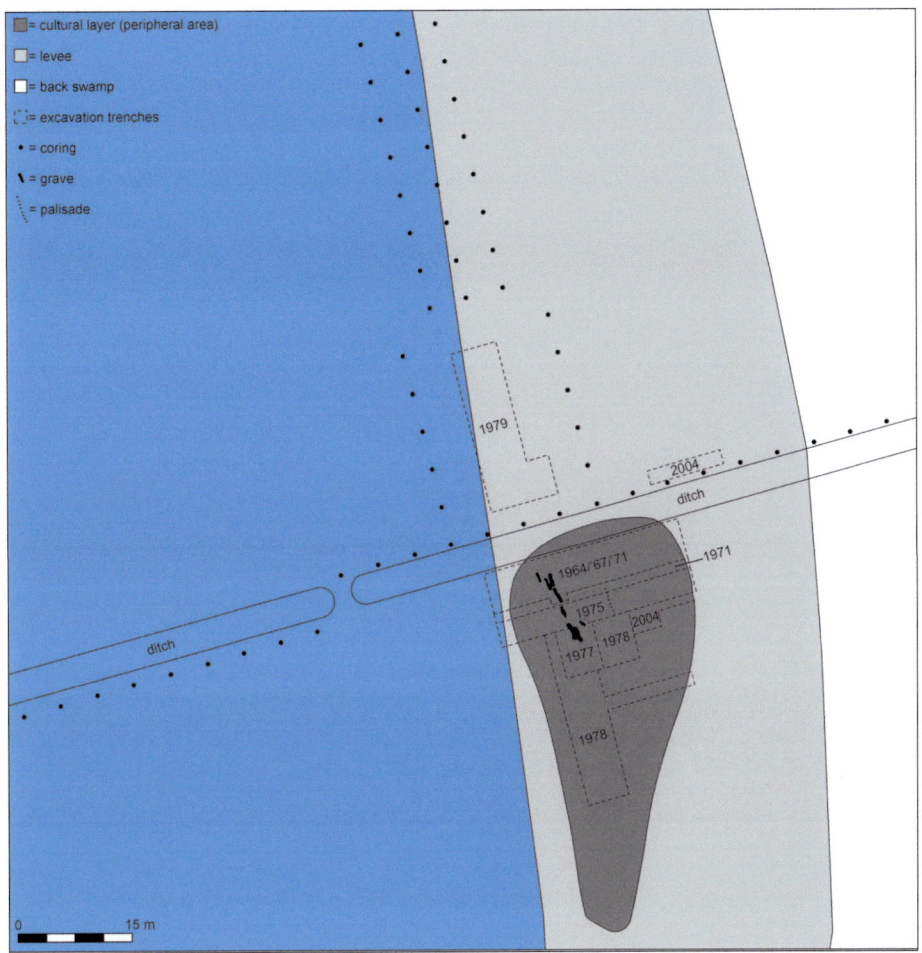

Figure I.19: Overview of Swifterbant-S2 excavation (source: Devriendt 2014)

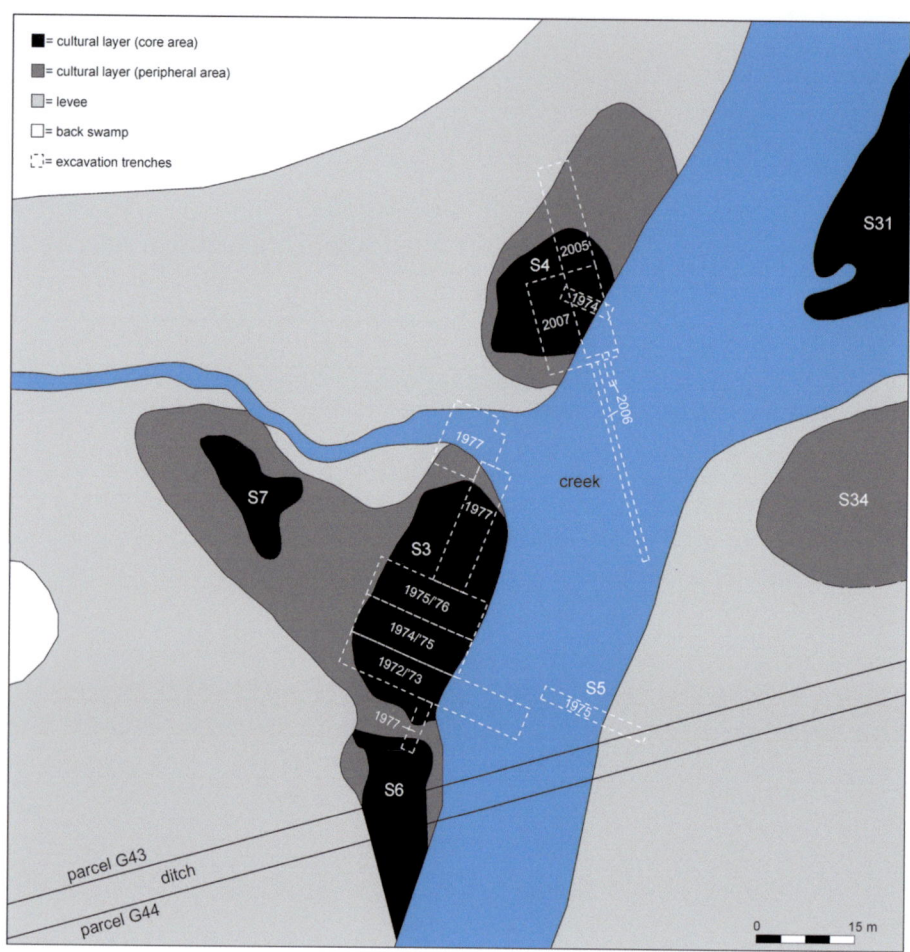

Figure I.20: Overview of Swifterbant-S3-S7 excavation (source: Devriendt 2014).

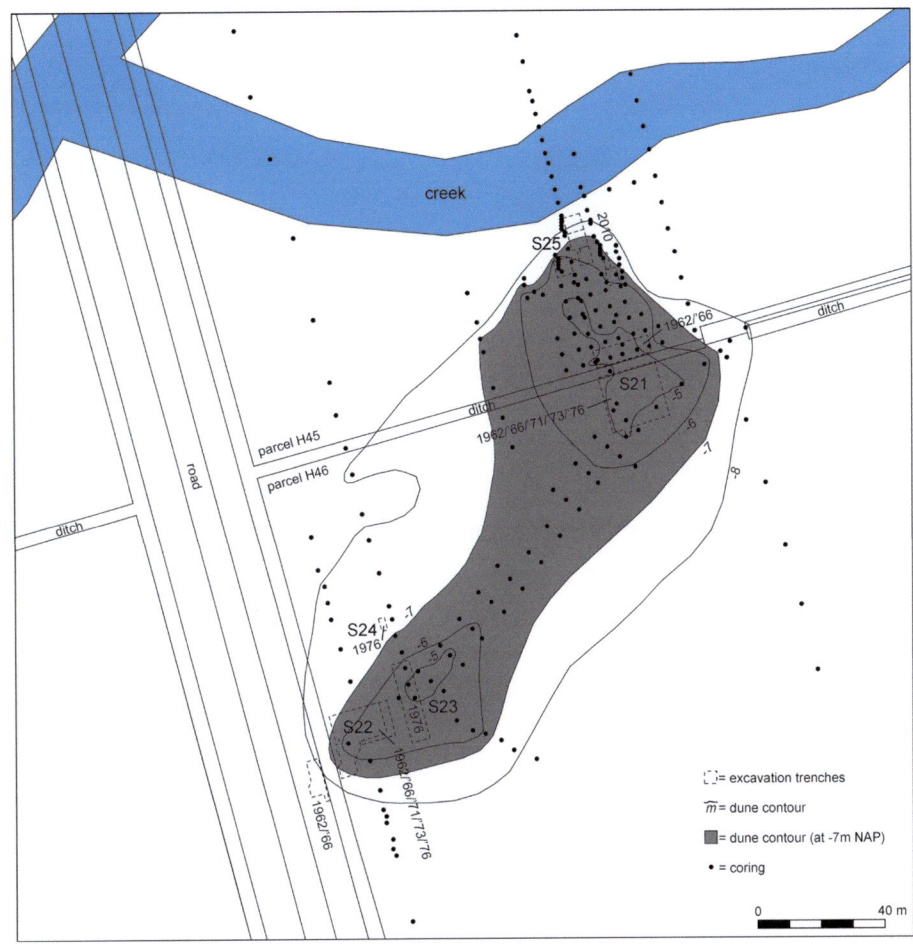

Figure I.21: Overview of Swifterbant-S21-S25 excavation (after Devriendt 2014 and Raemaekers *et al.* 2014).

I.4 Noordoostpolder

I.4.1 Emmeloord J97

References
Publications: Palarzcyk 1984 and 1986; Gehasse 1995; Van der Heijden 2000; Bulten *et al.* 2002; Raemaekers 2005; Ten Anscher 2012.
Data: urn:nbn:nl:ui:13-ans-q87 and urn:nbn:nl:ui:13-0cq-tot

Geographical location and research
The site Emmeloord-J97 is located beside the A6 motorway south of Emmeloord.[771] The first evidence for an archaeological site at this location was found by the State Service for the IJsselmeerpolders (RIJP) when ditches were dug in 1950.

To ascertain the size of the site and map the palaeolandscape context, the Institute of Prehistory and Protohistory, University of Amsterdam (IPP)[772] performed research at the location in 1984 and 1988.[773] It consisted of boreholes and small trial trenches.[774]

The site was further investigated between 1999 and 2000 in trenches and larger excavation pits.[775] During this investigation, a levee with an adjacent gully were found to contain archaeological remains. In the gully immediately beside the levee, part of a fish weir and fish trap were found. The results of this investigation led to an open area excavation of the gully, performed over two campaigns (2000 and 2001) during which a large portion of the gully was studied.[776] The first campaign began with the digging of a narrow trench alongside a row of posts that were found, and a sieving programme whereby part of the archaeological layer was excavated in units of 100 x 100 x 5 cm and sieved over a 5 mm mesh. More fish weirs and fish traps were found in the extension of the first trench. In the second campaign, it was decided that a large part of the gully should be excavated more systematically in trenches approx. 10 m wide, and varying in length. The trenches were dug perpendicular to the orientation of the gully so that cross-sections of the gully could be documented at several points, as well as one longitudinal profile in the deepest part of the gully.

Geological and pedological context
The subsurface of the site consists of sandy bed deposits of the Kreftenheye Formation covered by a layer of clay from the Wijchen Layer. A layer of Basal Peat (Nieuwkoop Formation) was deposited on this. Creeks formed in this peat under the influence of the river Vecht. The site was first used by humans when the creeks began to silt up around 3500 cal. BC, in the Subboreal. Around 3000 cal BC there was a period of renewed peat formation, in which the creek acted as a drainage channel, where conditions were nutrient-poor.

From around 2000 cal BC the creek briefly connected two lakes. In this period, too, the area was used by humans. One of the lakes must have broken through or flooded during this period, depositing a layer of sand on the deposits in the residual channel of the creek. This sand, which presumably consists of redeposited coversand or river sand, contains displaced artefacts from several periods.

The narrowing and eventual closure of the coastal inlet at Bergen around 1500 BC, the drainage of the IJssel and Vecht stagnated, causing large lakes to form. From that point until the draining of the polder, the site was covered by a layer of seabed deposit attributed to the Flevomeer, Almere and Zuiderzee deposits (Naaldwijk Formation).

Dating/phasing
The dating of the site is based on a large number of ^{14}C dates, dendrochronology and typological dating of finds. On this basis, use of the site appears to divide into two phases:
- Phase 1: 3650/3350-3000 BC (Middle Neolithic: Pre-Drouwen and Funnel Beaker)
- Phase 2: 2400-1500 BC (Late Neolithic / Middle Bronze Age: Bell Beaker, Barbed Wire and Hilversum)

There appears to be a hiatus of some 600 years between these two phases, which seems to coincide with a period when the creek acted as a drainage channel for the peat.

Finds, features and cultural context
Postholes were found on the highest part of the levee, some of which contained the preserved remains of the pointed wooden post. The postholes formed a cluster in which no recognisable structure could be discerned.

Several linear structures were found on the flank of the levee and in the channel, each of which consisted of dozens of vertically positioned wooden posts. At least ten fish weirs were found, some of which actually consisted of several phases. The rows of posts were perpendicular to the gully (fig. I.22).

Parts of the original weir were found at a number of fish weirs, against and between the vertically positioned posts. They consisted of screens, woven from thin branches, approx. 3-4 m in length and approx. 1 m high. The original height of the screens is unknown, given the fact that only the bottom part has been preserved. As well as these woven screens, horizontal bundles of branches

---

[771] J97 refers to the plot number.
[772] University of Amsterdam.
[773] Palarzcyk 1984 and 1986.
[774] The results of this investigation were never published, see however Gehasse (1995) and Ten Anscher (2012).
[775] Van der Heijden 2000.
[776] Bulten *et al.* 2002.

and vertical bundles of reeds were also found, which presumably had the same function.

Close to the fish weirs 44 fish traps were found, both fragments and complete examples. The fish weirs and traps date from both phases of use.

A large quantity of other material was retrieved, including pottery, worked flint and stone, and worked bone and antler. The pottery includes Funnel Beaker, Bell Beaker, Potbeaker, Barbed Wire and Hilversum pottery. One remarkable find is an almost complete potbeaker (without a bottom) which had been inverted and placed over one of the posts of a fish weir. The flint assemblage includes 180 identifiable tools, including arrowheads, scrapers and borers. The rest of the material (97%) consisted of processing waste. The stone was highly varied in terms of its type, and included axes, querns, hammerstones and whetstones. The bone objects included awls, needles, fishhooks, chisels, spatulas and jewellery. Alongside worked bone, unworked bone fragments were also found, as well as a large variety of fish remains.

Apart from a few finds (such as an almost complete potbeaker) the material does not appear to have been preserved *in situ*. Most of the material comes from the find-rich layer of redeposited coversand or river sand.[777] It includes finds from all periods represented at the site.

Palaeoecological and palaeogeographical context

Analysis of molluscs and macrobotanical remains (mainly seeds) around the fish weirs indicate a brackish environment that was almost always wet. Some seeds of freshwater species may have washed in from upstream, suggesting there was a supply of fresh water, possibly under tidal influence. The pollen suggests that there was an alder grove on the levee, with hazel and brambles in the higher parts, and willow in the lower-lying parts. Waterside plants grew along the gully and there were also plants (or water plants) in the creek. Sedge grew in the surrounding marshes. Pollen from sheep's sorrel and black bindweed (*Rumex acetosella* and *Polygonum convulvulus*) indicate human activity on the levees.

Taphonomy

When Flevomeer lake formed the top of the levee was exposed to erosion, causing the original archaeological layer to disappear. Based on the preservation of the remaining features, the researchers assume that some 50 cm of the levee eroded, and was then covered by a layer of detritus gyttja (Flevomeer Deposits). The other (lower-lying) parts of the site were exposed to only a small degree of erosion. A small proportion of the bone appears weathered, but the majority shows barely any signs of weathering. Several bone fragments from juvenile animals have for example been found. They are particularly vulnerable to erosion, and are only rarely encountered. As a result of settling, the vertical parts of the fish weirs and fish traps had concertinaed (fig. I.23). Some of the fish traps were found in their original three-dimensional form, while others had been flattened. This difference in preservation might be explained by different rates of sedimentation.

---

777 Only a small percentage of the finds have traces of wear, indicating that the material was not transported a long distance.

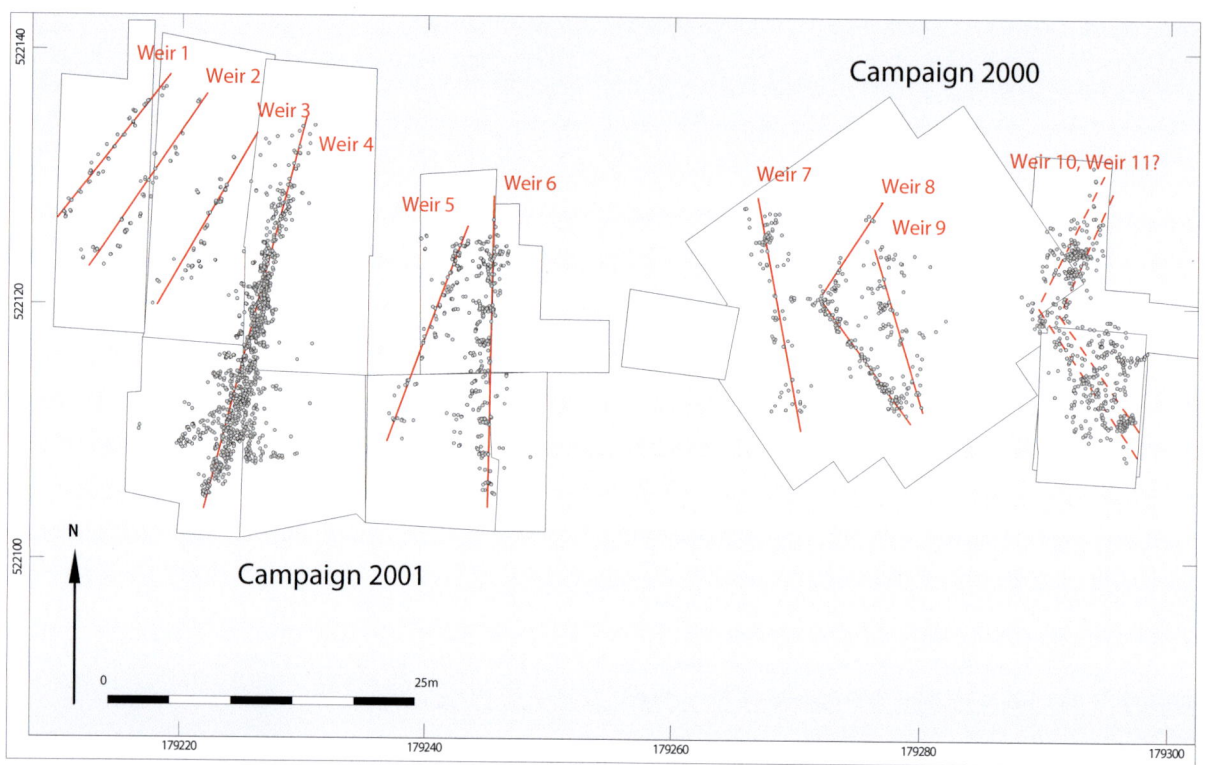

Figure I.22: Overview of the fish weirs (source: Bulten *et al.* 2002).

Figure I.23: Effect of settling/compression on wooden stakes (source: Bulten *et al.* 2002).

I.4.2 Schokkerhaven-E170

References
Publications: Gehasse 1995; Raemaekers 2005; Raemaekers *et al.* 2010; Van Heeringen *et al.* 2004; Out 2009; Hogestijn 1990, 1991; Weijdema *et al.* 2011; Ten Anscher 2012, 2013. Data: not publically available

Geographical location and research
The Schokkerhaven-E170 site is part of the UNESCO World Heritage Site of Schokland, an inundated island on the Noordoostpolder. Archaeological finds were first reported there in 1953, discovered on the surface and in the flank of a recently dug ditch. A number of additional surface finds were subsequently recovered in 1967 and 1969. The University of Amsterdam (IPP) performed a thorough investigation in the 1980s. During this investigation, ditch sides were discovered and a systematic study was launched, consisting of a borehole survey in 1985, surface prospection in 1986 and a trial trench survey in 1988. During this last survey, two trenches 3 m wide and 150 and 63 m long were dug. The trenches were orientated north-south, and were positioned in line along the river dune (fig. I.24).

In 1990 and 1991 surface prospection was performed, completing the archaeological fieldwork at this location. Regular inspections were carried out from 1994 to monitor the effect of any fall or rise in the water table on the archaeological remains (1994-2003).[778] Despite these multiple phases of research, the results have never been published in their entirety.[779]

Geological and pedological context
The site lies on part of a river dune 6 km long, which is orientated east-west. The top of the dune is almost immediately below the ploughsoil, and to the south it is bordered by a riverbed. The stratigraphy of the overlying layers has remained partially intact on this south side. The base of the profile consists of the river dune, covered with peat. Alternating clay and detritus deposits were found in the peat (Wormer Member, Holland Peat). The dune is partly covered by a layer of bog peat (1.3 m), and this in turn is covered by recently ploughed Zuiderzee Deposits (Naaldwijk Formation).

Dating/ phasing
The oldest archaeological remains date from the Early Neolithic, although some probable pit hearths in the ROB excavation pits of 1988 might suggest the site was used in the Mesolithic. Neolithic use of the dune was determined on the basis of typological features of the pottery and flint.[780] The periods identified range from the first phase of Swifterbant via TRB to the Late Neolithic. There is no concrete evidence for use in the Bronze Age, apart from a few decorated pottery sherds that might be attributable to the Early Bronze Age. It has been determined on the basis of several surface finds (pottery) that the site was also used in the Iron Age.

Finds, features and cultural context
Only a small number[781] of features were found in the top of the river dune during the 1988 trial trench survey. The description suggests they are pit hearths that might date to the Mesolithic. The only features from any of the other periods of use are several oak posts that might have been part of a palisade. This palisade has been dated to the Middle Neolithic Funnel Beaker Culture, on the basis of several ^{14}C dates.

The finds were recovered from the top of the dune sand and on the flanks, in the alternating clay and detritus layer. The Neolithic finds consist of pottery, lithics, flint, burnt and unburnt bone, antler and amber. Few numbers are given in the various publications.

Palaeoecological and palaeogeographical context
Pollen and macroremains were analysed on only a limited scale for this site. The analysis focused on use of the dune in the Neolithic. It was found that in this period there was deciduous woodland on the high (and dry) parts of the river dune, surrounded by reed peat, and there was open water in the immediate vicinity. A large variety of species were found in the samples analysed, indicating that the natural vegetation was significantly disturbed. This might have been caused by the activities and presence of humans. Remains of emmer wheat and barley were found in the samples.

Taphonomy
The presence of finds in the ploughsoil proofs that the top of the dune has eroded to some extent due agricultural activities. The archaeological layers appear to be completely intact on the south side of the river dune. Given the presence of features and organic remains (including wooden posts from a palisade) the preservation conditions can be regarded as very good. Changes in the water table and the acidity of the soil do however pose a threat to the archaeological remains still present. There are plans to artificially raise the water table to protect the remains.

---

778 Van Heeringen 2004.
779 Ten Anscher (2013) reviews the results of several investigations, though he notes that most of the basic processing of finds and features has yet to be take place.
780 Rijksdienst voor het Oudheidkundig bodemonderzoek (then known in English as the 'State Service for Archaeological Investigations', now the Cultural Heritage Agency of the Netherlands – RCE)
781 The number of features is unknown.

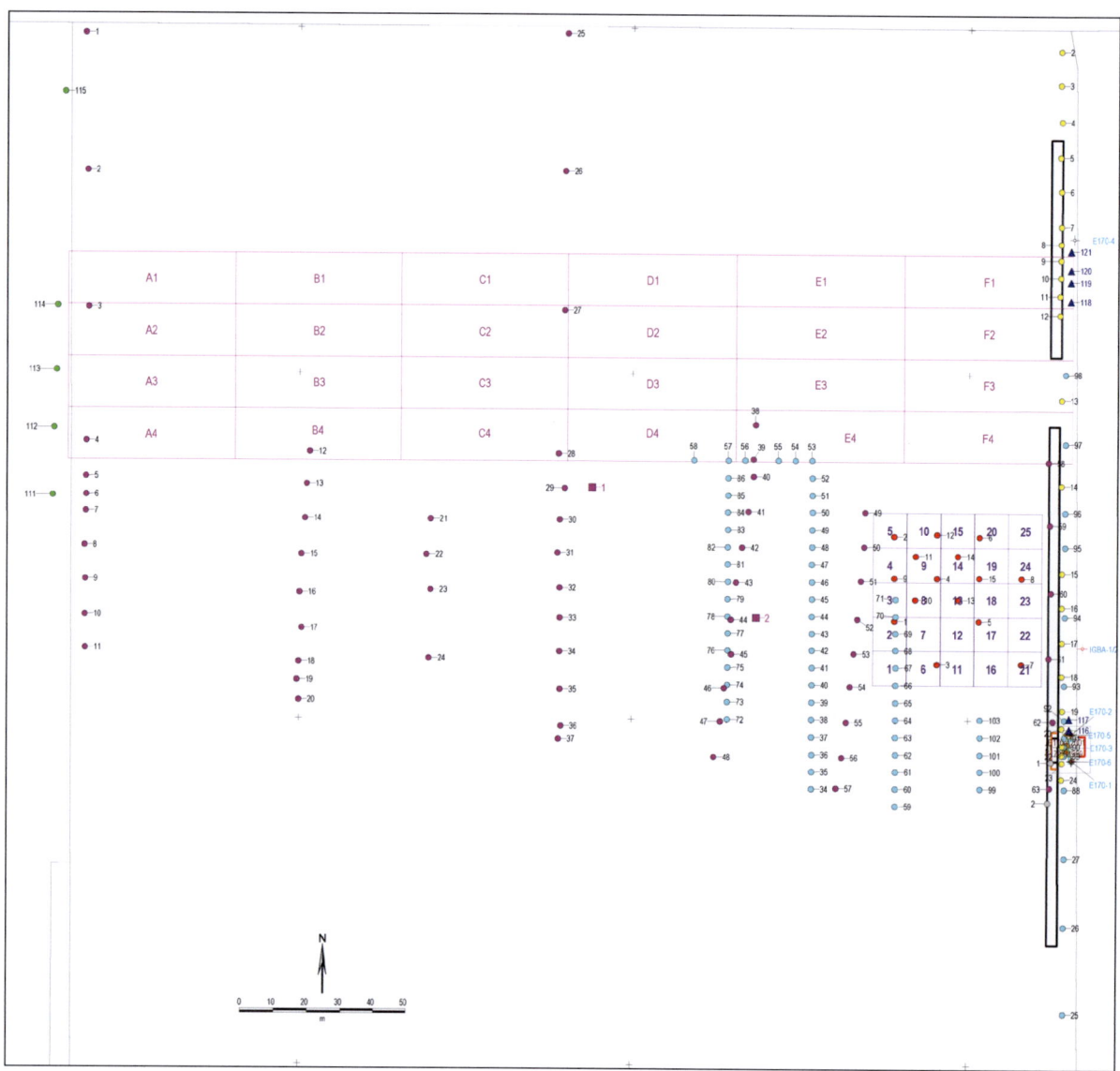

Figure I.24: Overview of excavation pits and borehole locations from all campaigns (NB. A row of boreholes was made over the entire length of the plot, towards the south (along the eastern edge), which are only partially visible in the figure because of the scale of the map) (source: Ten Anscher 2013).

I.4.3 Schokland P14

References
Publications: Ten Anscher 2012; Gehasse 1995
Data: not publicly available.

Geographical location and research
The site is in the northern half of the former island of Schokland on the Noordoostpolder (fig. I.25). The first artefact recovered – an axe (a *Fels-Ovalbeil*) – was found at this site in 1955. But the precise location was not identified until two years later, in 1957, by staff of the RIJP who found pottery and flint in the soil dug out of a drainage ditch dug during a levelling operation. After these first finds, a larger excavation pit was dug to survey the archaeological remains.

From 1982 onwards the University of Amsterdam (IPP) spent several years researching the site.[782] The size of the site and nature of the subsurface was determined by digging small (1 x 1 m) test pits and making boreholes along the flanks of the sandy ridge.

During investigations in the years that followed, several large excavation pits were dug on the sandy ridge and along the flank. The research in 1987 focused on the northeast part of the site, where several larger pits and a number of trial trenches were dug, and in 1989 and 1990 large excavation pits were dug at the highest point of the sandy ridge. The final archaeological campaign was in 1991, when several transects of boreholes were made to address a number of questions about the stratigraphy.

Geological and pedological context
The site is situated at the eastern end of a Late Pleistocene sandy ridge with an east-west orientation. The base of the soil profile is boulder clay deposits that date from the ice age before last (the Saalian). During the Late Glacial coversand was deposited over this, consisting of fairly fine, clay-poor sand that is light yellow in the high parts, transitioning to greyish on the flank. This layer varies in thickness from 0.20 m to 1.30 m. A podzol formed in the top of the coversand which has not been fully preserved everywhere due to bioturbation and erosion. Immediately to the east of the site there was a navigable river (the Vecht) when the site was in use in prehistory, which made it easily accessible.

It proved possible to record the stratigraphy of the various covering layers on the flank facing the river (fig. I.26). The layers consist of clay, gyttja, detritus gyttja and bog peat which were deposited over a long time during and after the period when the site was in use.

Dating and phasing
The traces of human activity can be dated to periods ranging from the Palaeolithic to the Iron Age. However, the majority of the archaeological remains date to the Neolithic and the Bronze Age. The different phases of human occupation at the site have been distinguished by combining lithostratigraphic data and pottery typochronology, from the Neolithic Swifterbant culture to the Middle Bronze Age (possibly Late Bronze Age).

Ten Anscher divided the Neolithic to Bronze/Iron Age occupation/use history of P14 into five phases. No older or younger archaeological remains have been included in this phasing.
- Phase 1: Swifterbant and pre-Drouwen Funnelbeaker
- Phase 2: Drouwen and Havelte Funnelbeaker
- Phase 3: Corded Ware and Bell Beaker
- Phase 4: Barbed Wire
- Phase 5: Middle Bronze Age – Early Iron Age

Finds, features and cultural context

Phase 1: Swifterbant and Pre-Drouwen Funnelbeaker
Having yielded tens of thousands of finds, P14 is one of the richest sites from this phase. The finds attributed to this phase include the following: approx. 10,000 pottery sherds, flint, lithics, artefacts of antler and bone, zoological material (mammal, bird and fish), botanical remains (grain, arable weeds, charcoal, gathered plants (hazelnuts, acorns, blackberries, hawthorn)).

The most striking features from this phase are 12 graves or possible graves. They include single individual graves and also skull burials and collective graves. The other features from this phase are a pit containing a complete beaker (*Ösenbecher*), many postholes (including two house plans), several pits and cattle hoofmarks.

Phase 2: Drouwen and Havelte Funnelbeaker
Only 153 pottery sherds can be attributed with any degree of certainty to this phase. Just two features have been attributed to this phase. The first is a disturbed grave and the second a small, shallow pit containing a possible wooden ladle (or a human skull).

Phase 3: Corded Ware and Bell Beaker
The finds from this phase consists of pottery, flint, worked wood, bone and antler artefacts, zoological material (mammal, bird and fish) and botanical remains (grain, arable weeds, charcoal, gathered plants (hazelnuts, acorns, blackberries, hawthorn)). Several features have been attributed to this phase: five graves, ard marks, various ditch systems, bank reinforcements (in the form of a concentration of stone), postholes, stakeholes, charcoal

---
782 Ten Anscher 2012.

concentrations, surface hearths (shallow dish-shaped depressions), large pits and cattle hoofprints.

Phase 4: Barbed Wire
Only a few finds could be linked with certainty to this phase. They include pottery, flint, lithics, bone and antler artefacts, zoological material (mammal, bird and fish), worked wood, botanical remains (grain, arable weeds, gathered plants (hazelnuts, acorns, blackberries, hawthorn)) and charcoal.

The features attributed to this phase are post-/stakeholes, cooking pits (recognisable by stones at the bottom of the pit that show signs of heating), hearths, charcoal concentrations, cattle hoofprints and footprints.

Phase 5: Middle Bronze Age – Early Iron Age
No features from this phase have been found. Large pieces of two earthenware pots were however found (one Middle Bronze Age Laren pot and an Early Iron Age *Schräghalspot*). Both are regarded as ritual depositions in wet marshy zones.

Palaeoecological and palaeogeographical context
The long use history of the site coincided with major changes in the landscape. Logically, therefore, the phases of activity distinguished should also be placed in different palaeoecological and palaeogeographical contexts. As a result of the rising water table, the landscape at and around the site gradually changed, and the available (habitable) land area shrank.

Phase 1 6000-4650 BP – 4900-3400 cal BC)
During this phase the average water level rose by 3.30 metres, which had a major impact on the landscape. During the initial period of this phase the sandy ridge was still relatively large and covered with mixed deciduous woodland, surrounded by a zone of alder and birch groves that transitioned to sedge marshland and a narrow reed zone along the banks of the Vecht (fig. I.27).[783]. During the rest of this phase the rising water table caused the alder and birch groves to gradually spread (shrinking the size of the higher, dry, parts of the sandy ridge), until an extensive area of alder grove had formed by the end.

Phase 2 (4650-4100 BP – 3400-2950 cal BC)
During this phase the average water level rose by one metre. The landscape around P14 underwent relatively little change, though the sandy ridge did continue to shrink. The higher parts of the ridge were generally dry and covered in mixed deciduous woodland in this phase. During the first 200 years the sandy ridge lay in the middle of expanding alder groves which included many areas of open water. Between the alder groves and the Vecht there was a strip of reed-rich sedge vegetation. However, during this period a wetter type of alder grove vegetation came to dominate. In the final 250 years of this phase the sedge vegetation expanded further, at the expense of the wetter alder groves.

Phase 3 (4400-3650 BP – 2800-2000 cal BC)
During this phase the average water level rose by less than a metre, so the landscape around P14 saw very little change. The sandy ridge shrank further. The higher parts were generally dry and covered with mixed deciduous woodland. Varying drainage conditions caused the surrounding vegetation to change from sedge to extensive birch groves. There was a zone of sedge vegetation along the Vecht.

Phase 4 (3600-3400 BP – 2000-1700 cal BC)
During this phase the average water level rose by 0.4 metres. Strongly fluctuating water levels and tidal influences in the first half of this phase caused both sedimentation and erosion on the flanks and the top of the sandy ridge. Around the ridge, which was still covered with mixed deciduous woodland, birch groves made way for reed and sedge vegetation.

Phase 5 (3400 -2000BP – 1700 cal BC-*start of Common Era*)
No regional water level data is available for this period. There were no strong fluctuations during the first 400 years. Drainage in the area gradually worsened during this period (as a result of the closure of the sea inlet at Bergen). Around the deciduous woodland on the sandy ridge, birch and alder groves developed, and beyond that, an extensive area of raised bog. After 3000 BP the sandy ridge shrank even further, and was gradually overgrown by birch.

Taphonomy
The river Vecht caused a lot of deposition and erosion at and around the site. As a result of this, virtually no intact archaeological layers were found, and finds from different periods were mixed. This palimpsest situation meant that the human occupation history of the site could be described in general outline, though much remains uncertain in terms of detail.[784]

The preservation of the features (postholes, ard marks, cattle hoofprints, footprints, ditches, burials) is very good, generally speaking. However, unburnt organic remains are very poorly preserved (including the burials), as a result of exposure to air in the higher parts, though they have been preserved in the lower parts.

---

[783] Ten Anscher 2012.

[784] Ten Anscher 2012, 23.

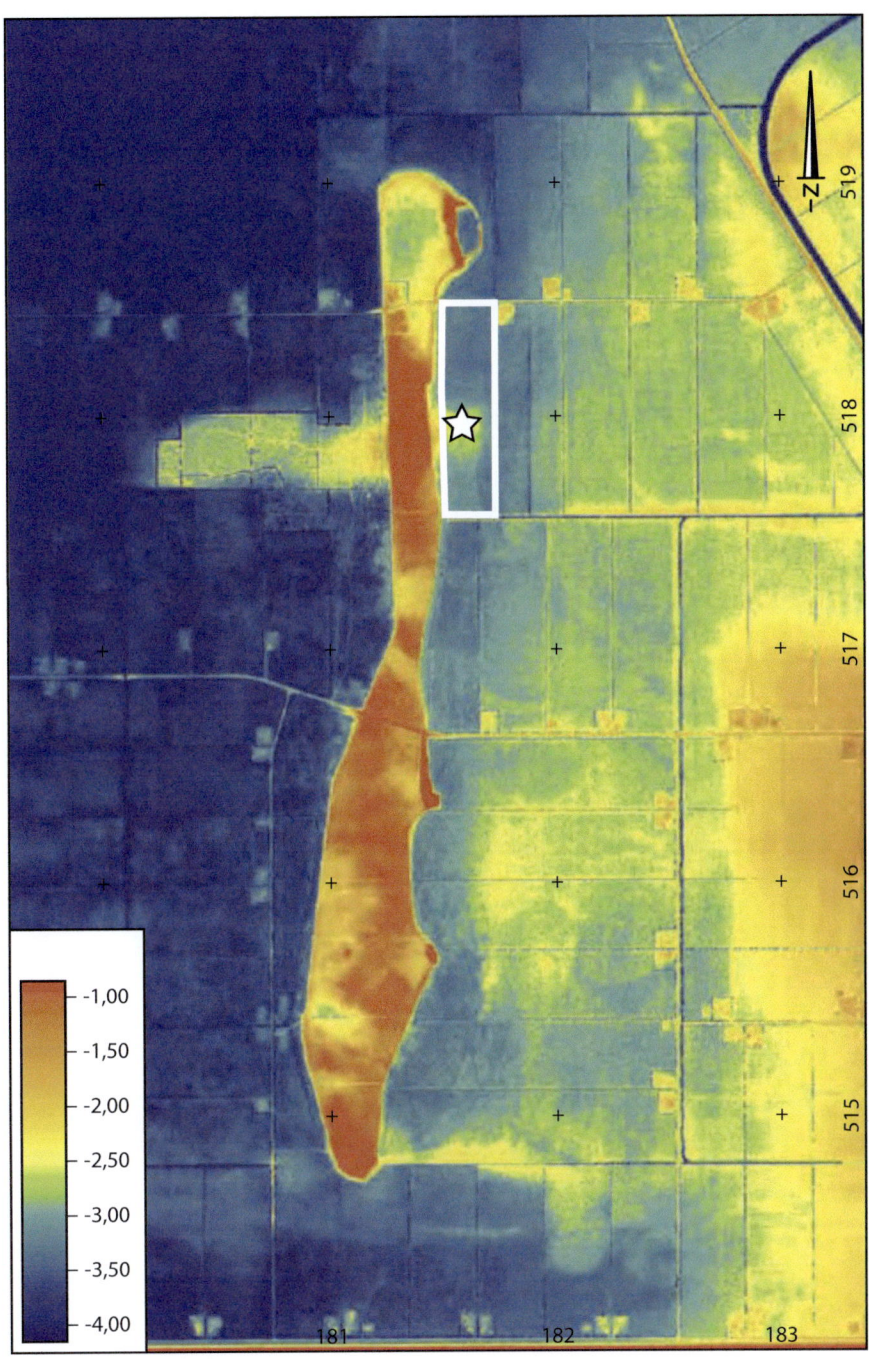

Figure I.25: Location map of Schokland-P14 plotted on the AHN (elevation) map of Schokland.

metres relative to NAP

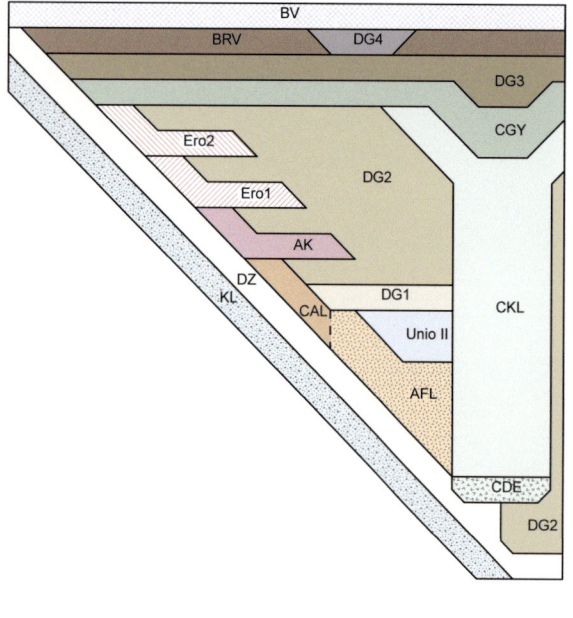

Figure I.26: Schematic geological profile of Schokland-P14. Legend KL / KZ) Boulder clay and boulder sand (moraine deposits), DZ) coversand, OAL) Old Archaeological Layer, AFL) refuse layer, UN-II) Unio II, DG1) Detritus gyttja 1, DG2) Detritus gyttja 2, AK) Arable layer, Ero1) Erosion layer 1, Ero2) Erosion layer 2, CDE) Cardium detritus, CKL) Cardium clay, CGY) Cardium gyttja, DG3) Detritus gyttja 3, BRV) Bog peat, DG4) Detritus gyttja 4 and BV/Ploughsoil) Almere and Zuiderzee deposits (source: Ten Anscher 2012).

Figure I.27: Landscape around Schokland-P14 during phase 1 (c. 4150-3700 cal. BC) (source: Ten Anscher 2012).

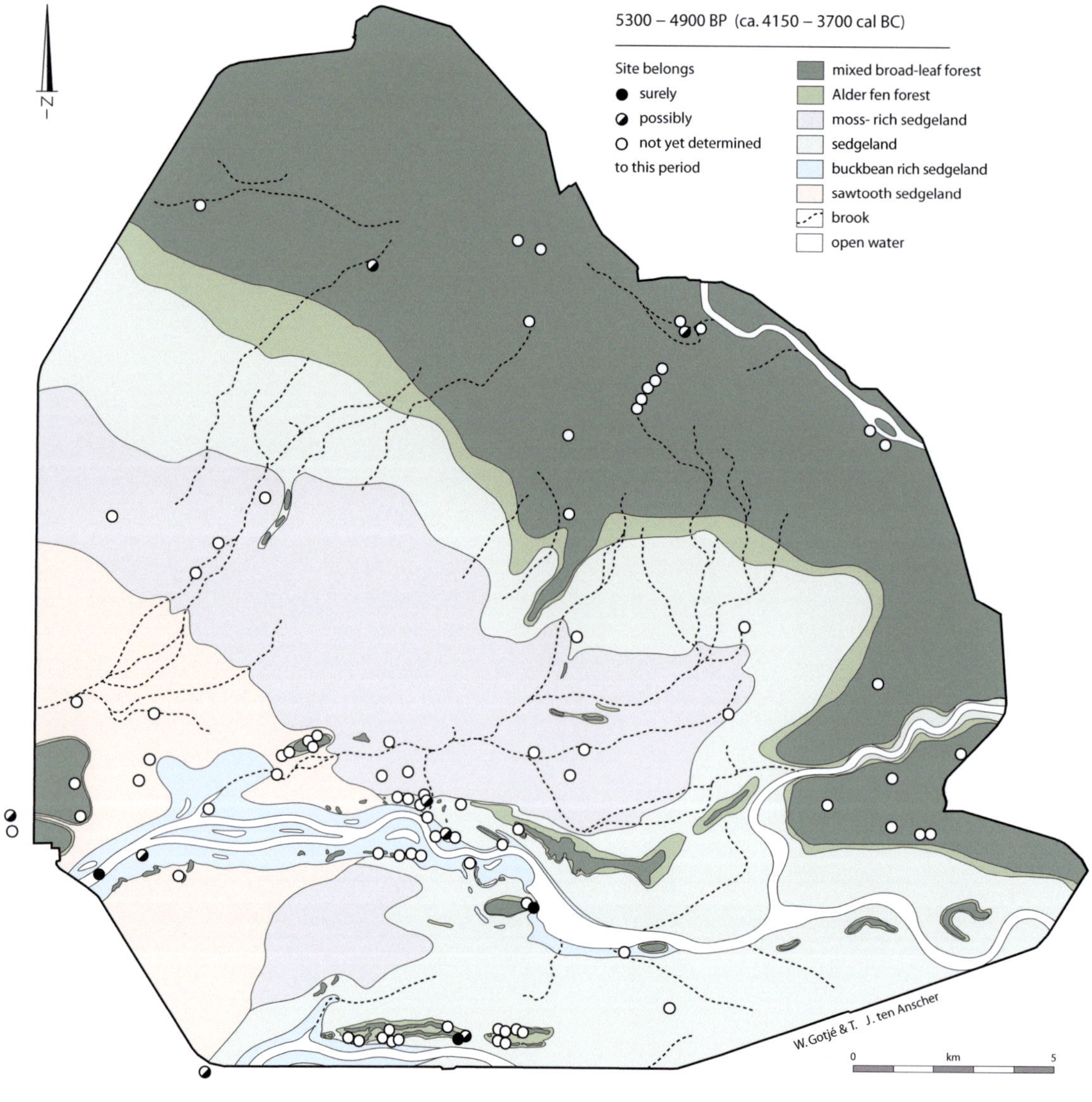

I.4.4 Urk – E4

References
Publications: Hogestijn 1993; Van Oorsouw 1998; Peters & Peeters 2001; Raemaekers 2005; Ten Anscher 2012.
Data: urn:nbn:nl:ui:13-2sc-36m

Geographical location and research
The Urk-E4 site is on the Noordoostpolder to the south of present-day Urk. It was discovered in 1991 during surface prospection by Flevoland provincial authority, when pottery and worked flint were found. In response to these finds borehole surveys were conducted in 1991 and 1992, in which two transects of boreholes were made over the river dune (still unknown at the time) using a 20 cm diameter hand auger (*megaboringen*). The top of the dune was sampled and sieved over a 4 x 4 mm mesh.

In 1996 further investigations were carried out at the site, involving boreholes made with a 20 cm diameter auger and three small trial trenches (this time 2 x 1 m). In 1997 the site was excavated in its entirety in view of construction plans, and because the preservation of the archaeological remains had deteriorated significantly since 1991.

To compile a detailed contour map and to determine whether any archaeological layer was present, the excavation work began with a borehole survey. Several rows of boreholes were made over the site, at a distance of 10 m between both the rows and the boreholes. In a number of places the grid was densified to 5 x 5 m. On the basis of the data from the boreholes and the infrastructural constraints, it was decided that the top and the eastern half of the river dune should be excavated. A total area of 880 m² was excavated.[785]

The majority of the features were sectioned or excavated in quadrants, with the exception of the grave pits (see below), which were excavated layer by layer down to the first skeletal remains. After a physical anthropologist had documented and sampled this level, the graves were removed *en bloc*, together with the surrounding sediments. These were then largely excavated at the repository in a laboratory setting, cleaning the skeletal remains which were subsequently preserved, along with some of the surrounding sediments Finally, the entire area was machine-excavated to identify any deeper-lying features.

Geological and pedological context
The Urk-E4 site is located on a river dune in the valley of the Overijsselse Vecht river. At the time of occupation the dune must have protruded several metres above the surrounding landscape. An archaeological layer formed at the top of the dune, fanning out onto the flanks to some extent. Clay, peaty clay and peat were deposited against the flank in different phases. Sea-level rise caused the river dune to be overgrown with peat during the course of the Neolithic. The top of the dune will have been completely covered by around 3400 BC. The entire thing is covered with a clayey ploughsoil into which some of the peat and part of the archaeological layer have been incorporated.

Dating/ phasing
Two main phases of occupation have been distinguished at Urk-E4:
- Phase 1: 7300-5050 cal. BC (Middle and Late Mesolithic)
- Phase 2: 4200-3400 cal. BC (Neolithic: Classic Swifterbant and Pre-Drouwen)

The dating of the first phase is based on ^{14}C dating of the pit hearths and typological features of two artefacts. However, the majority of the finds can be attributed to the Neolithic phase on the basis of typological features. The available ^{14}C datings suggest the site was not occupied in the intervening period. The end date of the second phase of occupation was determined on the basis of the probable 'inundation' of the dune (see above) and the absence of finds associated with the later Funnel Beaker Culture.

Finds, features and cultural context
On the higher parts of the dune there was an archaeological layer in the top of the dune. This layer was at the surface in both phases, so the finds have become mixed. They include pottery, bone, flint, lithics and charcoal.

On the higher parts of the dune, 35 Mesolithic pit hearths from the first phase of use have been found. The pits are 20 to 60 cm deep and have a diameter ranging from 40 to 100 cm; they are filled with dark-grey to black sand containing a large amount of charcoal. There are few finds from this period. The two artefacts from this period are a microlith (C-point) and a blade with steep retouche.

Phase 2 (Neolithic) is represented by four surface hearths or shallow pit hearths, 18 pits that cannot be more closely interpreted, four stake-/postholes, ard marks and five burial pits (fig. I.28). The remains of ten individuals were found in the burial ground; they were buried in different ways, in varying orientations and positions.[786] Interestingly, several individuals were interred in two of the burial pits (one burial pit contains two and the other three individuals) and only the skull of two of the

---

785 Of this, 182 m² was excavated in grid cells of 50 x 50 cm in stratigraphic units with a maximum thickness of 4 cm and the soil was sieved over a 2 x 2 mm mesh. In addition, an area of 176 m² was manually excavated in grid cells of 1 x 1 m and 2.5 x 2.5 m, again in stratigraphic units with a maximum thickness of 4 cm. Finds were collected by hand during the excavation. The rest of the area (522 m²) was machine-excavated to the level were features could be recognized, and finds were collected by hand.

786 Individual supine position, double (possibly) supine position, crouched position and skull burials.

individuals has been found. Six amber beads were found beside the head in a grave containing one individual. No grave gifts were found in the other graves.

The pottery belongs to the later phase of the Swifterbant culture/Pre-Drouwen. The tempering is dominated by stone grit. The decoration consists of fingernail impressions and fingerprints or grooves, but was made mainly by simple (one- or two-pronged) spatulas. The largest category of finds is flint. The majority of these artefacts are flakes, though scraping and cutting tools and a small quantity of arrowheads were also found.

Palaeoecological and palaeogeographical context

The available palaeoecological data apply only to the period of use in the Neolithic (phase 2). Samples are only available from the flank of the dune. There appears to have been a very open form of vegetation there, with virtually no shade-loving plants. A large concentration of common nettle seeds was found in the archaeological layer. This is a nitrophilous plant that might indicate human activity. Conditions were fairly moist at the base of the dune. There is evidence of a few black elder trees, but the majority of the vegetation consisted of plants growing along the banks and waterside. This wet environment was mainly fresh water, with only scant evidence of species that favour saltwater or brackish conditions.

Taphonomy

A large proportion of the site was exposed to a high degree of erosion, and the highest part of the dune has been disturbed by deep ploughing, which mixed the archaeological layer into the ploughsoil. As a result, there are no shallow features, and only deeper features like pit hearths and graves have survived. The lowering of the water table caused organic remains to oxidise. This might also explain the limited quantity of fish and bird bones found. Taphonomic processes (such as treading, bioturbation and local water erosion) were also taking place while the site was occupied, so the Mesolithic and Neolithic finds have become mixed.

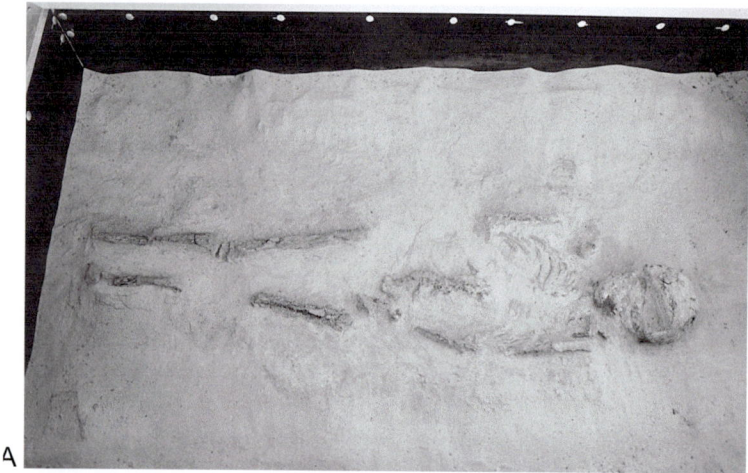

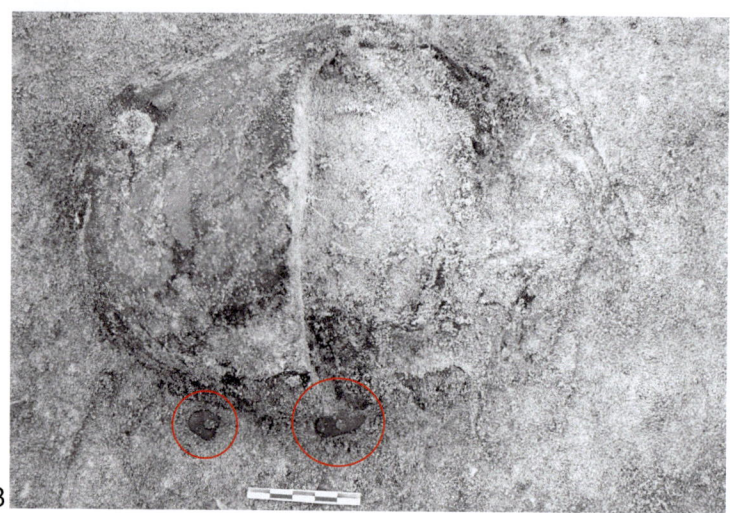

Figure I.28: One of the inhumation graves of Urk-E4. A. photograph showing entire grave as recovered. B. detail of two amber beads beside the head (source: Peters *et al.* 2001).

Appendix II. Glossary plant species

English name	Scientific name	Dutch name
Adder's Tongue	Ophioglossum vulgatum	Addertong
Alternate-flowered Water-milfoil	Myriophyllum alterniflorum	Teer vederkruid
Arctic Willow	Salix arctica	Arctische wilg
Ash	Fraxinus excelsior	Es
Aspen	Populus tremula	Ratelpopulier
Beaked Tasselweed	Ruppia maritima	Snavelruppia
Bedstraw	Galium	Walstro (G)
Birch	Betula	Berk (G)
Bogbean	Menyanthes trifoliata	Waterdrieblad
Bog-moss	Sphagnum	Veenmos (G)
Bracken	Pteridium aquilinum	Adelaarsvaren
Branched Bur-reed	Sparganium erectum	Grote egelskop
Broad Buckler-fern	Dryopteris dilatata	Brede stekelvaren
Bulrush	Typha	Lisdodde (G)
Bur-reed	Sparganium	Egelskop (G)
Buttercup	Ranunculus	Boterbloem (G)
Celery	Apium	Moerasscherm (G)
Charophytes	Charophyta	Kranswieren
Clubmoss	Lycopodium	Wolfsklauw (G)
Common Bulrush	Typha latifolia	Grote lisdodde
Common Club-rush	Schoenoplectus lacustris	Mattenbies
Crab Apple	Malus sylvestris	(Wilde) Appel
Crested Buckler-fern	Dryopteris cristata	Kamvaren
Crowberry	Empetrum nigrum	Kraaihei
Downy Birch	Betula pubescens	Zachte berk
Dropwort	Filipendula	Spirea (G)
Dryas	Dryas	Dryas (G)
Dwarf Birch	Betula nana	Dwergberk
Dwarf Willow	Salix herbacea	Kruidwilg
Elder	Sambucus	Vlier (G)
Elm	Ulmus	Iep (G)
Grass Family	Poaceae	Grassenfamilie
Grey Club-rush	Schoenoplectus tabernaemontani	Ruwe bies
Hare's-tail Cottongrass	Eriophorum vaginatum	Eenarig wollegras
Hawthorn	Crataegus	Meidoorn (G)

English name	Scientific name	Dutch name
Hazel	Corylus avellana	Hazelaar
Heath Family	Ericaceae	Heifamilie
Heather	Calluna vulgaris	Struikhei
Hypnaceae Family	Hypnaceae	Slaapmossen/Klauwtjesmossen
Juniper	Juniperus communis	Jeneverbes
Lesser Bulrush	Typha angustifolia	Kleine lisdodde
Lime	Tilia	Linde (G)
Malaceae	Malaceae	Appelachtigen
Male-fern	Dryopteris filix-mas	Mannetjesvaren
Male-fern / Buckler-fern	Dryopteris	Niervaren (G)
Marsh Cinquefoil	Comarum palustre	Wateraardbei
Marsh Fern	Thelypteris palustris	Moerasvaren
Marsh Marigold	Caltha palustris	Dotterbloem
Meadow Rue	Thalictrum	Ruit (G)
Mint	Mentha	Munt (G)
Mountain Ash	Sorbus aucuparia	Wilde lijsterbes
Mountain Pine	Pinus mugo	Dwergden
Mugwort	Artemisia	Alsem (G)
Narrow Buckler-fern	Dryopteris carthusiana	Smalle stekelvaren
Net-leaved Willow	Salix reticulata	Netwilg
Oak	Quercus	Eik (G)
Pine	Pinus	Den (G)
Polar Willow	Salix polaris	Poolwilg
Pondweed	Potamogeton	Fonteinkruid (G)
Poplar	Populus	Populier (G)
Rockrose	Helianthemum	Zonneroosje (G)
Rush	Juncus	Rus (G)
Saw-sedge	Cladium mariscus	Galigaan
Scots Pine	Pinus sylvestris	Grove den
Sea-buckthorn	Hippophae rhamnoides	Duindoorn
Sedge Family	Cyperaceae	Cypergrassenfamilie
Silver Birch	Betula pendula	Ruwe berk
Spiked Water-milfoil	Myriophyllum spicatum	Aarvederkruid
Spikemoss	Selaginella	Selaginella (G)
Stag's-horn Clubmoss	Lycopodium clavatum	Grote wolfsklauw
Water Chesnut	Trapa natans	Waternoot
Water Soldier	Stratiotes aloides	Krabbenscheer
Water-crowfoor	Ranunculus subgen. Batrachium	Waterranonkel (SG)
White Water-lily	Nymphaea alba	Witte waterlelie
Wild Pear	Pyrus pyraster	Peer
Willow	Salix	Wilg (G)
Yellow Iris	Iris pseudacorus	Gele lis
Yellow Water-lily	Nuphar lutea	Gele plomp

Literature

Albrechtsen, S.E. & E. Brinch Petersen 1976: Excavations of a Mesolithic Cemetery at Vedbaek, Denmark, *Acta Archaeologica* 47, 1-26.

Albarella, U., K. Dobney & P. Rowley-Conwy 2009: Size and shape of the Eurasian wild boar (Sus scrofa), with a view to the reconstruction of its Holocene history, Environmental Archaeology 14, 103-136.

Amkreutz, L.W.S.W., 2013: *Persistent Traditions. A long-term persepctive on communities in the process of Neolithisation in the Lower Rhine Area (5500-2500 cal BC)*, Leiden.

Andersen, S.H., 1974: Ringkloster: en jysk inlandsboplad med Ertebøllekultur, *Kuml* 1974, 11-108.

Andersen, S.H., 2000: 'Køkkenmøddinger' (Shell Middens) in Denmark: a Survey, *Proceedings of the Prehistoric Society* 66, 361-384.

Andersen, S.H., 2010: The first pottery in South Scandinavia, in: B. Vanmontfort, L. Louwe Kooijmans, L. Amkreutz & L. Verhart (eds), *Pots, farmers and foragers. Pottery traditions and social interaction in the earliest Neolithic of the Lower Rhine Area*, Leiden (Archaeological Series Leiden University 20), 167-176.

Arts, N., 1988: A survey of Final Palaeolithic archaeology in the southern Netherlands, in: M. Otte (ed.), *De la Loire à l'Oder. Les civilisations du Paléolithique final dans le nord-ouest européen*, Oxford (BAR International Series 444), 287-356.

Arts, N. & J. Deeben 1981: *Prehistorische jagers en verzamelaars te Vessem: een model*, Eindhoven.

Arts, N. & M. Hoogland 1987: A Mesolithic settlement area with a human cremation grave at Oirschot V, Municipality of Best, the Netherlands, *Helinium* 27, 172-189.

Aveling, E. M. & C. Heron 1998: Neolithic glue from the Sweet Track, Somerset, England. NewsWARP 23. Vol 23, 5-8.

Baetsen, S., 2008: Het grafveld, in: H. Koot, L. Bruning & R.A. Houkes (eds), *Ypenburg-locatie 4. Een nederzetting met grafveld uit het midden-neolithicum in het West-Nederlandse kustgebied*, Leiden, 119-188.

Baetsen, S. & L. Kootker 2011: Graf, in: T. Hamburg, A. Müller & B. Quadflieg (eds), *Mesolithisch Swifterbant. Mesolithisch gebruik van een duin ten zuiden van Swifterbant (8300-5000 v.Chr.). Een archeologische opgraving in het tracé van de N23/N307, provincie Flevoland*, Leiden/Amersfoort (Archol Rapport 174/ADC Rapport 3250), 147-156.

Bailey, G., 2007: Time perspectives, palimpsests and the archaeology of time, *Journal of Anthropological Archaeology* 26, 198-223.

Bakels, C.C. & J.T. Zeiler 2005: De vruchten van het land. De neolithische voedselvoorziening, in: L.P. Louwe Kooijmans, P.W. van den Broeke, H. Fokkens & A. van Gijn (red.), *Nederland in de prehistorie*, Amsterdam, 311-335.

Bakker, J.A., 1979: *The TRB West Group. Studies in the Chronology and Geography of the Makers of Hunebeds and Tiefstich Pottery*, Amsterdam (Cingula 5).

Bakker, J.A. & W.A.B. van der Sanden 2005: Trechterbekeraardewerk uit natte context: de situatie in Drenthe, *Nieuwe Drentse Volksalmanak* 112, 132-148.

Bakker, R. 2003: The emergence of agriculture on the Drenthe Plateau: a palaeobotanical study supported by high-resolution 14C-dating, Bonn (Archäologische Berichte 16).

Baumbartner, A., Sampol-Lopez, M., Cemeli, E., Schmidt, T.E., Evans, A.A., Donahue, R.E., Anderson, D., 2012: Genotoxicity Assessment of Birch-Bark Tar – A Most Versatile Prehistoric Adhesive. *Advances in Anthropology* 2(2), 49-56.

Beckerman, S.M., 2015: *Corded Ware coastal communities: using ceramic analysis to reconstruct third millennium BC societies in the Netherlands*, Groningen (PhD dissertation University of Groningen).

Berendsen, H.J.A., 2011: *De vorming van het land*, Assen.

Bettinger, R.L., R. Garvey & S. Tushingham 2015: *Hunter-gatherers. Archaeological and evolutionary theory*, New York.

Binford, L.R., 1982: The Archaeology of Place, *Journal of anthropological archaeology* 1 (1), 5-31.

Björck, S., M.J.C Walker, L.C. Cwynar, S. Johnsen, K-L Knudsen, J.J. Lowe & B. Wohlfarth 1998: An event stratigraphy for the Last Termination in the North Atlantic region based on the Greenland ice-core record: a proposal by the INTIMATE group, *Journal of Quaternary Science* 13(4), 283-292.

Bloo, S.B.C., 2002: Aardewerk, in: E.E.B. Bulten, F.J.G. van der Heijden & T. Hamburg (eds), *Emmeloord, Prehistorische visweren en fuiken*, Bunschoten (ADC Rapport 140), 78-87.

Bokelmann, K., 1994: Frühboreale Mikrolithen mit schäftungspech aus dem Heidmoor im Kreis Segeberg, Offa 51:7-47.

Bos, J.A.A. & R. Urz 2003: Late Glacial and Early Holocene environment in the middle Lahn River valley (Hessen, central-west Germany) and the local impact of Early Mesolithic man – pollen and macrofossil evidence, Vegetation History and Archaeobotany 12, 19-36.

Bos, J.A.A., B. van Geel, B.J. Groenewoudt & R.C.G.M. Lauwerier 2005: Early Holocene environmental change, the presence and disappearance of Early Mesolithic habitation near Zutphen (the Netherlands), Vegetation History and Archaeobotany 15, 27-43.

Bos, J.A.A., M.T.I.J. Bouman & C. Moolhuizen 2009: Analyse pollen en botanische macroresten/zaden, in: N.M. Prangsma & D.A. Gerrets (red.), *Hanzelijn Tunnel Drontermeer: verbinding tussen Oude en Nieuwe Land (Een Archeologische Begeleiding bij de Sallanddijk en een compenserend archeologisch onderzoek in gebied XVI)*, Amersfoort (ADC Rapport 1601), 34-37.

Bos, J.A.A. & F. Verbruggen 2012: Pollen onderzoek, in: M. Opbroek & E. Lohof (red.), *Tijd in centimeters. Een kijkje in het landschap van een dekzandrug te Almere*, Amersfoort (ADC Rapport 2662), 17, 32-45.

Bouman, M.T.I.J. & J.A.A. Bos 2012: Paleoecologie, in: T. Hamburg, A. Müller & B. Quadflieg (red.), *Mesolithisch Swifterbant. Mesolithisch gebruik van een duin ten zuiden van Swifterbant (8300-5000 v.Chr.). Een archeologische opgraving in het tracé van de N23/N307, provincie Flevoland*, Leiden/Amersfoort (Archol Rapport 174/ ADC Rapport 3250), 299-340.

Bourillet, J.F., J.Y. Reynaud, A. Baltzer & S. Zaragosi 2003: The 'Fleuve Manche': the submarine sedimentary features from the outer shelf to the deep-sea fans, *Journal of Quarternary Science* 18(3/4), 261-282.

Bouwmeester, H.M.P., H.A.C. Fermin & M. Groothedde (red.) 2006: *Geschapen landschap: tienduizend jaar bewoning en ontwikkeling van het cultuurlandschap op de Looërenk in Zutphen*, Deventer (BAAC report 00.068),

Braat, W.C., 1932: De archaeologie van de Wieringermeer, *Oudheidkundige Mededelingen van het Rijksmuseum te Leiden* 13, 15-56.

Bradley, R., 1984: *The Social Foundation of Prehistoric Britain*, London/New York.

Brinch Petersen, E. & C. Meiklejohn 2003. Three cremations and a Funeral: aspects of Burial Practice in Mesolithic Vedbaek, in: L. Larsson, H. Kindgren, K. Knutsson & A. Åkerlund (eds), *Mesolithic on the Move. Papers presented at the Sixth International Conference on the Mesolithic in Europe, Stockholm 2000*, Oxford, 485-493.

Brinkhuizen, D.C., 1976: De visresten van Swifterbant, *Westerheem* 25, 246-52.

Brinkkemper, O., W.J. Hogestijn, H. Peeters, D. Visser & C. Whitton 1999: The early Neolithic site at the Hoge Vaart, Almere, the Netherlands, with particular reference to non-diffusion of crop plants, and the significance of site function and sample location, *Vegetation History and Archaeobotany* 8. 79-86.

Brongers, J.A. & P.J. Woltering 1978: De prehistorie van Nederland, economisch-technologisch, Bussum.

Bronk Ramsey, C., 2013: *OxCal v.4.2 software*, Oxford.

E.E. Bulten & A. Clason 2001: The antler, bone and tooth tools of Swifterbant, The Netherlands (c. 5500 – 4000 cal. BC) compared with those from other Neolithic sites in the western Netherlands, in: A. M. Choyke & L. Bartosiewicz (eds) *Crafting Bone: Skeletal Technologies through Time and Space. Proceedings of the 2nd Meeting of the (ICAZ) Worked Bone Research Group, Budapest, 31 August – 5 September Meeting of the (ICAZ)WorkedBoneResearchGroup, Budapest, 31August – 5September nd 1999*. British Archaeological Reports, International Series 937. Oxford, 297-320.

Bulten, E.E.B., F.J.G. van der Heijden & T. Hamburg (red.) 2002: *Emmeloord, Prehistorische visweren en fuiken*, Bunschoten (ADC Rapport 140).

Bunnik, F.P.M. & F. Verbruggen 2011: *Palynologisch en archeobotanisch onderzoek aan monsters uit de boring Almere busbaan i.o.v. Becker & Van de Graaf*, Utrecht (TNO-rapport TNO-060-UT-2011-00069/B).

Çakirlar, C., B. Rianne, F. Koolstra, K. Cohen & D.C.M. Raemaekers 2020: Dealing with domestic animals in the fifth millennium cal BC Dutch wetlands: New insights from old Swifterbant assemblages, in: K.J. Gron, L. Sorensen & P. Rowley-Conwy (eds), *Farmers at the Frontier: A Pan-European Perspective on Neolithisation*, Oxbow, 263-287.

Cappers, R.T.J. & D.C.M. Raemaekers 2008: Cereal Cultivation at Swifterbant? Neolithic Wetland Farming on the North European Plain, *Current Anthropology* 49-3, 385-402.

Casparie, W.A. & J.P. De Roever, 1992: Vondsten van hout uit de opgravingen bij Swifterbant. *Bulletin van de Stichting Prehistorische Nederzetting Flevoland* 5 (3), 10-13.

Casparie, W.A., B. Mook-Kamps, R.M. Palfenier-Vegter, P.C. Struijk & W. van Zeist 1977: The palaeobotany of Swifterbant: A preliminary report, *Helinium* 17, 28-55.

Casparie, W.A. & J.G. Streefkerk 1992: Climatology, stratigraphy and paleoecological aspects of mire development, in: J.T.A. Verhoeven (ed.), *Fens and bogs in the Netherlands. Vegetation, history, nutrient dynamics and conservation*, Dordrecht, 81-129.

Champion, T., C. Gamble, S. Shennan & A. Whittle 1984: *Prehistoric Europe*, London.

Clevis, H., & A.D. Verlinde (red.) 1991: *Bronstijdboeren in Ittersumerbroek: opgraving van een bronstijdnederzetting in Zwolle-Ittersumerbroek*, Kampen.

Chapman, J., 2000: *Fragmentation in archaeology. People, places and broken objects in the prehistory of south-eastern Europe*, London.

Clarke D., 1976: Mesolithic Europe: the economic basis. in: G.G. Sieveking, I.W. Longewroth, K.E. Wilson (eds) Problems in Economic and Social Archaeology, London, 449-481

Clason, A.T., 1983: Worked and unworked antlers and bone tools from Spoolde, De Gaste, The IJsselmeerpolders and adjacent area, *Palaeohistoria* 25, 77-130.

Clason, A.T. & D.C. Brinkhuizen 1978: Swifterbant, Mammals, Birds, Fishes. A preliminary report, *Helinium* 18, 69-82.

Coe, A.L. (ed.) 2003: *The Sedimentary Record of Sea-level Change*, Cambridge.

Cohen, K.M. & M.P. Hijma 2008: Het Rijnmondgebied in het vroegholoceen: inzichten uit een diepe put bij Blijdorp (Rotterdam), *Grondboor & Hamer*, 64-71.

Cohen Stuart, C.D.R., J.J. Huisman, H.C.J. Visscher & S.A.D.S. Post 2006: *Basisrapportage vooronderzoek waardestelling, selectieadvies tekst bestemmingsplan. Plangebied 5B3, Vogelhorst*, Almere (Archeologische Rapporten Almere 3).

Constandse Westermann, T.S. & C. Meiklejohn 1979: The human remains from Swifterbant (Swifterbant Contributions 12), *Helinium* 19, 237 266.

Crombé, P. 2015: Forest fire dynamics during the Early and Middle Holocene along the southern North Sea basin as shown by charcoal evidence from burnt ant nests, *Vegetation History and Archaeobotany* 25(4), 311-321.

Crombé, Ph., Langohr, R., Louwagie, G., 2015: Mesolithic hearth-pits: fact or fantasy? A reassessment based on the evidence from the sites of Doel and Verrebroek (Belgium). Journal of Archaeological Science. 61, 158-171.

Crombé, Ph. & Langohr, R., 2020a: On the origin of Mesolithic charcoal-rich pits: a comment on Huisman et al. Journal of Archaeolological. Science, 119: 105058

Crombé, Ph. & Langohr, R., 2020b: Mesolithic charcoal-rich pits: "pit hearths"or "ant nests" A short response to Huisman et al. (2020), *Notae Prehistoricae* 40, 51-60.

Crombé, P., J. Verhegge, K. Deforce, E. Meylemans & E. Robinson 2015: Wetland landscape dynamics, Swifterbant land use systems, and the Mesolithic-Neolithic transition in the southern North Sea Basin, *Quarternary International* 378, 119-133.

Crombé, P., J. Deeben & M. van Strydonck 2014: Hunting in a changing environment: the transition from the Younger Dryas to the (Pre-)Boreal in Belgium and the southern Netherlands, in: J. Jaubert, N. Fourment & P. Depaepe (eds), *Transitions, ruptures et continuité en préhistoire / Transitions, rupture and continuity in prehistory, 2: paléolithique et mésolithique*, Paris (Actes XXVIIe congrès préhistorique de France, Bordeaux-Les Eyzies, 31 mai-5 juin 2010), 583-604

Culleton, B.J., 2006: Implications of a freshwater radiocarbon reservoir correction for the timing of late Holocene settlement of the Elk Hills, Kern County, California, *Journal of Archaeological Science* 33 (9), 1331-1339.

Cunliffe, B., 2008: *Europe between the Oceans. 9000 BC – AD 1000*, Yale.

Cunningham, P., 2012: The hazelnut assemblage from vindplaats 5, Netherlands, in: T. Hamburg, A. Müller & B. Quadflieg (red.), *Mesolithisch Swifterbant. Mesolithisch gebruik van een duin ten zuiden van Swifterbant (8300-5000 v.Chr.). Een archeologische opgraving in het tracé van de N23/N307, provincie Flevoland*, Leiden/Amersfoort (Archol Rapport 174/ ADC Rapport 3250), 495-500.

Dark, P, 2004: Plant remains as indicators of seasonality of site-use in the Mesolithic period, *Environmental Archaeology* 9 (1), 39-45.

Deckers, P.H., 1979: The flint material from Swifterbant, Earlier Neolithic of the Northern Netherlands. I. Sites S 2, S 4 and S 5. Final Reports on Swifterbant II, *Palaeohistoria* 21, 143 180.

Deckers, P.H., 1982: Preliminary notes on the Neolithic flint material from Swifterbant (Swifterbant Contribution 13), *Helinium* 22, 3339.

Degerbol M. & B. Fredskild 1970: The Urus (Bos primigenius Bojanus) and Neolithic domesticated cattle (Bos taurus domesticus Linné) in Denmark, *Det Kongelige Danske Videnskaberned Selskab Biolgiske Skrifter*, 17 : 1-234.

De Jong, J., 1966: *Uitkomsten van een pollenanalytisch onderzoek en een C14-ouderdomsbepaling van een sectie bij Swifterbant (Oostelijk Flevoland)*, Haarlem (Intern Rapport Rijksgeologische Dienst afdeling Kenozoïcum).

De Jong, J., 1974: *Pollenanalytisch onderzoek van een aantal boringen uit Oostelijk en Zuidelijk Flevoland*, Haarlem (Intern Rapport Rijksgeologische Dienst, Afdeling Palaeobotanie no. 705).

De Mulder, E.F.J., M.C. Geluk, I. Ritsema, W.E. Westerhoff & Th.E. Wong, 2003: De ondergrond van Nederland. Geologie van Nederland, deel 7: 247-352.

Delcourt, P.A. & H.R. Delcourt 2004: *Prehistoric native Americans and ecological change. Human ecosystems in eastern North America since the Pleistocene*. Knoxville.

Demirci, Ö., A. Lucquin, O.E. Craig & D.C.M. Raemaekers 2020: First lipid analysis of Early Neolithic pottery from Swifterbant (the Netherlands, ca. 4300-4000 BC), *Archaeological and Anthropological Sciences* 12, article number 105. https://doi.org/10.1007/s12520-020-01062-w

De Moor, J.J.W., J.A.A. Bos, M.T.I.J. Bouman, C. Moolhuizen, R. Exaltus, F.P.A. Maartense & T.J.M. van der Linden 2009: *Definitief Archeologische Onderzoek in het tracé van de Hanzelijn in het Nieuwe Land. Een interdisciplinaire geo-archeologische waardering van het begraven landschap van Oostelijk Flevoland*, Delft (Deltares-rapport 1001311-000-GEO-0005).

De Moor, J.J.W., A.M. Maurer, D. Fritzch & I. Devriendt 2013: *Almere Poort. Vindplaats 4E_17 "de Geest" – Nederlandstraat*, Amersfoort (Earth Integrated Archaeology Rapporten 45).

De Roever, J.P., 1976: Excavations at the river dune sites S21 S22 (Swifterbant Contribution 4), *Helinium* 16, 209-221.

De Roever, J.P., 1979: The Pottery from SwifterbantDutch Ertebølle? (Swifterbant Contribution 11), *Helinium* 19, 1336.

De Roever, J.P., 2004: *Swifterbant-aardewerk. Een analyse van de neolithische nederzettingen bij Swifterbant, 5e millennium voor Christus*, Groningen (Groningen Archaeological Studies 2).

Descola, Ph., 2010: *La fabrique des images*, Paris.

Devriendt, I., 2008: "Diamonds are a girl's best friend". Neolithische kralen en hangers uit Swifterbant, *Westerheem* 57 (6), 384-397.

Devriendt, I., 2013: *Swifterbant Stones. The Neolithic Stone and Flint Industry at Swifterbant (the Netherlands): from stone typology and flint technology to site function*, Groningen (Groningen Archaeological Studies 25).

De Wolf, H. & P. Cleveringa 2005: *Swifterbant: nat maar begroeid, wonen in een estuarium*, Utrecht (TNO-rapport NITG 05-014-B).

De Wolf, H. & P. Cleveringa 2006: *Palaeoecologisch diatomeënonderzoek van twee ontsluitingsprofielen te Swifterbant, Oostelijk Flevoland*, Utrecht (Intern Rapport TNO 2006-U-R0176/B).

De Wolf, H. & P. Cleveringa 2009a: *Palaeoecologisch diatomeënonderzoek van drie ontsluitingsprofielen genomen in 2006 te Swifterbant, Oostelijk Flevoland*, Beverwijk (WMC Rapport D11).

De Wolf, H. & P. Cleveringa 2009b: *Palaeoecologisch diatomeënonderzoek van S4-put 1 (bak 1 en 2) en van het lakprofiel te Swifterbant, Oostelijk Flevoland*, Beverwijk (WMC Rapport D29).

D'Hollosy, M. & S. Baetsen 2001: Fysische antropologie, in: F.J.C. Peters & J.H.M. Peeters (red.), *De opgraving van de mesolithische en neolithische vindplaats Urk-E4 (Domineesweg, gemeente Urk)*, Amersfoort (Rapportage Archeologische Monumentenzorg 93), 48-60.

Dimbleby, G.W., 1985: *The Palynology of Archaeological Sites*, London.

Drenth, E. & E. Lohof 2005: Mounds for the dead. Funerary and burial ritual in Beaker period, Early and Middle Bronze Age, in: L.P. Louwe Kooijmans, P.W. van den Broeke, H. Fokkens & A.L. van Gijn (eds), *The Prehistory of the Netherlands*, Amsterdam, 433-454.

Drenth, E. & L. Meurkens 2011: Laat-neolithische graven in: Hamburg, T., E. Lohof & B. Quadflieg (red.) *Bronstijd opgespoord. Archeologisch onderzoek van prehistorische vindplaatsen op Bedrijvenpark H2O – plandeel Oldebroek (Provincie Gelderland)*, Leiden/Amersfoort (Archol Rapport 142/ ADC Rapport 2627), 197-276

Drenth, E. & M.J.L.Th. Niekus 2009a: Stone Mace-heads and Picks: a Case-study from the Netherlands, in: P. Crombé, M. Van Strydonck, J. Sergeant, M. Boudin & M. Bats (eds), *Chronology and Evolution within the Mesolithic of North-West Europe. Proceedings of an International Meeting, Brussels, May 30th-June 1st 2007*, Newcastle upon Tyne, 747-766.

Drenth, E. & M.J.L.Th. Niekus 2009b: 14C-datierte Geröllkeulen aus den Niederlanden, *Archäologische Informationen* 32, 91-94.

Dresscher, S. & D.C.M. Raemaekers 2010: Oude geulen op nieuwe kaarten. Het krekensysteem bij Swifterbant (Fl.), *Paleo-aktueel* 21, 31-38.

Dunnell, R.C., 1992: The notion site, in: J. Rossignol & L. Wandsnider (eds), *Space, time and archaeological landscapes*, New York, 21-41.

Ente, P.J., J. Koning & R. Koopstra 1986: *De bodem van oostelijk Flevoland*, Lelystad (Flevobericht 258).

Ebbinge, B., 2007: *In de greep van Siberische lemmingen*, Zeist.

Esser, E., J. van Dijk & M. Rijkelijkhuizen 2010: *Dierlijke resten van laat- en postmiddeleeuwse (kapittel) boerderijen langs de Hogeweide (gem. Utrecht)*, Delft (Ossicle 184).

Exaltus, R.P., 1993: *Archeologisch onderzoek in het tracé van Rijksweg 27 (Zuidelijk Flevoland)*, Amsterdam (RAAP-rapport 83).

Exaltus, R., 2001: Deel 14: Micromorfologie: onderzoek aan slijpplaatmonsters van grondsporen, in: J.W.H. Hogestijn & J.H.M. Peeters (eds), *De mesolithische en vroeg-neolithische vindplaats Hoge Vaart-A27 (Flevoland)*, Amersfoort (Rapportage Archeologische Monumentenzorg 79).

Fokkens, H., 1998: *Drowned landscape. The Occupation of the Western Part of the Frisian-Drentian Plateau, 4400 BC-AD 500*, Assen.

Fokkens, H., B.J.W. Steffens & S.F.M. van As 2016: *Farmers, fishers, fowlers, hunters. Knowledge generated by development-led archaeology about the Late Neolithic, the Early Bronze Age and the start of the Middle Bronze Age (2850 – 1500 cal BC) in the Netherlands*, Amersfoort, (Nederlandse Archeologische Rapporten 53).

Foley, R., 1981: Off-site archaeology; an alternative approach for the short-sited, in: I. Hodder, G. Isaac & N. Hammond (eds), *Patterns of the past, studies in honour of David Clarke*, Cambridge, 157-183.

Fontijn, D.R., 2002: *Sacrificial landscapes. Cultural Biographies of persons, objects and 'natural' places in the Bronze Age of the southern Netherlands*, Leiden/Leuven (Analecta Praehistorica Leidensia 33).

Funk, C., 2011. Yupik Eskimo Gendered Information Storage Patterns. In: R. Whallon, W.A. Lovis & R.K. Hitchcock (eds), *Information and its role in hunter-gatherer bands*, Los Angeles, 29-48.

Geerts, R.C.A., A. Müller, M.J.L.Th. Niekus & F.J. Vermue (red.) 2019: *Mesolithische kampen onder de oever van het Reevediep, een archeologische opgraving van vindplaats 9 in het tracé van de hoogwatergeul in het Reevediep te Kampen*, Amersfoort (ADC Monografie 28/ADC Rapport 4750).

Gehasse, E.F., 1995: *Ecologisch-archeologisch onderzoek van het Neolithicum en de Vroege Bronstijd in de Noordoostpolder met de nadruk op P14 gevolgd door een overzicht van de bewoningsgeschiedenis en de bestaanseconomie binnen de Holocene delta*, Amsterdam (Ph.D thesis University of Amsterdam).

Geurts, A.J., 1991: *Schokland. De historie van een weerbarstig eiland*, Zutphen.

Geuverink, J., D.C.M. Raemaekers & I. Devriendt 2009: *Op zoek naar archeologie bij Doug's duin, Kamperhoekweg, Swifterbant, gemeente Dronten. Inventariserend veldonderzoek door middel van boringen*, Groningen (Grondsporen 4).

Gotjé, W., 1993: *De Holocene laagveenontwikkeling in de randzone van de Nederlandse kustvlakte (Noordoostpolder)*, Amsterdam (Ph.D. thesis VU University Amsterdam).

Gotjé, W., 1997a: *De vegetatie op en rond een Mesolithische en Vroeg Neolithische vindplaats. Een ecologisch onderzoek aan drie kernen op de vindplaats Hoge Vaart*, Amsterdam (BIAXiaal 36).

Gotjé, W., 1997b: *Het landschap in Zuidelijk Flevoland tussen 9500 en 4300 BP. Een landschapsreconstructie in het gebied Wet Bodembescherming*, Amsterdam (BIAXiaal 40).

Gotjé, W., 2001: Deel 9, Bodemkunde en landschapsecologie III: vegetatieontwikkeling en diatomeeën, in: J.W.H. Hogestijn & J.H.M. Peeters (eds), *De mesolithische en vroeg-neolithische vindplaats Hoge Vaart-A27 (Flevoland)*, Amersfoort (Rapportage Archeologische Monumentenzorg 79).

Groenendijk, H.A., 2004: Middle Mesolithic occupation on the extensive site NP3 in the peat reclamation district of Groningen, the Netherlands, in: P. Crombé & P. Vermeersch (eds), *7: Le mésolithique / The Mesolithic*, Oxford (Acts of the XIVth congress of the UISPP, University of Liège, Belgium, 2-8 September, 2001) (BAR International Series 1302), 19-26.

Groenendijk, H.A., 1997. *Op zoek naar de Horizon. Het landschap van Oost-Groningen en zijn bewoners tussen 8000 voor Chr. en 1000 na Chr.*, Groningen (Regio- en landschapsstudies nr. 4.).

Groenewoudt, B.J., P.A.C. Schut, F.J.G. van der Heijden, J.H.M. Peeters & M.H. Wispelwey 2006: *Een inventariserend veldonderzoek bij de Hunneschans (Uddel, Gelderland): nieuwe gegevens over de steentijdbewoning bij het Uddelermeer en een beknopt overzicht van de onderzoeksgeschiedenis van de Hunneschans*, Amersfoort (Rapportage Archeologische Monumentenzorg 143).

Groenewoudt, B.J., J.H.C. Deeben & H.M. van der Velde 2000: *Raalte-Jonge Raan 1998. De afronding van de opgraving en het esdekonderzoek*, Amersfoort (Rapportage Archeologische Monumentenzorg 73).

Haakanson, S. & P. Jordan 2011: 'Marking' the land: sacrifices, cemeteries and sacred places among the Iamal Nenetses. In: P. Jordan (ed.), *Landscape and culture in northern Eurasia*, New York, 161-178.

Habermehl, K.-H., 1975: Die Alterbestimmung bei Haus- und Labortieren, Berlijn/Hamburg.

Haanen, P.L.P. & J.W.H. Hogestijn, 2001: Deel17 Aardewerk: morfologische en technologische aspecten, in J.W.H. Hogestijn, & J.H.M. Peeters (red.) 2001: *De mesolithische en vroeg-neolithische vindplaats Hoge Vaart-A27 (Flevoland)*, Amersfoort (Rapportage Archeologische Monumentenzorg 79).

Hamburg, T., E. Lohof & B. Quadflieg (red.) 2011: *Bronstijd opgespoord. Archeologisch onderzoek van prehistorische vindplaatsen op Bedrijvenpark H2O – plandeel Oldebroek (Provincie Gelderland)*, Leiden/Amersfoort (Archol Rapport 142/ ADC Rapport 2627).

Hamburg, T., A. Müller & B. Quadflieg (red.) 2012: *Mesolithisch Swifterbant. Mesolithisch gebruik van een duin ten zuiden van Swifterbant (8300-500 BC). Een archeologische opgraving in het tracé van N23/N307, provincie Flevoland*, Leiden (Archol Rapport 174 / ADC Rapport 3250).

Hamburg, T., A. Tol, J. de Moor & Y. Lammers-Keijsers 2014: *Afgedekt verleden. Opsporing, waardering en selectie van prehistorische archeologische vindplaatsen in Flevoland. Programma Kennisontwikkeling Archeologie Hanzelijn (Thema 1B)*, Leiden/Amersfoort (Archol Rapport 244/Earth Integrated Archaeology rapporten 49).

Harsema, O.H. 1982: Settlement site selection in Drenthe in later prehistoric times: criteria and considerations, *Analecta Praehistorica Leidensia* 15, 145-160.

Havinga, A.J., 1963: *A palynological Investigation of Soil Profiles developed in Cover Sand*, Wageningen (Mededelingen van de Landbouwhogeschool te Wageningen, Nederland 63 (1)).

Havinga, A.J., 1974: Problems in the Interpretation of Pollen Diagrams of Mineral Soils, *Geologie en Mijnbouw* 53 (6), 449-453.

Hermsen, I., M. van der Wal & H. Peeters, 2015: *Afslag Olthof. Archeologisch onderzoek naar de vroegprehistorische vindplaatsen op de locaties Olthof-Noord en Olthof-Zuid in Epse-Noord*, Deventer, (Rapportage archeologie Deventer 34).

Hijma, M., 2009: *From river valley to estuary. The early-mid Holocene transgression of the Rhine-Meuse valley, The Netherlands*, Utrecht (Netherlands Geographical Studies 389).

Hijma, M.P., K.M. Cohen, G. Hoffmann & A.J.F. van der Spek 2009: From river valley to estuary: the evolution of the Rhine mouth in the early to middle Holocene (western Netherlands, Rhine-Meuse delta), *Netherlands Journal of Geosciences*, 13-53.

Hijma, M.P. & K.M. Cohen 2010: Timing and magnitude of the sea-level jump preluding the 8200 yr event, *Geology* 38(3), 275-278.

Hoek, W.Z., 1997a: *Palaeogeography of Lateglacial Vegetations. Aspects of Lateglacial and Early Holocene vegetation, abiotic landscape, and climate in The Netherlands*, Utrecht/Amsterdam (Netherlands Geographical Studies 230).

Hoek, W.Z., 1997b: *Atlas to Palaeogeography of Lateglacial Vegetations. Maps of Lateglacial and Early Holocene landscape and vegetation in The Netherlands, with an extensive review of available palynological data*, Utrecht/Amsterdam (Netherlands Geographical Studies 231).

Hogestijn, J.W., 1990: From Swifterbant to TRB in the IJssel-Vecht Basin-some suggestions, in: D. Jankowska (ed.), *Die Trichterbecherkultur. Neue Forschungen und Hypothesen. Teil I*, Poznan, 163-180.

Hogestijn, J.W., 2019: De Tweeling in Almere Stichtsekant, Laat-neolithische visweren in een verdwenen meer, *Archeologie in Nederland* 3 (5), 2-9.

Hogestijn, J.W.H., 1993: Flevoland, *Jaarverslag ROB 1992*, 206-209.

Hogestijn, J.W.H., & E. Drenth 2000/2001: In Slootdorp stond een trechterbeker-huis? Over midden- en laat-neolithische huisplattegronden uit Nederland, Archeologie 10, 42-79.

Hogestijn, J.W.H. & J.H.M. Peeters (red.) 2001: *De mesolithische en vroeg-neolithische vindplaats Hoge Vaart-A27 (Flevoland)*, Amersfoort (Rapportage Archeologische Monumentenzorg 79).

Hogestijn, J.W.H. & J.H.M. Peeters 2001: Deel 20 Op de grens van land en water: jagers-vissers-verzamelaars in een verdrinkend landschap, in Hogestijn, J.W.H. & J.H.M. Peeters (red.) *De mesolithische en vroeg-neolithische vindplaats Hoge Vaart-A27 (Flevoland)*, Amersfoort, (Rapportage Archeologische Monumentenzorg 79).

Hogestijn, W.J.H. & H.C.J. Visscher 2011: *Inventariserend Veldonderzoek Plangebied 4E Europakwartier West, Almere Poort*, Almere (Archeologische Rapporten Almere 80).

Holst D (2010) Hazelnut economy of early Holocene hunter-gatherers: a case study from Mesolithic Duvensee, northern Germany. Journal of Archaeological Science 37:2,871-2,880

Horreüs de Haas, J. & R. Horreüs de Haas 1984: *Als in het stenen tijdperk. Verslag van het spel in de Flevopolder*. Den Haag.

Hübner, E., 2011: *Jungneolithische Gräber auf der Jütischen Halbinsel. Typologische und chronologische Studien zur Einzelgrabkultur*, Copenhagen (Nordiske Fortidsminder, Serie B 24, 1-3).

Huisman, D.J., Niekus, M.J.LT, J.H.M. Peeters, R. Geerts & A. Müller, 2020: Arguments in favour of an anthropogenic origin of mesolithic pit hearths. A reply to Crombé and Langohr (2020), *Journal of Aracheological Science*, 119: 105144.

Huisman, D.J., Niekus, M.J.LT, J.H.M. Peeters, R. Geerts & A. Müller 2019: Deciphering the complexity of a 'simple' mesolithic phenomenon: Indicators for construction, use and taphonomy of pit hearths in Kampen (the Netherlands), Journal of Archaeological Science 119: 104987.

Huisman, D.J., A.G. Jongmans & D.C.M. Raemaekers 2009: Investigating Neolithic land use in Swifterbant (NL) using micromorphological techniques, *Catena* 78, 185-197.

Huisman, D.J. & D.C.M. Raemaekers 2014: Systematic cultivation of the Swifterbant wetlands (The Netherlands). Evidence from Neolithic tillage marks (c. 4300-4000 cal. BC), *Journal of Archaeological Science* 49, 572-584.

Huisman, H. 1977: Over het voorkomen van bruinhout en barnsteen in de ondergrond van Noord-Nederland en Noord Duitsland, grondboor & Hamer, 154-160

Hullegie, A.G.J., 2009: *Swifterbant S4: Archeo-ichtyologisch Onderzoek Vlak 3 en 9*, Groningen (unpublished Bachelor's thesis).

Ingold, T., 1993: The temporality of the landscape, World Archaeology, 25:2, 152-174.

Innes, J., J. Blackford & I. Simmons 2010: Woodland disturbance and possible land-use regimes during the Late Mesolithic in the English uplands: pollen, charcoal and non-pollen palynomorph evidence from Bluewath Beck, North York Moors, UK, *Vegetation History and Archaeology* 19, 439-452.

Isaac, G., 1981: Stone Age visiting cards: approaches to the study of early land-use patterns, in: I. Hodder, G. Isaac & N. Hammond (eds), *Patterns of the Past*, Cambridge, 206-227.

Innes, J.B., & J.J. Blackford 2003: The ecology of Late Mesolithic woodland disturbances: model testing with fungal spore assemblage data, *Journal of Archaeological Science* 30, 185-194.

Jansma, M.J., 1990: Diatoms from a Neolithic excavation on the former island of Schokland, IJsselmeerpolders, the Netherlands, *Diatom Research* 5, 301-309

Janssen, C.R., 1974: *Verkenningen in de palynologie*, Utrecht

Jelgersma, S., 1979: Sea-level changes in the North Sea Basin, in: E. Oele, R.T.E. Schüttenhelm & A.J. Wiggers (eds), *The Quaternary History of the North Sea*, Upsala (Acta Universitatis Upsaliensis. Symposia Universitatis Upsaliensis Annum Quingentesimum Celebrantis 2), 233-248.

Jordan, P., 2015: *Technology as human social tradition: cultural transmission among hunter-gatherers*, Oakland.

Jordan, P., 2003: *Material culture and sacred landscape. The anthropology of the Siberian Khanty*. Lanham.

Kampffmeyer, U., 1991: *Die Keramik der Siedlung Hüde I am Dümmer. Untersuchungen zur Neolithisierung der nordwestdeutschen Flachlands*, Göttingen.

Kerkhoven, A., 2011: *Compacte analyse verzamelde gegevens onderdeel Archeologie. Interne memo ihkv Programma Kennisontwikkeling Archeologie Hanzelijn (PAKH)*, 21 oktober 2011.

Kiden, P., L. Denys & P. Johnston 2002: Late Quaternary sea-level change and isostatic and tectonic land movements along the Belgian-Dutch North Sea coats: geological data and model results, *Journal of Quarternary Science* 17, 535-546.

Kintingh, K.W., 1989: The Effectiveness of Subsurface Testing: A Simulation Approach, *American Antiquity* 53/4, 686-707.

Kleijne, J.P., O. Brinkkemper, R.C.G.M. Lauwerier, B.I. Smit & E.M. Theunissen (eds), 2013: *A matter of life and death at Mienakker (the Netherlands): Late Neolithic behavioural variability in a dynamic landscape*, Amersfoort (Nederlandse Archeologische Rapporten 45),

Koch, E., 1998: *Neolithic bog pots from Zealand, Møn, Lolland and Falster*, Kopenhagen (Nordiske fortidsminder Serie B 16).

Kooistra, L.I., 2012a: Houtskool onderzoek, in: M. Opbroek & E. Lohof (red.), *Tijd in centimeters. Een kijkje in het landschap van een dekzandrug te Almere*, Amersfoort (ADC Rapport 2662), 18-19; 45-53. (Also published as: Kooistra, L.I., 2010: *Houtskoolonderzoek aan lagen op de overgang van Pleistoceen en Holoceen in Almere Haven (Maatweg – Meesweg – Meentweg)*, Zaandam (BIAXiaal 481).

Kooistra, L.I., 2012b: Hout, in: T. Hamburg, A. Müller & B. Quadflieg (red.), *Mesolithisch Swifterbant. Mesolithisch gebruik van een duin ten zuiden van Swifterbant (8300-5000 v.Chr.). Een archeologische opgraving in het tracé van de N23/N307, provincie Flevoland*, Leiden/Amersfoort (Archol Rapport 174/ ADC Rapport 3250), 369-374.

Kooistra, L.I., 2012c: Houtskool uit mesolithische kuilen, in: T. Hamburg, A. Müller & B. Quadflieg (red.), *Mesolithisch Swifterbant. Mesolithisch gebruik van een duin ten zuiden van Swifterbant (8300-5000 v.Chr.). Een archeologische opgraving in het tracé van de N23/N307, provincie Flevoland*, Leiden/Amersfoort (Archol Rapport 174/ ADC Rapport 3250), 375-387.

Kooistra, L.I., 2014: 9 De vegetatiegeschiedenis vanaf het Midden-Neolithicum tot in de Romeinse tijd, in: R.M. van Heeringen, W.A.M. Hessing, L.I. Kooistra, S.

Lange, B.I. Quadflieg, R. Schrijvers & W. Weerheijm, *Archeologisch landschapsonderzoek in het kader van het project Kwaliteitsverbetering Kotterbos (locatie Natuurboulevard) in de gemeente Lelystad, provincie Flevoland. Menselijke activiteit in natte landschappen in de Steentijd en de (Vroeg-) Romeinse tijd*, Amersfoort (Vestigia Rapport V1132), 63-84.

Kooistra, L.I., L. Kubiak-Martens & J.J. Langer 2009: Archeobotanisch onderzoek in: N.M. Prangsma & D.A. Gerrets (red.), *Hanzelijn Tunnel Drontermeer: verbinding tussen Oude en Nieuwe Land (Een Archeologische Begeleiding bij de Sallanddijk en een compenserend archeologisch onderzoek in gebied XVI)*, Amersfoort (ADC Rapport 1601), 52-57. (Also published as: Kooistra, L.I., L. Kubiak-Martens & J.J. Langer 2009: *Mesolithische haardkuilen van vindplaats Hanzelijn, Tunnel Drontermeer op houtskool onderzocht*, Zaandam (BIAXiaal 417).

Kooistra, M.J. & M.M. Pulleman 2010: Features Related to Faunal Activity, in: G. Stoops, V. Marcelino & F. Mees (eds), *Interpretation of Micromorphological Features of Soils and Regoliths*, 397-418.

Kooistra, L. & H. van Haaster 2011: *Inventarisatie van bio-archeologisch onderzoek in het kader van het Programma Kennisontwikkeling Archeologie Hanzelijn (PAKH)*, Zaandam (BIAX-rapport 554).

Koopstra, R., G. Lenselink & U. Menke 1993: *Geologische en bodemkundige atlas van het IJsselmeer*, Lelystad (Rijkswaterstaat Directie Flevoland).

Krakker, J.J., M.J. Shott & P.D. Welch 1983: Design and evaluation of shovel-test sampling in regional archaeological survey, *Journal of Field Archaeology* 10, 469-480.

Kroezenga, P., J.N. Lanting, R.J. Kosters, W. Prummel & J.P. de Roever 1991: Vondsten van de Swifterbantcultuur uit het Voorste Diep bij Bronneger (Dr.), *Paleo-aktueel* 2, 32-36.

Kubiak-Martens, L., Langer, J.J., Kooistra, L.I., 2012: Plantenresten en teer in haardkuilen. In: Hamburg, T., Müller, A., Quadflieg, B. (eds), MesolithischSwifterbant: mesolithisch gebruik van een duin ten zuiden van Swifterbant(8300-5000 v. Chr.): een archeologische opgraving in het tracé van de N23/N307,provincie Flevoland, Leiden &Amersfoort (Archol report 174 & ADC report 3250), 341-360.

Kubiak-Martens, L., O. Brinkkemper & T. Oudemans 2014: What's for dinner? Processed food in the coastal area of the northern Netherlands in the Late Neolithic, Vegetation History and Archaeobotany 24, 47-62.

Kubiak-Martens, L., 2006: Botanical remains and plant food subsistence, in: L.P. Louwe Kooijmans & P.F.B Jongste (eds), *Schipluiden, a Neolithic settlement on the Dutch North Sea Coast c. 3500 cal BC*, Leiden (Analecta Praehistorica Leidensia 37/38), 317-336.

Laarman, F.J., 2001: Deel 16 Archeozoölogie: aard en betekenis van de dierlijke resten, in: J.W.H. Hogestijn & J.H.M. Peeters (red.), *De mesolithische vindplaats Hoge Vaart-A27 (Flevoland)*, Amersfoort (Rapportage Archeologische Monumentenzorg 79).

Lane, P.J., 2014. Hunter-gatherer-fishers, ethnoarchaeology, and analogical reasoning, in: V. Cummings, P. Jordan & M. Zvelebil (eds), *The Oxford handbook of the archaeology and anthropology of hunter-gatherers*, Oxford, 104-150.

Lanting, J.N., 2001: Dating of cremated bones, *Radiocarbon* 43, 249-254.

Lanting, J.N. & J. van der Plicht 1997/1998: De 14C-chronologie van de Nederlandse pre- en protohistorie. II: Mesolithicum, *Palaeohistoria* 39/40, 99-162.

Lanting, J.N. & J. van der Plicht 1999/2000: De 14C-chronologie van de Nederlandse pre- en protohistorie. III: Neolithicum, *Palaeohistoria* 41/42, 1-110.

Larsson, L., 1983: Ageröd V: an Atlantic bog site in central Scania, Lund, *Acta Archaeological Lundensia* 12.

Larson, G., K. Dobney, U. Albarella, M. Fang, E. Matisoo-Smith, J. Robbins, S. Lowden, H. Finlayson, T. Brand, E. Willerslev, P. Rowley-Conwy, L. Andersson & A. Cooper 2005: Worldwide phylogeography of wild boar reveals multiple centers of pig domestication, *Science* 307, 1618-1621

Lauwerier, R.C.M.G. & J. Deeben 2011: Burnt animal remains from Feddermesser sites in the Netherlands, *Archäologisches Korrespondenzblatt* 41, 1-20.

Lauwerier, R.C.G.M., Th. Van Kolfschoten & L.H. van Wijngaarden-Bakker 2005: De archeozoologie van de steentijd, in: J. Deeben, E. Drenth, M.-F. van Oorsouw & L. Verhart (red.), *De steentijd van Nederland*, Zutphen (Archeologie 11/12), 39-66.

Lavrillier, A., 2011. The creation and persistence of cultural landscapes among the Siberian Evenkis: two conceptions of 'sacred' space. In: P. Jordan (ed.), *Landscape and culture in northern Eurasia*, New York, 215-231.

Leijnse, K., 2006: *Hanzelijn, tracédeel Nieuwe Land; archeologisch vooronderzoek: een inventariserend veldonderzoek-IVO fase 1 (afronding en IVO fase 2)*, Amsterdam (RAAP-rapport 1305).

Lenselink, G. & U. Menke 1995: *Geologische en bodemkundige atlas van het Markermeer*, Lelystad.

Little, A., B. Elliott, C. Conneller, D. Pomstra, A.A. Evans, L.C. Fitton., A. Holland, R. Davis, R. Kershaw, S. O'Connor, T. O'Connor, T. Sparrow, A.S. Wilson, P. Jordan, M.J. Collins, A. Carlo Colonese, O.E. Craig, R. Knight, A.J.A. Lucquin, B. Taylor & N. Milner

2016: Technological analysis of the world's earliest shamanic costume: a multi-scalar, experimental study of a red deer headdress from the Early Holocene site of Star Carr, North Yorkshire, UK, *PLoS ONE* 11(4): e0152136. (doi:10.1371/journal.pone.0152136).

Lohof, E., T. Hamburg & J. Flamman (red.) 2011: *Steentijd opgespoord. Archeologisch onderzoek in het tracé van de Hanzelijn-Oude Land*, Leiden/Amersfoort (Archol Rapport 138 & ADC Rapport 2576).

Louwe Kooijmans, L.P., 1976 Local Developments in a Borderland. A survey of the Neolithic at the Lower Rhine, *Oudheidkundige Mededelingen uit het Rijksmuseum van Oudheden te Leiden* 57, 227297.

Louwe Kooijmans, L.P. (red.), 2001a: *Archeologie in de Betuweroute. Hardinxveld-Giessendam Polderweg. Een mesolithisch jachtkamp in het rivierengebied (5500-5000 v. Chr.)*, Amersfoort (Rapportage Archeologische Monumentenzorg 83).

Louwe Kooijmans, L.P. (red.), 2001b: *Archeologie in de Betuweroute. Hardinxveld-Giessendam De Bruin. Een kampplaats uit het Laat-Mesolithicum en het begin van de Swifterbant-cultuur (5500-4450 v. Chr.)*, Amersfoort (Rapportage Archeologische Monumentenzorg 88).

Louwe Kooijmans, L.P., 2003: The Hardinxveld sites in the Rhine/Meuse Delta, the Netherlands, 5500-4500 cal BC, in: L. Larsson, H. Kindgren, K. Knutsson, D. Loeffler & A. Åkerlund (eds), *Mesolithic on the Move. Papers presented at the Sixth International Conference on the Mesolithic in Europe, Stockholm 2000*, Oxford, 608-624.

Louwe Kooijmans, L.P., 2007: Multiple choices. Mortuary practices in the Low Countries during the Mesolithic and Neolithic, 9000-3000 cal BC, in: L. Larsson, F. Luth & T. Terberger (eds), *Innovation and continuity – non-megalithic mortuary practices in the Baltic. New methods and research into the development of stone age society*, Mainz am Rhein (Berichte der Romisch Germanische Kommission 88), 551-580.

Louwe Kooijmans, L.P., 2010: De VL-pot van Kootwijk en enkele andere potdeposities uit de tweede helft van het vierde millenium v. Chr., in: T. de Ridder, L. Verhart, J. Coenraadts, G. Groeneweg, M. Lockefeer & L. Kruyf (eds), *Special 2010 Vlaardingen-cultuur*, (Westerheem Special nr. 2), 194-207.

Louwe Kooijmans, L.P., 2013: Reflections on the Mesolithic burial pits at Mariënberg (province of Overijssel), the Netherlands, in: M.J.L.Th. Niekus, R.N.E. Barton, M. Street & Th. Terberger, *A mind set on flint studies in honour of Dick Stapert*, Groningen (Groningen Archaeological Studies 16), 401-424.

Lovis, W.A. & R.E. Donahue 2011: Space, information and knowledge: Ethnocartography and North American boreal forest hunter-gatherers, in: R. Whallon, W.A. Lovis & R. Hitchcock, *Information and its role in hunter-gathrerer bands*, Los Angeles, 59-84.

Luijten, H., 1986: *Zadenanalyse van de prehistorische vindplaats P14 (Noordoostpolder)*, Amsterdam (unpublished student report).

Luijten, H., 1987: *Palaeobotanische analyse van enkele Vroeg-Neolithische afvallagen van Zuiderbuurt I (Noordoostpolder)*, Amsterdam (doctoraal scriptie UvA).

Makaske, B., D.G. van Smeerdijk, J.R. Mulder & T. Spek 2002a: De stijging van de waterspiegel nabij Almere in de periode 5300-2300 v. Chr, Wageningen (Alterra-rapport 478).

Makaske, B., D.G. van Smeerdijk, M.J. Kooistra, R.M.K. Haring, E.C. Verbauwen & A. Smit 2002b: *Een verkenning van begraven dekzandbodems in een bodembeschermingsgebied ten zuidoosten van Almere. Een interdisciplinair onderzoek naar de kwaliteit van het bodemarchief met implicaties voor archeologische waarden*, Wageningen (Alterra-rapport 486).

Makaske, B., D.G. van Smeerdijk, H. Peeters, J.R. Mulder & T. Spek 2003: Relative water-level rise in the Flevo Lagoon (The Netherlands), 5300 – 2000 cal. yr BC.: an evaluation of new and existing basal peat time-depth data, *Netherlands Journal of Geosciences / Geologie en Mijnbouw* 82, 115-131.

Margl, H. & K. Zukrigl 1981: Die Standorts- und Vegetationskartierung der Donau-Auen bei Wien, *Angewandte Pflanzensoziologie* 26, Wien.

Marguerie, D. & Hunot, J.-Y. 2006: Charcoal analysis and dendrology: data from archaeological sites in north-western France. *Journal of Archaeological Science* 34, 1417-1433

Mason, S. & J.G. Hather 2000. Parenchymatous plant remains from Staosnaig. In: S. Mithen (ed.), Hunter-gatherer landscape archaeology. The Southern Hebrides Mesolithic Project 1988-1998, Cambridge (McDonald Institute Monographs), 415-425.

Meiklejohn, C. & T.S. Constandse Westermann 1978: The human skeletal material from Swifterbant, Earlier Neolithic of the Northern Netherlands, I. Inventory and demography, Final Reports on Swifterbant I, *Palaeohistoria* 20, 3989.

Mellars, P., 1976: Fire ecology, animal populations and man: a study of some ecological relationships in prehistory, *Proceedings of the Prehistoric Society* 42, 15-45.

Mellars, P. & P. Dark (eds) 1998: *Star Carr in context: new archaeological and palaeoecological investigations at the Early Mesolithic site of Star Carr*, North Yorkshire, Oxford.

Ménot, G., E. Bard, F. Rostek, J.W.H. Weijers, E.C. Hopmans, S. Schouten, J.S.S. Damsté 2006: Early Reactivation of European Rivers during the Last Deglaciation, *Science* 313(5793), 1623-1625.

Menke, U. & G. Lenselink 1991: *Bodemkundig-geologisch onderzoek langs een oude loop van de Vecht op kavel D133 en D134, Noordoostpolder*, Lelystad (Intern rapport Rijkswaterstaat directie Flevoland 1991-23).

Menke, U., E. van de Laar & G. Lenselink 1998: *De geologie en bodem van Zuidelijk Flevoland*, Lelystad (Flevobericht 415).

Merlo, S., 2010: *Contextualising intra-site spatial analysis: the role of three-dimensional GIS modelling in understanding excavation data*, Cambridge (PhD dissertation Cambridge University).

Meyer-Orlac, R., 1982: *Mensch und Tod: Archäologischer Befund, Grenzen der Interpretation*, Freiburg.

Mietes, E.K. & R. Schrijvers 2005: *N23 Lelystad- Dronten. Een archeologisch bureauonderzoek en advies t.b.v. bestemmingsplanwijzigingen*, Amersfoort (Vestigia Rapportnummer V257).

Mischka, D., 2011: The Neolithic burial sequence at Flintbek LA 3, North Germany, and its cart tracks: a precise chronology, *Antiquity*, 85, 742-758.

Mithen, S., Finlayson, B., Najjar, M., Jenkins, E., Smith , S., Hemsley, S., Maricevic, D., Pankhurst, N., Yeomans, L. & H. Al-Almarat 2010: Excavations at the PPNA site of WF16: a report on the 2008 season. Annual of the Department of Antiquities of Jordan (53), 115-126.

Modderman, P.J.R., 1945: *Over de wording en de beteekenis van het Zuiderzeegebied*, Groningen/Batavia.

Moree J.J. & M.M. Sier (eds) 2015: *Interdisciplinary Archaeological Research Programme Maasvlakte 2, Rotterdam, 1: twenty meters deep! The Mesolithic period at the Yangtze Harbour site – Rotterdam Maasvlakte, the Netherlands: Early Holocene landscape development and habitation*, Rotterdam (BOORrapporten 566),

Müller, A., T. Hamburg, S. Knippenberg, L. Kubiak-Martens, L. Kooistra, J. de Moor & M. Niekus 2011: Synthese, in: T. Hamburg, A. Müller & B. Quadflieg (eds), *Mesolithisch Swifterbant. Mesolithisch gebruik van een duin ten zuiden van Swifterbant (8300-5000 v.Chr.). Een archeologische opgraving in het tracé van de N23/N307, provincie Flevoland*, Leiden/Amersfoort (Archol Rapport 174/ADC rapport 3250), 387-401.

Müller, A. & K. Leijnse 2003: *Hanzelijn, tracédeel Nieuwe Land; een inventariserend archeologisch onderzoek*, Amsterdam (RAAP-rapport 932).

Müller, A., H. Meerten, R.B.J. Brinkgreve, & D.J.M. Ngan-Tillard 2014: *Flevoland kennisontwikkeling programma archeologie hanzelijn: Mogelijkheden tot in-situ conservering van begraven archeologische landschappen, Deelonderzoek 2, De invloed van tijdelijke en permanente afdekkingen of ophoging op maaiveld op de conservering van archeologische vindplaatsen in de ondergrond*, Amersfoort.

Müller, D.W. 1987: gräber von Metallwerken aus der Glochenbecherkultur des Mittelelbe-Saale-Gebietes, *Ausgrabungen und Funde* 32, 175-179.

Nales,T., 2010: *Inventariserend veldonderzoek (IVO), fase 3. Waarderende fase d.m.v. boringen Stichtsekant, 1-R Bon-Eindhoven gemeente Almere*, Noordwijk (Becker & Van de Graaf rapport 862).

Newell, R.R., 1973: The post-glacial adaptations of the indigenous population of the northwest European plain, in: S.F. Kozlowski (ed.), *The Mesolithic in Europe*, Warsaw, 399-440.

Newell, R.R., 1980: Mesolithic dwelling structures: fact and fantasy, *Veröffentlichendes Museums für Ur- und Frühgeschichte Potsdam* 14/15, 235-284.

Newell, R.R., C. Constandse-Westermann & C. Meiklejohn 1979: The skeletal remains of Mesolithic Man in Western Europe: An evaluative catalogue, *Journal of Human Evolution* 8, 1-228.

Newell, R.R. & A.P.J. Vroomans 1972: *Automatic artefact registration and system for archaeological analysis with the Philips P1100 computer: a Mesolithic test case*, Oosterhout.

Niekus, M.J.L. Th., 2006: A geographically referenced 14C database for the Mesolithic and the early phase of the Swifterbant culture in the northern Netherlands, *Palaeohistoria* 47/48 (2005/2006), 41-99.

Niekus, M.J.L.Th., 2014: Between appearance and reality: the excavation of Bergumermeer S-64B (Province of Friesland) as a milestone of Stone Age research in the Netherlands, *Mesolithic Miscellany* 22-2, 45-55.

Niekus, M.J.L.T., D.C. Brinkhuizen, A.A. Kerkhoven, J.J. Huisman & D.E.P. Velthuizen 2012: An Early Atlantic Mesolithic site with micro-triangles and fish remains from Almere (the Netherlands), in: D.C.M. Raemaekers, E. Esser, R.C.G.M. Lauwerier & J.T. Zeiler (red.), *A bouquet of archaeozoological studies. Essays in honour of Wietske Prummel* (Groningen Archaeological Studies 21), Groningen, 61-76.

Niekus, M.L.Th., J. Jelsma & C. Luinge 2018: Bergumermeer S-64B (the Netherlands) revisited: some critical remarks on the interpretation of an extensive Late Mesolithic site complex with alleged dwelling structures, *Journal of Archaeological Science: reports* 18, 946-959.

Niekus, M.J.L.Th., S. Knippenberg & I.I.J.A.L.M. Devriendt 2011: Vuursteen, in: T. Hamburg, A. Müller & B. Quadflieg (eds), *Mesolithisch Swifterbant. Mesolithisch gebruik van een duin ten zuiden van Swifterbant (8300-5000 v.Chr.). Een archeologische opgraving in het tracé van de N23/N307, provincie Flevoland*, Leiden/Amersfoort (Archol Rapport 174/ADC Rapport 3250), 157-241.

Niekus, M.J.L.Th. & B. Smit, 2006: Wie het kleine niet eert... Micro-driehoeken in het mesolithicum van Noord Nederland, *Paleo-aktueel* 17, 46-55.

Nikitin, A.G., P. Stadler, N. Kotova, M. Teschler-Nicola, T.D. Price, J. Hoover, D.J. Kennett, I. Lazaridis, N. Rohland, M. Lipson & D. Reich 2019: Interactions between earliest Linearbandkeramik farmers and central European hunter gatherers at the dawn of European Neolithization, *Nature Scientific Reports* 9, 1-10 (doi.org/10.1038/s41598-019-56029-2).

Noens, G. 2011: *Een afgedekt mesolithisch nederzettingsterrein te Hempens/N31 (gemeente Leeuwarden, provincie Friesland, Nl.)*, Ghent (Archaeological Reports Ghent University 7).

Opbroek, M. & E. Lohof (eds) 2012: *Tijd in centimeters. Een kijkje in het landschap van een dekzandrug te Almere. Een inventariserend Veldonderzoek in de vorm van proefsleuven en een hoogwaardig booronderzoek*, Amersfoort (ADC Rapport 2662).

Out, W.A., 2009: *Sowing the seed? Human impact and plant subsistence in Dutch wetlands during the Late Mesolithic and Early and Middle Neolithic (5500-3400 cal BC)*, Leiden.

Overeem, M.J., 2005: *Trechterbekeraardewerk uit een bouwput op het terrein van het Universitair Medisch Centrum Groningen*, Groningen (Stadse fratsen 6).

Oversteegen, J., 2001: Archeozoölogie, in: F.J.C. Peters & J.H.M. Peeters (red.), *De opgraving van de mesolithische en neolithische vindplaats Urk-E4 (Domineesweg, gemeente Urk)*, Amersfoort (Rapportage Archeologische Monumentenzorg 93), 43-48.

Palarczyk, M.J., 1984: *Slootkantverkenningen in de Noordoostpolder en Oostelijk Flevoland*, Amsterdam (IPP).

Palarczyk, M.J., 1986: *Slootkantverkenning in de Noordoostpolder*, Amsterdam (IPP-scriptie).

Peeters, J.M.P., 2004: *Wetenschappelijke uitgangspunten voor de archeologische monumentenzorg in het kader van de aanleg van de Hanzelijn (Lelystad-Zwolle)*, Amersfoort (Rapportage Archeologische Monumentenzorg 110).

Peeters, J.H.M., 2007: *Hoge Vaart-A27 in context: towards a model of Mesolithic-Neolithic land use dynamics as a framework for archaeological heritage management*, Amersfoort (PhD thesis University of Amsterdam).

Peeters, J.H.M., 2008: *Almere: Stad met cultureel erfgoed van allure. Een 'biografie' van prehistorische jagers-verzamelaars in een verdrinkend landschap. Hazelnootlezing 8 mei 2007*, Almere (Hazelnootreeks 2).

Peeters, J.H.M., 2010a: Breaking down an early Neolithic palimpsest site: some notes on the concept of Percolation Theory and the understanding of spatial pattern formation, in: F. Niccolucci & S. Hermon (eds) Beyond the Artifact Digital Interpretation of the Past Proceedings of CAA2004, Prato 13-17 April 2004, 423-428.

Peeters, J.H.M., 2010b: Early Swifterbant pottery from Hoge Vaart-A27 (Almere, the Netherlands), in: B. Vanmontfort, L. Louwe Kooijmans, L. Amkreutz, & L. Verhart (eds) Pots, Farmers and Foragers. Pottery traditions and social interaction in the earliest Neolithic of the Lower Rhine Area, Leiden, 151-160.

Peeters, J.H.M., E. Hanraets, J.W.H. Hogestijn & E. Jansma 2001: Deel 12 Dateringen: 14C-analyse en dendrochronologie, in: J.W.H. Hogestijn & J.H.M. Peeters (red.), *De mesolithische vindplaats Hoge Vaart-A27 (Flevoland)*, Amersfoort (Rapportage Archeologische Monumentenzorg 79).

Peeters, J.H.M., J.W.H. Hogestijn & Th. Holleman 2004: *De Swifterbantcultuur. Een nieuwe kijk op de aanloop naar voedselproductie*, Abcoude.

Peeters, H., A. Makaske, J. Mulder, A. Otte-Klomp, D. van Smeerdijk, S. Smit & Th. Spek 2002: Elements for archaeological heritage management: exploring the archaeological potential of drowned Mesolithic and Early Neolithic landscapes in Zuidelijk Flevoland, *Berichten Rijksdienst voor het Oudheidkundig Bodemonderzoek* 45, Amersfoort, 81-123.

Peeters, H. & M.J.L.Th. Niekus 2017: Mesolithic pit hearths in the northern Netherlands: function, time-depth and behavioural context, in: N. Achard-Corompt, E. Ghesquière & V. Riquer (eds), *Creuser au Mésolithique / Digging in the Mesolithic: actes de la séance de la Société préhistorique française de Châlons-en-Champagne (29-30 mars 2016)*, Paris (Séances de la Société préhistorique française 12), 225-239.

Peeters, J.H.M., D.C.M. Raemaekers, I.I.J.A.L.M. Devriendt, P.W. Hoebe, M. Niekus, G.R. Nobles & M. Schepers 2017: *Paradise Lost? Insights into the early prehistory of the Netherlands from development-led archaeology*, Amersfoort (Nederlandse Archeologische Rapporten 62).

Peeters, J.H.M. & J-W. Romeijn 2016: Epistemic considerations about uncertainty and model selection in computational archaeology: A case study on exploratory modelling, in: M. Brouwer Burg, H. Peeters & W. Lovis (eds), *Uncertainty and Sensitivity Analysis in Archaeological Computational Modeling*, Springer (Interdisciplinary Contributions to Archaeology), 37-58.

Peters, F.J.C. & J.H.M. Peeters (red.) 2001: *De opgraving van de mesolithische en neolithische vindplaats Urk-E4 (Domineesweg, gemeente Urk)*, Amersfoort (Rapportage Archeologische Monumentenzorg 93).

Philippsen, B., 2012: *Variability of freshwater reservoir effects. Implications for radiocarbon dating of prehistoric pottery and organisms from estuarine environments*, Aarhus (PhD thesis).

Phillippsen, B., 2013: The freshwater reservoir effect in radiocarbon dating, *Heritage Science* 1,24.

Polak, B., 1936: Pollen- und torfanalytische Untersuchungen im künftigen nord-östlichen Polder der Zuidersee, *Recueil des Travaux botaniques néerlandais* 33, 313-332.

Pomstra, D. & D. Olthof 2006: Jager-verzamelaars in de Flevopolder. Verslag van een mesolitisch leefproject, *Westerheem* 55, 306-311.

Pomstra, D. & A. van Gijn 2013: The reconstruction of a Late-Neolithic House combing Primitive Technology and Science, *Bulletin of Primitive Technology* 45, 45-54.

Pons, L. J., 1992: Holocene peat formation in the lower parts of the Netherlands, in: J.T.A. Verhoeven (ed.), *Fens and bogs in the Netherlands. Vegetation, history, nutrient dynamics and conservation*, Dordrecht, 7-79.

Prangsma, N.M., & D.A. Gerrets (red.) 2009: *Hanzelijn Tunnel Drontermeer: verbinding tussen Oude en Nieuwe Land (Een Archeologische Begeleiding bij de Sallanddijk en een compenserend archeologisch onderzoek in gebied XVI)*, Amersfoort (ADC Rapport 1601).

Price, T.D., 1978: Mesolithic settlement systems in the Netherlands, in: P. Mellars (ed.), *The early postglacial settlement of Northern Europe*, Londen, 81-113.

Price, T.D., 1980: The Mesolithic of the Drents Plateau, *Berichten Rijksdienst voor het Oudheidkundig Bodemonderzoek* 30, Amersfoort, 11-63.

Price, T.D., 1981: Swifterbant, Oost Flevoland, Netherlands: excavations at river dune sites S21-S24, 1976, *Palaeohistoria* 23, 75-104.

Price, T.D., R. Whallon & S. Chappel 1974: Mesolithic sites near Havelte, province of Drenthe (Netherlands), *Palaeohistoria* 16, 7-61.

Prummel, W. & W.A.B. van der Sanden 2002: Een oeroshoren uit het Drostendiep bij Dalen, *Nieuw Drentse Volksalmanak* 119, 217-221.

Prummel. W., D.C.M. Raemaekers, S.M. Beckerman, N. Bottema, R. Cappers, P. Cleveringa, I. Devriendt & H. de Wolf 2009: Terug naar Swifterbant. Een kleinschalige opgraving te Swifterbant-S2 (gemeente Dronten), *Archeologie* 13, 17-45.

Quadflieg, B.I., 2007: *Nota van wijziging met betrekking tot het Programma van Eisen ten behoeve van de archeologische begeleiding in het project Hanzelijn. Deeltracé Tunnel Drontermeer, gebied XVI*, Amersfoort.

Raemaekers, D.C.M., 1997: The history of the Ertebølle parallel in Dutch Neolithic Studies and the curse of the point-based pottery, *Archaeological Dialogues* 4, 220-234.

Raemaekers, D.C.M., 1999: *The Articulation of a 'New Neolithic'. The meaning of the Swifterbant Culture for the process of Neolithization in the western part of the North European Plain*, Leiden (Archaeological Series Leiden University 3).

Raemaekers, D.C.M., 2000: *Plangebied Hout, gemeente Almere; Fase IB archeologische begeleiding: veldtoetsing archeologische verwachtingskaart*, Amsterdam (RAAP-rapport 601).

Raemaekers, D.C.M., 2001a: Aardewerk en verbrande klei, in: L.P. Louwe Kooijmans (ed.), *Hardinxveld-Polderweg: een woonplaats uit het Late mesolithicum in de Rijn/Maas-delta, 5500-5000 v. C.*, Amersfoort (Rapportage Archeologische Monumentenzorg 83), 105-117.

Raemaekers, D.C.M., 2001b: Aardewerk en verbrande klei, in: L.P. Louwe Kooijmans (ed.), *Hardinxveld-De Bruin: een kampplaats uit het Late mesolithicum en de vroege Swifterbant-cultuur in de Rijn/Maas-delta, 5500-4450 v. C.*, Amersfoort (Rapportage Archeologische Monumentenzorg 88), 117-152.

Raemaekers, D.C.M., 2003: Over benen werktuigen en deposities van runderhoorns. De betekenis van de categorieën wild en gedomesticeerd voor de Swifterbant-cultuur, *Paleo-aktueel* 14, 74-77.

Raemaekers, D.C.M., 2005: An outline of Late Swifterbant pottery in the Noordoostpolder (province of Flevoland, the Netherlands) and the chronological development of the pottery of the Swifterbant culture, *Palaeohistoria* 45/46, 11-36.

Raemaekers, D.C.M., 2015: Rethinking Swifterbant S3 ceramic variability. Searching for the transition to the Funnel Beaker culture before 4000 calBC, in: J. Kabaciński, S. Hartz, D.C.M. Raemaekers & T. Terberger (eds), *The Dąbki Site in Pomerania and the Neolithization of the North European Lowlands (c. 5000-3000 calBC)*, Rahden (Archäologie und Geschichte im Ostseeraum 8/Archaeology and History of the Baltic 8), 321-334.

Raemaekers, D., 2018: A Neolithic backwater? Dutch developments in the 4th millennium BC, in: H. Meller & S. Friedrich (eds), *Salzmünde – Regel oder Ausnahme? / Salzmünde – rule or exception? Internationale Tagung vom 18. bis 20. Oktober 2012 in Halle (Saale)*, Halle (Saale) (Tagungen; Vol. 16), 487-498.

Raemaekers, D.C.M., 2019: Taboo? The process of Neolitisation in the Dutch wetlands 91 re-examined (5000-3400 cal BC), in: R. Gleser & D. Hofmann (eds), *Contacts, Boundaries & Innovation: Exploring developed Neolithic societies in central Europe and beyond*, Sidestone press, 91-102.

Raemaekers, D.C.M., T.A. Abelen, E.F.A. Anker, L.A.B. Cardamone, J. Geuverink, E.J. Hensbroek & N.M.G. Vukkink 2010: *Schokkerhaven-E170 (gemeente Noordoostpolder). Vondsten AWN-veldverkenningen*

2002-2009 en ROB-opgraving 1988, Groningen (Grondsporen 9).

D.C.M. Raemaekers & J. P. de Roever 2010: The Swifterbant pottery tradition (5000-3400 BC): matters of fact and matters of interest, in: B. Vanmontfort, L. Louwe Kooijmans, L. Amkreutz, & L. Verhart (eds) Pots, Farmers and Foragers. Pottery traditions and social interaction in the earliest Neolithic of the Lower Rhine Area, Leiden, 135-150.

Raemaekers, D.C.M., J. Geuverink, M. Schepers, B.P. Tuin, E. van der Lagemaat & M. van der Wal 2011: *A Biography in stone. Typology, age, function and meaning of Early Neolithic perforated wedges in the Netherlands*, Groningen, (Groningen Archaeological studies 14).

Raemaekers, D.C.M., A. Maurer, J. van der Laan & I. Woltinge 2013/2014: Swifterbant-S25 (gemeente Dronten, provincie Flevoland). Een bijzondere vindplaats van de Swifterbant-cultuur (ca. 4500-3700 cal. BC), Palaeohistoria 55/56, 1-56.

Raemaekers, D.C.M., A.J. Borsboom & A. Müller 2003: *Plangebied 4E-Euroquartier, Almere Poort, gemeente Almere; een inventariserend archeologisch onderzoek*, Amsterdam (RAAP-rapport 528).

Raemaekers, D.C.M. & P. de Roever 2020: *Swifterbant S4 (the Netherlands): Occupation and exploitation of a Neolithic levee site (c. 4300-4000 cal. BC)*, Groningen (Groningen Archaeological Studies 36).

Raemaekers, D.C.M., I. Devriendt, R.T.J. Cappers & W.Prummel 2005: Het Nieuwe Swifterbant Project. Nieuw onderzoek aan de mesolithische en neolithische vindplaatsen nabij Swifterbant (provincie Flevoland, Nederland), *Notae Praehistorica* 25, 119-127.

Raemaekers, D.C.M., J. Geuverink, A. Mauer, E. Scheele & J. van der Laan 2010: *Van Swifterbant naar TBR (4300-3700 v. Chr.). Een archeologisch onderzoek van een midden-neolithische oeverzone*, Groningen (Intern Rapport Universiteit Groningen / Grondsporen 14).

Raemaekers, D.C.M., J. Geuverink, I. Woltinge, J. van der Laan, A. Maurer, E.E. Scheele, T. Sibma & D.J. Huisman 2014: Swifterbant-S25 (gemeente Dronten, provincie Flevoland). Een bijzondere vindplaats van de Swifterbant-cultuur (ca. 4500-3700 cal. BC), *Palaeohistoria* 55/56, 1-56.

Raemaekers, D.C.M. & W.J.H. Hogestijn 2008: Weg met de Klokbekerweg? De interpretatie van vondsten van de Klokbeker-cultuur in Swifterbant en de provincie Flevoland, *Westerheem* 57, 409 – 417.

Raemaekers, D.C.M., L. Kubiak-Martens, T.F.M. Oudemans 2013: New food in old Pots – charred organic residues in Early Neolithic ceramic vessels from Swifterbant, the Netherlands (4300-4000 cal. BC), *Archäologisches Korrespondenzblatt* 43(3), 315-334.

Raemaekers, D.C.M., A. Maurer, J. van de Laan & I. Woltinge 2014: Swifterbant-S25 (gemeente Dronten, provincie Flevoland). Een bijzondere vindplaats van de Swifterbant-cultuur (ca. 4500-3700 cal. BC), *Paleohistoria* 55/56, 1-56.

Raemaekers, D.C.M., H.M. Molthof & E. Smits 2009: The textbook 'dealing with death' from the Neolithic Swifterbant culture (5000-3400 BC), the Netherlands, *Berichte Römisch-Germanische Kommission* 88, 479-500.

Regnel, M., 2012: Plant subsistence and environment at the Mesolithic site Tagerup, southern Sweden; new insights on the "Nut Age", *Vegetation History and Archaeobotany* 21, 1-16.

Regnell M, M.J. Gaillard, T.S. Bartholin & P. Karsten 1995: Reconstruction of environment and history of plant use during the late Mesolithic (Ertebølle culture) at the inland settlement of Bökeberg III, southern Sweden. Vegetation Historical Archaeobotany 4:67-91

Reimer, P.J. & R.W. Reimer 2001: A Marine Reservoir Correction Database and on-line Interface, *Radiocarbon*. 43, nr. 2A, 461-463.

Reimer, P.J., E. Bard, A. Bayliss, J.W. Beck, P.G. Blackwell, C. Bronk Ramsey, C.E. Buck, H. Cheng, R.L. Edwards, M. Friedrich, P.M. Grootes, T.P. Guilderson, H. Haflidason, I. Hajdas, C. Hatté, T.J. Heaton, D.L. Hoffmann, A.G. Hogg, K.A. Hughen, K.F. Kaiser, B. Kromer, S.W. Manning, M. Niu, R.W. Reimer, D.A. Richards, E.M. Scott, J.R. Southon, R.A. Staff, C.S.M. Turney & J. van der Plicht 2013: INTCAL13 and MARINE13 Radiocarbon Age Calibration Curves 0-50,000 Years CAL BP, *Radiocarbon* 55:4, 1869-1887.

Roeleveld, W. & W. Gotjé 1993: Holocene waterspiegelontwikkeling in de Noordoostpolder in relatie tot zeespiegelbeweging en kustontwikkeling, in: W. Gotjé, *De Holocene laagveenontwikkeling in de randzone van de Nederlandse kustvlakte (Noordoostpolder)*, Amsterdam (Ph.D. thesis), 76-86.

Rompelman, E., 2003: *Macro-botanisch onderzoek van een Laat-Swifterbant jacht- en viskamp*, Amsterdam (Scriptie Universiteit van Amsterdam).

Roovers, H.P.A., D.E.P. Velthuizen & M. de Boer 2011: *Zwaanpad 2003. Verslag van de AWN opgravingscursus*, Lelystad (AWN Afdeling 21 Rapport 2011-1).

Schwabedissen, H., 1979: Der Beginn des Neolithikums imnordwestlichen Deutschland. In: H. Schirnig (ed.) *Großsteingräberin Niedersachsen*, Hildesheim, 203- 222.

Schelvis, J., 2001: Deel 10 Bodemkunde en landschapsecologie IV: arthropoden, in: J.W.H. Hogestijn & J.H.M. Peeters (red.), *De mesolithische vindplaats Hoge Vaart-A27 (Flevoland)*, Amersfoort (Rapportage Archeologische Monumentenzorg 79).

Schepers, M., 2014a: *Reconstructing vegetation diversity in coastal landscapes*, Groningen.

Schepers, M., 2014b: Wet, wealthy worlds: The environment of the Swifterbant river system during the Neolithic occupation (4300-4000 cal BC), *Journal of Archaeology in the Low Countries* 5.1, 80-103.

Schepers, M., J.F. Scheepens, R.T.J. Cappers, O.F.R. van Tongeren, D.C.M. Raemaekers & R.M. Bekker 2013: An objective method based on assemblages of subfossil plantmacro-remains to reconstruct past natural vegetation: a casestudy at Swifterbant, The Netherlands, *Vegetation History and Archaeobotany* 22(3), 243-255.

Schepers, M. & I. Woltinge 2020: *Chapter 2. Landscape development and stratigraphy,* in: D.C.M. Raemakers & J.P. de Roever (eds), *Swifterbant S4 (the Netherlands): Occupation and exploitation of a Neolithic levee site (c. 4300-4000 cal. BC),* Groningen (Groningen Archaeological Studies 36), 15-24.

Scherjon, F., C. Bakels, K. MacDonald & W. Roebroeks, 2015: Burning the land. An ethnographic study of off-site fire use by current and historically documented foragers and implications for the interpretation of past fire practices in the landscape. *Current Anthropology*, 56-3, 299-326.

Shipman, P., G.F. Foster, M. Schoeniger 1984: Burnt bones and teeth: an experimental study of colour, morphology, crystal structure and shrinkage. *Journal of Archaeological Science* 11:307-325

Siebelink, M. A. van Gijn, D. Pomstra & Y. Lammers-Keijsers met een bijdrage van J.J. Langer 2012: gebruiksporenanalyse van vuursteen, in: T. Hamburg, A. Müller & B. Quadflieg (eds) *Mesolithisch Swifterbant. Mesolithisch gebruik van een duin ten zuiden van Swifterbant (8300-500 BC). Een archeologische opgraving in het tracé van N23/N307, provincie Flevoland*, Leiden (Archol Rapport 174 / ADC Rapport 3250), 243-268.

Simmons, I.G., 1996: *The environmental impact of later Mesolithic cultures: the creation of moorland landscape in England and Wales*, Edinburgh.

Smit, B.I., O. Brinkkemper, J.P. Kleijne, R.C.G.M. Lauwerier & E.M. Theunissen (eds) 2012: *A kaleidoscope of gathering at Keinsmerbrug (the Netherlands): Late Neolithic behavioural variability in a dynamic landscape*, Amersfoort (Nederlandse Archeologische Rapporten 43)

Smit, B.I., 2010: *Valuable flints: research strategies for the study of early prehistoric remains from the Pleistocene soils of the Northern Netherlands*, Groningen (PhD thesis Groningen University) (Groningen Archaeological Studies 11).

Smit, A., G. Mol & R.M. van Heeringen 2005: *Natte voeten voor Schokland, Inrichting hydrologische zone Archeologische monitoring 2003-2004*, Amersfoort (Rapportage Archeologische Monumentenzorg 124).

Smith, C., 1989: British antler mattocks. In: C. Bonsall (ed.), *The Mesolithic in Europe*, Edinburgh, 272-283.

Smith, W., 2013: *Een integrale opsporingsformule voor prospectief booronderzoek*, Almere (Archeologische Rapporten Almere 97).

Smith, W. & J.W.H. Hogestijn 2013: De invloed van variatie in vondstdichtheden op de Vindkans van vuursteenvindplaatsen. Poissonverdeling versus de negatieve binomiale verdeling, Almere (Archeologische Rapporten Almere 92).

Smits, L. & L.P. Louwe Kooijmans 2006: Graves and human remains, in: L.P. Louwe Kooijmans & P.F.B. Jongste (eds), *Schipluiden. A neolithic settlement on the Dutch North Sea Coast, c. 3500 cal BC*, Leiden (Analecta Praehistorica Leidensia 37/38), 91-112.

Spek, Th., E.B.A. Bisdom & D.G. van Smeerdijk 2001a: Deel 7 Bodemkunde en landschapsarcheologie I: veranderingen in bodem en landschap, in: J.W.H. Hogestijn & J.H.M. Peeters (red.), *De mesolithische vindplaats Hoge Vaart-A27 (Flevoland)*, Amersfoort (Rapportage Archeologische Monumentenzorg 79).

Spek, Th., E.B.A. Bisdom & D.G. van Smeerdijk 2001b: Deel 8 Bodemkunde en landschapsarcheologie II: veranderingen in bodem en landschap, in: J.W.H. Hogestijn & J.H.M. Peeters (red.), *De mesolithische vindplaats Hoge Vaart-A27 (Flevoland)*, Amersfoort (Rapportage Archeologische Monumentenzorg 79).

Stapel, B., 1991: *Die geschlagene Steingeräte der Siedlung Hüde I am Dümmer*, Münster (Veröffentlichen der Urgeschichtlichen Sammlungen des Landesmuseum zu Hannover 38).

Stapert, D., 1985: A small Creswellian site at Emmerhout (province of Drenthe, the Netherlands), *Palaeohistoria* 27, 1-65.

Strahl, E., 1990: *Das Endneolithikum im Elb-Weser-Dreieck*, Hannover (Veröffentlichungen der urgeschichliche Sammlungen des Landesmuseums zu Hannover 36).

Stortelder, A.H.F., J.H.J. Schaminée & M. Hermy 1999: Querco-Fagetea, in: A.H.F. Sortelder, J.H.J. Schaminée & P.W.F.M. Hommel (red.), *De vegetatie van Nederland 5, Plantengemeenschappen van ruigten, struwelen en bossen*, Uppsala-Leiden.

Ten Anscher, T.J., 2012: *Leven met de Vecht. Schokland-P14 en de Noordoostpolder in het Neolithicum en de Bronstijd*, Amsterdam.

Ten Anscher, T.J., 2013: *Bureaustudie ten behoeve van de instandhouding van de archeologische waarden in het werelderfgoedgebied Schokland, kavels E170/E171, gemeente Noordoostpolder*, Amsterdam (RAAP-rapport 2723).

Ten Anscher, T.J. & E.F. Gehasse 1993: Neolithische en Vroege Bronstijd-bewoning langs de benedenloop van de Overijsselse Vecht, in: J.H.F. Bloemers, W. Groenman-van Waateringe & H.A. Heidinga (eds), *Voeten in de aarde. Een kennismaking met de moderne Nederlandse archeologie*, Amsterdam, 25-44.

Ten Anscher, T.J., E.F. Gehasse & J.A. Bakker 1993: A pre-megalithic TRB and Late Swifterbant complex at P14-Schokland, gemeente Noordoostpolder, the Netherlands, in: J. Pavúk (ed.), *Actes du XIIe Congrès International des Sciences Préhistorique et Protohistorique, Bratislava 1991 (part II)*, Bratislava, 460-466.

Ter Wal, A., 1995/1996: Een onderzoek naar de depositie van vuurstenen bijlen, *Palaeohistoria* 37/38, 127-158.

Theunissen, E.M., O. Brinkkemper, R.C.G.M. Lauwerier, B.I. Smit & I.M.M. van der Jagt (eds) 2014: *A mosaic of habitation at Zeewijk (the Netherlands): Late Neolithic behavioural variability in a dynamic landscape*, Amersfoort (Nederlandse Archeologische Rapporten 47).

Thrane, H., 1989: Danish Plough-Marks from the Neolithic and Bronze Age, *Journal of Danish Archaeology* 8, 111-125.

Tobolski, K., 2000: *Vademecum Geobotanicum. Przewodnik do oznaczania torfów i osadów Jeziornych*, Warszawa.

Tol, A.J., 2007: *Aanleg N23 tussen Lelystad en Dronten Archeologisch vooronderzoek: een inventariserend veldonderzoek (verkennende en karterende fase)*, Amsterdam (RAAP-rapport 1469).

Tol, A.J., J.W.H.P. Verhagen, A. Borsboom & M. Verbruggen 2004: *Prospectief Boren. Een studie naar de betrouwbaarheid en toepasbaarheid van booronderzoek in de prospectiearcheologie*, Amsterdam (*RAAP-rapport* 1000).

Tol, A.J., J.W.H.P. Verhagen & M. Verbruggen 2006: Leidraad inventariserend veldonderzoek. Deel: karterend booronderzoek, SIKB.

Ufkes, A., 1997: Edelhertgeweien uit natte context in Drenthe, *Nieuwe Drentse Volksalmanak* 114, 142-170.

Ufkes, A., 1993: Vroeg-Neolitische votiefgaven ?: edelhertgeweien uit Drenthe en Groningen, *Paleo-aktueel* 4, 28-30.

Van Beek, R., 2009: *Reliëf in Tijd en Ruimte. Interdisciplinair onderzoek naar bewoning en landschap van Oost-Nederland tussen vroege prehistorie en middeleeuwen*, Wageningen.

Van Betuw-Demon, S., 1997: Stichting Prehistorische Nederzetting Flevoland, in: G.H.L. Tiesinga (ed.), *Dwarsliggers komen in het IJsselmeergebied niet voor*, Lelystad, 110-112.

Vajda, E.J., 2010: Ket shamanism, *Shaman* 18 (1-2), 125-143.

Van Beurden, L., 2007: *Vegetatie en ontwikkelingen in het rivierengebied in de Bronstijd*, Zaandam (BIAXiaal 331).

Van de Geer, P., 2013: *Steentijd op de Stichtsekant. Definitieve opgraving van drie vindplaatsen in Stichtsekant, gemeente Almere*, Leiden (Archol Rapport 212).

Van de Plassche, O., 1982: Sea-level change and water movements in the Netherlands during the Holocene, *Mededelingen Rijks Geologische Dienst* 36-1, 1-93.

Van de Plassche, O., S.J.P. Bohncke, B. Makaske & J. van der Plicht 2005: Water-level changes in the Flevo area, central Netherlands (5300-1500 BC): implications for relative mean sea-level rise in the Western Netherlands, *Quaternary International* 133-134, 77-93.

Van der Heide, G.D., 1950: Enkele resultaten van het oudheidkundig bodemonderzoek in het Zuiderzeegebied, met name in de Noordoostpolder, *Bulletin van de Koninklijke Nederlandse Oudheidkundige Bond* 3, 79-94.

Van der Heide, G.D., 1951. Die Archäologie des Zuiderzeegebietes, *Berichten van de Rijksdienst voor het Oudheidkundig Bodemonderzoek* 1, 42-46.

Van der Heide, G.D., 1955: *Aspecten van het archeologisch onderzoek in het Zuiderzeegebied*, Zwolle.

Van der Heide, G.D., 1965a: Opgravingen bij Swifterbant, *Fibula* 7, 86-89.

Van der Heide, G.D., 1965b: Enkele aantekeningen betreffende prehistorische bewoning van het oostelijk deel van het Zuiderzeegebied, *Kamper Almanak* 1965/1966, 200-214.

Van der Heide, G.D. & A.J. Wiggers 1954: Enkele resultaten van het geologische en archeologische onderzoek betreffende het eiland Schokland en zijn naast omgeving, in: A.J. Zuur (ed.), *Langs gewonnen velden. Facetten van Smedings werk*, Wageningen, 96-113.

Van der Heijden, F.J.G., 2000: *Gemeente Noordoostpolder. Aanvullend archeologisch onderzoek Vindplaats Rijksweg A6 – Kavel J97*, Bunschoten (ADC Rapport 69).

Van der Heijden, F.J.G. & T. Hamburg 2002: Weren en fuiken, in: E.E.B. Bulten, F.J.G. van der Heijden & T. Hamburg (red.), *Emmeloord, prehistorische visweren en fuiken*, Amersfoort (ADC Rapport 140), 34-56.

Van der Kroft, P. 2012: Het vuursteen van P14 werkput 1989-17, in: T. Ten Anscher Leven met de Vecht, Schokland-P14 en de Noordoostpolder in het Neolithicum en de Bronstijd, Zutphen, 617-676.

Van der Lijn, P., 1973: Het keienboek: Mineralen, gesteenten en fossielen in Nederland, Zutphen.

Van der Linden, M., 2008: *Mesolithische en Vroeg-Neolithische bewoningssporen bij Swifterbant? Pollenonderzoek op vindplaats 5 in het kader van de aanleg van de N23 tussen Lelystad en Dronten*, Zaandam (BIAXiaal 365).

Van der Linden, M., 2010: *Palynologisch onderzoek aan een veen- en kleipakket uit het Laat-Mesolithicum bij Almere-De Vaart*, Zaandam (BIAXiaal 501).

Van der Linden, M., 2012: Pollenanalyse aan haardkuilen, in: T. Hamburg, A. Müller & B. Quadflieg, *Mesolithisch Swifterbant. Mesolithisch gebruik van een duin ten zuiden van Swifterbant (8300-5000 v.Chr.). Een archeologische opgraving in het tracé van de N23/N307, provincie Flevoland*, Leiden/Amersfoort (Archol Rapport 174/ ADC Rapport 3250), 361-367.

Van der Linden, M. & L.I. Kooistra 2019: Hoofdstuk 7 Hoogveenontwikkeling in Nederland gedurende het Holoceen, in: A. Jansen & A. Grootjans (red.), *Hoogvenen. Landschapsecologie, behoud, beheer, herstel*, Gorredijk (Noordboek Natuur), 91-100.

Van der Plicht, J., L.W.S.W. Amkreutz, M.J.L.Th. Niekus, J.H.M. Peeters & B.I. Smit 2016: Surf'n turf in Doggerland: dating, stable isotopes and diet of Mesolithic human remains from the southern North Sea, *Journal of Archaeological Science Reports* 10, 110-118.

Van der Sanden, W.A.B., 2002: Runderhoorns, wagens en andere Drentse veenvondsten, *Nieuwe Drentse Volksalmanak* 119, 128-167.

Van der Sanden, W.A.B., 1997: Aardewerk uit natte context in Drenthe: het vroeg- en laat-neoliticum en de vroege bronstijd, *Nieuwe Drentse Volksalmanak* 114, 127-141.

Van der Sanden, W.A.B. & E. Taayke 1995: Aardewerk uit natte context in Drenthe: 1100 v.Chr. tot 500 na Chr., *Nieuwe Drentse Volksalmanak* 112:149-186.

Van der Veen, Y., 2008: *Palynologisch onderzoek aan de akker van Swifterbant S4*, Groningen (unpublished Bachelors thesis).

Van der Velde, H.M., N. Bouma & D.C.M. Raemaekers 2019: A monumental burial ground from the Funnel Beaker Period at Oosterdalfsen (the Netherlands), in: J. Müller, M. Hinz & M. Wunderlich (eds), *Megaliths – Societies – Landscapes Early Monumentality and Social Differentiation in Neolithic Europe. Proceedings of the international conference »Megaliths – Societies – Landscapes. Early Monumentality and Social Differenzierung in Neolithic Europe« (16th-20th June 2015) in Kiel*, Bonn (Frühe Monumentalität und soziale Differenzierung 18), 319-328.

H.M. van der Velde, N. Bouma, D.C.M. Raemaekers (eds) 2021: *Making a Neolithic non-megalithic monument. A TRB burial ground at Dalfsen (the Netherlands), c. 3000-2750 cal. BC*, Leiden.

Van der Waals, J.D., 1964: Neolithic disc wheels in the Netherlands; with a note on the early Iron Age disc wheels from Ezinge, *Palaeohistoria* 10, 103-146.

Van der Waals, J.D., 1972: Die durchlochten rössener Keile und das frühe Neolithikum in Belgien und in den Niederlanden, *Fundamenta* A3, 154-184.

Van der Waals, J.D. & H.T. Waterbolk 1976: Excavations at Swifterbantdiscovery, progress, aims and methods (Swifterbant Contribution 1), *Helinium* 16, 414.

Van der Werf, S., 1991: *Bosgemeenschappen, Natuurbeheer in Nederland 5*, Wageningen.

Van Diepen, R., 2013: Een Zeeuws steentijdhuis in het Horsterwold, in: R. van Diepen, W. van der Most & H. Pruntel (eds), *Nieuw land in vogelvlucht*, Lelystad, 135-143.

Van Geel, B., S.J.P. Bohncke & H. Dee 1981: A Palaeoecological Study from an Upper Late Glacial and Holocene Sequence from "De Borchert", The Netherlands, *Review of Palaeobotany and Palynology* 31, 347-448.

Van Gijn, A.L., 2010: *Flint in focus. Lithic Biographies in the Neolithic and Bronze Age*, Leiden.

Van Gijssel, K. & B. van der Valk 2005: Aangespoeld, gestuwd en verwaaid: de wording van Nederland, in: L.P. Louwe-Kooijmans, P.W. van den Broeke, H. Fokkens & A. van Gijn, *Nederland in de Prehistorie*, Amsterdam, 45-74.

Van Haaster, H., 2010: *Waardering van botanische macroresten van vier archeologische vindplaatsen bij Schokland*, Zaandam (BIAXiaal 496).

Van Heeringen, R.M., R. Schrijvers, K.E. Waugh, 2018: *Handreiking prospectief onderzoek in Flevoland voor het opsporen en waarderen van vindplaatsen uit de vroege prehistorie*, Amersfoort (Vestigia rapport V1372).

Van Heeringen, R.M., W.A.M. Hessing, L.I. Kooistra, S. Lange, B.I. Quadflieg, R. Schrijvers & W. Weerheim 2014: *Archeologisch landschapsonderzoek in het kader van het project Kwaliteitsverbetering Kotterbos (locatie Natuurboulevard in de gemeente Lelystad, Provincie Flevoland. Menselijke activiteit in natte landschapen in de Steentijd en de (vroeg-) Romeinse tijd*, Amersfoort (Vestigia-rapport 1132).

Van Heeringen, R.M., G.V. Mauro & A. Smit (eds) 2004: *A pilot study on the monitoring of the physical quality of three archaeological sites at UNESCO world heritage site at Schokland, Province of Flevoland, the Netherlands*, Amersfoort (Nederlandse Archeologische Rapporten 26).

Van Lil, R., 2008: *Aanleg N23 tussen Lelystad en Dronten. Een Inventariserend Veldonderzoek in de vorm van een waarderend booronderzoek van vindplaats 5*, Amersfoort (ADC Rapport 1577).

Van Mourik, J.M., 2001: Pollen and spores, preservation in ecological settings, in: E.G. Briggs, P.R. Crowther (eds), *Palaeobiology* II, Blackwell Science, Oxford, 315-318.

Van Mourik, J.M., 2003: Life cycle of pollen grains in mormoder humus forms of young acid forest soils: a micromorphological approach, *Catena* 54, 651-663.

Van Mourik, J.M. & B. Jansen 2013: The added value of biomarker analysis in palaeopedology; reconstruction of the vegetation during stable periods in a polycyclic driftsand sequence in SE-Netherlands, *Quaternary International* 306, 14-23.

Van Mourik, J.M., K.G.J. Nierop & D.A.G. Vandenberghe 2010: Radiocarbon and optically stimulated luminescence dating based chronology of a polycyclic driftsand sequence at Weerterbergen (SE Netherlands), *Catena* 80, 170-181.

Van Oorsouw, M.F., 1998: Flevoland 1995 & 1996, *Jaarverslag ROB 95/96*, 159-166.

Van Regteren Altena, J.F., J.A. Bakker, A.T. Clason, W. Glasbergen, W. Groenmanvan Waateringe & L.J. Pons, 1962, The Vlaardingen Culture (II), *Helinium* 2, 97-103.

Van Rijn, P., 2002: Houtonderzoek, in: E.E.B. Bulten, F.J.G. van der Heijden (red.), *Prehistorische visweren en fuiken bij Emmeloord*, Amersfoort (ADC Rapport 140), 57-77.

Van Rijn, P. & L.I. Kooistra 2001: Deel 15 Hout en houtskool: het gebruik van hout als constructiemateriaal en brandstof, in: J.W.H. Hogestijn & J.H.M. Peeters (red.), *De mesolithische vindplaats Hoge Vaart-A27 (Flevoland)*, Amersfoort (Rapportage Archeologische Monumentenzorg 79).

Van Rooij, J.A.G., 2007: *Botanische resten uit Swifterbant S4*, Groningen (unpublished Bachelor thesis).

Van Smeerdijk, D.G., 1989: Alder car, growth and drowning in the IJsselmeer region, an aspect of Dutch coastal development, *Acta Botanical Neerlandica* 38, 477-491.

Van Smeerdijk, D.G., 2001: Palynologie, in: F.J.C. Peters & J.H.M. Peeters (red.), *De opgraving van de mesolithische en neolithische vindplaats Urk-E4 (Domineesweg, gemeente Urk)*, Amersfoort (Rapportage Archeologische Monumentenzorg 93), 70-76.

Van Smeerdijk, D.G., 2002: *Palaeo-ecologisch onderzoek aan een bodemprofiel uit de locatie Almere Kasteel, Gemeente Almere*, Zaandam (BIAXiaal 138).

Van Smeerdijk, D.G., 2003: *Pollenonderzoek aan materiaal uit de top van een Pleistocene dekzandrug in Almere-Hout ten behoeve van de Cursus Archeologie*, Zaandam (BIAXiaal 179).

Van Smeerdijk, D.G., 2006: *Palynologisch onderzoek en datering van de overgang van het Pleistocene zand naar het afdekkende veen bij de Noorderplassen-West in Almere*, Zaandam (BIAXiaal 283).

Van Vuure, C., 2005: *Retracing the Aurochs – History, Morphology and Ecology of an extinct wild Ox*, Sofia-Moscow.

Van Zeist, W. & R.M. Palfenier-Vegter 1983: Seeds and fruits from the Swifterbant S3 site (Final reports on Swifterbant IV), *Palaeohistoria* 23, 105-168.

Van Zijverden, W., 2002: Fysische geografie, in: E.E.B. Bulten, F.J.G. van der Heijden & T. Hamburg (red.), Emmeloord, Prehistorische visweren en fuiken, Bunschoten (ADC-rapport 140).

Vera, F.W.M., 1997: *Metaforen voor de wildernis: eik, hazelaar, rund en paard*, Wageningen.

Verbaas, A., M.J.L.Th. Niekus, A.L. van Gijn, S. Knippenberg, Y.L. Lammers-Keijsers & P.C. van Woerdekom 2011: Vuursteen uit de Hanzelijn, in: E. Lohof, T. Hamburg & J. Flamman (red.), *Steentijd opgespoord: archeologisch onderzoek in het tracé van de Hanzelijn-Oude Land*, Leiden/Amersfoort (Archol report 138/ADC report 2576), 335-393.

Verhagen, J.W.H.P., E. Rensink, M. Bats & Ph. Crombé 2011: *Optimale strategieën voor het opsporen van Steentijdvindplaatsen met behulp van booronderzoek. Een Statistisch perspectief*, Amersfoort (Rapportage Archeologische Monumentenzorg 197).

Verhagen, P., E. Rensink, M. Bats & P. Crombé 2013: Establishing discovery probabilities of lithic artefacts in Palaeolithic and Mesolithic sites with core sampling, *Journal of Archaeological Science* 40, 240-247.

Verhart, L., 2012: Contact in stone: adzes, Keile Spitzhauen in Lower Rhine Basin Neolithic stone tools and the transition from Mesolithic to Neolithic in Belgium in the Netherlands, 5300-4000 cal BC, *Journal of Archaeology in the Low Countries*, 4(1), 5-35.

Verhart, L., 2010: *De geur van veen. Vlaardingen en de ontdekking van de Vlaardingen-cultuur*, Utrecht.

Verhart, L.B.M. 2000: Times fade away: the neolithization of the southern Netherlands in an anthropological and geographical perspective, Leiden (PhD dissertation Leiden University).

Verlinde, A.D., 1974: A Mesolithic Settlement with Cremation at Dalfsen, *Berichten van de Rijksdienst voor het Oudheidkundig Bodemonderzoek* 24, 113-117.

Verlinde, A.D. & R.R. Newell 2006: A multicomponent complex of mesolithic settlements with late mesolithic grave pits at Mariënberg in Overijssel, in: B.J. Groenwoudt, R.M. Van Heeringen & G.M. Scheepstra (eds), *Het zandeilandenrijk van Overijssel. Bundel verschenen ter gelegenheid van de pensionering van A.D. Verlinde als archeoloog in, voor en van Overijssel*, Amersfoort (Nederlandse Archeologische Rapporten 22), 83-270.

Verlinde, A.D. & R.R. Newell, 2013: *The Mesolithic cemetery at Mariënberg (NL), a rebuttal to alternative interpretations*, Amersfoort (Nederlandse Archeologische Rapporten 42).

Verneau, S.M.J.P., 2001 : Vuursteen, in: F.J.C. Peters & J.H.M. Peeters (eds), *De opgraving van de mesolithische en neolithische vindplaats Urk-E4 (Domineesweg, gemeente Urk)*, Amersfoort (Rapportage Archeologische Monumentenzorg 93), 93-109.

Vernimmen, T.J.J., 1999: *Archeobotanisch waarderingsonderzoek aan de hand van monsters van een drietal vindplaatsen te Schokland*, Amersfoort (Interne Rapporten Archeobotanie ROB 19).

Vernimmen, T.J.J., 2001: Macroresten, in: F.J.C. Peters & J.H.M. Peeters (red.), *De opgraving van de mesolithische en neolithische vindplaats Urk-E4 (Domineesweg, gemeente Urk)*, Amersfoort (Rapportage Archeologische Monumentenzorg 93), 60-70.

Vernimmen, T.J., 2004: Preservation of botanical remains, in: R.M. van Heeringen, G.V. Mauro & A. Smit (red.), *A Pilot Study on the Monitoring of the Physical Quality of Three Archaeological Sites at the UNESCO World Heritage Site at Schokland, Province of Flevoland, the Netherlands*, Amersfoort (Nederlandse Archeologische Rapporten 26), 73-86.

Visser, C.A., C. Gaffney & W.A.M. Hessing 2011: *Het gebruik van geofysische prospectietechnieken in de Nederlandse archeologie*, Amersfoort (Vestigia-rapport V887).

Visser, D., C. Whitton, O. Brinkkemper & J.W.H. Hogestijn 2001: Deel 11 Archeobotanie: de analyse van botanische macroresten, in: J.W.H. Hogestijn & J.H.M. Peeters (red.), *De mesolithische vindplaats Hoge Vaart-A27 (Flevoland)*, Amersfoort (Rapportage Archeologische Monumentenzorg 79).

Vissers, M.J., S. van Asselen & J.J. Hekman 2014: *Programma Kennisontwikkeling Archeologie Hanzelijn, Thema 2A: veranderingen in de waterhuishouding gerelateerd aan bodemeigenschappen en de gevolgen daarvan voor de conservering van afgedekte archeologische vindplaatsen in Flevoland*, De Bilt (Grontmij Archeologische Rapport 1314).

Vlierman, K., 1985: Neolithische en middeleeuwse vondsten op de kavels oz 35 en oz 36 in zuidelijk Flevoland, Rijksdienst voor de IJsselmeerpolders, Lelystad (RIJP-rapporten 1).

Von Brandt, A., 1984: *Fish catching methods of the world*, London.

Vorenhout, M., 2011a: *Evaluatie invoer database PKAH. Korte notitie MVH Consult n.a.v. literatuur database bestaande onderzoeken binnen in situ behoud* (Interne memo ihkv Programma Kennisontwikkeling Archeologie Hanzelijn (PAKH), 15 december 2011).

Vorenhout, M., 2011b: *Onderzoeksplannen PKA. Korte notitie MVH Consult n.a.v. literatuur database bestaande onderzoeken binnen in situ behoud* (Interne memo ihkv Programma Kennisontwikkeling Archeologie Hanzelijn (PAKH), 15 december 2011).

Vos, P.C., 2015: *Origin of the Dutch Coastal Landscape. Long-term landscape evolution of the Netherlands during the Holocene described and visualized in national, reginal and local palaeogeographical map series*, Utrecht.

Vos, P. & S. de Vries 2018: *Tweede generatie palaeogeografische kaarten van Nederland (versie 2.0)*, Utrecht (Deltares).

Vos, P. & P. Kiden 2005: De landschapsvorming tijdens de steentijd, in: J. Deeben, E. Drenth, M.-F. van Oorsouw & L. Verhart (eds), *De steentijd van Nederland*, Zutphen (Archeologie 11/12), 7-38.

Vos, P., M. van der Meulen, H. Weerts & J. Bazelmans 2020: *Atlas of the Holocene Netherlands, landscape and habitation since the last ice age*, Amsterdam.

Wandsnider, L., 1997: The Roasted and the Boiled: Food Composition and Heat Treatment with Special Emphasis on Pit-Hearth Cooking, Journal of Anthropological Archaeology, 1-48.

Wansleeben, M. & L.B.M. Verhart 1990: The Meuse valley project: the transition from the Mesolithic to the Neolithic in the Dutch Meuse Valley, in: P.M. Vermeersch & P. Van Peer (eds), *Contributions to the Mesolithic in Europe*, Leuven, 389-402.

Wansleeben, M. & W. Laan 2012a: Ruimtelijke analyse, in: T. Hamburg, A. Müller & B. Quadflieg (eds) *Mesolithisch Swifterbant. Mesolithisch gebruik van een duin ten zuiden van Swifterbant (8300-500 BC). Een archeologische opgraving in het tracé van N23/N307, provincie Flevoland*, Leiden (Archol Rapport 174 / ADC Rapport 3250) 85-112.

Wansleeben, M. & W. Laan 2012b: The archaeological practice of discovering Stone Age sites, *Analecta Praehistorica Leidensia* 43/44, 253-261.

Warning, S., B.I. Smit & H.C.J. Visscher 2009: *Vindplaats 1R-2: Onderzoeksgebied 1R2 (Stichtse Kant), Gemeente Almere een inventariserend veldonderzoek: fase 3 (waardering)*, Weesp (RAAP-rapport 2012).

Waterbolk, H.T., 1957: De Ertebölle-cultuur in Nederland?, *Westerheem* 6, 86-90.

Waterbolk, H.T. 1960: Preliminary report on the excavations at Anlo in 1957 and 1958, *Palaeohistoria* 8, 59-90.

Waterbolk, H.T., 1985: The Mesolithic and Early Neolithic settlement of the Northern Netherlands in the light of radiocarbon evidence, in: R. Fellmann, G. Germann & K. Zimmermann (eds), *Jagen und Sammeln. Festschrift für Hans-Georgi Bandi zum 65. Geburtstag*, Bern (Jahrbuch des Bernische Historische Museum 63-64), 273-281.

Waterbolk, H . J . & H . T . Waterbolk 1992 : Barnsteen in het waddengebied , Waddenbulletin 27 , 7077.

Weeda, E.J., R. Westra, Ch. Westra & T. Westra 1985: *Nederlandse oecologische flora. Wilde planten en hun relaties 1*, Deventer.

Weeda, E.J., R. Westra, Ch. Westra & T. Westra 1991: *Nederlandse oecologische flora. Wilde planten en hun relaties 4*, Deventer.

Weeda, E.J., R. Westra, Ch. Westra & T. Westra 1994: *Nederlandse oecologische flora. Wilde planten en hun relaties 5*, Deventer.

Weijdema, F., O. Brinkkemper, H. Peeters & B. van Geel 2011: Early Neolithic human impact on the vegetation in a wetland environment in the Noordoostpolder, central Netherlands, *Journal of Archaeology in the Low Countries* 3 1/2, 31-46.

Wentink. K., 2006 : *Ceci n'est pas une hache. Neolithic Depositions in the Northern Netherlands*, Leiden.

Wentink, K., 2020: *Stereotype, the role of grave sets in Corded Ware and Bell Beaker funerary practices*, Leiden.

Whallon, R., 2011. An introduction to information and its role in hunter-gatherer bands. In: R. Whallon, W.A. Lovis & R.K. Hitchcock (eds), *Information and its role in hunter-gatherer bands*, Los Angeles, 1-28

Whallon, R. & T.D. Price 1976: Excavations at the river dune sites S11S13 (Swifterbant Contribution 5), *Helinium* 16, 222229.

Waugh, F. W., 1916: *Iroquois Foods and Food Preparation*. Canada Department of Mines Geological Survey, Memoir 86, No. 12, Anthropological Series, Ottawa.

Whitley, T.G., 2016: Archaeological simulation and the testing paradigm, in: M. Brouwer Burg, H. Peeters & W. Lovis (eds), *Uncertainty and sensitivity analysis in archaeological computational modeling*, New York (Interdisciplinary Contributions to Archaeology), 131-156.

Whittle, A., 1996: *Europe in the Neolithic. The creation of new worlds*, Cambridge.

Wiggers, A.J., 1955: *De wording van het Noordoostpoldergebied*, Zwolle (Van Zee tot Land 14). (also published as: A.J. Wiggers, (1955): *De wording van het Noordoostpoldergebied. Een onderzoek naar de physisch-geografische ontwikkeling van een sedimentair gebied*, Zwolle (PhD-thesis).

Wolf, R.J.A.M., A.H.F. Stortelder & R.W. de Waal (red.) 2001: *Ooibossen*, Utrecht (Bosecosystemen van Nederland 2).

Woldring, H., M. Schepers, J. Mendelts & R. Fens 2012: Camping and foraging in Boreal hazel woodland – the environmental impact of Mesolithic hunter-gatherers near Groningen, in: M.J.L.Th. Niekus, R.N.E. Barton, M. Street & T. Terberger (eds), *A mind set on flint: studies in honour of Dick Stapert*, Groningen (Groningen Archaeological Studies 16), 381-392.

Woltinge, I., 2009: *Almere Lage Vaart: op zoek naar de Oude Eem. Booronderzoek naar Oude Getijde Afzettingen aan de Trekweg en Kievitsweg in Almere*, Almere (Grondsporen 6).

Woltinge, I., M. Opbroek, L.A. Tebbens, I. Devriendt & E. Drenth 2019: *Mesolitisch verblijf en maretakspitsen aan de Staringlaan te Soest, De opgraving van een mesolitische 'persistent place'*, 's Hertogenbosch (BAAC Rapport A-15.0124).

Zeiler, J.T., 1986: Swifterbant: Dwelling place for a season or throughout the whole year? An archaeozoological contribution, in: H. Fokkens, B. Banga & M. Bierma (eds), *Op zoek naar de mens en materiële cultuur*, Groningen, 85-95.

Zeiler J.T., 1997: *Hunting, fowling and stock-breeding at Neolithic sites in the western and central Netherlands*, Groningen (Ph.D. thesis).

Zeiler, J.T. & D.C. Brinkhuizen, 2012: The faunal remains, in: B.I. Smit, O. Brinkkemper, J.P. Kleijne, R.C.G.M. Lauwerier & E.M. Theunissen (eds), A kaleidoscope of gathering at Keinsmerbrug (the Netherlands). Late Neolithic behavioural variability in a dynamic landscape, Amersfoort (Nederlandse Archeologische Rapporten 43), 131-147.

Zeiler, J.T. & D.C. Brinkhuizen 2013: Faunal remains, in: J.P. Kleijne, O. Brinkkemper, R.C.G.M. Lauwerier, B.I. Smit & E.M. Theunissen (eds), A Matter of Life and Death at Mienakker (the Netherlands) Late Neolithic Behavioural Variability in a Dynamic Landscape, Amersfoort (Nederlandse Archeologische Rapporten 45), 155-173.

Zeiler, J.T. & D.C. Brinkhuizen 2014: Faunal remains, in: E.M. Theunissen, O. Brinkkemper, R.C.G.M. Lauwerier, B.I. Smit & I.M.M. van der Jagt (eds), A Mosaic of Habitation at Zeewijk (the Netherlands). Late Neolithic Behavioural Variability in a Dynamic Landscape, Amersfoort (Nederlandse Archeologische Rapporten 47), 177-196.

Zeiler, J.T. & A.T. Clason 1993: Fowling in the Dutch Neolithic at inland and coastal sites, Archaeofauna 2, 67-74.

Zvelebil, M. 1994: Plant use in the Mesolithic and its role in the transition to farming, *Proceedings of the Prehistoric Society* 60, 35 74.

Zvelebil, M., S.W. Green & M.G. Macklin 1992: Archaeological Landscapes, Lithic Scatters, and Human Behavior, in: J. Rossignol & L. Wandsnider (eds), *Space, time and archaeological landscapes*, New York, 193-226.